Tetraspanins

PROTEINS AND CELL REGULATION

Volume 9

Aims and Scope

Our knowledge of the ways in which a cell communicates with its environment and how it responds to information received has reached a level of almost bewildering complexity. The large diagrams of cells to be found on the walls of many a biologist's office are usually adorned with parallel and interconnecting pathways linking the multitude of components and suggest a clear logic and understanding of the role played by each protein. Of course this two-dimensional, albeit often colourful representation takes no account of the three-dimensional structure of a cell, the nature of the external and internal milieu, the dynamics of changes in protein levels and interactions, or the variations between cells in different tissues.

Each book in this series, entitled "*Proteins and Cell Regulation*", will seek to explore specific protein families or categories of proteins from the viewpoint of the general and specific functions they provide and their involvement in the dynamic behaviour of a cell. Content will range from basic protein structure and function to consideration of cell type-specific features and the consequences of diseaseassociated changes and potential therapeutic intervention. So that the books represent the most up-to-date understanding, contributors will be prominent researchers in each particular area. Although aimed at graduate, postgraduate and principal investigators, the books will also be of use to science and medical undergraduates and to those wishing to understand basic cellular processes and develop novel therapeutic interventions for specific diseases.

For further volumes:
http://www.springer.com/series/6442

Fedor Berditchevski • Eric Rubinstein
Editors

Tetraspanins

Springer

Editors
Fedor Berditchevski
School of Cancer Sciences
University of Birmingham
Birmingham, UK

Eric Rubinstein
Inserm/Universite Paris-Sud UMRS_1004
14 Avenue Paul Vaillant Couturier
Villejuif Cedex, France

ISBN 978-94-007-6069-1 ISBN 978-94-007-6070-7 (eBook)
DOI 10.1007/978-94-007-6070-7
Springer Dordrecht Heidelberg New York London

Library of Congress Control Number: 2013934542

Printed on acid-free paper

Springer is part of Springer Science+Business Media (www.springer.com)

Fedor Berditchevski dedicates this work to the memory of his father, Benor Berditchevski.

Foreword

Since arriving at the molecular scene more than 20 years ago tetraspanins continue to capture imagination of researchers in various fields of biology. Initially described as targets for tumour-specific mAbs (CO-029/TSPAN8, ME491/CD63), the anti-proliferative mAb (TAPA-1/CD81/TSPAN28), and the antibodies recognising "cluster differentiation" (CD) antigens (CD9/TSPAN29, CD37/TSPAN26 and CD53/TSPAN25), tetraspanins later came into prominence as regulators of membrane dynamics which play an important role in cell-cell fusion, cell adhesion and endocytic trafficking.

The first review article describing tetraspanins as a distinct family of four-transmembrane domain proteins (then called "tetraspans" or TM4SF proteins) was written in 1991 and over the years which followed a dozen of more reviews focussing on various aspects of tetraspanin function have been published. This volume represents a collection of 15 up-to-date articles covering the whole tetraspanin field.

The presence of four transmembrane domains separating two extracellular regions of unequal size, a number of conserved amino acids (including polar residues in the transmembrane domains) and a characteristic Cystein-Cystein-Glycine (CCG) triplet constitute the hallmarks of tetraspanins. High-resolution crystal structure of the large extracellular domain of CD81, and subsequent modeling indicated that despite extensive sequence divergence this domain is structurally conserved, and highly characteristic of tetraspanins. Structural features of tetraspanins are described in Chap. 1.

With completion of various genome sequencing projects it became clear that tetraspanins are present in all multicellular organisms: there are 33 identified tetraspanins in mammals, 20 in *C. elegans* and 37 in *Drosophila*. The intricacies of the evolution of the tetraspanin superfamily are discussed in Chap. 2.

What do tetraspanins do? Many researchers addressed this question through the identification of tetraspanin interacting proteins. This approach yielded nearly 60 tetraspanin-associated proteins. The interaction of many of these proteins with several tetraspanins, the interaction of tetraspanins with one another and other findings contributed to the idea that tetraspanins were organizing a network of interactions at the membrane referred to as "tetraspanin web", forming discrete tetraspanin-enriched microdomains (TERM). Chapters 3 and 4 will discuss the organisation and molecular dynamics of tetraspanins in the light of our idea of the "web" and TERMs.

Tetraspanins directly influence the function of the molecules they associate with, including binding of the associated receptors to their ligands, receptor oligomerisation and signal transduction (Chap. 4). Tetraspanins were also shown to regulate various aspects of endocytic trafficking of the associated proteins (Chap. 5). One of the preeminent interactions is that with a subset of integrins. This interaction is discussed in details in Chap. 6, in relation with the ability of tetraspanins to modulate adhesion and migration.

The function of tetraspanins in the context of the whole organism was investigated using knock-out mice. These experiments established the role of tetraspanins in maintaining the kidney structure, platelet aggregation, retinal vascularization and pathological angiogenesis (Chap. 7). They also revealed the requirement of tetraspanins for sperm-egg fusion (Chap. 9) and normal immunity (Chap. 10). The importance of tetraspanins in the life cycle and development of invertebrates, plants and fungi is discussed in Chap. 8.

Several tetraspanins have a distribution restricted to particular organs. Uroplakins (TSPAN20 and TSPAN21), principal components of urothelial plaques that cover almost the entire apical surface of the mammalian bladder urothelium, are expressed in urothelial cells. Their functions in urothelial biology and disease are discussed in Chap. 12. Peripherin/RDS (TSPAN22) and Rom-1 (TSPAN23) are retinal specific tetraspanins. These proteins are key regulators of the photoreceptor architecture and mutations RDS and Rom-1 were linked with retinal degenerative diseases (Chap. 13).

A number of tetraspanins have been described as tumour-specific antigens whose expression is deregulated during cancer development and progression. Recent studies have shown that the role of tetraspanins in cancer is more complex than previously thought: not only do the changes in expression of tetraspanins affect growth and invasive behaviour of tumour cells, but this also modifies tumour microenvironment. Chapter 11 summarizes these studies.

Tetraspanins have been hitchhiked by several pathogens. Cell infection by several viruses and bacteria can be affected by targeting certain tetraspanins. CD81 is special because it is absolutely required for the infection by two major human pathogens, the hepatitis C virus and the malaria parasite, as described in Chap. 15. The involvement of several tetraspanins in viral life cycles, especially HIV, is reviewed in Chap. 14.

Contents

Chapter 1
Structural Bases for Tetraspanin Functions

Michel Seigneuret, Hélène Conjeaud, Hui-Tang Zhang, and Xiang-Peng Kong

Abstract The tetraspanin transmembrane glycoproteins are considered as "molecular facilitators" which simultaneously interact with, and thereby bring into close proximity specific proteins involved in cellular activation and transduction processes. Elucidation of the 3D structure of tetraspanins is an essential step in understanding of their facilitator function and of the molecular basis of their partner specificity. Although there are currently no experimental atomic resolution structures of a whole tetraspanin molecule, recent information gained from three different approaches has led to a rather comprehensive picture of the structural organization of tetraspanins. These include: (1) crystallographic structures of the main extracellular domain of the ubiquitous tetraspanin CD81; (2) a 6 Å-resolution cryo-EM structure of the tetraspanins uroplakin UPIa and UPIb in the urothelial plaque of mammalian urothelium; (3) molecular modeled-structures of the complete CD81 tetraspanin. On the basis of such structural data, a qualitative view of tetraspanin structure-function relationship is emerging, including a delineation of regions of the molecule involved in specific interactions with partners, as well as an understanding of the structural basis of the multilevel partner specificity of tetraspanins and of the tetraspanin network organization.

M. Seigneuret (✉) • H. Conjeaud
Matière et Systèmes Complexes – CNRS UMR 7057, Université Paris-Diderot,
10 rue Alice Domon et Léonie Duquet, Paris cedex 13 75205, France
e-mail: michel.seigneuret@univ-paris-diderot.fr

H.-T. Zhang • X.-P. Kong
Department of Biochemistry and Molecular Pharmacology, New York University
School of Medicine, 550 First Avenue, New York, NY 10016, USA

F. Berditchevski and E. Rubinstein (eds.), *Tetraspanins*, Proteins and Cell Regulation 9,
DOI 10.1007/978-94-007-6070-7_1,

Abbreviations

AUM	Asymmetric unit membrane (AUM)
Cryo-EM	Cryo-electron microscopy
EC1	First tetraspanin extracellular region
EC2	Second tetraspanin extracellular region
EM	Electron microscopy
IC	Intracellular
TEM	Tetraspanin-enriched microdomain
TM	Transmembrane
UP	Uroplakin
UPEC	Uropathogenic E. coli
UPIa	Uroplakin Ia
UPIb	Uroplakin Ib
UPII	Uroplakin II
UPIIIa	Uroplakin IIIa
UTI	Urinary tract infection

1.1 Introduction

1.1.1 Functions of Tetraspanins

Tetraspanins constitute a superfamily of transmembrane glycoproteins that are involved in various aspects of the regulation of cellular development, proliferation, activation, and motility. The best characterized human members include CD81, CD9, CD53, CD82, CD151, CD37 and CD63 which are expressed in various cell types, as well as the tetraspanins with more specialized function and restricted distribution such as uroplakins Ia and Ib, found in the asymmetric unit membranes of the urothelium and RDS/peripherin and ROM, located in photoreceptor outer segment discs. Many studies suggest that the role of tetraspanins is related to their ability to interact with other proteins such as adhesion molecules, receptor and co-receptor molecules, major histocompatibility complex antigens, cytoplasmic kinases as well as other tetraspanins. Current hypotheses view tetraspanins as "molecular facilitators", the function of which would be to simultaneously interact with, and thereby bring into close proximity specific proteins involved in cellular processes (for reviews see Hemler 2003; Levy and Shoham 2005). Such properties lead to the formation of large membrane complexes involved in specific activation and transduction of signaling processes. Tetraspanins themselves undergo homologous and heterologous associations, which may form the basis of a tetraspanin web (Charrin et al. 2009a). Tetraspanins have been shown to interact with lipid rafts (Delaguillaumie et al. 2004; Xu et al. 2009) and have been suggested to be involved in microdomains called tetraspanin-enriched microdomain having composition properties and detergent-solubilization different from rafts but still involving cholesterol as a major lipid component (Berditchevski et al. 2002, for a recent review see Yanez-Mo et al. 2009).

1.1.2 Partner Specificities of Tetraspanin

A striking feature of tetraspanin associations with their partners is their multilevel specificity (for recent reviews on tetraspanin molecular interactions see Stipp et al. 2003; Charrin et al. 2009a; Yanez-Mo et al. 2009). Many interactions with partners appear to be specific to a single tetraspanin (e.g. CD19-CD81, PSG17-CD9, Pro-HB-EGF-CD9, UPIa with UPII, UPIb with UPIIIa). On the other hand, several tetraspanins have been shown to share common partners (CD9P1/EWI-F, EWI-2 or Claudin-1 with CD81 and CD9, MHC-I and MHC-II with CD81, CD82 and CD53). A particular case is constituted by integrins which appear to be privileged partners of tetraspanins. On the other hand, each specific integrin appear to interact only with one or two specific tetraspanin with variable affinity (e.g. $\alpha 3\beta 1$, $\alpha 6\beta 1$ and $\alpha 6\beta 4$ with CD151, $\alpha 4\beta 1$ with CD81, $\alpha 3\beta 1$, $\alpha 6\beta 1$ and $\alpha 1\beta 1$ with CD9, LFA-1 with CD82).

1.1.3 Structural Studies of Tetraspanins

Tetraspanins are characterized by four transmembrane segments (TM1-4) linked by one short extracellular (EC1), one short intracellular (IC) and one large extracellular (EC2) stretches (12). Tetraspanins also possess a number of conserved residues. The most conserved residues are an ubiquitous CCG motif as well as 2–6 other cysteines located on the EC2 stretch. Also significantly conserved are a number of very polar residues (Asn, Glu, Gln) and small size residues (Gly, Ala) in the transmembrane domain.

Elucidation of the 3D structure of tetraspanins is essential for understanding of their facilitator function and the molecular basis of partner specificity. A system of choice for such structural studies is naturally occurring two-dimensional crystals involving uroplakins UPIa and UPIb in the urothelial plaque of mammalian urothelium that has led to extensive investigation by cryoelectron microscopy (Min et al. 2006, see Sect. 1.5). Apart from this particular case, experimental structural studies of full-length tetraspanin molecules are lacking. This is presumably due to the usual difficulties associated with preparation of relatively large quantities of suitable samples of transmembrane proteins for X-ray crystallography, cryoelectron microscopy or high resolution NMR. However the recent reports describing high level expression systems for tetraspanins represent an encouraging step toward this goal (Jamshad et al. 2008; Takayama et al. 2008).

A usual approach to circumvent difficulties in production of full length transmembrane proteins for structural experiments is to study isolated aqueous domains of transmembrane proteins. This was done by Kitadokoro et al. (2001) who reported the crystallographic structure of a soluble form of the human tetraspanin CD81 EC2 domain (Fig. 1.1). The structure appears mushroom-shaped and consists of a

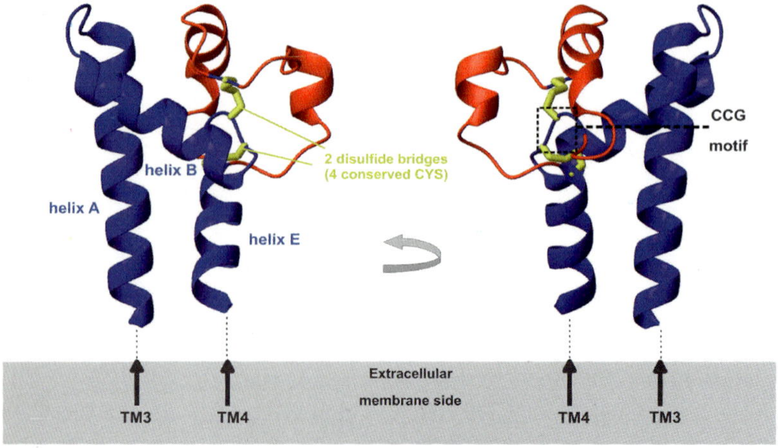

Fig. 1.1 Ribbon representation of the crystallographic structure of the human CD81 second extracellular domain EC2 according to Kitadokoro et al. (2001). The lower conserved membrane proximal subdomain is drawn in *blue* and the upper hypervariable subdomain is drawn in *red*

five-helix bundle stabilized by two disulfide bridges involving the CCG motif and two other conserved cysteines. The EC2 appears to be organized in two subdomains. The first, membrane proximal, subdomain involves two antiparallel helices (A and E), that form the stalk of the mushroom as well as a third helix (B) which is connected to helix A by a short loop. The second subdomain is sequentially inserted within the first subdomain and located on its top. It is composed of two shorter helices (C and D). The two disulfide bridges maintain the two subdomains in a defined orientation. Seigneuret et al. (2001) found that the structural features of the CD81 EC2 are conserved only partially among tetraspanins, with the membrane proximal subdomain being structurally conserved and the upper subdomain structurally hypervariable (see Sect. 1.3.3).

1.2 The CD81 Molecular Model and Its Structural Relevance for the Tetraspanin Superfamily

Considering the current shortcomings of experimental methods for structure determination of complete tetraspanins other than uroplakins, an alternate approach is structure prediction and molecular modeling. Seigneuret proposed in 2006 a molecular model of the complete structure of the human tetraspanin CD81 (Seigneuret 2006). While the crystallographic structure of the CD81 EC2 (Kitadokoro et al. 2001) was used as a starting point, the rest of the molecule was built from various prediction methods, including studies of the periodicity of sequence conservation of various properties, secondary structure prediction, protein docking and homology modeling (Fig. 1.2).

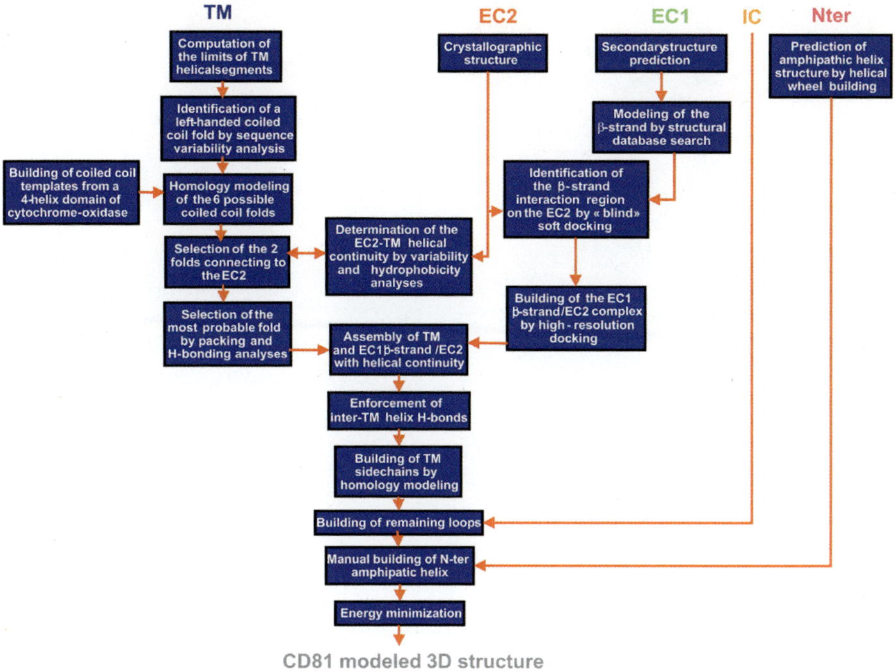

Fig. 1.2 Flowchart of the procedure used for molecular modeling of the human CD81 complete structure (Seigneuret 2006)

1.2.1 Description of Modeled Structure

The modeled structure of the complete CD81 molecule is shown in Fig. 1.3. The transmembrane domain is organized as a four-stranded left handed antiparallel coiled coil square bundle of helices. Two adjacent helices of this coiled coil directly connect to two helices of the larger extracellular domain EC2. Namely, after emerging from the bilayer, TM3 and TM4 become helices A and E of the EC2 without interruption of the helical conformation and remain packed, although departing from their coiled coil geometry due to the constraints of the EC2 tertiary structure. The smaller extracellular domain EC1 extents from the two other helices of the transmembrane coiled coil, TM1 and TM2, and packs against the conserved subdomain of the EC2 mainly on helices A, B and E. The EC1 contains a small partially hydrophobic β-strand which inserts into a conserved hydrophobic groove of the EC2 and runs roughly antiparallel to helix B. The EC2 hypervariable subdomain (Seigneuret et al. 2001), which for CD81 contains the two small helices C and D, is located on the side of the extracellular domain opposite to the EC1. It is linked to the conserved EC2 subdomain by two disulfide bridges. At the intracellular side, the N-terminal segment forms an amphipathic membrane-parallel helix when palmitoylated (palmitoyl residues are not shown on the figure). The IC loop, connecting transmembrane helices TM2 and TM3 is very short (four residues). The C-terminal segment is disordered.

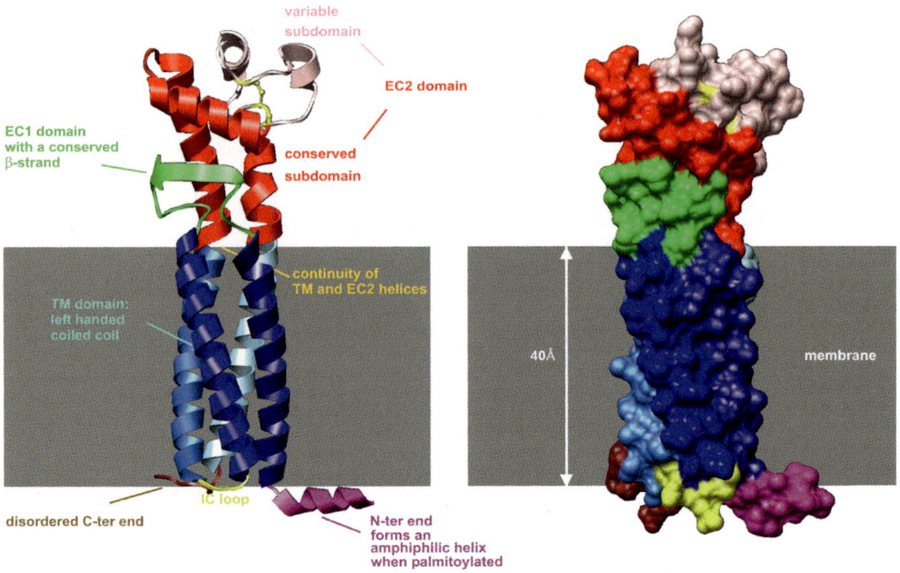

Fig. 1.3 The modeled CD81 3D structure. Ribbon (*left*) and surface (*right*) representation of the CD81 tertiary structure and topology. TM1-TM4, the conserved and variable subdomains of the EC2, the EC1, the IC loop and the N-terminal and C-terminal regions are respectively represented in *marine blue*, *blue*, *royal blue*, *light blue*, *red*, *pink*, *green yellow*, *magenta* and *brown*. Disulfide bridges are in *yellow*. The lipid bilayer (40 Å thickness) is depicted in *gray*

The structure of the modeled CD81 is very compact due to the tight packing of the left-handed coiled coil transmembrane domain and its continuity with the EC2 as well as to the packing of the EC1 on the EC2. Due to this tight packing of the EC1 in the EC2 groove, the whole extracellular domain more or less retains its mushroom shape. It protrudes out of the bilayer by 3.3 nm, i.e. about the size of an Ig domain. On the IC side, the N-terminal amphipathic helix, the IC loop and the C-terminal disordered domain emerge in the aqueous phase at similar levels, presumably interacting with the membrane surface. The modeled CD81 structure therefore assumes a cylindrical shape which is asymmetrically implanted in the bilayer, emerging much more on the extracellular than on the intracellular side.

1.2.2 Intrinsic Validity of Modeled Structure

Several arguments suggest that the above CD81 molecular model is plausible. Firstly, the model was built from predictions that, although mostly derived independently from each other, were found to constitute a self-consistent ensemble. The conclusion that the transmembrane domain of CD81 adopts a coiled coil fold was derived from analyses of the periodicity of sequence and residue size conservation in alignments of transmembrane regions of tetraspanin. This gave rises to six possible folds corresponding to permutations of four helices within a square bundle. Several criteria were used to select one particular fold corresponding to an antisymmetric four-stranded coiled coil. Independently, it was concluded from periodicity analyses of sequence and residue hydrophobicity conservation in the regions linking TM3 to EC2 helix A and TM4 to EC2 helix B, that helical continuity occurred in both cases. It then appeared that the selected transmembrane antisymmetric four-stranded coiled coil could be connected with such helical continuity to the EC2 crystallographic structure with adequate stereochemistry and sequence continuity. Furthermore, such connectivity imposed an orientation to the EC2 in which an existing hydrophobic groove, formed within helices A, B and E, was adequately positioned for interaction with the shorter EC1 domain. It was further predicted that the CD81 EC1 contains a short β-strand enriched in hydrophobic residues. Again independently, it was demonstrated by docking simulations that the EC2 hydrophobic groove was the preferred binding site for a β strand peptide corresponding to that predicted for the EC1. In all, molecular modeling of the CD81 structure was like assembling a puzzle from independently crafted pieces that were found to fit together.

Another argument in favor of the modeled CD81 structure is that its overall structural properties are similar to those of experimental transmembrane protein structures. Its membrane spanning transmembrane domain has a hydrophobic surface devoid of polar residues (Fig. 1.4a). Besides, this transmembrane domain is flanked on each side by so-called "aromatic belts" of external tyrosines and tryptophans found in many experimental protein structures and that are thought to provide anchoring to membrane interfaces (Lee 2003). Exterior phenylalanine sidechains are also present but occupy less superficial positions, as also found for experimental protein structures (Ulmschneider and Sansom 2001) (Fig. 1.4b). Besides, interactions between transmembrane helices in the modeled CD81 structures are similar to those found in experimental protein structure. Indeed, stability of transmembrane helical bundles result mainly from two factors (Liang 2002; Schneider 2004): (1) size complementarity between adjacent transmembrane helices mainly resulting from contact between small size and bulky residues leading to efficient Van der Waals interactions; (2) hydrogen bonding between polar residues located at helix interfaces. Both types of interactions exist in the CD81 modeled structure (see Sect. 1.3.2).

1.2.3 Comparison with Experimental Data

A first way of evaluating the accuracy of the CD81 model is to validate that residues subjected to post-translational modification in CD81 and other tetraspanin modifications are accessible and not buried in the molecule.

The structure was found to be in agreement with these data in that all such residues have a significant exposed surface area (see details in Seigneuret 2006). CD81 contains six palmitoylatable cysteines which are all found to be accessible in the

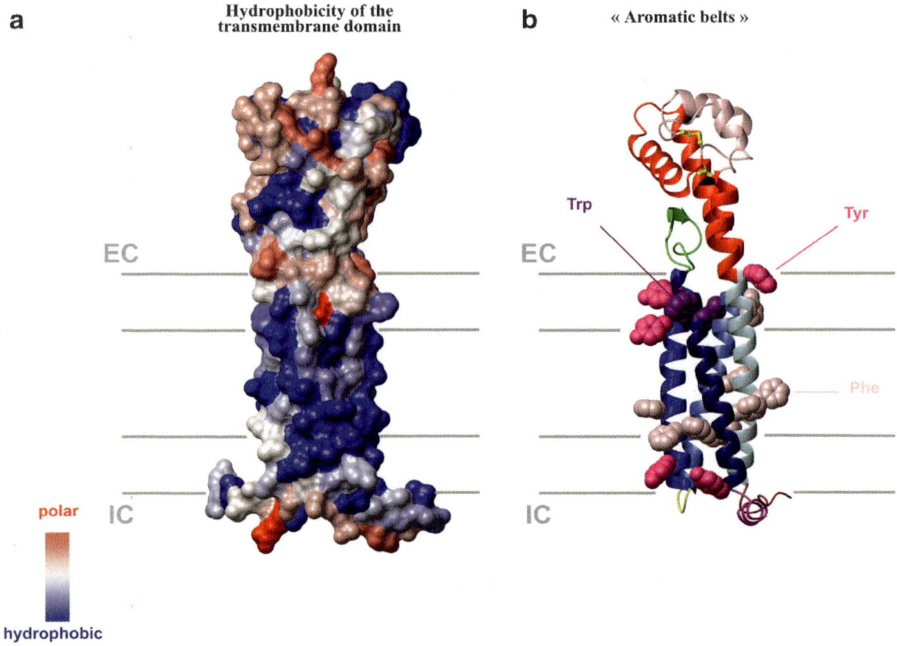

Fig. 1.4 Polarity and organization of aromatic residues in the modeled CD81 3D structure. (**a**) Surface representation of residue polarity calculated according to Eisenberg et al. (1984). The surface color ranges linearly from *red* to *blue*, corresponding respectively to polar and hydrophobic. (**b**) Mixed CPK/ribbon representation of the modeled CD81 structure highlighting exterior aromatic residues of the transmembrane domain. Phe, Tyr and Trp residues which are respectively colored *pink*, *magenta* and *purple*. The rest of the molecule is color-coded as in Fig. 1.3. The expected limits of the lipid bilayer (40 Å thickness) and of the hydrophobic region (30 Å thickness) are indicated as *thin gray lines*

CD81 structural model and located in regions of the modeled protein close to the intracellular lipid headgroup regions of the membrane. CD81 residues in the position similar to the positions of cysteines in CD151 are also accessible. Besides, unlike CD81, most tetraspanins are also glycosylated. In particular CD9 is glycosylated at two positions located in the EC1 and the corresponding CD81 residues are largely solvent accessible in the modeled structure. Finally, in the modeled CD81 structure, no EC2 residue known to be accessible either in CD81 itself or in other tetraspanins was found to be masked by the EC1. These include CD81 residues known to be involved in the interaction with HCV glycoprotein E2, as well as residues corresponding to glycosylation sites in CD53 or CD63 and a residue corresponding to a RDS/peripherin cysteine involved in disulfide-mediated dimer formation.

A stronger confirmation of the overall validity of the CD81 modeled structure appeared when a cryo-EM structure of the uroplakin complexes at 6 Å resolution was published (Min et al. 2006; see also Sect. 1.5). The experimental structures of uroplakin tetraspanins UPIa and UPIb were found to share the following features with the CD81 modeled structure: (1) cylindrical rod shape of the whole molecule; (2) arrangement of the four transmembrane helices as a square bundle with a left handed tilt; (3) similar arrangement and contacts of the four helices TM1-TM4 within the transmembrane bundle; (4) continuous extension of the transmembrane helices into the extracellular domains EC1 and EC2 yielding a similar orientation of the extracellular region relative to the transmembrane region; (5) close packing of the EC1 on an hydrophobic region of the EC2. While there are however a few differences between the two structures, these can be straightforwardly explained. In particular, Min et al. (2006) found that the cross angles between adjacent helix pairs are ~10° and ~25°. In the modeled CD81 structure, all such cross angles have an identical value of ~20° (unpublished). While this later value is in between the experimental values for UPIa/UPb, this difference likely represents an approximation related to the modeling procedure for which a very regular left-handed antisymmetric coiled coil was used as a template for the CD81 transmembrane domains. Indeed coiled coil arrangements of transmembrane helices have been reported to be less regular than for soluble coiled coils (Langosch and Heringa 1998; Walshaw and Woolfson 2001). On the other hand, both the UPIa/b experimental structure and the CD81 modeled structure have similar interaxis distances between adjacent transmembrane helices (~10 Å). Another difference is that the UPIa/b EC1, although closely packed to the EC2 hydrophobic surface as in the CD81 model, does not appear to contain an extended β strand. However, Seigneuret (2006) found that, according to secondary structure prediction methods, only ~70% of tetraspanins contain a β strand within the EC1. UPIa and UPIb were among those for which no β strand was found. It is probable that for several tetraspanins, insertion of the EC1 into the specific geometry of the hydrophobic groove of the EC2 is associated with distortion from the β strand geometry. In all, it appears that there is a good agreement between the UPIa and UPIb cryo-EM structure and the CD81 modeled structure.

1.3 General Consequences of the CD81 Model for Tetraspanin Structure

1.3.1 Overall Structure

The availability of the CD81 modeled structure raises the question as to which aspects of the structure are common to all tetraspanins. The striking resemblance found with the cryo-EM structure of UPIa/UPIb (Min et al. 2006), although these are relatively distant members (only 12–13% sequence identity), suggests that several features are ubiquitous. Those probably include the arrangement of the transmembrane domains as a relatively square bundle with a left handed tilt, the relative arrangement of transmembrane helices TM1-TM4 within the bundle, the continuity of transmembrane helices with the extracellular regions and the tight packing of the EC1 with the EC2.

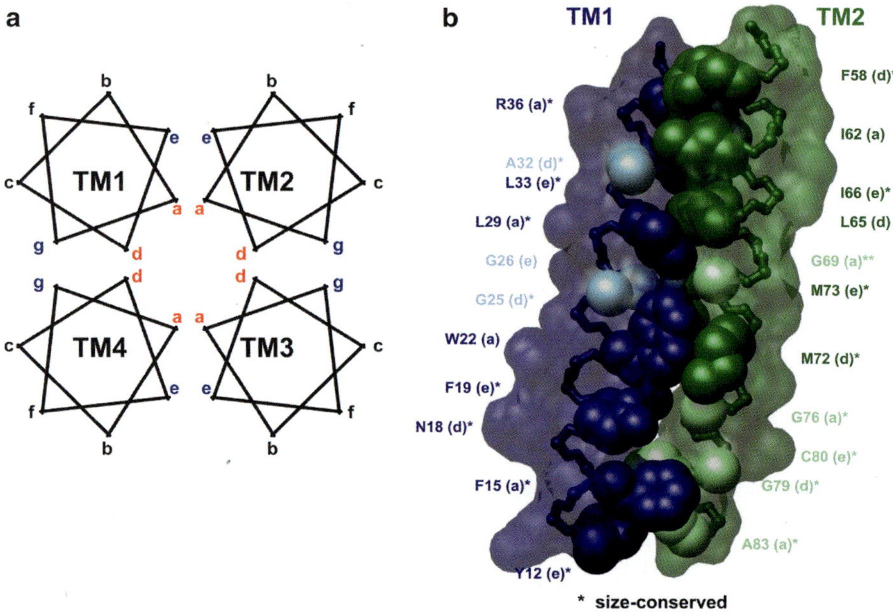

Fig. 1.5 Transmembrane helix interactions in the modeled CD81 3D structure: (**a**) Helical wheel representation of the four stranded antisymmetric coiled coil organisation of the TM domain of CD81. Residues at positions *a* and *d* are buried and contribute mainly to helix packing. Residues at positions *e* and *g* are semi-exposed and contribute to both helix packing and the molecular surface. (**b**) Interhelix packing with complementarity of size-conserved bulky and small interior sidechains for the TM1-TM2 interface. The peptide backbone of both helices and the C_αs plus side chains of residues contributing to the helix-helix interface (coiled coil residues *a*, *d* and *e*) are shown respectively as bond and atom representation. Small and large residues are colored in *light* and *dark blue* for TM1 and in *light* and *dark green* for TM2, respectively. Size-conserved residues among tetraspanins with more than 30% sequence homology with CD81 are indicated by an *asterisk*

1.3.2 Coiled Coil Interactions and Hydrogen Bonding in the Transmembrane Domain

It appears likely that the arrangement of the transmembrane domain as a four-stranded antisymmetric left-handed coiled coil is common to all tetraspanins. Due to a more regular pitch of individual helices (3.5 residues per helix turn), the packing of coiled coils (Fig. 1.5a) involves specific core residues positions termed *a* and *d* for the more internal and *e* and *g* for the more external with repetitive spacing in the sequences (heptad repeats) (Lupas 1996). Helix interactions in soluble (Lupas 1996) or transmembrane (Langosch and Heringa 1998; Javadpour et al. 1999; Liu et al. 2004) coiled coils are known as “knobs-into-holes” type of interactions between sidechains of residues located at the critical *a*/*d* and *e*/*g* helix positions. For transmembrane coiled coils, the importance of residue sidechain type or volume conservation among homologous proteins in such packing has been emphasized

Table 1.1 Interhelix proximities between small and bulky core residue pairs in the transmembrane domain of the modeled CD81 structure

TM1-TM2	TM2-TM3	TM3-TM4	TM4-TM1
Y12(*e*)*-A83(*a*)*	G61(*g*)-W111(*d*)	L91(*a*)*-C227(*e*)	A208(*g*)-L35(*g*)
Y12(*e*)*-C80 (*e*)	I62(*a*)-A108(*d*)	F94(*d*)*-S223(*a*)*	I215(*g*)*-G25(*g*)*
F15(*a*)*-A83(*a*)*	L65(*d*)-A108(*d*)	F95(*e*)-C227(*e*)	V212(*d*)*-A32(*d*)*
F15(*a*)*-C80 (*e*)	L65(*d*)-C104(*g*)	E105(*a*)*-A213(*e*)*	S223(*a*)*-N18(*d*)*
F15(*a*)*-G79(*d*)*	G69(*a*)*-L101(*d*)*	G112(*a*)-I205(*d*)*	
F19(*e*)-C80 (*e*)	M72(*d*)*-C91(*a*)		
W22(*a*)-G69 (*a*)*	G76(*a*)*-F94(*d*)*		
G26(*e*)-M73(*e*)*	G79(*d*)*-L90(*g*)*		
L29(*a*)*-G69(*a*)*			

Coiled coil positions of the residues are indicated in *italic*. Proximities were measured using a 1 Å distance cutoff between sidechains (or C_α for glycines) to account for the uncertainties of the modeling. Residues that are size-conserved are indicated by an *asterisk*

(Eilers et al. 2002). Seigneuret (2006) analyzed sequences alignments of each transmembrane helix of tetraspanins with at least ~30% homology with CD81 and found definite conserved heptad repeats for both residue types and residue size (although not as regular as for soluble coiled coils). Indeed, many transmembrane left-handed coiled coils are known to depart locally from a regular knobs-into-holes arrangement (Langosch and Heringa 1998; Walshaw and Woolfson 2001). For tetraspanins with lower homology, some deviations from size-conserved heptad repeats occur for a limited number of positions. This likely arises from correlated residue substitutions (Gobel et al. 1994; Kovalenko et al. 2005) in which concerted residue size variations occur. These data suggest a left-handed antisymmetric coiled coil organization for TM1-4 of all tetraspanins. On the other hand, actual tetraspanin transmembrane coiled coils are probably less regular than in the CD81 model, as suggested by the cryo-EM UPIa/UPIb structure (Min et al. 2006).

Such coiled coil organization gives rise to contacts of small and bulky core residues at the interface between adjacent helices that promote shape complementarity and efficient Van der Waals interactions. In Fig. 1.5b, as an example, the interface between TM1 and TM2 is depicted. Although modeling of the sidechain conformation and packing is likely to be approximate (so that atomic contacts are not always effective), there is a visible size complementarity of several small and bulky core residues of the two helices, most of which are size-conserved. In particular, near the intracellular end of TM2, there is a cluster of small residues (G69, G76, G79, C80, G82 and A83). These residues mainly interact with another cluster of very bulky residues of TM1 (Y12, F15, N18, F19 and W22) in a size-complementary manner. Table 1.1 lists proximities between small and bulky core residues in all adjacent transmembrane helix pairs. More than half of such proximities occur between pairs of size-conserved residues, suggesting that size complementarity may indeed play a role in the stability of the CD81 transmembrane domain. This stabilizing role of residue size complementarity provides an explanation for the conservation of many small and bulky residues in the transmembrane domain of tetraspanins.

Kovalenko et al. (2005) also evidenced the role of contact between small and bulky residues in the TM1-TM2 interaction using an original Monte Carlo modeling procedure taking into account correlated substitutions. These authors, using alignments of 28 tetraspanin sequences, reported the occurrence of heptad repeats only for TM1, TM2 and TM3. It must be noted that their TM4 alignment of CD81 and CD9 with other sequences is different from that of Seigneuret (2006). Kovalenko et al. (2005) also found that mutations of conserved TM1 and TM2 interior glycines in human CD9 caused aggregation of mutant proteins inside the cell suggesting misfolding of the transmembrane domain.

The CD81 model also suggests that hydrogen bonds can be formed between interior polar residues of different transmembrane helices involving conserved very polar residues such as asparagine, glutamic acid and glutamine. Three hydrogen bonds were proposed for CD81: N18 (TM1)—S223 (TM4); W22 (TM1)—E219 (TM4) and E109 (TM3)—A209 peptide carbonyl (TM4) (Seigneuret 2006). The hydrogen bonding network may be in part different for other tetraspanins, particularly since the position of the E219 residue is specific to CD81. While there is also a very polar (usually Q or E) residue in TM4 in many other tetraspanins, it is located farther from the cytoplasmic side. In a recent modeling study, Bari et al. (2009) suggested the following interhelix hydrogen bonds for human CD82: N17 (TM1)—S249 (TM4), as in CD81, and Q99 (TM3)—E242 (TM4). This represents probably the most common situation since the corresponding very polar (Q, E, D or N) interior residue positions are shared by the majority of tetraspanins, including CD151, CD53 and CD37. On the other hand, other tetraspanins lack some of these very polar residues such as CD63 (in TM1) or CD9 (in TM4). Here other hydrogen bonding schemes involving interior serine or threonine residues or peptide carbonyls must occur. It is interesting to emphasize that, in almost all tetraspanins, TM2 is devoid of interior polar residue and, therefore, it is likely to interact with adjacent helices only through Van der Waals contact interactions. On the other hand, while the exact interhelix hydrogen-bonding network may be variable in tetraspanins, it always involves TM1-TM4 and TM3-TM4 interaction.

In fact, the framework of transmembrane helix interactions in tetraspanins seems to be similar to that of GPCRs. Liu et al. (2004) have emphasized the importance of the size conservation of interior small residues often located at *a* and *d* coiled coil positions in the packing of GPCR transmembrane helices. They found that while a specific position may be size-conserved among the whole GPCR superfamily, each subfamily may have a preference for a specific residue type. A similar situation is indeed found for tetraspanins. Also, while the transmembrane helices of GPCR are linked by hydrogen bonds, the exact pattern of such bonds depends on the subfamily (Lomize et al. 1999), as is also likely for tetraspanins.

Interestingly, Kovalenko et al. (2005) found that mutations of interior small residues involved in coiled coil interhelix interactions diminished intermolecular CD9 interactions. Bari et al. (2009) found that mutation of all three very polar interior residues of CD82 involved in interhelix hydrogen bonds affected its interaction with CD9 and CD151 (Bari et al. 2009). This suggests that interactions with other transmembrane partners are critically dependent upon intramolecular interactions within transmembrane helices of tetraspanins.

1.3.3 Organization of the Extracellular Domain

In the human CD81 EC2 crystal structure, a surface hydrophobic patch contributed by residues from helices A, B and E is apparent and corresponds to crystallographic contact between adjacent EC2 molecules. The conservation of some of these residues among tetraspanins led to the suggestion that the hydrophobic patch might be involved in tetraspanin interactions with themselves (Kitadokoro et al. 2001). Here, in the modeled CD81 structure, these residues are in part masked by their interaction with the EC1, a fact that is also found in the UPIa/UPIb cryo-EM structure. This suggests that such masked residues are actually conserved internal hydrophobic residues of the extracellular domain that become unmasked in the soluble EC2. The idea that the EC2 hydrophobic patch is not involved in tetraspanin-tetraspanin interaction is consistent with the report (Berditchevski et al. 2001) that the removal of the EC2 by mutagenesis of CD151 does not affect its association with itself and other tetraspanins (CD9, CD81, CD63). Moreover, a study by Drummer et al. (2005) reports that mutations of two residues involved in the CD81 EC2 hydrophobic patch (F150 and V146), although decreasing EC2 oligomerization in solution, has no detectable effect on CD81 homodimerization in situ (Drummer et al. 2005). This suggests that other structural factors are involved in tetraspanin-tetraspanin interactions.

Secondary structure prediction (with ~75% accuracy) suggests that ~70% of tetraspanins possess a β-strand region in the EC1. These include among others CD81, CD82, CD9, CD151, CD37 and CD63. As stated above for UPIa/b, this may not be the case for other tetraspanins for which the corresponding EC1 region while still packed in the EC2 groove may deviate significantly from the β-strand conformation. In all cases, this EC1 region is enriched in hydrophobic residues, which interact with conserved hydrophobic residues of the EC2 groove. In a previous study, the structural conservation and variability pattern of the EC2 was characterized among the tetraspanin superfamily. It was found that, in spite of limited sequence similarity, helices A, B and E of the EC2 form a structurally conserved subdomain among tetraspanins (Seigneuret et al. 2001). The present data suggests that, for ~70% of tetraspanins, the EC1 shares similar features, i.e. it has a largely conserved structure in spite of significant sequence divergence. Since the conserved EC2 subdomain and the EC1 are packed together, these appear to constitute a structurally-conserved extracellular subdomain. This conserved subdomain is topped by a smaller structurally hypervariable subdomain from the EC2 (Seigneuret et al. 2001, see Sect. 1.4.1).

1.3.4 Organization of the Tetraspanin Intracellular Region

Structural conservation on the intracellular side of tetraspanin appears to be more contrasted. The small IC loop connecting TM2 and TM3 is four residues long in the CD81 structural model. It corresponds to a sequence pattern which is found in

~60% of tetraspanins. It is therefore likely that the loop adopts a conformation similar to that of CD81 in these species. The N-terminal intracellular stretch of CD81, which contains two palmitoylatable cysteines, has been modeled as a membrane-parallel amphipathic helix and corresponding regions of ~40% tetraspanins are also predicted to have a comparable amphipathic pattern. Although many tetraspanins contain a single cysteine in this region, the missing cysteine is often replaced by a hydrophobic residue so that the amphipathic character is retained. The idea that this amphipathic helix is formed only upon palmitoylation suggests a possible mechanism for regulation of tetraspanin interactions. The heterologous interactions between tetraspanins have been shown to depend upon palmitoylation (Charrin et al. 2002). It is tempting to propose that interactions between the amphipathic N-terminal helices are involved in such interactions. Finally, the intracellular C-terminal stretch is among the most divergent regions in tetraspanins. While it is suggested to be disordered in CD81, it may adopt specific conformations in other members, especially since it is often involved in very specific functions (for review see Stipp et al. 2003; Charrin et al. 2009a).

1.4 Conservation and Variability in Human Tetraspanins and Specificity of Partner Associations

The multilevel specificity of interactions of tetraspanins with their partners raises two complementary questions: (1) what is the structural origin of the high diversity of tetraspanins partners specific for each member? (2) what is the structural origin of the occurrence of partners common to several tetraspanins? The results described above suggest that a large part of the tetraspanin molecule is overall structurally conserved for a majority of the family members and at least for the more common human members (CD81, CD82, CD9, CD151, CD37 and CD63). This corresponds to the transmembrane domains and the region of the extracellular domain directly connected to the transmembrane domains, namely the conserved part of the EC2 (including helices A, B and E) and the EC1 (including its β strand). For CD81 the conserved regions amount to ~75% of the molecule. In the large extracellular domain (hypervariable region of the EC2) and both intracellular extremities of the tetraspanin molecule, one finds regions that are structurally non-conserved (or less conserved) among the family members. This suggests that specificity for a partner in tetraspanins is dictated by two distinct types of variability at the molecular level: the occurrence of such structurally variable regions and variability of surface residues in structurally conserved regions.

1.4.1 Structural Variability in Tetraspanins

The best documented occurrence of structural variability in tetraspanins concerns the hypervariable subdomain of the EC2. Seigneuret and colleagues (2001) performed multiple sequence alignments and secondary structure predictions the EC2 of 43

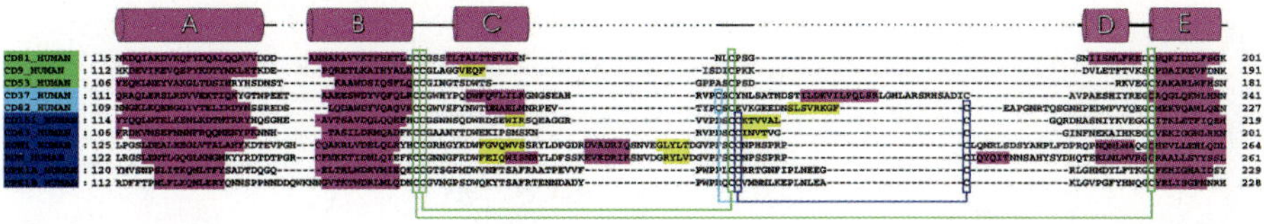

Fig. 1.6 Secondary structure prediction of the tetraspanin EC2 of common human tetraspanins. Sequences regions highlighted in *magenta* and *yellow* correspond, respectively, to helices and strands. The background colors of sequence names correspond to the different tetraspanin groups: *green*: group1, *light blue*: group 2a, *blue*: group 2b. The positions of the experimental helical regions of the CD81 EC2 structure are indicated as *magenta* tubes at the *top* of the figure. The *lines* connecting conserved cysteine residues indicate disulfide bridges with color coding corresponding to the tetraspanin groups. Secondary structure prediction was performed using the Jpred method (Cuff et al. 1998)

different tetraspanins types and compared the results to the crystallographic structure of the soluble CD81 EC2 domain (Kitadokoro et al. 2001). They found that the three helices A, B and E of the membrane proximal subdomain could be adequately predicted for all tetraspanins and possess similar lengths. On the other hand, the region located between the CCG motif and the last conserved cysteine, corresponding to helices C and D in CD81, is extremely variable in size and is composed of stretches separated by a variable number of cysteines and each yielding a predicted secondary structure variable from one tetraspanin to another (helix, strand, or loop). A multiple sequence alignment with predicted secondary structure is shown in Fig. 1.6 for the most common human tetraspanins. It was inferred from this study that the tetraspanin EC2 contains both conserved and hypervariable subdomains. The conserved subdomain retains a three helix bundle organization similar to that found in CD81 with helices A, B and E, the former and the later linking the EC2 to the transmembrane domain. The hypervariable subdomain is maintained in a defined orientation and topology with regards to the conserved subdomain by the two canonical disulfide bridges involving the CCG motif and the two other cysteines conserved in all tetraspanins. This variable subdomain is made of peptide stretches having secondary structure specific to each tetraspanin and substituting for helices C and D of CD81. In addition, this variable domain may contain one or two additional disulfide bridges. This led to a classification of tetraspanin according to the number of EC2 disulfide bridges (Fig. 1.6): group 1 (e.g. CD81, CD9, CD53) contains only the two canonical disulfide bridges; group 2 contain three disulfide bridges, it is further divided in group 2a (e.g. CD82, CD37) and group 2b (e.g. CD151, CD63, RDS/peripherin, ROM1, UPIa/UPIb) depending on the location of the third disulfide; group 3 contains four disulfide bridges and includes five tetraspanins (Tspan5, Tspan10, Tspan14, Tspan15, Tspan17 and Tspan33). Figure 1.7 shows, together with the experimental CD81 EC2 structure, homology models of two tetraspanins EC2 in which the variable stretches are non regular loops. The constant orientation and topology of the hypervariable subdomain is due not only to the two canonical disulfides (the relative orientation of which is due to the fact that each one involves successive cysteines in the CCG motif) but also to the conserved

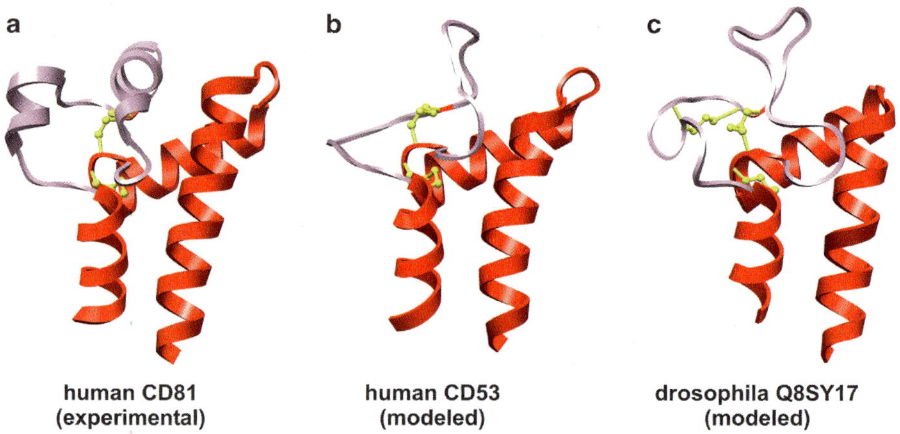

Fig. 1.7 Homology modeling of the tetraspanin EC2. A, ribbon representation of the experimental structure (Kitadokoro et al. 2001) of the human CD81 EC2 (**a**) and the predicted structures of the human CD53 (**b**) and *Drosophila Melanogaster* Q8SY17 (**c**) EC2. Conserved and variable subdomains are colored in *red* and *pink*, respectively. Disulfide bridges are in *yellow*

glycine and a conserved proline that impose specific conformational constraints to the main chain. A recent modeling study appears to confirm the occurrence of conserved and variable subdomains in peripherin/RDS (Vos et al. 2010). Using circular dichroism, secondary structure prediction and a "threading" molecular modeling approach, these authors predicted a structure with three helices at positions similar to helices A, B and E of CD81 and a region between helices B and E containing helices, strands and coils, structurally different from that of CD81.

The hypervariable subdomain in various tetraspanins is known to play an important role in their interaction with specific partners or ligands as well as in cellular processes for which tetraspanin partners still have to be identified (reviewed in (Stipp et al. 2003; Charrin et al. 2009a; Yanez-Mo et al. 2009)). Association of CD151 with α3 and α6 integrin was found to be critically dependant on a 194–196 QRD sequence located in this subdomain. Recently, the 185–192 region located upstream was also found important for the interaction with α3 integrins (Yamada et al. 2008). For CD9, known to interact laterally with proHB-EGF, residues G158, V159 and 175, also located in the variable subdomain appear to have essential roles in the upregulation of this receptor for binding of diphteria toxin (Hasuwa et al. 2001). Again for CD9, a 173–175 SFQ sequence appears essential for oocyte-sperm fusion (Higginbottom et al. 2003). Interestingly, one of these later CD9 residues is required for PSG17 binding. Also for CD81, almost all residues important for HCV E2 glycoprotein binding have been mapped in the hypervariable region (Kitadokoro et al. 2001; Drummer et al. 2002).

The hypervariable region of the EC2 represents the most salient example of how structural variability in tetraspanins may promote specificity for interaction with partners. However, structural variability at the cytoplasmic side may also contribute to such specificity. The amphipathic helix conformation of the palmitoylated N-terminal region proposed for CD81, as well as the sequence pattern responsible for the conformation of the IC loop appear to occur in ~40% and ~60% of tetraspanins, respectively. Therefore, other tetraspanins may possess different structures in these regions that may contribute to selectivity for partners. Furthermore, the C-terminal stretch is the most heterogenous region in tetraspanins both in sequence and length. Although disordered in CD81, it may adopt specific structures in other tetraspanins and seem to be associated with very specific interactions (Stipp et al. 2003). A likely role for such cytoplasmic regions is the binding of cytoplasmic signaling enzymes such as PKCs as suggested for CD9 (Zhang et al. 2001).

1.4.2 Surface Residue Variability and Conservation in the Structurally Conserved Regions of Tetraspanins

Apart from the occurrence of structurally variable and hypervariable domains, a likely origin for the partner specificity of tetraspanins lies in the variability of surface residues in the overall structurally conserved part common to most tetraspanins. This corresponds to a large median region corresponding to the whole transmembrane domain, and the structurally conserved part of the extracellular domain proximal to the transmembrane domain (i.e. the structurally conserved subdomain of the EC2 and the EC1). Recently, Conjeaud and Seigneuret (to be published) have used the CD81 modeled structure to delineate surface and buried residues in this structurally conserved region of the molecule. Multiple sequence alignments were then used to assign surface residues in the structurally conserved regions for the best characterized human tetraspanins (i.e. the CD's, RDS/peripherin, ROM1 and UPIa/b). Pairwise sequence conservation between such surface residues was then computed for each couple of tetraspanins, separately for the transmembrane domain and the structurally conserved part of the extracellular domain. The results are displayed in Fig. 1.8.

The first observation is that the pairwise homology between surface residues of the structurally conserved part of the extracellular domain is low, sometimes close to zero and rarely reaching above 20%. In addition, when mapped onto the CD81 modeled structure, the conserved residues are scattered over the molecular surface. This suggests that although this portion of tetraspanins is structurally conserved, it retains a potential for tetraspanin-specific interactions. Therefore aside the occurrence of structurally variable subdomains, a second possible origin for the diversity and specificity of tetraspanin partners is the high variability of the surface residues of the structurally conserved part of the extracellular domain. Whilst this idea will require further experimental support in the future, recent findings from Yalaoui and colleagues indicate that a 22 residue stretch encompassing the end of EC2 helix A, the loop connecting helices A and B and helix B are all important for the ability of CD81 to support infection of hepatocytes by Plasmodium yoeli (Yalaoui et al. 2008). It was inferred that this CD81 region is mandatory for its interaction with an unidentified partner that could function as a sporozoite receptor.

A second observation drawn from Fig. 1.8 is that the pairwise homology between surface residues of the transmembrane domain can reach much higher values, (from

	Pairwise homology of transmembrane domain surface residues (%)											
	CD81	CD9	CD82	CD151	CD53	CD37	CD63	UP1a	UP1b	ROM1	RDS	
CD81		54	39	34	38	29	27	20	20	18	13	CD81
CD9	20		43	44	29	29	28	23	22	10	15	CD9
CD82	8	5		29	35	44	28	15	11	16	10	CD82
CD151	10	10	21		30	25	27	23	18	14	11	CD151
CD53	15	3	23	23		32	31	10	17	16	10	CD53
CD37	10	15	22	10	7		19	14	15	14	11	CD37
CD63	7	5	18	8	13	13		15	18	12	12	CD63
UP1a	5	10	12	7	10	7	17		48	10	9	UP1a
UP1b	5	10	12	7	10	7	22	27		10	7	UP1b
ROM1	5	5	14	11	9	2	7	11	11		55	ROM1
RDS	9	9	11	7	9	7	7	20	9	25		RDS
	CD81	CD9	CD82	CD151	CD53	CD37	CD63	UP1a	UP1b	ROM1	RDS	
Pairwise homology of EC2 domain structurally conserved part surface residues (%)												

Fig. 1.8 Pairwise homology between surface-exposed residues of the structurally conserved portion of tetraspanins. Portions framed in *red* and *blue* correspond respectively to the surface of the transmembrane domain and to the surface of the EC2 domain structurally conserved portion. Pairwise homologies are indicated in % and also background color-coded in a linear *grayscale* (*white*: 0%, *black* 100%). Surface residues were selected as those having a side chain molecular surface exposure of more that 15% as measured with the Naccess program (http://www.bioinf.manchester.ac.uk/naccess/)

30% to more than 50%). However, such high values occur only for tetraspanin pairs involved in similar functional processes (e.g. peripherin/RDS and ROM1, UPIa/b) or for some of the more ubiquitous CD tetraspanins. The fact that tetraspanin pairs without similar functional involvement do not yield such a high homology indicate that the result is not due to a bias associated with the majoritary occurrence of hydrophobic residues on transmembrane surfaces (i.e. the average random residue homology at the surface of transmembrane domains is not higher than at the surface of soluble domains). This indicates a high conservation of the transmembrane domain molecular surface for some specific tetraspanin pairs. To study this property in more details, a planar representation of tetraspanin molecular surfaces was developed (Fig. 1.9) allowing one their direct comparison. Figure 1.10a shows the result of such comparison for CD81 and CD9 for which the pairwise identity between transmembrane surface residues reaches 54%. The mapping of conserved residues on the surface of the CD81 modeled structure is shown in Fig. 1.10b. It appears that the conserved residues are organized as continuous patches of variable sizes on the tetraspanin transmembrane domain molecular surface. This leads to the proposal that such conserved patches are interaction motifs for common partners of tetraspanins. Interestingly, these observations appear to be consistent with results concerning the regions implicated in binding of EWI-2 and CD9-P1/EWI-F to tetraspanins (common partners for both CD9 and CD81). The interaction of EWI-2 with CD9 was mapped to two regions: the EC2 and the part encompassing the second half of

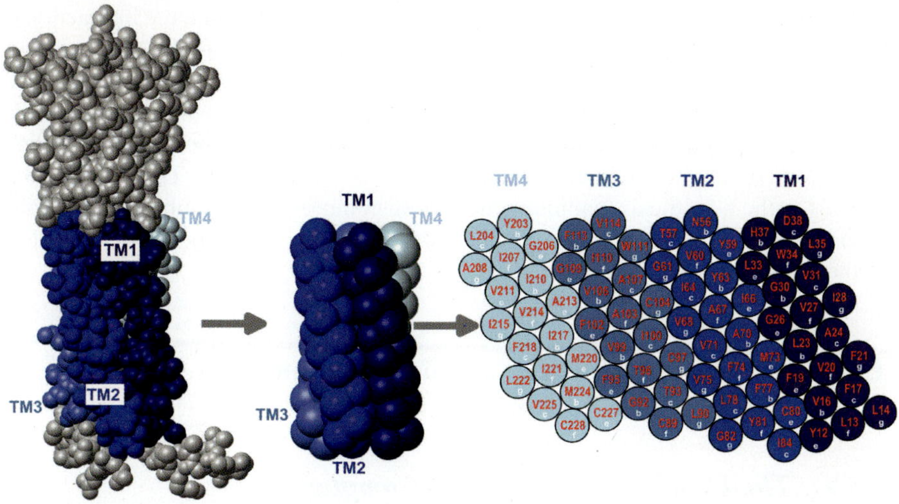

Fig. 1.9 Principle of the planar representation of tetraspanin molecular surfaces. Each residue contributing to the molecular surface of the transmembrane domain of the modeled CD81 structure (i.e. positions *b,c,f* and *g* in Fig. 1.5) was replaced by a sphere centered on its C_α. The rest of the molecule was discarded. The resulting cylinder was "cut" between TM4 and TM1, unfolded and projected onto a planar surface. A similar representation was then performed for each tetraspanin by matching the surface residues with those of CD81 using sequence alignments of the transmembrane domains. TM1-TM4 are respectively represented in *marine blue*, *blue*, *royal blue* and *light blue*

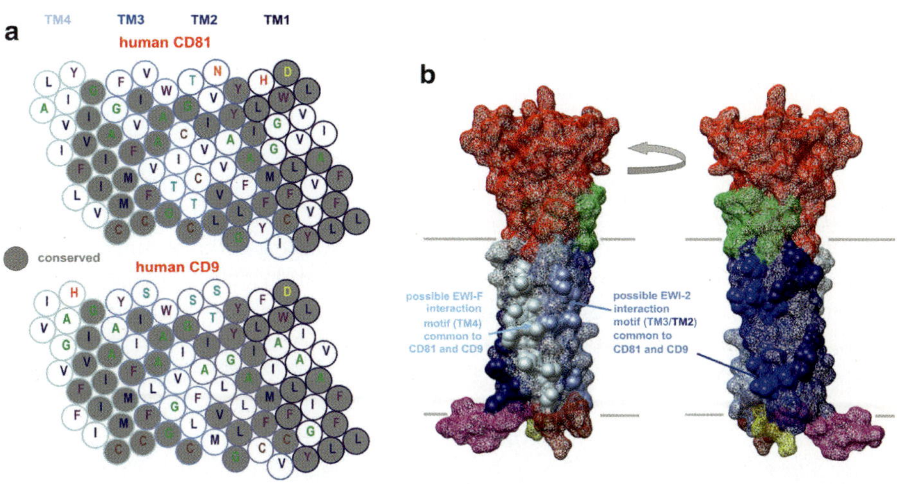

Fig. 1.10 (**a**) Detection of homologies between residues of the molecular surfaces of the transmembrane domains of CD81 and CD9. Conserved residues are on *gray* background. (**b**) Mapping of the conserved transmembrane domain surface residues between CD81 and CD9 on the molecular surface of the CD81 modeled structure. Contributions of the conserved residues are in solid surface, the rest of the molecule is in mesh surface. Colors are as in Fig. 1.3

TM2 and the whole TM3 (Charrin et al. 2003). On the other hand, the interaction of EWI-F with both CD9 and CD81 was found to involve TM4 and/or the C-terminus (although this latter region bears no homology in both tetraspanins) (Charrin et al. 2001, 2009a, b). As emphasized in Fig. 1.10b, there are patches of conserved transmembrane residues between CD81 and CD9 located at such positions that could correspond to interaction motifs with these two common partners. This lends further support to the hypothesis that interactions with common transmembrane partners of tetraspanins are due to the occurrence of conserved interaction motifs at the surface of the transmembrane domain.

1.5 Uroplakin Complex as a Unique Example of Tetraspanin Networks

1.5.1 The Uroplakin Tetraspanin Complex

Mammalial uroplakin (UP) tetraspanin complex constitutes two tetraspanins, UPIa and UPIb, and each is paired with a single transmembrane partner, UPII and UPIIIa, respectively (Fig. 1.11). It can serve as a unique example of tetraspanin complexes as it naturally forms a crystalline array (network) covering almost the entire apical surface of the epithelium of the lower urinary tract. The tetraspanin UPIa and UPIb are highly homologous, with 34% identities in amino acid sequence, and they belong to group 2b (see Sect. 1.4.1) of tetraspanins as their have three pairs of cysteine residues in their extracellular domains. The uroplakin crystalline protein array has a rather stiff concave appearance and it has been named urothelial plaques. It has also been called asymmetric unit membrane (AUM) as it was noticed, when it was discovered in the late 60s, that the extracellular leaflet is roughly twice thicker than the cytoplasmic leaflet when viewed in thin section EM (Koss 1969; Hicks 1975). The urothelial plaques are assembled in the cytoplasm and mature into discoidal fusiform vesicles before getting delivered to the apical plasma membrane. The lining of urothelial surface by urothelial plaques contributes to its functions as a permeability barrier keeping the urine separated from the cellular contents (Hicks 1966; Negrete et al. 1996), and it may in addition contribute to maintaining the integrity of the apical membrane during the dynamic cycles of micturition (Staehelin et al. 1972). The uroplakin complex can also serve as the surface receptor for uropathogenic E. coli (UPEC) (Zhou et al. 2001; Xie et al. 2006), which is the major causative agent for urinary tract infection (UTI), one of the most common infectious diseases. The type 1-piliated E. coli, which accounts for more than 85% of UPEC, gains a foothold in the urinary tract by binding the mannose moiety of UPIa via the bacterial adhesin lectin FimH located at the distal ends of its filamentous pili. This binding further triggers a signaling cascade in the urothelial umbrella cells, leading to a cytoskeletal rearrangement and internalization of the bacteria, a first step in establishing an infection (Mulvey et al. 1998; Wang et al. 2008, 2009). Hence, the uroplakin tetraspanin complex can serve as an example of how the tetraspanin

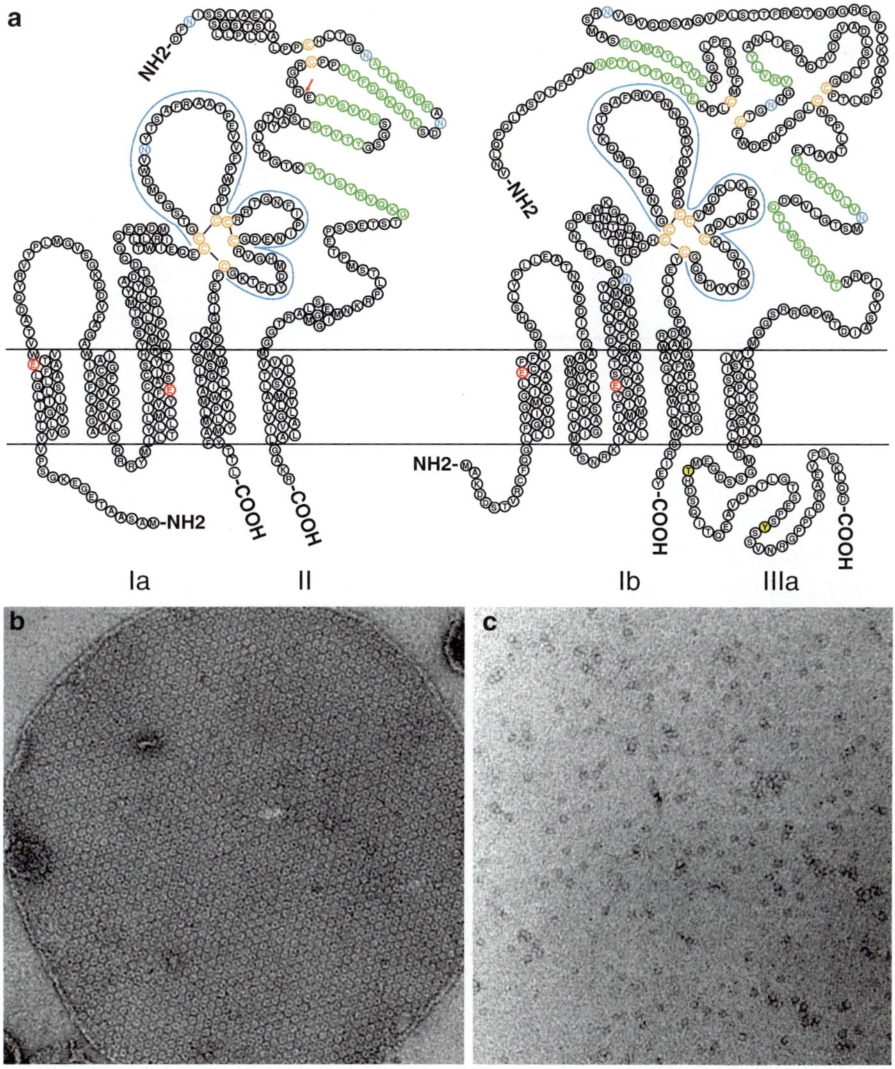

Fig. 1.11 Uroplakins and the urothelial plaques. (**a**) A schematic representation of the uroplakin heterodimers (mouse sequences). Tetraspanin UPIa and UPIb form heterodimers with the single transmembrane UPII and UPIIIa, respectively. The prosequence of UPII before the furin-cleavage site (*arrow*) is kept here to indicate that this region may be retained in the mature UP particles (Hu et al. 2008). The cysteine residues are arranged into proximity to indicate potential disulfide bonds, and the extracellular variable loops of the tetraspanin UPs are highlighted by *blue arches*. *Blue colored* Asn residues are potential glycosylation sites, and green residues are regions predicted to form beta strands. Like many other tetraspanins, the transmembrane domains of UPIa and UPIb contain Glu residues (*red*). Two potential phosphorylation sites (*yellow*) of UPIIIa are indicated (Mahbub Hasan et al. 2005; Thumbikat et al. 2009). (**b**) A mouse urothelial plaque visualized by negative staining EM. (**c**) Individual 16-nm uroplakin particles separated from the plaques by detergent wash

network regulates signaling pathways in pathogen invasion (for a recent review of uroplakin functions, see Wu et al. 2009).

1.5.2 Molecular Structure of the 16-nm Uroplakin Particles

The naturally occurring 2D crystalline urothelial plaques, with size up to 1 μm in diameter, can be isolated from animal tissue (Fig. 1.11b), allowing structural studies of the uroplakins by various biophysical methods. Low resolution techniques, including negative staining coupled with image processing (Hicks and Ketterer 1969; Vergara et al. 1969; Brisson and Wade 1983; Taylor and Robertson 1984; Walz et al. 1995; Min et al. 2002), quick-freeze deep-etch (Severs and Warren 1978; Kachar et al. 1999), and atomic force microscopy (Min et al. 2002) showed that the plaques are actually formed by protein particles of ~ 16 nm in diameter and a hexagonal stellate shape. Each 16 nm particle can be resolved into six inner and six outer subdomains forming two concentric rings. One inner and one outer subdomain interconnect to form a subunit—one sixth of the hexagonal particle. More recent cryo-electron microscopy (Cryo-EM) 3D reconstructions revealed that each subdomain contains a total of five transmembrane helices, corresponding to one tetraspanin and a single-pass uroplakin (Fig. 1.12) (Oostergetel et al. 2001; Min et al. 2003, 2006). The EM localization using the E. coli FimH adhesin, which specifically binds to UPIa, indicates that the UPIa/UPII pair occupies the inner subdomain, whereas the UPIb/UPIIIa pair occupies the outer subdomain (Min et al. 2002). The 6 Å resolution cryo-EM structure of the 16 nm particle allowed a visualization of the transmembrane helices of the uroplakins, and the configuration of the helices permitted the identification of how the uroplakins are paired (Min et al. 2006). The four transmembrane helices of tetraspanin UPIa and UPIb are tightly packed into bundles which extend to the extracellular domains thus giving these tetraspanins an overall cylindrical shape. This is consistent with the structural model of CD81 (see Sect. 1.2). The single-pass UPII and UPIIIa adopt an inverted 'L'-shape, anchored by its transmembrane helix packed against the four transmembrane helices of their specific tetraspanin partner tetraspanin, thus forming a five-helix bundle within each subdomain. The long arm of the inverted 'L' continues up against the cylindrical UP tetraspanins. The short arm of the 'L' extends to join the short arm of the other inverted 'L' from the paired subdomain within the same subunit, thus forming a 'joint'. Interestingly, the joint provides the only contact between the two subdomains within a subunit, while the two tetraspanins, UPIa and UPIb, do not appear to have any direct contact. This type of loose connection between the two subdomains within a subunit suggests a flexible interaction between the inner and outer subdomains, thus providing a basis for possible structural changes of UPs upon binding to the E. coli FimH adhesin (see below). The four transmembrane helix bundles of UPIa/Ib are left-handed, although the cross-angles (as defined according to Bowie 1997a), between these TMs are somewhat variable (Fig. 1.12c). In this regard, the four transmembrane helices of the UP tetraspanins can be

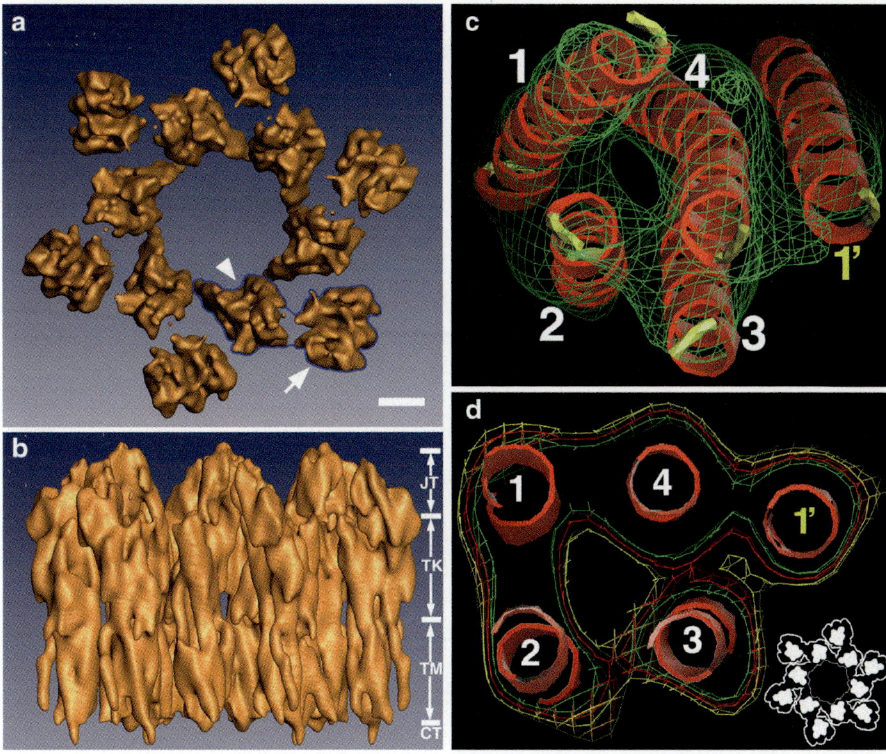

Fig. 1.12 Cryo-EM structure of the 16 nm uroplakin particle. (**a**) The *top* view of the hexagonal 16 nm UP particle. A subunit is indicated by an outline, which consists of an inner domain (*arrowhead*) and out domain (*arrow*). (**b**) The four zones of the uroplakin particle: from the top down, the joint (*JT*), the trunk (*TK*), the transmembrane (*TM*), and the cytoplasmic domain (*CT*). (**c**) The transmembrane helices of the UPIa/UPII tetraspanin pair viewed from the cytoplasmic side. The helices 1–4 belong to the tetraspanin UPIa and helix 1' belongs to UPII. (**d**) The section at the middle of the transmembrane region of UPIa/UPII pair. The inset indicates the position of the 5-helix bundle in the outline of the 16 nm particle

grouped into two pairs, the TM1–TM2 pair and the TM3–TM4 pair (Fig. 1.12c). The cross-angle between the two helices within each pair is ~10°, which is relatively small, whereas the cross-angle between the two pairs is ~25°, which is slightly larger than the 20° common cross-angle in transmembrane proteins in general (Bowie 1997b). The small packing cross-angles between the helices in the two pairs allow the helices within a pair to be closely associated with each other along the entire length (Eilers et al. 2002).

The molecular structure of the 16 nm particle may allow it to transmit a transmembrane signal upon bacterial attachment (Min et al. 2003; Wang et al. 2009). A key question of bacterial invasion in UTI is how an urothelial umbrella cell senses bacterial attachment and how the signal of bacterial attachment is transmitted transmembranously through the rather impermeable apical membrane. It has been shown recently that binding of the UPEC adhesin FimH can induce large conformational changes to the top joint region of the 16 nm UP particle, formed by the extracellular domains of UPII and UPIIIa, and that these conformational changes propagate through the transmembrane helices of the uroplakins to the cytoplasmic side of the membrane (Wang et al. 2009). These transmembranous conformational changes of the uroplakins may trigger, through linkers to cytoskeletons, the downstream cytoplasmic signaling cascades leading to cytoskeletal rearrangement and bacterial invasion. A recent study has shown that FimH binding can induce a phosphorylation of a Thr residue in the C-terminal of UPIIIa (Fig. 1.11a) (Thumbikat et al. 2009). Hence the mechanic/conformational signal induced by UPEC binding of the extracellular domains of the UP tetraspanin complex may be converted into a chemical signal that further triggers the downstream signaling cascade.

1.5.3 Network Formation of Uroplakin Tetraspanin Complex

The molecular structure of the uroplakin crystalline plaques provides an example of how tetraspanin networks can be formed (Min et al. 2006). There are three levels of interactions in the UP tetraspanin network (Fig. 1.13): (1) the primary interaction between a tetraspanin (UPIa or UPIb) and its single transmembrane partner (UPII or UPIIIa); (2) the secondary interaction between the primary complexes, and (3) the tertiary interaction between these secondary complexes (Hemler 2003; Levy and Shoham 2005; Martin et al. 2005). The primary interaction between UPIa/Ib and their partners is very extensive, involving both TM helices and extracellular domains, and the primary complexes—the subdomains in the 16 nm particle—can be isolated separately by ion exchange chromatography (Liang et al. 2001). There are two types of secondary interactions between the primary complexes in the UP tetraspanin network. One is between the UPIa/UPII and UPIb/UPIIIa primary complexes (i.e., the inner and outer subdomains), via the contact of UPII and UPIIIa at the joint, to form a subunit (six of which form a 16 nm particle). The other secondary interaction is between the UPIa/UPII complexes (the inner subdomains), via the contact between UPIa of one primary complex and UPII of a neighboring primary complex; this secondary interaction is responsible for linking the six inner subdomains to form the inner ring of the 16 nm particle. The third level of interaction is between the outer subdomains of neighboring 16 nm particles. This interaction can be visualized only in the electron density map at very low contour levels, and it may be rather weak, allowing dynamic movement of the particles in the membrane (Kachar et al. 1999).

1.6 Conclusion

Although there is currently no experimental atomic resolution structure of a whole tetraspanin molecule the current structural information allowed us to present a rather comprehensive picture of various intramolecular interactions that stabilize

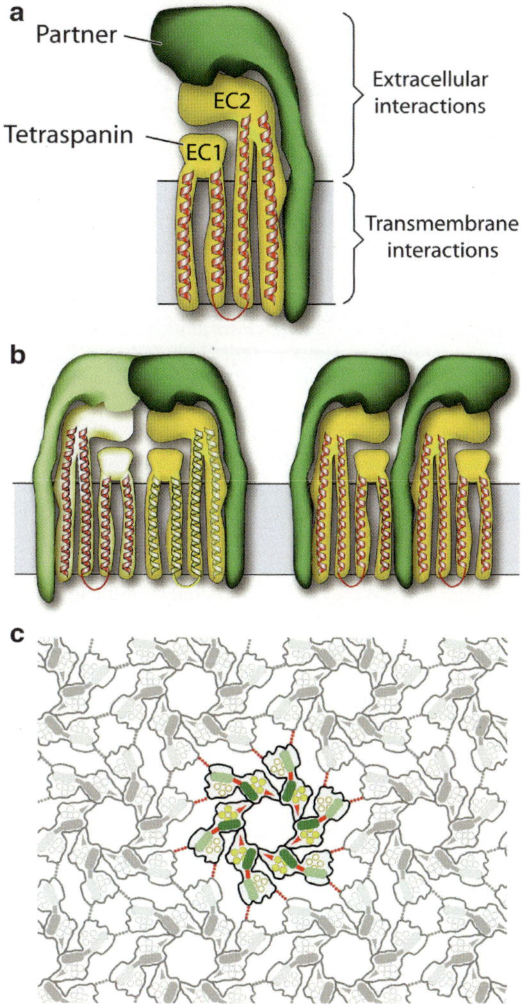

Fig. 1.13 Uroplakin tetraspanin network. (**a**) Primary interaction. UPIa or UPIb interacts with its partner UPII or UPIIIa. (**b**) Two types of secondary interactions. UPIa/II interacts with UPIb/UPIIIa primary complexes (*left*); UPIa/UPII interacts with another UPIa/UPII complexes (*right*). (**c**) Formation of the urothelial tetraspanin network. The outer subdomains of neighboring 16 nm particles interact each other (*dotted red lines*)

tetraspanin structure. The stable tetraspanin structure is ideal for docking other tall signaling transmembrane proteins, and help these proteins to pass messages into the cell. They are themselves sometimes the receptors and signaling molecules for some bacteria and viruses. The stable tetraspanins may thus act as stable pilings in a lipid sea for other floating proteins to dock and function.

Acknowledgments The work done in the authors' laboratories was supported by grants from the Centre National de la Recherche Scientifique, the Institut National de la Santé et de la Recherche Médicale, SIDACTION, the Association National pour la Recherche Contre le SIDA (to HC and MS) and NIH grant DK52206 (to HTZ and XPK).

References

Bari R, Zhang YH, Zhang F, Wang NX, Stipp CS, Zheng JJ, Zhang XA (2009) Transmembrane interactions are needed for KAI1/CD82-mediated suppression of cancer invasion and metastasis. Am J Pathol 174:647–660

Berditchevski F, Gilbert E, Griffiths MR, Fitter S, Ashman L, Jenner SJ (2001) Analysis of the CD151-alpha3beta1 integrin and CD151-tetraspanin interactions by mutagenesis. J Biol Chem 276:41165–41174

Berditchevski F, Odintsova E, Sawada S, Gilbert E (2002) Expression of the palmitoylation-deficient CD151 weakens the association of alpha 3 beta 1 integrin with the tetraspanin-enriched microdomains and affects integrin-dependent signaling. J Biol Chem 277(40):36991–37000, Oct 4

Bowie JU (1997a) Helix packing angle preferences. Nat Struct Biol 4:915–917

Bowie JU (1997b) Helix packing in membrane proteins. J Mol Biol 272:780–789

Brisson A, Wade RH (1983) Three-dimensional structure of luminal plasma membrane protein from urinary bladder. J Mol Biol 166:21–36

Charrin S, Le Naour F, Oualid M, Billard M, Faure G, Hanash SM, Boucheix C, Rubinstein E (2001) The major CD9 and CD81 molecular partner. Identification and characterization of the complexes. J Biol Chem 276:14329–14337

Charrin S, Manie S, Oualid M, Billard M, Boucheix C, Rubinstein E (2002) Differential stability of tetraspanin/tetraspanin interactions: role of palmitoylation. FEBS Lett 516:139–144

Charrin S, Le Naour F, Labas V, Billard M, Le Caer JP, Emile JF, Petit MA, Boucheix C, Rubinstein E (2003) EWI-2 is a new component of the tetraspanin web in hepatocytes and lymphoid cells. Biochem J 373:409–421

Charrin S, le Naour F, Silvie O, Milhiet PE, Boucheix C, Rubinstein E (2009a) Lateral organization of membrane proteins: tetraspanins spin their web. Biochem J 420:133–154

Charrin S, Yalaoui S, Bartosch B, Cocquerel L, Franetich JF, Boucheix C, Mazier D, Rubinstein E, Silvie O (2009b) The Ig domain protein CD9P-1 down-regulates CD81 ability to support plasmodium yoelii infection. J Biol Chem 284:31572–31578

Cuff JA, Clamp ME, Siddiqui AS, Finlay M, Barton GJ (1998) JPred: a consensus secondary structure prediction server. Bioinformatics 14:892–893

Delaguillaumie A, Harriague J, Kohanna S, Bismuth G, Rubinstein E, Seigneuret M, Conjeaud H (2004) Tetraspanin CD82 controls the association of cholesterol-dependent microdomains with the actin cytoskeleton in T lymphocytes: relevance to co-stimulation. J Cell Sci 117:5269–5282

Drummer HE, Wilson KA, Poumbourios P (2002) Identification of the hepatitis C virus E2 glycoprotein binding site on the large extracellular loop of CD81. J Virol 76:11143–11147

Drummer HE, Wilson KA, Poumbourios P (2005) Determinants of CD81 dimerization and interaction with hepatitis C virus glycoprotein E2. Biochem Biophys Res Commun 328:251–257

Eisenberg D, Schwarz E, Komaromy M, Wall R (1984) Analysis of membrane and surface protein sequences with the hydrophobic moment plot. J Mol Biol 179:125–142

Eilers M, Patel AB, Liu W, Smith SO (2002) Comparison of helix interactions in membrane and soluble alpha-bundle proteins. Biophys J 82:2720–2736

Gobel U, Sander C, Schneider R, Valencia A (1994) Correlated mutations and residue contacts in proteins. Proteins 18:309–317

Hasuwa H, Shishido Y, Yamazaki A, Kobayashi T, Yu X, Mekada E (2001) CD9 amino acids critical for upregulation of diphtheria toxin binding. Biochem Biophys Res Commun 289:782–790

Hemler ME (2003) Tetraspanin proteins mediate cellular penetration, invasion, and fusion events and define a novel type of membrane microdomain. Annu Rev Cell Dev Biol 19:397–422

Hicks RM (1966) The permeability of rat transitional epithelium. Kertinization and the barrier to water. J Cell Biol 28:21–31

Hicks RM (1975) The mammalian urinary bladder: an accommodating organ. Biol Rev Camb Philos Soc 50:215–246

Hicks RM, Ketterer B (1969) Hexagonal lattice of subunits in the thick luminal membrane of the rat urinary bladder. Nature 224:1304–1305

Higginbottom A, Takahashi Y, Bolling L, Coonrod SA, White JM, Partridge LJ, Monk PN (2003) Structural requirements for the inhibitory action of the CD9 large extracellular domain in sperm/oocyte binding and fusion. Biochem Biophys Res Commun 311:208–214

Hu CC, Bachmann T, Zhou G, Liang FX, Ghiso J, Kreibich G, Sun TT (2008) Assembly of a membrane receptor complex: roles of the uroplakin II prosequence in regulating uroplakin bacterial receptor oligomerization. Biochem J 414:195–203

Jamshad M, Rajesh S, Stamataki Z, McKeating JA, Dafforn T, Overduin M, Bill RM (2008) Structural characterization of recombinant human CD81 produced in Pichia pastoris. Protein Expr Purif 57:206–216

Javadpour MM, Eilers M, Groesbeek M, Smith SO (1999) Helix packing in polytopic membrane proteins: role of glycine in transmembrane helix association. Biophys J 77:1609–1618

Kachar B, Liang F, Lins U, Ding M, Wu XR, Stoffler D, Aebi U, Sun TT (1999) Three-dimensional analysis of the 16 nm urothelial plaque particle: luminal surface exposure, preferential head-to-head interaction, and hinge formation. J Mol Biol 285:595–608

Kitadokoro K, Bordo D, Galli G, Petracca R, Falugi F, Abrignani S, Grandi G, Bolognesi M (2001) CD81 extracellular domain 3D structure: insight into the tetraspanin superfamily structural motifs. EMBO J 20:12–18

Koss LG (1969) The asymmetric unit membranes of the epithelium of the urinary bladder of the rat. An electron microscopic study of a mechanism of epithelial maturation and function. Lab Invest 21:154–168

Kovalenko OV, Metcalf DG, Degrado WF, Hemler ME (2005) Structural organization and interactions of transmembrane domains in tetraspanin proteins. BMC Struct Biol 5:11

Langosch D, Heringa J (1998) Interaction of transmembrane helices by a knobs-into-holes packing characteristic of soluble coiled coils. Proteins 31:150–159

Lee AG (2003) Lipid-protein interactions in biological membranes: a structural perspective. Biochim Biophys Acta 1612:1–40

Levy S, Shoham T (2005) The tetraspanin web modulates immune-signalling complexes. Nat Rev Immunol 5:136–148

Liang J (2002) Experimental and computational studies of determinants of membrane-protein folding. Curr Opin Chem Biol 6:878–884

Liang FX, Riedel I, Deng FM, Zhou G, Xu C, Wu XR, Kong XP, Moll R, Sun TT (2001) Organization of uroplakin subunits: transmembrane topology, pair formation and plaque composition. Biochem J 355:13–18

Liu W, Eilers M, Patel AB, Smith SO (2004) Helix packing moments reveal diversity and conservation in membrane protein structure. J Mol Biol 337:713–729

Lomize AL, Pogozheva ID, Mosberg HI (1999) Structural organization of G-protein-coupled receptors. J Comput Aided Mol Des 13:325–353

Lupas A (1996) Coiled coils: new structures and new functions. Trends Biochem Sci 21:375–382

Mahbub Hasan AK, Sato K, Sakakibara K, Ou Z, Iwasaki T, Ueda Y, Fukami Y (2005) Uroplakin III, a novel Src substrate in Xenopus egg rafts, is a target for sperm protease essential for fertilization. Dev Biol 286:483–492

Martin F, Roth DM, Jans DA, Pouton CW, Partridge LJ, Monk PN, Moseley GW (2005) Tetraspanins in viral infections: a fundamental role in viral biology? J Virol 79:10839–10851

Min G, Stolz M, Zhou G, Liang F, Sebbel P, Stoffler D, Glockshuber R, Sun TT, Aebi U, Kong XP (2002) Localization of uroplakin Ia, the urothelial receptor for bacterial adhesin FimH, on the six inner domains of the 16 nm urothelial plaque particle. J Mol Biol 317:697–706

Min G, Zhou G, Schapira M, Sun TT, Kong XP (2003) Structural basis of urothelial permeability barrier function as revealed by Cryo-EM studies of the 16 nm uroplakin particle. J Cell Sci 116:4087–4094

Min G, Wang H, Sun TT, Kong XP (2006) Structural basis for tetraspanin functions as revealed by the cryo-EM structure of uroplakin complexes at 6-A resolution. J Cell Biol 173:975–983

Mulvey MA, Lopez-Boado YS, Wilson CL, Roth R, Parks WC, Heuser J, Hultgren SJ (1998) Induction and evasion of host defenses by type 1-piliated uropathogenic Escherichia coli. Science 282:1494–1497

Negrete HO, Lavelle JP, Berg J, Lewis SA, Zeidel ML (1996) Permeability properties of the intact mammalian bladder epithelium. Am J Physiol 271:F886–F894

Oostergetel GT, Keegstra W, Brisson A (2001) Structure of the major membrane protein complex from urinary bladder epithelial cells by cryo-electron crystallography. J Mol Biol 314:245–252

Schneider D (2004) Rendezvous in a membrane: close packing, hydrogen bonding, and the formation of transmembrane helix oligomers. FEBS Lett 577:5–8

Seigneuret M (2006) Complete predicted three-dimensional structure of the facilitator transmembrane protein and hepatitis C virus receptor cd81: conserved and variable structural domains in the tetraspanin superfamily. Biophys J 90:212–227

Seigneuret M, Delaguillaumie A, Lagaudriere-Gesbert C, Conjeaud H (2001) Structure of the tetraspanin main extracellular domain. A partially conserved fold with a structurally variable domain insertion. J Biol Chem 276:40055–40064

Severs NJ, Warren RC (1978) Analysis of membrane structure in the transitional epithelium of rat urinary bladder. 1. The luminal membrane. J Ultrastruct Res 64:124–140

Staehelin LA, Chlapowski FJ, Bonneville MA (1972) Lumenal plasma membrane of the urinary bladder. I. Three-dimensional reconstruction from freeze-etch images. J Cell Biol 53:73–91

Stipp CS, Kolesnikova TV, Hemler ME (2003) Functional domains in tetraspanin proteins. Trends Biochem Sci 28:106–112

Takayama H, Chelikani P, Reeves PJ, Zhang S, Khorana HG (2008) High-level expression, single-step immunoaffinity purification and characterization of human tetraspanin membrane protein CD81. PLoS One 3:e2314

Taylor KA, Robertson JD (1984) Analysis of the three-dimensional structure of the urinary bladder epithelial cell membranes. J Ultrastruct Res 87:23–30

Thumbikat P, Berry RE, Zhou G, Billips BK, Yaggie RE, Zaichuk T, Sun TT, Schaeffer AJ, Klumpp DJ (2009) Bacteria-induced uroplakin signaling mediates bladder response to infection. PLoS Pathog 5:e1000415

Ulmschneider MB, Sansom MS (2001) Amino acid distributions in integral membrane protein structures. Biochim Biophys Acta 1512:1–14

Vergara J, Longley W, Robertson JD (1969) A hexagonal arrangement of subunits in membrane of mouse urinary bladder. J Mol Biol 46:593–596

Vos WL, Vaughan S, Lall PY, McCaffrey JG, Wysocka-Kapcinska M, Findlay JB (2010) Expression and structural characterization of peripherin/RDS, a membrane protein implicated in photoreceptor outer segment morphology. Eur Biophys J 39:679–688

Walshaw J, Woolfson DN (2001) Socket: a program for identifying and analysing coiled-coil motifs within protein structures. J Mol Biol 307:1427–1450

Walz T, Haner M, Wu XR, Henn C, Engel A, Sun TT, Aebi U (1995) Towards the molecular architecture of the asymmetric unit membrane of the mammalian urinary bladder epithelium: a closed "twisted ribbon" structure. J Mol Biol 248:887–900

Wang H, Liang FX, Kong XP (2008) Characteristics of the phagocytic cup induced by uropathogenic *Escherichia coli*. J Histochem Cytochem 56:597–604

Wang H, Min G, Glockshuber R, Sun TT, Kong XP (2009) Uropathogenic E. coli adhesin-induced host cell receptor conformational changes: implications in transmembrane signaling transduction. J Mol Biol 392(2):352–361

Wu XR, Kong XP, Pellicer A, Kreibich G, Sun TT (2009) Uroplakins in urothelial biology, function, and disease. Kidney Int 75:1153–1165

Xie B, Zhou G, Chan SY, Shapiro E, Kong XP, Wu XR, Sun TT, Costello CE (2006) Distinct glycan structures of uroplakins Ia and Ib: structural basis for the selective binding of FimH adhesin to uroplakin Ia. J Biol Chem 281:14644–14653

Xu C, Zhang YH, Thangavel M, Richardson MM, Liu L, Zhou B, Zheng Y, Ostrom RS, Zhang XA (2009) CD82 endocytosis and cholesterol-dependent reorganization of tetraspanin webs and lipid rafts. FASEB J 23:3273–3288

Yalaoui S, Zougbede S, Charrin S, Silvie O, Arduise C, Farhati K, Boucheix C, Mazier D, Rubinstein E, Froissard P (2008) Hepatocyte permissiveness to Plasmodium infection is conveyed by a short and structurally conserved region of the CD81 large extracellular domain. PLoS Pathog 4:e1000010

Yamada M, Tamura Y, Sanzen N, Sato-Nishiuchi R, Hasegawa H, Ashman LK, Rubinstein E, Yáñez-Mó M, Sánchez-Madrid F, Sekiguchi K (2008) Probing the interaction of tetraspanin CD151 with integrin alpha 3 beta 1 using a panel of monoclonal antibodies with distinct reactivities toward the CD151-integrin alpha 3 beta 1 complex. Biochem J 415(3):417–427

Yanez-Mo M, Barreiro O, Gordon-Alonso M, Sala-Valdes M, Sanchez-Madrid F (2009) Tetraspanin-enriched microdomains: a functional unit in cell plasma membranes. Trends Cell Biol 19:434–446

Zhang XA, Bontrager AL, Hemler ME (2001) Transmembrane-4 superfamily proteins associate with activated protein kinase C (PKC) and link PKC to specific beta(1) integrins. J Biol Chem 276:25005–25013

Zhou G, Mo WJ, Sebbel P, Min G, Neubert TA, Glockshuber R, Wu XR, Sun TT, Kong XP (2001) Uroplakin Ia is the urothelial receptor for uropathogenic *Escherichia coli*: evidence from in vitro FimH binding. J Cell Sci 114:4095–4103

Chapter 2
The Evolution of Tetraspanins Through a Phylogenetic Lens

Rob DeSalle, Tung-Tien Sun, Tjard Bergmann, and Antonio Garcia-España

Abstract The tetraspanin superfamily of proteins provides an excellent system for examining many important evolutionary phenomena at the level of gene and protein sequences. Because dozens of eukaryotic organisms now have their full genomes sequenced, tetraspanins from these genomes can be compared and placed into a phylogenetic context. The whole genome information allows for researchers to trace with great precision the evolutionary events that have molded the broad array of tetraspanins found in eukaryotic genomes. We first demonstrate that phylogenetic analysis of tetraspanins from the fully sequenced genomes of an exemplar set of eukaryotes can give a fairly complete picture of the relationships of the families and subfamilies of tetraspanins. We can use the phylogenetic analysis of these tetraspanins to classify the various families and subfamilies of tetraspanins and use the

R. DeSalle
Sackler Institute for Comparative Genomics, American Museum of Natural History, New York, NY 10024, USA

T.T. Sun
Department of Cell Biology, New York University School of Medicine, New York, NY 10016, USA

Department of Dermatology, New York University School of Medicine, New York, NY 10016, USA

Department of Pharmacology, New York University School of Medicine, New York, NY 10016, USA

Department of Urology, New York University School of Medicine, New York, NY 10016, USA

T. Bergmann
ITZ, Ecology and Evolution, Tierä rztliche Hochschule Hannover, Hannover, Germany

A. Garcia-España (✉)
Research Unit, Universitat Rovira i Virgili, Hospital Joan XXIII, Pere Virgily Institute (IISPV), Tarragona 43007, Spain
e-mail: agarciae.hj23.ics@gencat.cat

F. Berditchevski and E. Rubinstein (eds.), *Tetraspanins*, Proteins and Cell Regulation 9, DOI 10.1007/978-94-007-6070-7_2,

sequence information as diagnostics for identifying novel tetraspanins. By using a phylogenetic perspective we also examine several important evolutionary processes in the tetraspanins such as intron evolution and the evolution of small protein motifs. We also describe a website for researchers who are interested in tetraspanin evolution, classification, identification and information called TSPAN4.web (http://research.amnh.org/users/desalle/data/tspan/).

2.1 Tetraspanins

An obligatory (but short) introduction: Tetraspanins are members of a large group of integral membrane proteins (Maecker et al. 1997; Hemler 2001, 2003; Boucheix and Rubinstein 2001). Humans have 33 tetraspanin members scattered throughout the genome (see Table 2.1) whose functions are distributed widely in cells and tissues. The structure of tetraspanins is widely conserved across large phylogenetic with the typical tetraspanin being 200–350-amino-acid-long with four transmembrane (TM) domains. In addition two extracellular loops exist in these proteins, one being small (SEL—about 13–30 amino acids long) and the other large (LEL—up to 150 amino acids long). Many tetraspanin proteins were originally identified as human tumor antigens while others are associated with several forms of retinal degeneration. Still others have been associated with mental retardation syndromes (Zemni et al. 2000). Tetraspanin-enriched microdomains can form through primary associations with a variety of transmembrane and intracellular signaling/cytoskeletal proteins and secondary associations (Levy and Shoham 2005a, b). The conserved structure of tetraspanins over such extreme functional diversity and phylogenetic time makes them an ideal subject for evolutionary analysis. This chapter examines the superfamily of proteins through a phylogenetic "looking glass", by first explaining the caveats of phylogenetic analysis of tetraspanins. Next, we examine nomenclatural issues that arise as a result of having a phylogenetic framework for this superfamily. We also examine two important aspects of protein evolution using the tetraspanins—intron and short amino acid motif evolution. We conclude by demonstrating how a close up view of the phylogenetics of specific tetraspanins can enhance our understanding of the structure and function of these proteins.

2.2 Phylogenomic Methods

The ins and outs of protein family trees: Many gene families have been analyzed using phylogenetic approaches. Often the methodology and limitations of such analyses are unclear. While an exhaustive explanation of the phylogenetic approaches is beyond the scope of this chapter, we present here a critical discussion of why we chose our particular approaches to analyze the phylogenetic evolution of tetraspanins. First and foremost to keep in mind when analyzing gene families is the concept

Table 2.1 Tetraspanin superfamily nomenclature, family designations and divergence times

Protein	Gene	Aliases	Family	DT
TSPAN1	TSPAN1	TSP-1	CD	V
TSPAN2	TSPAN2	TSP-2	CD	T
TSPAN3	TSPAN3	TSP-3	CD63	V
TSPAN4	TSPAN4	TSP-4/NAG2	CD	V
TSPAN5	TSPAN5	TSP-5	RD	V
TSPAN6	TSPAN6	TSP-6	CD63	T
TSPAN7	TSPAN7	CD231/TALLA-1/A15	CD63	V
TSPAN8	TSPAN8	CO-029	CD	T
TSPAN9	TSPAN9	NET-5	CD	V
TSPAN10	TSPAN10	OCULOSPANIN	RD	V
TSPAN11	CD151-like	CD151-like	RD	V
TSPAN12	TSPAN12	NET-2	Uroplakin	C
TSPAN13	TSPAN13	NET-6	CD63	V
TSPAN14	TSPAN14		RD	V
TSPAN15	TSPAN15	NET-7	RD	D
TSPAN16	TSPAN16	TM4-B	CD	M
TSPAN17	TSPAN17		RD	M
TSPAN18	TSPAN18		CD	V
TSPAN19	TSPAN19		CD	V
TSPAN20	UPK1B	UP1b, UPK1B	Uroplakin	V
TSPAN21	UPK1A	UP1a, UPK1A	Uroplakin	V
TSPAN22	RDS	RDS, PRPH2	RD	V
TSPAN23	ROM1	ROM1	RD	V
TSPAN24	CD151	CD151	CD	V
TSPAN25	CD53	CD53	CD	M
TSPAN26	CD37	CD37	CD	M
TSPAN27	CD82	CD82/KAI-1	CD	V
TSPAN28	CD81	CD81	CD	V
TSPAN29	CD9	CD9	CD	V
TSPAN30	CD63	CD63	CD63	V
TSPAN31	TSPAN31	SAS	CD63	V
TSPAN32	TSPAN32	TSSC6	Uroplakin	M
TSPAN33	TSPAN33		CD	M

M mammal divergence at 100 MYA, *T* tetrapod divergnce at 370 MYA, *V* veterbrate divergence at 450 MYA, *C* chordate divergence at 535 MYA and *D* deuterstome divergence at 570 MYA

of homology of genes and proteins. Genes in a gene family can be **orthologous** or **paralogous** with each other (Thornton and DeSalle 2000). Orthologous genes are those that are in different organisms as a result of common ancestry via speciation. Paralagous genes are those that are in genomes (the same or different genomes) as a result of gene duplication. For example, human UP1A and human UP1B are both considered uroplakins, they are in reality paralogs of each other. By the same token, chimpanzee UP1A and human UP1B share some common ancestry they are also paralogs of each other. On the other hand though, chimpanzee UP1A and human UP1A are considered orthologs of each other. A first approximation of homology is usually made using similarity via a BLAST score. The determination of orthology can then be made by optimizing some aspect of the similarity scoring or through phylogenetic analysis (Chiu et al. 2006).

A major issue in the analysis of gene families that is also a consideration with tetraspanins, is to decide whether the analysis should be done on protein or DNA sequences. Since gene families that include Bacteria and Archaea and Eukarya will span the entire time that life has existed on this planet—3.5 or so billion years, considerable sequence change has occurred amongst the genes in the gene family. Using DNA sequences at this level is problematic because third positions in the genes will have evolved much more rapidly and the extreme amount of change that has occurred is difficult to compensate for even by modeling nucleotide sequence change. On the other hand, amino acid coding of the sequences evolves at a slower rate making such sequences more amenable to models that have been developed to compensate for such sequence change. Extreme sequence divergence also means that sequence alignment becomes a problem, and the alignment of amino acid sequences is simpler than alignment of DNA sequences at this degree of sequence change (although amino acids can be aligned first and used as a guide for DNA sequence alignment). Alignment and choice of model to compensate for extreme sequence change are two major initial problems to consider when examining gene families. Since tetraspanins appear to be present in all eukaryotic life this means that the common ancestor of the members of this gene family are at least the age of eukaryotes—more than 1.5 billion years. This observation means that amino acids are perhaps an appropriate source of data for phylogenetic studies of this large group of genes.

Another major issue has to do with how to generate phylogenetics, once an alignment of the gene family members has been produced. There are two main approaches to generating phylogenies both with their advantages and detractions. The most commonly used by molecular biologists are what are called distance or phenetic approaches. The linear sequence information in this method are condensed into a distance (or similarity) measure based on a model of sequence change for each pair of genes (or proteins) in the data set. The pairwise distances are then used in an algorithm that generates a phenogram that represents the distance information in the condensed matrix. The advantage of this kind of approach is its computational ease and rapidity. A second category of approaches leaves the sequence information intact as unitary characters and utilizes a character by character methods to generate a phylogenetic hypothesis. In this approach, the DNA sequence positions in the gene or the amino acid positions in the protein are assessed with optimality criteria for their fit onto a phylogenetic hypothesis. What this means is that the data for the genes or proteins in the analysis needs to be assessed for optimality with respect to each tree that can be generated for the genes or proteins in the analysis. For instance, for three proteins, three trees need to be assessed for optimality [if the three proteins are A, B and C, then the three trees are ((A,B)C), ((A,C),B) and ((B,C)A)]. When the number of proteins or genes in an analysis is over 15 or so, the ability of computers to compute exact solutions is prohibitive (the NP complete problem) and heuristic approaches to get a best estimate of optimality are used (Felsenstein 2004). These character-based approaches can use parsimony or likelihood methods for assessing optimality of a tree topology, given the assumptions of the approaches. The choice one makes as to which method to use is often based on accessibility and speed.

A third major concern regarding phylogenetic analysis of gene and protein families has to do with the robustness of the inferences made when using small numbers of characters. Methods such as bootstrapping (Felsenstein 1985) and jackknifing (Farris et al. 1996) can be used to assess the robustness of inference at nodes in the tree. These methods are resampling techniques that can be applied to both distance and character-based analyses. For most phylogenetic comparisons, the robustness of inference at nodes is roughly correlated with the amount of sequence information for each taxon, so inferences made with single genes or proteins for each taxon are not necessarily robust. As a general rule of thumb, any bootstrap or jackknife value above 65% is credible (Hillis and Huelsenbeck 1992), but bootstrap and jackknife values of gene and protein phylogenies should probably be viewed differently from the same measures for organismal trees. This is because these measures, when used in organismal studies, can tell the systematist where further work is needed and where more sequence information needs to be collected from the genomes of the organisms being analyzed, to converge on a robust inference. In gene and protein family trees no new data can be added. Other approaches for assessing robustness that are based on character-by-character analysis exist that place phylogenetic analysis in more of a statistical context such as Bayesian Phylogenetic analysis that estimates a posterior probability for each node in a phylogenetic tree (Huelsenbeck and Ronquist 2001). Since the Bayesian posterior is a probability, researchers have used the classical p value cutoff of 0.05 as an indicator of significance for these statistics.

A fourth aspect of gene and protein family phylogenetics concerns sampling. Some phylogenetic studies take the approach of including just the genes from a single group, like the uroplakins, to obtain as many representatives of the genes from as many organisms as possible regardless of whether a whole genome for the organism exists. Other researchers have limited their analyses to those organisms with fully sequenced genomes to examine all of the genes in a gene family from existing full genomes. We have argued elsewhere (Garcia-España et al. 2008) that analysis of organisms with fully sequenced genomes is the most efficient and informative approach, because in this case the *absence* of a gene in a subgroup of species can infer special significance. If genes from an organism without a fully sequenced genome are used then no inference about its absence can be made, thus imposing severe limitations on the interpretations of the data.

Finally, it is critical when thinking about protein and gene family phylogenetics to define the role of rooting or choice of outgroups. Of course, results of gene and protein family phylogenies can be presented as unrooted networks, and these can be quite informative, but being able to root the network renders polarity to the changes in the tree that can be inferred from the topology of the tree. Choice of outgroups in gene and protein family analysis can come from two sources. Firstly, if one is clear that a group of genes is orthologous, a closely related gene family that is not part of that group can be used as the outgroup. Secondly, if one has a clearly defined group of orthologous genes in a gene family, then the gene or protein from the most primitive organism in the analysis can be used as an outgroup. While it is easy to feed a lot of tetraspanin data through a phylogenetic analysis program, the nuances discussed above concerning choice of characters (i.e. DNA sequences or amino acid sequences), orthology, choice of algorithm or optimality criteria (i.e. distance analysis, parsimony or likelihood), robustness of inference (bootstrap, jackknife or Bayesian posteriors) and outgroup choice can all have a huge impact on interpreting results.

2.3 Classification System for Tetraspanins

A tree-based nomenclature for tetraspanins: We begin this discussion with a note on nomenclature. This large group of genes (proteins) includes 33 members in the human genome (Table 2.1). The members of this group of proteins are also sometimes called the transmembrane 4 superfamily (TM4SF) proteins. The nomenclature of the genes and proteins within this large group of proteins is most clearly articulated by the HUGO (Human Genome Organization) Gene Nomenclature Committee (http://www.genenames.org/index.html). According to the HUGO nomenclature system, there are 33 genes that exist in the human genome that encode the tetraspanin proteins and some of their "aliases" (Table 2.1). Note that some of the tetraspanin proteins produced by these 33 genes have been annotated as TSPAN followed by a number. Still others in the large group of genes are named uroplakins (UPK), Retinal degeneration slow (RDS) and the well known CD proteins (followed by a number) because of their specific cell expression pattern and function. These proteins are found in a wide range of living species and present in plants, animals, fungi and protists. Because of the breadth of organismal range and functionality of these proteins, we have adopted the convention of calling the entire group of tetraspanins a superfamily as in the TM4SF tradition. We then divide this tetraspanin superfamily into families, which are then divided into groups based on the existing annotations of genes and proteins in this superfamily.

Several research groups have used tetraspanins as the subject of gene family analysis (Huang et al. 2005; Todres et al. 2000; Garcia-España et al. 2008). In general, their results are congruent with respect to the monophyly of members in the major groups of tetraspanins. Some differences occur between the phylogenies when deeper nodes are examined, and hence relationships of families of genes within the superfamily may be different between the two studies. However, these differences in interpretation are due to the lack of robustness at nodes at the base of the trees in all of the studies accomplished so far. In this chapter, we discuss the classification system of Garcia-Espana et al. 2008 as a framework for tetraspanin evolution. A detailed phylogeny of the tetraspanins from this study can be found at the TSPAN4 website (see last section of this chapter for a full discussion of the website as a research tool). Since both studies that use large

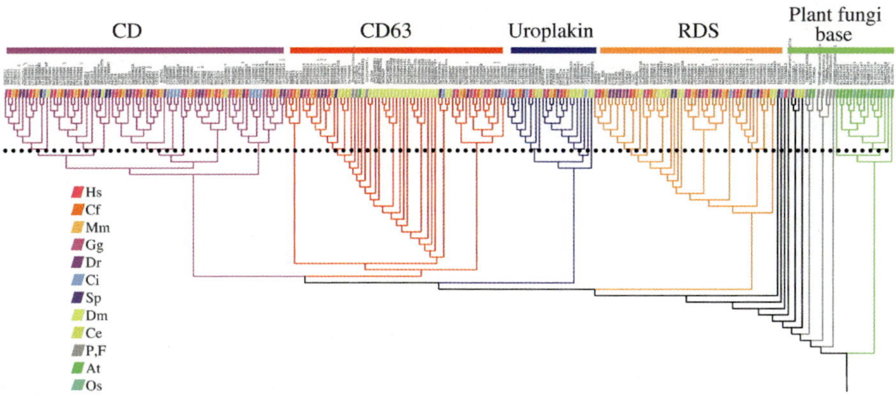

Fig. 2.1 Phylogenetic tree from Garcia-Espana et al. (2008). Species are designated by *colored boxes* with a legend for the species designation given (species abbreviations are given in Garcia-Espana and at http://research.amnh.org/users/desalle/data/tspan/). More detailed "close-ups" of the four major groups of tetraspanins designated here are available on the TSPAN4 website http://research.amnh.org/users/desalle/data/tspan/. The tetraspanin superfamily can be subdivided into four major monophyletic families (the CD family, the CD63 family, the uroplakin family, and the RDS family) and a group of nonmonophyletic tetraspanins at the base of the tree that comprises fungal, plant, and protist tetraspanins. The *black dotted line* represents the general area of the tree below which bootstrap and jackknife values drop below 60% and Bayes proportions below 90%

sampling of tetraspanin genes (Huang et al. 2005; Garcia-España et al. 2008) tend to group genes similarly, the groupings can serve as a basis for the classification of tetraspanins.

The phylogenetic analyses summarized in the tree in Fig. 2.1 (all trees discussed in this chapter can also be found at http://research.amnh.org/users/desalle/data/tspan/) shows four major clades that we have given the rank of family (called the **CD family**, the **CD63 family**, the **uroplakin family**, and the **RDS family**; Table 2.1). The largest cluster of tetraspanins, i.e., the CD family, comprises proteins annotated in existing genome databases as vertebrate CD and Tsp proteins with several invertebrate tetraspanins. This group includes all of the previously annotated tetraspanins with the designation CD in their name (151, 53, 9, 81, 82, 37) except for CD63. This latter CD tetraspanin is placed into its own family, the second largest of the four families with respect to members. This family contains the CD63 orthologs from several vertebrates and cluster of genes at chromosome location 42E in the Drosophila genome. The CD63 family of tetraspanins is highly divergent with several previously annotated vertebrate TSPAN proteins (TSPAN13, TSPAN31, TSPAN3, TSPAN6, and TSPAN7). The uroplakin family is made up of the classically named vertebrate uroplakin (UP) genes and several invertebrate tetraspanins (represented by the well-characterized Drosophila tetraspanin expansion group) as well as TSPAN32 and TSPAN12. The final large family of animal tetraspanins is called the RDS family, because it includes the RDS-ROM tetraspanins and this family also includes the orthologs of Human TSPAN10, TSPAN14, TSPAN5, TSPAN17, TSPAN15, TSPAN33.

2.4 The Origin of the Tetraspanins

Superfamilies, families and groups: By examining the clustering of tetraspanin orthologs and assaying the taxonomic representation within the ortholog groups, we can estimate the times of origin and divergence of the various groups. For instance, each of the four major families, i.e., CD, CD63, Uroplakin and RDS, have both vertebrate and invertebrate representatives of the Bilateria, but no fungal, plant or protist members. In addition, recent analysis of Cnidarian, Placozoan and Poriferan tetraspanins indicates that these phyla also have representatives of all four families of tetraspanins and that, while the Choanoflagellate, Monosiga also has a tetraspanin (data not shown), it is not orthologous to any of the tetraspanins in the CD, CD63, Uroplakin or RDS families of tetraspanins. The distribution of tetraspanins in the genomes of all of these animals and the choanoflagellate indicates that the expansion of this superfamily into the four large families we describe above was an "invention" in the genome of the ancestor of the Metazoa. These observations suggest that the origin of the expanded tetraspanin superfamily into four families corresponds to a divergence of more than about 540–650 million years in the ancestor of all metazoans that most likely existed prior to the Vendian period (Hedges and Kumar 2002; Doolittle et al. 1996).

Table 2.1 lists the human tetraspanin genes and their approximate time of origin using this approach. Using this approach we can designate certain tetraspanins as "inventions" of particular ancestors in the history of animals. For instance, there appears to have been a burst of tetraspanin "invention" in the ancestor of vertebrates. This burst also corresponds with well-known genome duplications in the ancestor of particular lineages of vertebrates. Another significant "burst" of tetraspanin origin also occurred in the ancestor of mammals, where CD37, CD53, TSPAN16, TSPAN17, TSPAN32 and TSPAN33 arose (Garcia-España et al. 2008). Future work using this approach should incorporate the newly emerging mammalian genomes to determine whether any of the tetraspanins are specific to orders of mammals such as the primates.

2.5 Introns and Cysteines

Evolution of intron junctions and protein motifs: The tetraspanin superfamily offers an excellent system for examining specific aspects of genome and protein evolution. In this section, we examine two evolutionary phenomena specific to tetraspanins. The first concerns the evolution of introns (Garcia-España et al. 2009; Garcia-España and DeSalle 2009) and the second concerns the evolution of repeated motifs in proteins (DeSalle et al. 2010). Using the robust phylogeny of the tetrapanins these interesting aspects of the gene family can be examined in precise detail. While several elegant studies of intron evolution using whole genome approaches have been useful in detecting genome-wide intron evolutionary trends, taking a

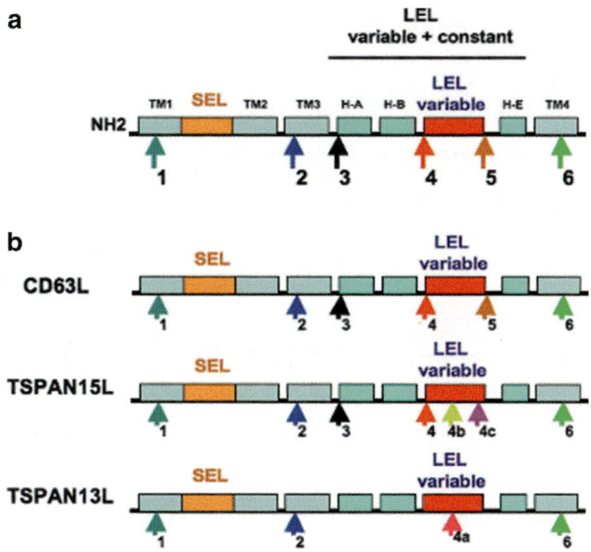

Fig. 2.2 Cartoons of intron positions in the tetraspanin genes. (**a**) The small (SEL; *orange*) and large (LEL; *red*) extracellular loops are indicated. *Light blue* represents the four transmembrane domains while no color, represents the intracellular regions. Ancestral intron positions *1–6* are indicated on the protein by *colored arrows* of the same color that will be used through all of the figures. Abbreviations are *TM-1–4* transmembrane domains; *H-A, H-B* and *H-E* constant helices in the LEL. (**b**) Animal CD63L (*top*), TSPAN15L *(middle*) and TSPAN13L (*bottom*) tetraspanins' consensus intron structure. *Purple red* (intron 4a), *yellow* (intron 4b) and (intron 4c) indicate four new intron junctions discussed in the text

gene family-specific approach can also be useful. The analysis of gene family, again, in fully sequenced genomes, can reveal patterns of intron gain and loss more precisely. In addition, if a gene family is used where specific function of the gene products is known, more precise interpretation of the gain and loss patterns can be made. The intron structure of tetraspanins is also interesting because of the relatively large number of introns in the genes in this superfamily. While an examination of the range of tetraspanin genes for intron position reveals that there are at least 105 unique intron positions in the tetraspanins of fungi, plants, protists and animals, the most common intron structure of tetraspanins in animals is a six intron scheme (Fig. 2.2a). More precisely, there are three major intron patterns (Fig. 2.2b) from which all other animal intron patterns are derived. As with the appearance of new tetraspanins in the genomes of animals, we can use the phylogenomic approach to give dates to the gain and loss of new introns in tetraspanins. These data show that there is a strong correlation of the appearance of tetraspanins with novel functions with the insertion of introns in new positions in the overall tetraspanin gene structure (Garcia-España et al. 2009). For instance, as we discussed above, there was a burst of appearance of novel tetraspanins in the ancestor of vertebrates, and this burst of new tetraspanins is accompanied by the appearance of six new introns in

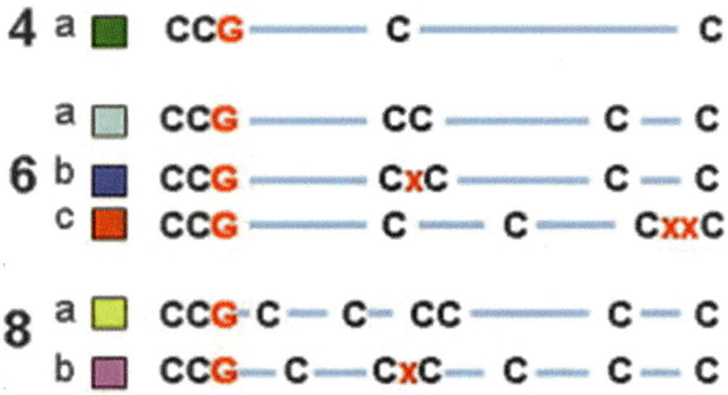

Fig. 2.3 The LEL patterns of cysteines according to their number and relative position to each other. *Blue lines* indicate any number of residues between adjacent cysteines. Each small x (*x*) indicates a single residue. Capital *red* G (*G*) indicates the glycine residue of the CCG motif. Each of the six cyteine patterns is represented by a distinct *colored box*

these genes. Furthermore, the position of these new introns in tetraspanins is non-random, as nearly 50% of new introns appear in the small extracellular loop (SEL), which accounts for only 10% of the entire length of most tetraspanin proteins. When the large extracellular loop (LEL), which on average makes up only 25% of tetraspanin proteins, is examined for novel intron occurrence Garcia-Espana and colleagues (2009) observed that another 25% of novel introns accrue in this region. Clearly the two extracellular loops (totaling ~35% of the entire amino acid sequence of most tetraspanins) are accruing the vast majority (70–75%) of novel introns. Using the patterns of intron gain and loss, it was also estimated that there are 105 intron gain events (42 alone in C. elegans) and only four intron loss events (Garcia-Espana et al. 2009). Finally, using the phylogenomic approach, we were able to determine that indels (i.e. insertions or deletions) at the ends of DNA exonic sequences could have caused the appearance of two discordant intron positions between orthologous tetraspanins (Garcia-España and DeSalle 2009). These data suggest that an intron sliding mechanism (Tarrío et al. 2008) can be used to explain these observations. This intron-sliding mechanism could have been important in generating functional diversity in this superfamily of tetraspanin proteins.

Cysteine residues have been used to characterize tetraspanins in the past because these cysteine reside in distinct motifs and because these cysteine residues may play important roles in the secondary and tertiary structure of tetraspanin proteins. Most of these cysteine motifs are found in the large extracellular loop (LEL) and so the phylogenomic analysis used to examine them focused on this region of the tetraspanin structure. In general, the number of cysteines is even (four, six or eight) suggesting that they interact in pairs in disulfide bonding. The exceptions to this even number of cysteines are the RDS/ROM and plant tetraspanins. There are six easily recognized cysteine motifs that can be examined in a phylogenomic context (Fig. 2.3). The results of this analysis suggest that the cysteine motifs are correlated closely with phylogeny and hence novel cysteine motifs are correlated with the bursts of appearance of new tetraspanins. In addition, DeSalle et al. (2010) suggest that the four cysteine motif (see Fig. 2.3) is highly derived occurring at the tips of the tetraspanin tree. What this means is that the four cysteine motif is a new evolutionary innovation. This suggests that the reduction in number of cysteines in the LEL is a recurring and more recent event in the evolution of the animal tetraspanins. One of the more important results of this study concerns the examination of the highly conserved CCG motif in the LEL. A detailed phylogenetic analysis of this motif reveals that it originated in the common ancestor of Unikonts—animals, Fungi and Amoebozoa. The CCG motif does not appear to be in all tetraspanins of plants, Stramenophiles, Alveolata, Discicristata, or Excavata, suggesting that it did not exist in the common ancestor of Bikonts.

2.6 Up Close and Personal

The uroplakins: An examination of closely related tetraspanins within specific families and groups can shed light on the evolutionary steps leading to the structure and function of proteins in the superfamily. As an example, we discuss the analysis of uroplakins by a detailed phylogenetic analysis in Garcia-Espana et al. (2006). This family of proteins are the integral membrane subunits of urothelial plaques (also known as the Asymmetric Unit Membrane [AUM]) that line the specialized apical surface of the mammalian urinary bladder epithelium. While there are four major uroplakins (UP1a, UP1b, UPK2 and UPK3) only two of them are tetraspanins (UP1a and UP1b), while the other two (UP2, UP3a) span the membrane only once. UP1a and UP1b interact selectively with UP2 and UP3a, respectively, to form Ia/II and Ib/IIIa complexes that further assemble to make up the urothelial plaques. A detailed phylogenetic analysis of all four of these four major uroplakin proteins individually, revealed a general correlation of protein evolution with organismal evolution. By tracing the presence and absence of the genes for these proteins in the genomes of vertebrates, Garcia-Espana et al. (2006) demonstrated two major phenomena relevant to the evolution of these proteins: (1) the UPIa and UPIb genes co-evolved by gene duplication in the common ancestor of vertebrates, as did UPII and UPIIIa; and (2) uroplakins can be lost in different combinations in vertebrate lineages generating a great deal of variability in the functionality of the proteins. Specifically, duplication of an ancestral UPI gene into UPIa and UPIb occurred in the common ancestor of cartilaginous fish and other vertebrates. Concomitantly, the duplication of a UPII/UPIIIa gene occurred in the same ancestor to produce the UPII and UPIIIa proteins. In addition, using a coevolutionary approach (where the proteins were compared in the following pairs UPIa/UPII, UPIa/UPIII, UPIb/UPII and UPIb/UPIIIa), these authors also showed that only the UPIa/UPII and UPIb/UPIIIa coevolutionary pairs showed statistical correlation suggesting there is a strong co-evolutionary relationship between UPIa and UPIb and their partners UPII and UPIIIa/IIIb, respectively (Garcia-Espana et al. 2006). These results further support the biochemical analyses regarding the stoichiometric interactions (Tu et al. 2002; Hu et al. 2005) and strengthen the hypothesis that UPIa and UPII interact with each other and that UPIb and UPIIIa interact with each other in the formation of urothelial plaques.

2.7 TSPAN 4

A website for studying the evolution of tetraspanins: To facilitate broader use of the phylogenomic approach to the analysis of tetraspanins, we have developed a website for the use of tetraspanin researchers. The webiste can be accessed at http://research.amnh.org/users/desalle/data/tspan/. Figure 2.4 shows the homepage of the website that we have named T4NET. There are five functions that the website performs. First, the website holds the most recent tetraspanin tree (* in Fig. 2.4) using accessions from whole genomes. This tree has a pull down function that allows the user to click on one of the four major families of proteins to get an upclose view of the structure and organization of protein groups. Second, the website allows users to input a new putative tetraspanin sequence and identify it using the "gene identifier" (# in Fig. 2.4) function on the website. Figure 2.5 shows the webpage where gene identification can be accomplished. Users simply cut and paste their putative tetraspanin gene sequence into the query box (large arrow) and use the Submit button (small arrow) for the webpage to determine the identity of their putative tetraspanin. Input sequences have to be in amino acid FASTA format. The third function of the website allows the user to view the uroplakin 1a, 1b phylogenetic

Fig. 2.4 Homepage for the TSPAN4 website. There are five functions that the website performs. A tetrspanin tree (*) function, a "gene identifier" function a uroplakin tree function (@), access the figures in tetraspanin paper (&) and accession numbers for tetraspanin genes that are in fully sequenced genomes ($). The website can be accessed at http://research.amnh.org/users/desalle/data/tspan/

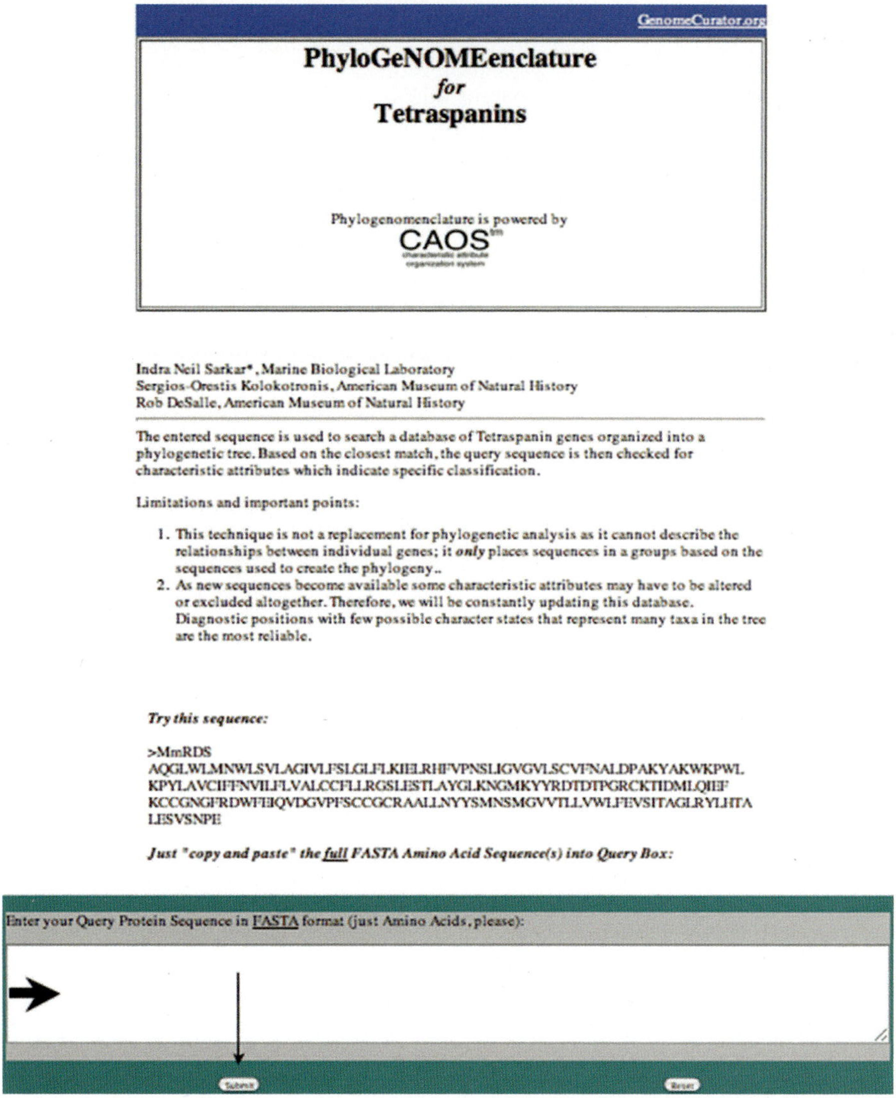

Fig. 2.5 Webpage showing the "gene identifier" on the TSPAN 4 website. The *large arrow* indicates the query box where a novel tetraspanin sequence can be pasted and the *small arrow* indicates the "Submit" button

tree (@ in Fig. 2.4) and to rapidly access the sequences that were used to generate the tree. The fourth function on the website allows users to access the figures (& in Fig. 2.4) produced for the various tetraspanin studies that were reviewed in this chapter. This page will also have the full references and link outs or pdfs for all of the papers relevant to tetraspanin evolution. The final function of the website is a list of all of the tetraspanin genes that are in fully sequenced genomes ($ in Fig. 2.4).

2.8 Conclusion

A phylogenetic context for tetraspanins allows for a precise and logical classification system for this large superfamily of genes. Placing the tetraspanins in a phylogenetic context and using fossil dates for divergence of the model organism fully sequenced genomes allows us to establish the dates of origin for the major tetraspanin groups. In addition, the phylogenetic approach allows for a detailed examination of two phenomena involved in the structure and function of tetraspanins. First we can use the phylogenetic approach to examine the evolution of intron position in the tetraspanin families. Second we can use the tree-based approach to examine the evolution of the structurally and functionally important cysteine residues in the proteins in this superfamily. By examining very closely the evolution of members of a well known group of tetraspanins, the uroplakins, we demonstrate how the phylogenetic approach can be useful in the reconstruction of evolutionary events for a functionally important protein family.

References

Boucheix C, Rubinstein E (2001) Tetraspanins. Cell Mol Life Sci 58:1189–1205

Chiu JC et al (2006) OrthologID: automation of genome-scale ortholog identification within a parsimony framework. Bioinformatics 22:699–707

DeSalle R, Mares R, Garcia-España A (2010) Evolution of cysteine patterns in the large extracellular loop of tetraspanins from animals, fungi, plants and single-celled eukaryotes. Mol Phylogenet Evol 56:486–491

Doolittle RF et al (1996) Determining divergence times of the major kingdoms of living organisms with a protein clock. Science 271:470–477

Farris JS, Albert VA, Källersjö M, Lipscomb D, Kluge AG (1996) Parsimony jackknifing outperforms neighbor-joining. Cladistics 12:99–124

Felsenstein J (1985) Confidence limits on phylogenies: an approach using the bootstrap. Evolution 39:783–791

Felsenstein J (2004) Inferring phylogenies. Sinauer Associates, Sunderland

Garcia-España A, DeSalle R (2009) Intron sliding in tetraspanins. Commun Integr Biol 2:1–3

Garcia-España A, Chung PJ, Zhao X, Lee A, Pellicer A, Yu J, Sun TT, DeSalle R (2006) Origin of the tetraspanin uroplakins and their co-evolution with associated proteins: implications for uroplakin structure and function. Mol Phylogenet Evol 41:355–367

Garcia-España A, Chung PJ, Sarkar IN, Stiner E, Sun TT, DeSalle R (2008) Appearance of new tetraspanin genes during vertebrate evolution. Genomics 91:326–334

Garcia-España A, Mares R, Sun TT, DeSalle R (2009) Intron evolution: testing hypotheses of intron evolution using the phylogenomics of tetraspanins. PLoS One 4:e4680

Hedges SB, Kumar S (2002) Genomics: vertebrate genomes compared. Science 297:1283–1285

Hemler ME (2001) Specific tetraspanin functions. J Cell Biol 155:1103–1107

Hemler ME (2003) Tetraspanin proteins mediate cellular penetration, invasion, and fusion events and define a novel type of membrane microdomain. Annu Rev Cell Dev Biol 19:397–422

Hillis DM, Huelsenbeck JP (1992) Signal, noise and reliability in molecular phylogenetic analyses. J Hered 83:189–195

Hu C et al (2005) Assembly of urothelial plaques: tetraspanin function in membrane protein trafficking. Mol Biol Cell 16:3937–3950

Huang S et al (2005) The phylogenetic analysis of tetraspanins projects the evolution of cell-cell interactions from unicellular to multicellular organisms. Genomics 86:674–684

Huelsenbeck JP, Ronquist F (2001) Bayesian, is a reference to Mr. Bayes (http://en.wikipedia.org/wiki/Bayesian_inference). Bioinformatics 17:754–755

Levy S, Shoham T (2005a) Protein-protein interactions in the tetraspanin web. Physiology (Bethesda) 20:218–224

Levy S, Shoham T (2005b) The tetraspanin web modulates immune-signalling complexes. Nat Rev Immunol 5:136–148

Maecker HT, Todd SC, Levy S (1997) The tetraspanin superfamily: molecular facilitators. FASEB J 11:428–442

Tarrío R, Ayala FJ, Rodríguez-Trelles F (2008) Alternative splicing: a missing piece in the puzzle of intron gain. Proc Natl Acad Sci USA 105:7223–7228

Thornton JW, DeSalle R (2000) Phylogenetics meets genomics: homology and evolution in gene families. Annu Rev Genomics Hum Genet 1:43–72

Todres E, Nardi JB, Robertson HM (2000) The tetraspanin superfamily in insects. Insect Mol Biol 9:581–590

Tu L, Sun TT, Kreibich G (2002) Specific heterodimer formation is a prerequisite for uroplakins to exit from the endoplasmic reticulum. Mol Biol Cell 13:4221–4230

Zemni R et al (2000) A new gene involved in X-linked mental retardation identified by analysis of an X;2 balanced translocation. Nat Genet 24:167–170

Chapter 3
Organisation of the Tetraspanin Web

Eric Rubinstein, Stéphanie Charrin, and Michael G. Tomlinson

Abstract Tetraspanins are currently hypothesized to promote membrane compartmentalization, through their ability to organize a network of molecular interactions termed the tetraspanin web or tetraspanin-enriched microdomains. In this chapter we will describe how the discovery of this unique ability of tetraspanins to interact with one another and with many other surface proteins led to this concept, and will discuss the hierarchical organization of these structures. We will also show how tetraspanins modulate the function of the proteins they associate with, including the regulation of trafficking, ligand binding, signal transduction and enzymatic activities.

3.1 A Brief Introduction to Tetraspanins

3.1.1 The Conserved Tetraspanin Structure

Tetraspanin proteins are composed of four membrane spanning regions, cytoplasmic amino and carboxy termini and two extracellular regions (Fig. 3.1). A number of features distinguish tetraspanins from other proteins with four transmembrane domains (Boucheix and Rubinstein 2001; Charrin et al. 2009a; Hemler 2003, 2005;

E. Rubinstein (✉) • S. Charrin (✉)
Inserm U1004, 14 Av Paul Vaillant Couturier, 94807 Villejuif Cedex, France

Institut André Lwoff, University Paris-Sud, 14 Av Paul
Vaillant Couturier, 94807 Villejuif Cedex, France
e-mail: eric.rubinstein@inserm.fr; stephanie.charrin@inserm.fr

M.G. Tomlinson (✉)
School of Biosciences, College of Life and Environmental Sciences,
University of Birmingham, Birmingham B15 2TT, UK
e-mail: m.g.tomlinson@bham.ac.uk

F. Berditchevski and E. Rubinstein (eds.), *Tetraspanins*, Proteins and Cell Regulation 9,
DOI 10.1007/978-94-007-6070-7_3,

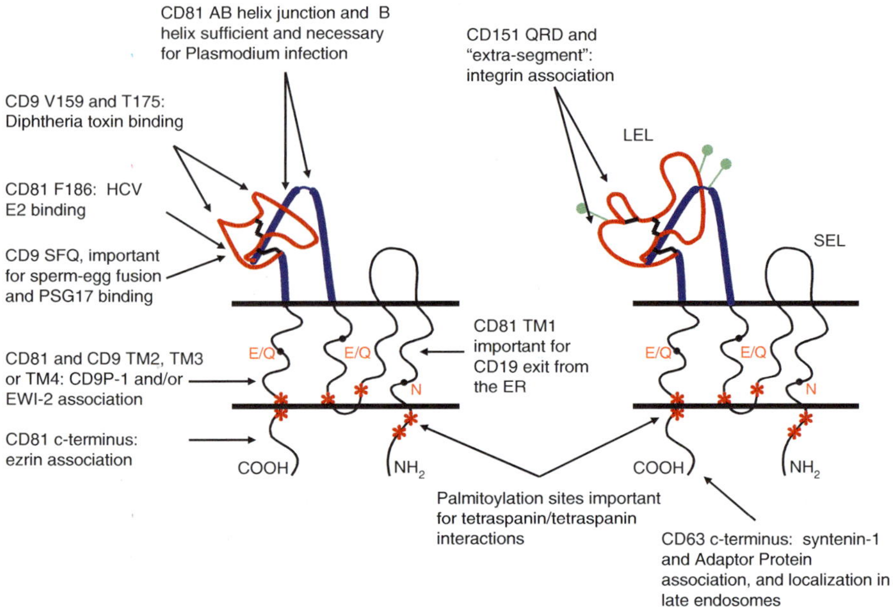

Fig. 3.1 A schematic view of tetraspanins with 2 (*left*) or 3 (*right*) disulfide bound in the large extracellular domain. The structurally variable region of this domain is in *red*, while the three conserved helices are in *bold blue*. Also highlighted are conserved polar residues of the transmembrane regions, as well as juxtamembrane cysteines that constitute potential palmitoylation sites. Known functional sites are indicated

Levy and Shoham 2005). Their two extracellular regions are of unequal size, with the first having less than 30 amino acids in the human proteins, while the second contains 76–131 amino acids. As such, the extracellular region of tetraspanins protrudes only 3.5–5 nm from the cell surface (Kitadokoro et al. 2001; Min et al. 2006), which is approximately the size of an immunoglobulin superfamily (IgSF) domain (Jones et al. 1992). The only high-resolution structural data to date concerns the large extracellular region of tetraspanin CD81, which consists of five α-helices (termed A-E) that form a 'mushroom-type' structure. The major extracellular region contains between four and eight conserved cysteine residues that form structurally important disulphide bonds, including a cysteine-cysteine-glycine (CCG) motif that is the major hallmark of tetraspanins. There is also a structurally variable region immediately downstream of the CCG motif, located within the C and D α-helices, which has been shown to mediate tetraspanin interactions with specific non-tetraspanin partner proteins. The transmembrane regions contain conserved charged or polar amino acids, namely an asparagine residue, within the first transmembrane, and glutamine or glutamic acid in the third and fourth. The cytoplasmic tails are short, containing generally, but not always, less than 20 residues. Tetraspanins can be subjected to three major post-translational modifications. These are N-linked glycosylation of the large extracellular region, which has up to three predicted sites (for review (Boucheix and Rubinstein 2001)), palmitoylation of several cysteine residues near the membrane/cytoplasm interface (Berditchevski et al. 2002; Charrin et al. 2002; Levy et al. 1991; Seehafer et al. 1988; Yang et al. 2002; Zhou et al. 2004), and ubiquitination of the cytoplasmic tails (Lineberry et al. 2008; Tsai et al. 2007). Finally, tetraspanins share up to 78% amino acid sequence identity for highly related tetraspanins, such as Tspan5 and Tspan17, and as low as 6% identity for more distantly related tetraspanins. Nevertheless, a conserved intron/exon organisation within tetraspanin genes suggests that this family has arisen by gene duplication from a common ancestor (Garcia-Espana et al. 2008).

3.1.2 Tetraspanin Expression Across the Tree of Life

Tetraspanins have been found in all metazoans, plants and some protozoans and multicellular fungi (Huang et al. 2005). A total of 33 tetraspanins have been identified in human and mouse, 46 in zebrafish, 36 in the fly *Drosophila melanogaster*, 20 in the roundworm *Caenorhabditis elegans*, at least 25 in the *Schistosoma* human parasitic flatworms, 17 in the model plant *Arabidopsis thaliana*, at least six in the parasite protozoan *Entamoeba histolytica*, and at least three in some multicellular pathogenic fungi. Unicellular fungi, such as *Saccharomyces cerevisiae* or *Schizosaccharomyces pombe*, do not express tetraspanins. The relatively large size of the tetraspanin superfamily and its widespread expression across the tree of life suggests a fundamental role in cell function.

3.1.3 Tetraspanin Function: Lessons from Knockouts

A fundamental role for tetraspanins is supported by some striking phenotypes that have been observed in tetraspanin-deficient humans, mice, flies, worms, fungi and plants. In humans, mutations in a retina-specific tetraspanin, perpherin/RDS, result in retinal degeneration leading to blindness as a result of disorganised outer segments (Goldberg 2006), which are the light sensing regions of rod and cone cells. Mutations in Tspan12 also lead to progressive loss of vision, but by causing exudative vitreoretinopathy (Nikopoulos et al. 2010; Poulter et al. 2010), a disease characterised by the failure of retinal vasculature to develop properly. In addition, a mutation in tetraspanin CD151 leads to severe clinical problems including kidney failure and a skin blistering disease known as pretibial epidermolysis bullosa, probably due to abnormal epithelial cell basement membranes (Karamatic et al. 2004). Finally, mutation of CD81 causes an antibody deficiency syndrome due to impaired B cell receptor activation (van Zelm et al. 2010).

Several tetraspanins have now been experimentally deleted from the mouse genome. The phenotypes for mice deficient for peripherin/RDS (Goldberg 2006), Tspan12 (Junge et al. 2009), CD151 (Baleato et al. 2008; Sachs et al. 2006) and

CD81 (Levy and Shoham 2005) are consistent with the human diseases, with additional phenotypes observed for CD151-deficient mice including impaired pathological angiogenesis (Takeda et al. 2007), impaired wound healing (Cowin et al. 2006) and platelet function (Lau et al. 2004; Orlowski et al. 2009), hyper-stimulatory dendritic cells (Sheng et al. 2009) and hyper-reactive T cells (Wright et al. 2004a). Interestingly, retinal degeneration is also a feature of mice deficient in ROM-1 (Goldberg 2006) which, like peripherin/RDS, is a retina-specific tetraspanin. Moreover, like CD151, platelet function is impaired in Tspan32-deficient mice (Goschnick et al. 2006) and T cells are hyper-reactive in the absence of CD37 (van Spriel et al. 2004) or Tspan32 (Tarrant et al. 2002). A particularly striking phenotype was observed in CD9-deficient mice, which are largely infertile due to impaired sperm-egg fusion (Kaji et al. 2000; Le Naour et al. 2000; Miyado et al. 2000), and indeed this phenotype is mimicked, though less severe, in CD81-deficient mice (Rubinstein et al. 2006). These findings that different tetraspanin knockouts display common phenotypes represents one of several clues to suggest that certain tetraspanins co-operate to function on the same pathway.

Tetraspanin knockouts in lower organisms have also yielded strong phenotypes. The fruit fly *Drosophila melanogaster* exhibits delayed synapse formation in the absence of tetraspanin late bloomer (Kopczynski et al. 1996) and light-dependent retinal degeneration in the absence of sunglasses (Xu et al. 2004). In addition, the nematode worm *Caenorhabditis elegans* suffers from epidermal blistering in the absence of TSP-15 (Moribe et al. 2004), pathogenic fungi fail to penetrate host leaves in the absence of tetraspanin Pls1 (Clergeot et al. 2001), and the plant *Arabidopsis thaliana* is developmentally impaired without tetraspanin Tornado2 (Cnops et al. 2006; Olmos et al. 2003).

Despite these striking phenotypes, tetraspanin knockout studies have not revealed tetraspanin functions to the extent that was originally hoped. Indeed, the mechanisms responsible for most of the observed phenotypes are not clear. In addition, there are many instances in which tetraspanin deletion has yielded phenotypes that are either very mild or not detectable at all. A good example is mouse CD9, a tetraspanin which is relatively highly expressed in a broad range of different cell types. With the exception of the sperm-egg fusion defect (Kaji et al. 2000; Le Naour et al. 2000; Miyado et al. 2000), and also paranodal junction defects and increased monocyte fusion (Ishibashi et al. 2004; Takeda et al. 2003), CD9-deficient mice appear remarkably normal. This is particularly apparent for platelets, a cell type on which CD9 is the second most highly expressed surface protein, which are only very mildly hyper-reactive, resulting in slightly larger clot formation in vivo (Mangin et al. 2009). It is hypothesized that in all these cases other tetraspanins may functionally compensate for the loss of a single tetraspanin protein. Although evidence for this is currently in short supply, this notion is supported by the observation that double knockout of the related tetraspanins CD9 and CD81 in mice results in lung abnormalities similar to chronic obstructive pulmonary disease. Importantly, this is not evident in either of the single knockouts (Takeda et al. 2008). In addition, the sperm-egg fusion defect in $CD9^{-/-}/CD81^{-/-}$ mice is even more profound than in either $CD9^{-/-}$ or $CD81^{-/-}$ animals (Rubinstein et al. 2006). Furthermore, over-expression of CD81 in the egg can rescue the fusion defect in the absence of CD9 (Kaji et al. 2002). The phenotypes of tetraspanin knock-out animals are described in more detail in other book chapters.

3.1.4 Tetraspanins as Membrane Organisers

The plasma membrane is crowded with transmembrane proteins that regulate cell communication with the extracellular environment. These proteins usually act either as receptors or co-receptors for ligands that transmit a signal across the membrane, or as transporters (channels and pumps) allowing certain molecules to cross the membrane, or as enzymes. The cloning of the first tetraspanins, in the late 1980s and the early 1990s, defined a new family of proteins that did not resemble any other membrane proteins. They lacked known motifs implicated in signal transduction and very few examples of physiological tetraspanin ligands have been reported. The best characterized ligand is the placenta-derived pregnancy-specific glycoprotein 17 (PSG17), which binds to CD9 and modulates cytokine production by macrophages (Ellerman et al. 2003; Ha et al. 2005; Waterhouse et al. 2002). Other potential ligands include the pro-inflammatory cytokine IL16 as a ligand for CD9 (Qi et al. 2006), the tissue inhibitor of metalloproteinase TIMP-1 as a ligand for CD63 (Jung et al. 2006), the matrix metalloprotease proMMP17 as a ligand for CD151 (Shiomi et al. 2005), and DARC (Duffy antigen receptor for chemokines) as a ligand for CD82 (Bandyopadhyay et al. 2006). It is important to note that the definitive demonstration that a protein is a ligand for a tetraspanin is difficult, since tetraspanins can regulate the surface expression of other transmembrane proteins which, in turn, could themselves be the actual receptors. Therefore novel functions have to be considered for tetraspanins. Their ability to interact with each other and with specific non-tetraspanin partner proteins has led to the hypothesis that they function as 'organisers' of cell membrane protein complexes by the formation of a tetraspanin 'web' or microdomains. This concept, confronted with the concept of lipid rafts, recently evolved towards a role of tetraspanins in membrane compartmentalization or in regulating the lateral segregation of a subset of membrane constituents. It is this relatively well-studied aspect of tetraspanin function that will be explored in the rest of the chapter.

3.2 Unravelling the Tetraspanin Web

3.2.1 Discovery of the Tetraspanin Web

In the early years following the cloning of the first tetraspanins, the only tools available to researchers were the anti-tetraspanin monoclonal antibodies that had been used in the expression cloning process. In search of clues towards their function, some researchers looked for molecules associating with their favourite tetraspanins. These studies revealed, for example, an interaction between CD81 and MHC class II antigens (Schick and Levy 1993), and between CD9 and the integrins α4β1 and α5β1 (Rubinstein et al. 1994). Conversely, the study of molecules associating with CD19, a B lymphocyte co-stimulatory IgSF molecule, and with β1 integrins, yielded the tetraspanins CD81 and CD63, respectively (Berditchevski et al. 1995; Bradbury et al. 1992; Matsumoto et al. 1993). In these early years, diphtheria toxin researchers identified the membrane precursor of the heparin-binding EGF growth factor as the toxin receptor, and demonstrated a functional complex between this receptor and CD9 (Iwamoto et al. 1994). Meanwhile, an interaction of the two tetraspanins CD81 and CD82 was shown with the T lymphocyte co-stimulatory IgSF molecules CD4 and CD8 (Imai and Yoshie 1993). In this study CD81 and CD82 were also shown to interact with each other; this initial demonstration of tetraspanin-tetraspanin interactions was confirmed by three laboratories, which collectively also demonstrated that MHC class II antigens and certain integrins associated with several tetraspanins (Angelisova et al. 1994; Berditchevski et al. 1996; Rubinstein et al. 1996). Further work has extended the number of molecules similarly associating with different tetraspanins, and mass-spectrometry analyses have provided an extensive list of proteins associating with tetraspanins (Andre et al. 2006; Charrin et al. 2003a, 2001; Clark et al. 2001; Kovalenko et al. 2007; Le Naour et al. 2006; Stipp et al. 2001a, b; Zhang et al. 2003). Major groups of tetraspanin-associated proteins include, among others, certain integrins (e.g. α3β1, α4β1, α6β1, α6β4), IgSF molecules (MHC molecules, CD9P-1/EWI-F/FPRP/CD317, EWI-2/PGRL/IgSF8/CD318, CD4, CD8, CD3, ICAM-1, VCAM-1), ectopeptidases (CD26), proteases (ADAM10, MT-MMP1) and intracellular signalling proteins such as heterotrimeric G-proteins, phosphatidylinositol 4-kinase (PI4K) and activated conventional protein kinase C enzymes (Tables 3.1 and 3.2).

Initial characterization of these tetraspanin-containing complexes revealed certain features. One single complex could contain at least three tetraspanins, two copies of a given tetraspanin, or two tetraspanins and either an integrin or an MHC class II molecule (Angelisova et al. 1994; Berditchevski et al. 1996; Rubinstein et al. 1996). Moreover, after stable transfection, CD9 did not seem to compete with the other tetraspanins for associating with integrins or MHC class II molecules, but seemed to incorporate the pre-existing complexes, suggesting a high plasticity of these structures (Rubinstein et al. 1996). Altogether these findings led to the idea that tetraspanins organize a network of molecular interactions at the cell surface that was referred to as the "tetraspanin web" (Rubinstein et al. 1996). It should be stressed that this term did not imply, in the authors' view, a spider's web of rigidly connected proteins across the entire cell surface, but rather a combination of various interactions (Boucheix and Rubinstein 2001).

3.2.2 Building the Web Through Different Levels of Interaction

Through the use of different detergents in tetraspanin co-immunoprecipitation experiments, three levels of interaction in the tetraspanin web have been proposed:

Table 3.1 Established and potential tetraspanin/partner pairs

Tetraspanins	Partners	Confirmed by XL	Reference	Comment	Other tetraspanins demonstrated to associate with this molecule using milder detergents
UP1a	UPII	Yes	For review (Sun 2006)	The UP1b/UPIII pair served as a control	
UP1b	UPIII	Yes	For review (Sun 2006)	The UP1a/UPII pair served as a control	
CD9	CD9P-1/EWI-F/FPRP/CD315	Yes	Charrin et al. (2001), Stipp et al. (2001b)	Specific digitonin-resistant complex. Partially resistant to Triton X-100 disruption	CD82, CD151, CD63, Tspan8 (Charrin et al. 2001; Claas et al. 2005)
	EWI-2/CD316	Yes	Charrin et al. (2003a), Stipp et al. (2001a)	Specific digitonin-resistant complex	CD53, CD82, CD151 (Charrin et al. 2003a; Zhang et al. 2003)
	EpCAM	Yes	Le Naour et al. (2006)	Specific digitonin-resistant complex	Tspan-8 (Claas et al. 2005)
	Pro-HB-EGF	Yes	Iwamoto et al. (1994)	Only CD9 among tetraspanins can upregulate its activity	CD63, CD81, CD82 (Nakamura et al. 2000)
	ICAM-1	No	Barreiro et al. (2005), (2008)	FRET analysis suggests a proximity of ICAM1 with CD9 but not CD151	CD151 (Barreiro et al. 2005)
	Claudin-1	Yes	Kovalenko et al. (2007)	Claudin-1 identified in a Triton X-100 CD9 immunoprecipitation by mass spectrometry	CD81 and CD151 (Harris et al. 2008; Kovalenko et al. 2007)
	Pro-TGFα	No	Shi et al. (2000)	Triton X-100 co-immunoprecipitation. No other tetraspanin tested	
	Integrin $\alpha_2\beta_1$	No	Cailleteau et al. (2010)	Detection identified using the split-ubiquitin assay; confirmation by co-immunoprecipitation in Brij58 buffer.	CD151 (Sincock et al. 1999)
CD81	CD19	No	Bradbury et al. (1992), Horváth et al. (1998), Matsumoto et al. (1993), Shoham et al. (2003)	Specific digitonin-resistant complex; CD21 is supposed to interact with CD81 through CD19	CD9, CD81 (Horváth et al. 1998)

(continued)

Table 3.1 (continued)

Tetraspanins	Partners	Confirmed by XL	Reference	Comment	Other tetraspanins demonstrated to associate with this molecule using milder detergents
	Integrin $\alpha_4\beta_1$	No	Serru et al. (1999)	Specific digitonin-resistant complex	CD9, CD53, CD63, and CD82 (Mannion et al. 1996; Rubinstein et al. 1996)
	CD9P-1/EWI-F/FPRP	Yes	Charrin et al. (2001), Stipp et al. (2001b)	Specific digitonin-resistant complex	CD82, CD151, CD63, Tspan8 (Charrin et al. 2001; Claas et al. 2005)
	EWI-2/CD316	Yes	Charrin et al. (2003a), Clark et al. (2001), Stipp et al. (2001a)	Specific digitonin-resistant complex	CD53, CD82, CD151 (Charrin et al. 2001; Claas et al. 2005)
	Claudin-1	No	Harris et al. (2008), Kovalenko et al. (2007)	FRET indicates proximity	CD9 CD151 (Kovalenko et al. 2007)
CD151	Integrin $\alpha_3\beta_1$	Yes	Serru et al. (1999), Sincock et al. (1999), Yauch et al. (1998), (2000)	Specific digitonin, NP-40 and Triton X-100-resistant complex	CD9, CD81, CD82, CD63, Tspan4 (Berditchevski et al. 1995; Berditchevski et al. 1996; Tachibana et al. 1997)
	Integrin $\alpha_6\beta_1$	Yes	Kazarov et al. (2002), Serru et al. (1999), Sincock et al. (1999)	Specific digitonin and NP-40 resistant complex	CD9, CD81, CD82, CD63, Tspan4 (Berditchevski et al. 1995, 1996; Tachibana et al. 1997)

	Integrin $\alpha_6\beta_4$	No	Chometon et al. (2006), Sterk et al. (2000), (2002)	A direct association is suggested by the targeting of CD151 to hemisdesmo somes by this integrin and the absence of staining of hemidesmosomes by a CD151 mAb which detects the fraction of CD151 not associated with integrins.	CD9 (Baudoux et al. 2000)
	Integrin $\alpha_{IIb}\beta_3$	+/–	Lau et al. (2004)	Interaction observed in Triton X-100 but no other tetraspanin tested in parallel. Cross-linking experiments did not provide any information on the size of the complex; altered integrin $\alpha_{IIb}\beta_3$ functions of CD151 null platelets	CD9, CD63 (Indig et al. 1997; Israels et al. 2001; Slupsky et al. 1989)
	VCAM-1	No	Barreiro et al. (2005), (2008)	FRET analysis suggests proximity of VCAM1 with CD151 but not CD9	CD9 (Barreiro et al. 2005)
	MT1-MMP	No	Yanez-Mo et al. (2008)	FRET analysis suggests proximity of MT1-MMP with CD151 but not CD9. Specific effect of CD151 silencing on MT1-MMP function	CD9, CD81, CD63, Tspan12 (Lafleur et al. 2009; Yanez-Mo et al. 2008)
CD53	ADAM10	No		Specific digitonin-resistant complex (our unpublished data)	CD9, CD63, CD81, CD82, CD151, Tspan12 (Arduise et al. 2008; Xu et al. 2009)
CD63	H^+K^+ATPase	No	Duffield et al. (2003)	Interaction observed in Triton X-100 but no other tetraspanin tested in parallel	
	CXCR4	No	Yoshida et al. (2008)	CD63 modifies trafficking of CXCR4	
	Synaptotagmin VII		Flannery et al. (2010)		CD63 regulates subcellular localization of synaptotagmin VII

(continued)

Table 3.1 (continued)

Tetraspanins	Partners	Confirmed by XL	Reference	Comment	Other tetraspanins demonstrated to associate with this molecule using milder detergents
Tspan12	Frizzled 4	No	Junge et al. (2009)	Tspan12 is a component of the norrin receptor complex, as determined by cross-linking, and modulates norrin-induced signalling. A number of other tetraspanins were shown not to stimulate norrin signalling	
Tspan8	E-cadherin		Greco et al. (2010)	Cross-linking reveals the existence of a complex the size of which is compatible with a proximal interaction; no detection of a CD9/E-cadherin interaction under the same conditions; E-cadherin association with tetraspanins is usually not observed in co-immunoprecipitation experiments	
TspanC8 tetraspanins (Tspan5, 10, 14, 15, 17, 33)	ADAM10	Yes	Prox et al. (2012), Dornier et al. (2012) Haining et al. (2012)	Specific digitonin-resistant complex; All 6 TspanC8 tetraspanins were shown to promote ADAM10 exit from the Endoplasmic Reticulum, and 4 of them to increase ADAM10 cell surface expression	CD9, CD63, CD81, CD82, CD151, Tspan12 (Arduise et al. 2008; Xu et al. 2009)

A tetraspanin's partner is defined as a molecule that interacts directly with this tetraspanin to form a primary complex. In co-immunoprecipitation experiments, these primary complexes are visualized under lysis conditions "disrupting" tetraspanin/tetraspanin interactions (Triton X-100, digitonin). Specific cross-linking experiments or FRET may also reveal such primary interactions.
XL cross-linking.

Table 3.2 Proteins interacting with tetraspanins, without strong tetraspanin partner candidate

Associated molecule	Identified associated tetraspanins	Most stringent detergent in which an interaction with a tetraspanin was observed and comment
Adhesion molecules		
α1β1 integrin	CD9	Brij97 (Lozahic et al. 2000)
α5β1 integrin	CD9,CD151	CHAPS; Brij58; interaction not observed in Brij97 (Rubinstein et al. 1994; Sincock et al. 1999)
α7β1 integrin	CD151, CD9	CHAPS (Sterk et al. 2000)
αLβ2 integrin	CD82	Brij96 (Shibagaki et al. 1999)
αMβ2 integrin	CD63	Brij58 (Skubitz et al. 1996)
αvβ5 integrin	CD81	Brij97 (Chang and Finnemann 2007)
CD42/gpIb	CD9	CHAPS + crosslinking (Longhurst et al. 1999; Slupsky et al. 1997)
CD44	CD9[a], Tspan8, Tspan12[a]	Brij97 (Le Naour et al. 2006; Xu et al. 2009); Crosslinking with CD9 after depalmitoylation (Kovalenko et al. 2007)
GPVI	Tspan9, CD9, CD151	Brij97 (Protty et al. 2009)
Syndecan	CD9	CHAPS (Jones et al. 1996)
Lu/B-CAM/CD239	CD9[a]	Brij97 (Andre et al. 2006)
CD36	CD9	Brij96 (Miao et al. 2001)
L1-CAM	CD9	An interaction had been observed in CHAPS (Schmidt et al. 1996). We could not observe it in Brij97 (our unpublished data)
Ig domain proteins		
MHC-I	CD82, CD81, CD53, CD9[a]	Brij 97 (CD9) (Lagaudriere-Gesbert et al. 1997b; Le Naour et al. 2006); Crosslinking with CD9 after depalmitoylation (Kovalenko et al. 2007); the interaction with CD53 and CD81 is only supported by mAb FRET experiments (Szöllösi et al. 1996)
MHC-II	CD9[a], CD53, CD81, CD82, CD37	CHAPS (Angelisova et al. 1994; Rubinstein et al. 1996); mAb FRET (Szöllösi et al. 1996); Based on the pattern of biotin-labelled proteins, the interaction is probably conserved in Brij 97 (Horváth et al. 1998)
CD2	CD9, CD53	CHAPS (Bell et al. 1992; Toyo-Oka et al. 1999); elsewhere, no interaction of CD2 with CD81 and CD82 could be observed in CHAPS and Brij97 (Imai et al. 1995; Imai and Yoshie 1993)
CD3	CD9, CD81, CD82	CHAPS (Toyo-Oka et al. 1999). Association not observed in Brij 96 (Imai and Yoshie 1993)
CD4	CD81,CD82	Brij96 (Imai and Yoshie 1993)

(continued)

Table 3.2 (continued)

Associated molecule	Identified associated tetraspanins	Most stringent detergent in which an interaction with a tetraspanin was observed and comment
CD5	CD9	CHAPS (Toyo-Oka et al. 1999); elsewhere, the interaction with CD81 and CD82 observed in CHAPS was variable and not observed in Brij96 (Imai and Yoshie 1993)
CD8	CD81, CD82	Brij96 (Imai and Yoshie 1993)
Other receptors		
EGFR	CD82, CD9, Tspan12[a]	Brij98 (Odintsova et al. 2000), CHAPS CD9 (Murayama et al. 2008), Brij 96/97 Tspan12 (Xu et al. 2009)
GPR56	CD9, CD81[a]	Brij96 (Little et al. 2004)
cKit/CD117	CD9, CD81, CD63	CHAPS (Anzai et al. 2002)
c-MET	CD82, CD151	Brij98 (Klosek et al. 2005; Takahashi et al. 2007; Todeschini et al. 2007). No interaction of CD82 with c-Met could be observed elsewhere (Sridhar and Miranti 2006)
CD71 (transferrin receptor)	CD81, Tspan12[a]	Brij 97 (Abache et al. 2007; Xu et al. 2009)
AMPA receptor	Tspan7	Bassani et al. (2012)
Membrane-anchored enzymes		
CD26/dipeptidyl peptidase IV	CD9[a]	Brij97 (Le Naour et al. 2006); based on the pattern of proteins co-immunoprecipitated with CD26, Tspan8 probably also associates with CD26
CD13/aminopeptidase N	CD9, CD81[a], Tspan12[a]	Mass-spectrometry (Xu et al. 2009). A mAb screened for its ability to co-immunoprecipitate tetraspanins recognized CD13 and was used to confirm the presence of CD13 in a CD9 IP (Rubinstein's lab, unpublished). Association with CD81 confirmed in (Xu et al. 2009)
CD38/cyclic ADP-ribose hydrolase	CD9	CHAPS (Zilber et al. 2005)
CD224/γ-glutamyltransferase 1	CD9[a], CD37, CD81, CD53, CD82	Brij97 (Le Naour et al. 2006; Nichols et al. 1998)
γ-secretase	CD9, CD81	CHAPSO (Wakabayashi et al. 2009)

Multipass membrane proteins		
Choline transporter like protein 1 (CTL1)/CDw92	CD9[a]	Brij97 (Andre et al. 2006; Le Naour et al. 2006)
CD20	CD53, CD81, CD82	mAb FRET (Szöllösi et al. 1996)
L6	CD81, CD63, CD151,	0.8% Brij 98/0.2% Triton X-100 (Lekishvili et al. 2008)
CD47	CD9	CHAPS (Longhurst et al. 1999)
Others		
CD46	CD9[a], CD81, CD82, CD151	Brij97 (Lozahic et al. 2000); interaction with tetraspanins probably indirect interaction through integrins
Leu-13/IFITM1	CD81	CHAPS (Takahashi et al. 1990)
Aggrus/podoplanin	CD9	Bri97. The expression of CD9 in a cancer cell line delayed platelet activation through podoplanin (Nakazawa et al. 2008)
SCIMP	CD37, CD53, CD81	CHAPS, Brij98 (Draber et al. 2011)

Results of the proteomic analyses are included only if the interaction has been confirmed at least for one tetraspanin

Many of the interactions observed in the mildest detergent (CHAPS, Brij 58 or 98) have not been tested in Brij97

In most cases only the tetraspanin cited has been tested

[a]Indicates identification of the associated protein in a proteomic analysis (but not necessarily the first identification). The interaction may have been observed in other experiments

primary, secondary and tertiary. Primary interactions are thought to be relatively strong, direct interactions between a tetraspanin and a so-called specific 'partner protein.' These interactions tend to be resistant, or partially resistant, to relatively stringent detergents such as 1% Triton X-100 and digitonin (Serru et al. 1999; Yauch et al. 1998). The latter is unusual in that it forms a 1:1 complex with cholesterol, which it precipitates with tetraspanins that are tightly associated with cholesterol and with each other, therefore potentially leaving in the supernatant some tetraspanin-partner pairs (Charrin et al. 2003c). Examples of such primary interactions include CD151 with the laminin-binding integrin α3β1, and CD9 with the IgSF protein CD9P-1 (Table 3.1). These primary interactions can be captured by chemical cross-linking, which is a gold standard in the field for confirming primary interactions. Commonly-used cross-linkers are the homo-bifunctional *N*-hydroxysuccinimide esters such as the water-insoluble disuccinnimidyl suberate (DSS), used for surface and intracellular cross-linking, and its water-soluble analogue Bis(sulphosuccinimidyl) suberate (BS_3), used for surface cross-linking. These reagents react with primary amine groups, present on lysine residues and the NH_2-termini of proteins, to form covalent amide bonds. The spacer arm length of 1.1 nm is sufficiently small such that proteins must be in very close proximity to be cross-linked and are likely to be interacting with each other. However, this approach is unlikely to reveal solely direct interactions because, in theory, a multi-molecular complex could be chemically cross-linked if there was appropriate juxtaposition of primary amine groups. An important control in these studies, therefore, is to determine whether the size of the interacting complex is consistent with a direct tetraspanin-partner protein interaction. This can be achieved most simply by western blotting but also by gel filtration.

Secondary interactions are the weaker interactions between different primary complexes and are thought to be the consequence of tetraspanin-tetraspanin interactions. These have been found to be preserved in detergent solutions such as 1% Brij97, which is less stringent than Triton X-100, and is thought to largely maintain tetraspanin-tetraspanin interactions (Berditchevski et al. 1996). Calcium and magnesium chloride are frequently added to Brij97 lysis buffer for better preservation of secondary interactions. The effects of the divalent cations is possibly due to their absorption onto the surface of membranes which increases their resistance to detergent solubilisation (Charrin et al. 2002). These secondary interactions thus provide a means by which primary complexes can be clustered together to form the tetraspanin web. A number of clever experiments have supported this key concept by investigating the interaction of the CD151 partner α3β1 with non-partner tetraspanins. For example, RNAi knockdown or germline deletion of CD151 dramatically reduced α3β1 interaction with other tetraspanins (Winterwood et al. 2006; Yamada et al. 2008a; Yang et al. 2008). In the reverse experiment in a CD151-deficient cell line, α3β1 only interacted with other endogenous tetraspanins when CD151 was transfected (Charrin et al. 2003b). Moreover, transfection of a non-palmitoylated mutant of CD151, which was known to interact normally with α3β1 but poorly with other tetraspanins, reduced the interaction of α3β1 integrin with endogenous tetraspanins (Berditchevski et al. 2002). Similar conclusions were reached from cell line experiments using the CD9 partner CD9P-1, since an interaction between the latter and other tetraspanins was only observed when CD9 was transfected (Charrin et al. 2003b).

Tertiary interactions in the tetraspanin web were defined as those which are preserved in Brij58, Brij98 and CHAPS, but not in Brij97. Some of the proteins co-immunoprecipitated with tetraspanins using only these detergents may be actual tetraspanin-associated proteins that form a complex dissociated with the more stringent conditions. However, because these detergents solubilise membranes relatively poorly, they may, in fact, describe proteins placed in the proximity to the tetraspanin-based structures on the plasma membrane. This chapter will therefore focus on primary and secondary interactions as the building blocks of the tetraspanin web. Indeed, strong support for this model has come from the only structural study of native tetraspanin complexes using cryo-electron microscopy (Min et al. 2006). Kong and colleagues took advantage of the fact that the bladder epithelium is almost entirely covered by a hexagonal, crystalline array of 16 nm urothelial plaques, which are particularly amenable to purification and subsequent cryo-electron microscopy (Wu et al. 2009). Importantly, the urothelial plaques contain two tetraspanins, uroplakin (UP) Ia and Ib, and their respective non-tetraspanin partner proteins UPII and UPIII (Wu et al. 2009). The 6Å structural resolution showed that the tetraspanins were compact, rod-shaped structures extending only about 5 nm from the plasma membrane. The larger partner proteins appeared to interact tightly with the tetraspanins via contacts between transmembrane and extracellular regions, and partially buried the tetraspanins. These UPIa-UPII and UPIb-UPIII primary interactions made less extensive secondary contacts with each other to form the individual plaques, composed of six molecules of each uroplakin, and the wider array of interconnected plaques across the epithelial surface (Min et al. 2006).

3.2.3 Interaction Mechanisms

The cryo-electron microscopy investigation of uroplakins by Kong et al. is the only study to date that demonstrates structurally how a tetraspanin interacts with a partner protein (Min et al. 2006). In this publication, the large extracellular region and the transmembrane regions (most likely transmembranes 3 and/or 4) of the uroplakin tetraspanins appeared to interact with the uroplakin partners. However, the 6Å resolution was too low to reveal the precise interaction mechanism. In future, X-ray crystallography may provide the higher resolution required to reveal such fine details, especially given recent advances in generating high resolution structures of multi-pass transmembrane proteins using this method (Ubarretxena-Belandia and Stokes 2010).

A number of studies have used biochemical approaches with chimeric tetraspanins to identify interaction mechanisms. In these experiments, the chimeras are generated by replacing regions of the tetraspanin in question with the equivalent regions of a tetraspanin that does not interact with the partner under investigation.

The CD151 interaction with laminin-binding integrins has been intensively studied using this approach, and researchers have largely focussed on the particularly tight binding with integrin α3β1. The regions of each protein that are essential for the interaction are a portion of the membrane-proximal stalk of α3β1 and the variable region of the large extracellular domain of CD151 (Berditchevski et al. 2001; Yauch et al. 2000). Essential amino acid residues within this variable region have been identified by mutagenesis as the QRD motif (where Q is glutamine, R is arginine and D is aspartic acid) within the D helix (Kazarov et al. 2002). Mutation of this site has been subsequently utilised by several groups to explore the functional consequences of disrupting the CD151-α3β1 interaction (Kazarov et al. 2002; Novitskaya et al. 2010; Sadej et al. 2009, 2010; Zuo et al. 2010). The inability of the QRD mutant to interact with integrins was, however, recently called into question (Zevian et al. 2011). In the initial studies, the analyses of the interaction of this mutant with integrins were performed in cells expressing WT CD151. The mutant was known to lose its interaction with the integrin α3β1 after cell lysis in Triton X-100, but not in Brij96/97. This was interpreted as the QRD mutant interacting indirectly with the integrins through interaction with WT CD151 (Kazarov et al. 2002). However, Zevian et al. found that the QRD mutant interacted with the integrins α3β1 and α6β1 in the absence of WT CD151. Because the QRD mutant had a similar ability as WT CD151 to restore an interaction of the integrin with other tetraspanins, a likely interpretation is that the QRD site makes the interaction of CD151 with integrins more stable, but without making an essential contribution. This interpretation is supported by functional studies that use the QRD mutant CD151 to rescue motility and adhesion defects in CD151-deficient cells (Zevian et al. 2011). It should however be noted that in other studies, the QRD mutant failed to restore the functional effects associated with depletion of CD151, such as the response to TGFβ1 (Sadej et al. 2010). Finally, the chimeric tetraspanin approach, combined with CD151 mAb epitope mapping, has identified regions of CD151 that are masked by α3β1 interaction and pointed to two additional short regions upstream of the QRD motif that appear to interact with the integrin (Yamada et al. 2008b). Together these studies suggest that the CD151 QRD mutation should be used with caution and carefully evaluated for α3β1 interaction in each experimental system.

The interactions of CD9 and CD81 with the related IgSF proteins EWI-2 and CD9P-1/EWI-F have also been studied using chimeric proteins and point mutants and revealed the important contribution of transmembrane domains (Andre et al. 2009; Charrin et al. 2003a; 2001, 2009b; Montpellier et al. 2011). The EWI-2 interaction with CD81 requires the membrane proximal fourth IgSF domain of EWI-2, in addition to a glycine-zipper motif in the transmembrane region and palmitoylation sites (Montpellier et al. 2011). For CD81, palmitoylation is not required, but the major extracellular region and transmembranes 3 and 4 are essential. It is possible that these regions form an extended binding surface for interaction with partners, similar to that predicted from the structural study of uroplakins (Min et al. 2006). Interestingly, these regions are all required for CD81-mediated HCV entry; the authors propose that forms of CD81 that cannot bind partners may homo-dimerize more easily and thus be less competent to facilitate entry of HCV (Montpellier et al. 2011). Somewhat surprisingly, the interaction mechanisms of EWI-2 with CD81 and CD9 appear to be different, since the fourth IgSF domain of EWI-2 is not required for CD9 binding (Montpellier et al. 2011) and two regions of CD9 are required for maximal EWI-2 interaction, namely the second half of the large extracellular region, which includes the variable region, and a region spanning most of transmembrane 2 and all of transmembrane 3 (Charrin et al. 2003a).

A similar study has addressed the interaction between CD81 and a second EWI family member, CD9P-1/EWI-F. The fourth transmembrane domain of the tetraspanin is the most important region and appears to interact with the transmembrane region of CD9P-1/EWI-F (Charrin et al. 2009b). Similar requirements were observed for the interaction of CD9P-1/EWI-F with CD9 (Charrin and Rubinstein unpublished data). One final interaction mapping example concerns CD81 and the IgSF protein CD19, which plays an important co-stimulatory role in B cell receptor signalling. The large extracellular region of CD81 is important in binding to the extracellular region of CD19, although a requirement for the first transmembrane domain for exit of CD19 from the Golgi, suggests that this domain might also interact with CD19 (Shoham et al. 2006).

In summary, these studies have identified both transmembrane domains and the variable part of the tetraspanin LEL as important regions for specific partner interactions. Interestingly the 3rd and 4th transmembrane regions of CD9 and/or CD81 that are required for the interaction with CD9P-1/EWI-F or EWI-2, are situated on the same side of the molecule as the variable subdomain of the LEL according to the modelling of CD81 (Seigneuret 2006). It therefore seems plausible that partner-binding interfaces could span the entire length of a tetraspanin, with particular specificity provided by the variable region. The other side of the molecule may be involved in the interaction with other tetraspanins, or alternatively with a simultaneous binding of other partner proteins. Such hypotheses will remain speculative until they can be tested experimentally by, for example, obtaining crystal structures of tetraspanins with their partners.

3.2.4 Several Levels of Interaction for the Interaction with Intracellular Molecules?

In comparison with the plethora of surface molecules associating with tetraspanins, only few cytoplasmic molecules have been shown to interact with tetraspanins. The analysis of the interaction of Syntenin-1, a scaffold protein with PDZ domains, with tetraspanins suggests that this molecule could interact directly with CD63 and indirectly with other tetraspanins, as it was shown to interact with several tetraspanins under conditions preserving the interaction of tetraspanins with the integrin $\alpha3\beta1$ (0.8% Brij98, 0.2% Triton X-100), but only with CD63 using a higher concentration of Triton X-100 (Latysheva et al. 2006). Other tetraspanins, such as CD9, CD81,

CD151 and Tspan7, have a PDZ-binding sequence which may be involved in the binding of other PDZ proteins. In this regard a peptide corresponding to the C-terminus of CD81 was shown to interact with several PDZ domains, and an interaction of CD81 with the PDZ domain proteins EBP50 and SAP97 was shown by co-immunoprecipitation (Pan et al. 2007). In addition, Tspan7 was shown to interact with the PDZ domain of "protein interacting with C kinase 1" (PICK1) (Bassani et al. 2012). Whether PDZ proteins contribute to the interaction of tetraspanins with other transmembrane proteins remains to be determined.

Another example of direct interactions is that involving ERM proteins, which interact with actin through their C-terminal domain and with the cytoplasmic domain of several adhesion molecules through their N-terminal domain (Charrin and Alcover 2006). The tetraspanins CD9 and CD81, but not CD151, were shown to co-immunoprecipitate ERM proteins. The interaction of CD81 with these proteins may be direct because the C-terminus of CD81 pulled down both ezrin and moesin (Sala-Valdes et al. 2006). However, there might also be an indirect interaction through CD9P-1/EWI-F and EWI-2, as the cytoplasmic domain of both proteins could also pull down ezrin and moesin (Sala-Valdes et al. 2006). Antibody-ligation of CD81 caused tyrosine-phosphorylation and threonine-dephosphorylation of ezrin (Coffey et al. 2009) and this required the C-terminal domain of CD81. Tetraspanins have been found in several studies to induce a reorganization of the actin cytoskeleton (Lagaudriere-Gesbert et al. 1998; Shigeta et al. 2003). Whether such reorganization is dependent on the interaction with ezrin remains to be determined.

Two kinases associate with several tetraspanins, namely the lipid kinase phosphatidylinositol 4-kinase (PI4K) (Berditchevski et al. 1997) and protein kinase C (PKCα and PKCβ2) (Andre et al. 2006; Zhang et al. 2001). So far the interactions were only observed in mild detergents. In particular the interaction with PI4K was not observed in Brij 96/97. Thus whether these kinases associate directly with one tetraspanin or one of their partners is not clear. In this regard, PI4K is palmitoylated (Barylko et al. 2009) and this post-translational modification may not only target this enzyme to the membrane but also its association with tetraspanins (see below). The use of chimeric molecules composed of CD9, which associates with PKC, and Tspan7 (TALLA-1), which does not, did not yield conclusive results on the role of CD9 cytoplasmic domains for this association (Zhang et al. 2001). Nevertheless, only the integrins associating with tetraspanins were found to interact with PI4K and PKCβ2, and this specificity was shown to be determined by the integrin ectodomain, which is responsible for the interaction with tetraspanins (Yauch et al. 1998; Zhang et al. 2001). Thus tetraspanins could link activated PKC and PI4K to the subset of integrins they interact with.

Proteomic analyses revealed the Brij97-resistant interaction of CD81 or CD9 with a subset of heterotrimeric G proteins (Andre et al. 2006; Le Naour et al. 2006; Little et al. 2004), but as a specificity control, neither CD63 nor CD151 were found to interact with Gαq/11 (Little et al. 2004). The functional relationship between tetraspanins and heterotrimeric G proteins is not yet elucidated. Because the entire fraction of the orphan heterotrimeric G protein–coupled receptor GPR56 associated with Gαq/11 is also associated with CD81, and expression of CD81 increases the interaction between GPR56 and Gαq/11, Little et al. have suggested that CD81 may function as a heterotrimeric G protein–coupled receptor scaffold (Little et al. 2004). Scaffold proteins are believed to contribute to GPCR signalling specificity by engaging additional signalling pathways or localizing GPCR signalling events to specific subcellular sites (Force et al. 2007).

Finally it was shown that the non-palmitoylated form of CD81 associates with the serine/threonine-binding signalling protein, 14-3-3. Both the palmitoylation of CD81 and association with 14-3-3 were modulated by oxidative stress (Clark et al. 2004).

3.2.5 The Role of Lipids in the Web

A prominent view of the membrane is that the lipid bilayer is not a structurally passive solvent, and that lipids contribute to the heterogeneity of the plasma membrane. In model membranes, long saturated lipids (typically sphingolipids) associate tightly with cholesterol and form, when mixed with phospholipids, a liquid-ordered phase. Such experiments provided a molecular basis for the existence of lipid rafts or membrane microdomains in the plasma membrane, which were initially defined as liquid-ordered phases in the lipid bilayer, dispersed in a liquid-disordered matrix of unsaturated glycerolipids (Simons and Toomre 2000). With the difficulties in visualizing membrane microdomains, and the introduction of advanced microscopy techniques that allowed the dynamics of membrane constituents to be studied, rafts are now viewed as fluctuating nanoscale assemblies of sphingolipids, cholesterol and proteins that can be stabilized to coalesce (Lingwood and Simons 2010).

A property of raft constituents is their resistance to detergent solubilization with the result that they float in low density fractions of sucrose gradients (Simons and Toomre 2000). A fraction of tetraspanins was found to partition in light fractions of a sucrose gradient after particular lysis conditions (Berditchevski et al. 2002; Charrin et al. 2002; Claas et al. 2001), which led to the suggestion that tetraspanin-based complexes were cell surface microdomains that were referred to as tetraspanin-enriched microdomains or "TERM" (Berditchevski et al. 2002). It should be pointed out that this appellation of tetraspanin-enriched microdomain is synonymous with that of "tetraspanin web", and that in the current view it corresponds, similarly to rafts, to dynamic nanoscale tetraspanin-based-assemblies (see next paragraph). Tetraspanins can sometimes be enriched in discrete regions of the plasma membrane. For example, CD81 concentrates in the immune synapse (Mittelbrunn et al. 2002), both CD9 and CD151 are enriched in endothelial docking structures (Barreiro et al. 2005), and both CD9 and CD81 are enriched in unknown structures at the basal surface of prostate cancer cells (Espenel et al. 2008). We believe that these tetraspanin-enriched structures should not be confounded with "tetraspanin-enriched microdomains," because as we will see in the next paragraph they are only a part of the story. Indeed, tetraspanin-based interactions characterizing the tetraspanin web occur in certain hematopoietic cells where there are no evident

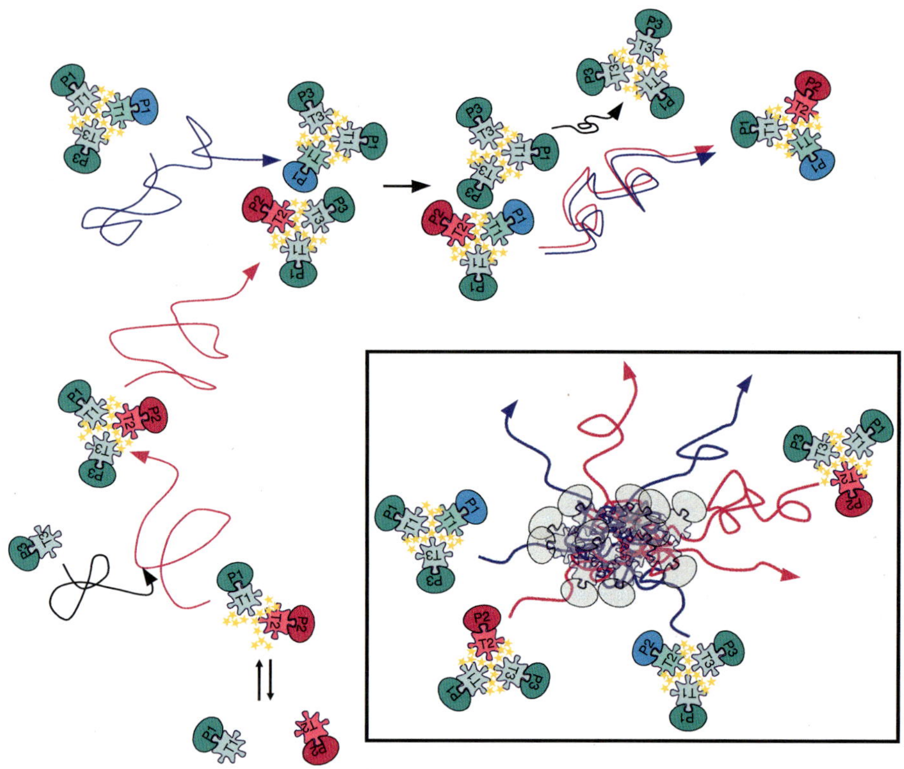

Fig. 3.2 A dynamic view of the tetraspanin web: this model is based on biochemical analysis of the tetraspanin web and the recent analysis of the dynamics of CD9 and CD151. One of these two studies was a tracking of CD9 at the single molecule level. To stay close to this technique, two tetraspanin/partner pairs are coloured in *red* and *blue*, and the traces correspond to their movement, as if they were the only molecules of this cartoon experimentally labeled. Left: basal level of interactions: small clusters of tetraspanins (T1, T2, …), each specifically associated with a molecular partner (P1, P2, …), would patrol in the plasma membrane, frequently interacting with other clusters and exchanging some of their constituents, contributing to the diversity of interactions within the tetraspanin web. Note that in this model, the clusters can contain several copies of a given tetraspanin/partner pair. Insert: upon particular stimulations, that remain largely to be defined, some tetraspanins become transiently confined within discrete areas of the plasma membrane where more stable interactions take place. This may favour specific enrichment of particular tetraspanin partners in this area

tetraspanin-enriched structures. In addition, recent advanced imaging techniques support the existence of tetraspanin/tetraspanin interactions outside discrete tetraspanin-enriched areas (Barreiro et al. 2008; Espenel et al. 2008). Moreover, Claas et al. have clearly shown that interactions between several components of the tetraspanin web occur outside detergent-resistant membranes, and thus the ability of tetraspanins to associate with each other is not directly linked to their resistance to solubilisation (Claas et al. 2001).

The ability of tetraspanins to partition into low density fractions of sucrose gradients has prompted researchers to determine whether tetraspanins are components of classical lipid raft microdomains. A number of differences between tetraspanins and lipid rafts have been highlighted. First, although they do partition into low density fractions of sucrose gradients under lysis conditions that maintain tetraspanin-tetraspanin interactions, such as Brij96/97/99 or CHAPS, tetraspanins are minimally resistant to solubilisation with 1% Triton X-100, in contrast to lipid rafts (Berditchevski et al. 2002; Charrin et al. 2003b; Claas et al. 2001; Delaguillaumie et al. 2004). Moreover, lysis in Brij detergents at 37°C maintains tetraspanin-tetraspanin interactions but disrupts lipid rafts (Charrin et al. 2003b; Claas et al. 2001). Finally, several studies, including proteomic studies, have demonstrated that tetraspanins do not associate with lipid raft components such as GPI-linked glycoproteins (Andre et al. 2006; Charrin et al. 2003c; Claas et al. 2001; Le Naour et al. 2006), and single-molecule tracking of CD9 and the GPI-linked CD55 showed different behaviors for these proteins (Espenel et al. 2008).

Certain lipids do, however, appear to have a role in the assembly of the tetraspanin web and are essential for tetraspanin function. Cholesterol, for example, appears to bind tightly to tetraspanins, as detected by labelling experiments with a cholesterol analogue (Charrin et al. 2003c). This finding is consistent with the capacity of the cholesterol-precipitating reagent digitonin to precipitate tetraspanins (Charrin et al. 2003c). Membrane cholesterol is required for normal tetraspanin functions, since extraction of this lipid with methyl-β-cyclodextrin (MβCD) inhibits CD81-mediated malaria sporozoite infection of liver cells (Silvie et al. 2006) and tetraspanin-regulated signalling events in T and B cells (Charrin et al. 2003c; Delaguillaumie et al. 2004). Moreover, MβCD alters CD9 diffusion, as determined by single molecule tracking (Espenel et al. 2008), and decreases binding of a CD81 antibody that recognises CD81 when it is associated with other tetraspanins (Silvie et al. 2006). Despite these positive findings, cholesterol extraction with MβCD from intact cells does not affect tetraspanin-tetraspanin interactions in subsequent co-immunoprecipitation experiments (Charrin et al. 2003c; Claas et al. 2001), although the presence of MβCD in the lysis buffer does disrupt such interactions (Charrin et al. 2003c). A possible explanation for this observation is that a fraction of cholesterol from the inner leaflet of intact cells is not extracted by MβCD, and is important for tetraspanin-tetraspanin interactions. It is currently unclear how cholesterol interacts with tetraspanins. One hypothesis, based on structural modelling of tetraspanins, proposes that membrane-exposed, asymmetrically clustered aromatic residues mediate cholesterol binding (Seigneuret 2006). A second hypothesis is that tetraspanin palmitate moieties bind cholesterol, which is supported by structural studies on the β2-adrenergic receptor which demonstrate palmitate-cholesterol interaction (Cherezov et al. 2007). Interestingly, mutation of the membrane-proximal cysteine residues, that are the sites for tetraspanin palmitoylation, partially reduces tetraspanin-tetraspanin association (Berditchevski et al. 2002; Charrin et al. 2003c; Yang et al. 2002; Zhou et al. 2004), abolishes the digitonin-precipitation of CD9 under certain lysis conditions (Charrin et al. 2003c) and modifies the dynamics of single-molecule interactions (Espenel et al. 2008). Consistent with this, over-expression of a protein acetyltransferase, DHHC2, that palmitoylates tetraspanins, enhances tetraspanin-tetraspanin interactions (Sharma et al. 2008). Certain membrane gangliosides also seem important in building of the web. In particular, ganglioside G_{M3} appears to interact with CD9 (Ono et al. 2001), while G_{D1a} regulates the interaction of CD82 with other tetraspanins (Odintsova et al. 2006). Moreover, CD82 promotes an increase in expression of G_{D1a} and G_{M1} (Odintsova et al. 2003). Finally, hydrolysis of sphingomyelin into ceramide by sphingomyelinase treatment was found to stimulate CD81 internalization and reduce HCV entry (Voisset et al. 2008). Further studies are now required to determine the full complexity of tetraspanin-lipid interactions and the extent to which this regulates the assembly and function of the tetraspanin web.

3.3 The Tetraspanin Web as a Platform for the Regulation of Membrane Proteins

3.3.1 A Two-Part Model for Web Function

The biochemical analysis of tetraspanin-containing complexes has been essential to demonstrate a multi-level organisation of the tetraspanin web. The recent studies of the dynamics of some of its constituents, together with in situ analysis of certain interactions, has helped to refine our view of the tetraspanin web, and of its potential function (Barreiro et al. 2008; Espenel et al. 2008). Indeed, two major insights from the recent studies of tetraspanins using advanced microscopic techniques are as follows. Firstly, tetraspanins shuttle between areas where they concentrate due to transient confinement, and areas that encompass the rest of the membrane. Secondly, tetraspanin/tetraspanin interactions occur both inside and outside tetraspanin-enriched areas.

We propose that the tetraspanin web is a dynamic entity that recruits certain molecules and regulates their function, potentially through their assembly into specialised structures in the membrane. This process can be envisaged through a two-part model (Charrin et al. 2009a). In the first part, relatively small clusters of tetraspanins, and their associated partners, would patrol the membrane and generate a succession of transient interactions that would contribute to the variety of interactions inside the tetraspanin web. In the second part, certain stimuli would induce the coalescence of multiple clusters to generate tetraspanin-enriched areas and specialised structures (Fig. 3.2).

In the first part of the model depicted in Fig. 3.2, tetraspanins are shown embedded in small clusters with their partners and possibly lipids. Evidence for this is twofold. Firstly, some anti-tetraspanin mAbs are able to induce, in the absence of secondary reagent, the regrouping of a major fraction of their target antigen in large patches (Arduise et al. 2008). This has been demonstrated in hematopoietic cell lines where there are no evident tetraspanin-enriched areas as determined by immunofluorescence microscopy. This suggests that a large fraction of a particular tetraspanin is localized in complexes containing at least two copies of the molecule.

This is because the bivalent anti-tetraspanin mAbs could not generate a large molecular aggregate of multiple clusters if only one tetraspanin was present in each cluster. In addition, other tetraspanins were also shown to concentrate in the patches induced by mAbs directed to a particular tetraspanin. This suggests that the complexes containing the two copies of the first tetraspanin also contain at least one copy of other tetraspanins. Secondly, in the tracking of single fluorescent CD9 molecules, that revealed the frequent and transient co-diffusion of two CD9 molecules undergoing Brownian trajectories, no abrupt change in CD9 velocity was observed when the molecules started to co-diffuse (Espenel et al. 2008). As one would expect the diffusion of transmembrane proteins in liquid membranes to decrease with increasing size of the diffusing protein (Gambin et al. 2006), this observation seems to exclude the interaction of two isolated CD9 molecules. Instead, it suggests that the co-diffusion represents a transient interaction between two small clusters, or the movement of a CD9 molecule from one cluster into another.

The second part of this model deals with the formation of tetraspanin-enriched areas which have been observed in various studies. In the single molecule tracking analysis of Espenel et al., dot-like tetraspanin-enriched areas were shown to be in permanent exchange with the rest of the membrane and to correspond to a decrease in diffusion and local confinement of the tetraspanins (Espenel et al. 2008). This result was consistent with the study of Barreiro et al., describing CD9- and CD151-enriched areas, generated through ICAM-1 and VCAM-1 engagement, where the diffusion was diminished (Barreiro et al. 2008). From these data, we hypothesize that tetraspanin-enriched areas are generated following the engagement of one of the tetraspanin partners, to create a nucleation point around which several tetraspanin clusters would aggregate. The ability of the HIV gag protein to create such tetraspanin-enriched areas suggests that the initial event might not only be induced from outside the cell, but also from the inside (Krementsov et al. 2010).

Therefore, in this two-step model, the relatively small tetraspanin-enriched clusters would, through the permanent exchange of their constituents, generate pre-assembled transient complexes. The pre-assembled complexes would facilitate the building of a complex molecular assembly upon proper stimulation and may contribute to regulating the stoichiometry and the dynamics of these complexes. In this regard, integrins have been shown to associate with other tetraspanin-associated molecules, such as MHC molecules, CD9P-1/EWI-F or EWI-2, under conditions preserving tetraspanin/tetraspanin interactions (Charrin et al. 2003b; Kolesnikova et al. 2004; Rubinstein et al. 1996; Winterwood et al. 2006). This model is also consistent with the finding that engagement of endothelial VCAM-1 or ICAM-1, by cells expressing either one of their ligands, induces the formation of a docking structure containing both adhesion receptors and tetraspanins (Barreiro et al. 2008). In addition, tetraspanins may favour the association of several copies of a given associated protein, as suggested by the CD9-dependant association of several MHC class II molecules (Unternaehrer et al. 2007). Such micro-clustering of certain proteins could have important consequences for their function. The tetraspanin web may also serve as a regulated adaptor platform for the communication of protein partners and intracellular signalling molecules such as PI4K or PKC (Berditchevski et al. 1997; Zhang et al. 2001). In one variation of this model, the interaction with tetraspanins may restrict other interactions of some surface proteins. This may be especially the case for ectoenzymes such as ADAM10, which was shown to be largely unable to cleave proTNFα while associated with tetraspanins (Arduise et al. 2008).

In this model, tetraspanins appear to play a role in optimizing the building of membrane structures. In this context, the role of these molecules could be revealed only under conditions where the optimization of a particular process is essential. This may in part explain why many tetraspanin knock-out mice have only mild phenotypes, or phenotypes restricted to particular organs. In this regard, the deficiency in sperm-egg fusion in the absence of CD9 may be so strong because the time window for fertilization is short, and optimization essential (Kaji et al. 2000; Le Naour et al. 2000; Miyado et al. 2000). This may also explain why the role of CD151 in angiogenesis is revealed only under "pathological" conditions (Takeda et al. 2007) or why the role of Tsp12 in *C. elegans* is revealed under conditions of Notch sensitization (Dunn et al. 2010).

3.3.2 The Tricky Issue of Identifying Tetraspanin-Specific Functions

Since individual mammalian cell types appear to express almost two thirds of the 33 tetraspanins in the genome (Protty et al. 2009), the tetraspanin web is clearly quite complex in nature. A prediction of such complexity, in combination with the two-part model described above, is that the experimental identification of functions for individual tetraspanins might be quite tricky. For example, if a particular tetraspanin was targeted by either knockdown, over-expression or with a mAb, this might be expected to yield a phenotype due to a general perturbation of the tetraspanin web, rather than due to a specific effect on the targeted tetraspanin. Indeed, targeting of a different tetraspanin by the same experimental technique might be expected to yield a similar phenotype.

There are a number of published reports in which this prediction has held true. For example, mice deficient for either CD9 or CD81 displayed reduced female fertility due to impaired sperm-egg fusion (Rubinstein et al. 2006), and mice deficient for either CD151 or Tspan32 had a mild bleeding phenotype due to impaired platelet function (Goschnick et al. 2006; Lau et al. 2004; Orlowski et al. 2009). In addition, gene knockout of CD37, CD81, CD151 or Tspan32 each yielded mice in which the T lymphocytes were hyper-proliferative in response to stimulation through the T cell receptor (Levy and Shoham 2005; Wright et al. 2004b). Similarly, a number of different anti-tetraspanin mAbs were shown to each induce T lymphocyte co-stimulation (Carter and Fearon 1992; Lagaudriere-Gesbert et al. 1997a; Lebel-Binay et al. 1995; Tai et al. 1997), and recombinant forms of the major extracellular region of different tetraspanins appeared to each inhibit HIV infection and macrophage fusion (Ho et al. 2006; Parthasarathy et al. 2009). However, there are also examples of studies that have revealed functions specific to individual tetraspanins. These have often employed biochemical approaches to identify specific partner proteins, which have helped to provide a mechanism behind observed phenotypes. These studies will be described in the following section.

3.4 Emerging Functional Mechanisms for Specific Tetraspanins

Tetraspanins have been implicated in a variety of cellular functions through studies that have employed gene knockouts, mRNA knockdown, over-expression, tetraspanin antibodies and/or recombinant extracellular regions. However, very few such studies have identified specific functional mechanisms because of the difficulties outlined in the previous section. This section will outline four major functions that have emerged for specific tetraspanins, namely the regulation of receptor-ligand binding, signalling, trafficking and proteolysis. The focus will be on studies that have most clearly defined the mechanism of tetraspanin action; where mechanism is less clear the reader is encouraged to read the most relevant later chapters of this book for a more thorough description.

3.4.1 Regulation of Receptor-Ligand Binding

Probably the most thoroughly studied tetraspanin-partner protein interaction is that of CD151 with the laminin-binding integrins $\alpha3\beta1$, $\alpha6\beta1$ and $\alpha6\beta4$. This is because the interaction is relatively stable and so historically was easily identified, because these integrin partners have a well-identified key function, and also because CD151 is over-expressed in many cancers and has potential as a therapeutic target (Romanska and Berditchevski 2011; Sadej et al. 2009; Yang et al. 2008; Zoller 2009). CD151-integrin interactions occur early in biosynthesis and appear to be largely mediated by regions in the variable domain of the major extracellular loop of CD151 (see Sect. 3.2.3). The intimacy of CD151 and laminin-binding integrins is underscored by defective integrin function in the absence of CD151. These defects are in integrin signalling, internalisation, glycosylation and ligand binding through the process of adhesion strengthening.

It has been recently proposed that integrin-mediated adhesion strengthening can be explained by formation of so called "catch bonds", which are generally defined as prolonged interactions between a receptor and its ligand when a mechanical force is applied (Thomas 2008). Integrins are subject to force because of their capacity to bind extracellular matrix proteins on one hand and simultaneous interaction with the cellular actin cytoskeleton on the other, which, in turn, applies inwards forces from the integrin-mediated adhesion site at the cell periphery (Puklin-Faucher and Sheetz 2009). Thus adhesion strengthening could be the result of force-induced increases in adhesion of integrins to the actin cytoskeleton and subsequent changes in integrin catch bonds, through a conformational change in the ligand-binding integrin headpiece (Kong et al. 2009). Interestingly, a C-terminal mutant of CD151 was found to impair α6β1 adhesion strengthening to laminin-coated magnetic beads subjected to a magnetic force (Lammerding et al. 2003). Importantly, there was no effect on the initial adhesion to laminin and adhesion strengthening to fibronectin. The authors proposed that their CD151 C-terminal mutant, that retained the capacity to interact with α6β1, was acting as a dominant negative by disrupting the function of endogenous CD151. Furthermore, it was proposed this mutant has a negative effect on adhesion strengthening by disrupting force-induced interactions of integrins with the cytoskeleton (Lammerding et al. 2003).

The idea that CD151 can promote adhesion strengthening of laminin-binding integrins was supported by a second study on α3β1 (Nishiuchi et al. 2005). The authors used two tools (a mAb to CD151 that resulted in dissociation of its interaction with α3β1 and knockdown of CD151 expression using RNAi) to show that in the absence of CD151, α3β1-mediated cellular adhesion to laminin was impaired. In addition, disruption of the CD151-α3β1 interaction resulted in a reduction in the amount of the active form of the integrin. Taken together, these studies support the idea that CD151 regulates the strength of adhesion of its laminin-binding integrin partners, and provide a potential explanation for the well-established effects of CD151 on cell motility, spreading and cable formation (Zhang et al. 2002). They also provide a potential explanation for the defects observed for CD151-deficient platelets (Lau et al. 2004; Orlowski et al. 2009). The initial activation events were normal in these platelets, but integrin αIIbβ3 post-occupancy events were impaired, as indicated by a delayed clot retraction, a reduction of platelet aggregation and a defect of spreading on fibrinogen. These results suggest that CD151 may also regulate strengthening of adhesion mediated by αIIbβ3 (Lau et al. 2004), an integrin with which it associates (Fitter et al. 1999).

Consistent with a major role for CD151 in the regulation of adhesion strengthening for laminin-binding integrins, CD151 in most epidermal cells is localised to the basal layer, where it can regulate the interaction of integrins with laminin in the basement membrane (Chometon et al. 2006; Sterk et al. 2000). In keratinocytes of the skin, CD151 and integrin α6β4 are localised to hemidesmosomes, which are stable, rivet-like structures that anchor the basement membrane to keratin intermediate filaments. A major consequence of CD151 mutation in patients is restricted (pretibial) epidermolysis bullosa (Karamatic et al. 2004), in which defective hemidesmosomes fail to properly anchor the epidermis to the dermis, resulting in fragile skin that is easily blistered. Interestingly, the stable attachment mediated by CD151-α6β4 complexes does not seem to require linkage to other tetraspanins (Zevian et al. 2011). Human CD151 mutation also results in hereditary nephritis, leading to end-stage kidney failure, due to glomerular and tubular basement membrane abnormalities (Karamatic et al. 2004). CD151-deficient mice display a similar kidney phenotype on certain genetic backgrounds, which is also similar to that observed in integrin α3-deficient mice (Baleato et al. 2008; Sachs et al. 2006). Possible explanations for these kidney defects in the absence of CD151 include defective

$\alpha 3\beta 1$ adhesion strengthening (Sachs et al. 2006) or defective integrin-mediated assembly of the glomerular basement membrane, since thickening of this membrane is the first observed morphological defect in CD151-deficient kidneys (Baleato et al. 2008).

A similar regulation of adhesion strengthening may explain the role of CD81 in leukocyte extravasation. Following the initial rolling phase, firm adhesion of leucocytes is accomplished through the interaction of integrins with endothelial cell ICAM-1 and VCAM-1. Leukocyte CD81 expression was shown to confer increased resistance to detachment from VCAM-1 by incremental shear forces, strongly suggesting that CD81 enhances integrin $\alpha 4\beta 1$ adhesion strengthening. It also conferred increased resistance to detachment from an integrin $\alpha 5\beta 1$ ligand, but not from ICAM-1, a ligand for the integrin $\alpha L\beta 2$ (Feigelson et al. 2003).

Another example of a tetraspanin regulating ligand binding of a partner protein comes from studies on CD9 and proHB-EGF, the transmembrane precursor of heparin-binding epidermal growth factor which is a receptor for diphtheria toxin (Iwamoto et al. 1994). Expression of CD9, but not other tetraspanins, substantially increased toxin binding in the absence of any change in receptor affinity or expression level. The authors suggested that CD9 might induce a conformational change in proHB-EGF to increase the number of binding sites for diphtheria toxin.

Finally, receptor clustering represents a mechanism by which tetraspanins can promote adhesive capacity of their partners through an avidity effect. An example of this phenomenon is found on endothelial cells activated by inflammatory stimuli. On these cells, the IgSF adhesion molecules ICAM-1 and VCAM-1 cluster with tetraspanins to form adhesive platforms for the efficient capture of blood leukocytes via their integrin ligands (Barreiro et al. 2008). The importance of tetraspanins in organising these docking structures is demonstrated by knockdown of CD9 or CD151, which result in impaired leukocyte adhesion and transmigration (Barreiro et al. 2005). More recently, CD63 was found to be required for clustering of the adhesion molecule P-selectin on endothelial cells, such that leukocyte recruitment was impaired in the absence of CD63 in both in vitro and in vivo models (Doyle et al. 2011).

3.4.2 Regulation of Receptor Signalling

Tetraspanins are known to interact with certain intracellular signalling proteins. These include heterotrimeric G proteins, activated conventional protein kinase C enzymes, PI 4-kinase and any number of PDZ-containing proteins that might interact with the COOH-terminal tails of some tetraspanins (Charrin et al. 2009a). However, since it does not appear that tetraspanins are receptor-type proteins with recognised ligands, it is unlikely that tetraspanins themselves induce intracellular signalling. Instead, tetraspanins appear to regulate the signalling function of certain partner proteins that truly are signal-initiating receptors. Indeed, the well documented capacity of many tetraspanin antibodies to induce signals is probably the consequence of cross-linking such tetraspanin-associated proteins. The variety of signalling pathways that are regulated by tetraspanins suggests that these molecules are not responsible for coupling associated molecules with a particular signalling pathway. The remainder of this section will focus on the regulation of signalling by tetraspanins where the partner proteins responsible for the signalling have been identified: CD151 and integrin signalling, CD82 and receptor tyrosine kinases, CD37 and dectin-1, and Tspan12 and Norrin/β-catenin signalling.

As in the previous section on receptor-ligand binding, CD151 has been well studied as a regulator of laminin-binding integrin signalling. In CD151-deficient lung endothelial cells, plated on the laminin-rich basement membrane extract Matrigel, activation of the PI3K/Akt pathway was reduced, along with its downstream target endothelial nitric oxide synthase (eNOS) (Takeda et al. 2007). These findings could explain in part the defective pathological angiogenesis observed in these mice (Takeda et al. 2007). In tumour epithelial cells, laminin-induced signalling through a second key mediator of integrin signalling, the non-receptor tyrosine kinase FAK, was impaired in the absence of CD151 (Yamada et al. 2008a; Yang et al. 2008). This would appear to be a direct effect of CD151 on the integrin, since an antibody that dissociates CD151 from α3β1 similarly resulted in impaired downstream FAK signalling (Nishiuchi et al. 2005).

The receptor tyrosine kinases EGFR and c-Met have been found to co-immunoprecipitate with CD82 under relatively mild lysis conditions, thus it is not clear whether these proteins are direct molecular partners or whether they interact indirectly within the tetraspanin web (Odintsova et al. 2000; Takahashi et al. 2007). Nevertheless, CD82 clearly inhibits signalling by these receptors in response to their growth factor ligands (Odintsova et al. 2000; Sridhar and Miranti 2006; Takahashi et al. 2007). A number of studies have attempted to determine the mechanisms responsible, and a rather complex picture has emerged. In the case of EGFR signalling in response to its ligand EGF, CD82 appears to impair signalling by reducing the initial dimerisation of the receptor (Odintsova et al. 2003), and by promoting downregulation of the receptor by endocytosis (Odintsova et al. 2000). These effects could be mediated by the capacity of CD82 to upregulate the surface expression of certain gangliosides, to promote lateral diffusion of the activated receptor, and/or to recruit protein kinase C-α and caveolin-1 (Danglot et al. 2010; Odintsova et al. 2006; Wang et al. 2007). Gangliosides, which consist of a ceramide moiety embedded in the membrane and an extracellular oligosaccharide, are also emerging as key players in CD82 inhibition of c-Met signalling (Takahashi et al. 2007; Todeschini et al. 2007). Indeed, it is becoming increasingly apparent that membrane lipids profoundly affect the function of membrane proteins, whether they are the annular lipids that directly contact the proteins, or other lipids in the immediate vicinity (Marsh 2008). If tetraspanin webs truly concentrate certain lipid species, regulation of partner proteins through the nature of the local lipid environment is a distinct possibility and may emerge as a general theme in tetraspanin research.

Dectin-1 is a C-type lectin-like receptor expressed by antigen presenting cells and is a key pattern recognition receptor for β-glucan on fungal cell walls. Dectin-1 is unusual in signalling via a so-called hemITAM motif (YxxL) which activates the tyrosine kinase Syk (Marakalala et al. 2010). CD37 co-immunoprecipitates with dectin-1 and is able to stabilise dectin-1 expression at the cell surface (Meyer-Wentrup et al. 2007). As such, CD37-deficient cells have reduced dectin-1 surface levels and yet dectin-1 signalling is dramatically enhanced (Meyer-Wentrup et al. 2007). This suggests that CD37 is a powerful negative regulator of dectin-1 signalling, although the mechanism has yet to be elucidated.

In a recent high-profile study, signalling by the Wnt receptor Frizzled 4, in response to the ligand Norrin, was shown to be dependent on Tspan12 (Junge et al. 2009). Like other Wnt receptors, Frizzled 4 associates with the co-receptors Lrp5 or Lrp6, and once activated by its ligands Wnt or the structurally unrelated Norrin, stabilises β-catenin. The latter translocates to the nucleus, where it functions as a co-activator of TCF/LEF family transcription factors, to initiate transcriptional programs that regulate important processes such as development and cancer progression (Rao and Kuhl 2010). Tspan12-deficient mice were found to phenocopy mice deficient for Frizzled 4, Lrp5 or Norrin, in exhibiting defective vascular development in the retina (Junge et al. 2009). Tspan12 over-expression promoted Norrin-induced signalling and Tspan12 knockdown impaired this pathway. This appeared to be specific to Tspan12 and Norrin, since other tetraspanins and the alternative Frizzled 4 ligand Wnt did not affect signalling in a cell line model. In biochemical experiments, Tspan12 was found to associate with Frizzled 4 and to promote clustering of the receptor (Junge et al. 2009). The importance of Tspan12 in this signalling pathway was highlighted more recently when mutations in Tspan12 were found to be the cause of familial exudative vitreoretinopathy, an inherited blinding disorder of the retinal vascular system for which mutations in Frizzled 4, Lrp5 or Norrin are also causative (Nikopoulos et al. 2010; Poulter et al. 2010).

3.4.3 Regulation of Trafficking

There are a number of examples of tetraspanins regulating the biosynthetic maturation of their specific partner proteins. A relatively well-characterised example is CD19, an immunoglobulin superfamily protein that is largely specific to B cells and which is a co-receptor for B cell receptor signalling. It forms a complex with CD81, its tetraspanin partner, and CD21, the receptor for the C3d component of complement (Bradbury et al. 1992; Matsumoto et al. 1993). Importantly, CD19 has a cytoplasmic tail which, when tyrosine phosphorylated, recruits intracellular signalling proteins such as Src family tyrosine kinases, PI3K and vav. The CD19/CD21/CD81 complex functions to bridge the innate and adaptive immune systems by enhancing B cell activation in response to complement-tagged antigens, since such antigens are able to co-ligate the complex with the B cell receptor (Fearon and Carter 1995; Tedder et al. 1994). In the absence of CD81 in mice, CD19 expression at the B cell surface was reduced by about 50% due to impaired trafficking from the endoplasmic reticulum to the Golgi; N-linked glycosylation of CD19 was also abnormal

(Shoham et al. 2003). Moreover, in a CD81-deficient human patient there was complete loss of surface CD19, again due to defective endoplasmic reticulum to Golgi trafficking (van Zelm et al. 2010). As a consequence of these effects on CD19, the mice and human deficient for CD81 had impaired B cell antibody responses (Levy and Shoham 2005; van Zelm et al. 2010).

The uroplakins represent an additional example of tetraspanin function in biosynthetic maturation of their partner proteins. The two tetraspanin uroplakins, UPIa and UPIb, form direct complexes with their partners UPII and UPIII, respectively. In a cell line model, UPII and UPIII fail to exit the endoplasmic reticulum and traffic to the plasma membrane in the absence of their tetraspanin partners (Tu et al. 2002). Similarly, expression of EWI-2, an immunoglobulin superfamily protein that partners with CD81, is increased at the cell surface upon CD81 expression (Stipp et al. 2003). Finally, ADAM10 was recently shown to directly associate with tetraspanins having eight cysteines in their extracellular domain (TspanC8), including Tspan5, Tspan14, Tspan15 and Tspan33. These tetraspanins promote ADAM10 expression by favoring, at least in part, exit from the endoplasmic reticulum ((Prox et al. 2012); Dornier et al. 2012; Haining et al. 2012).

Finally, the tetraspanin CD63, which is somewhat unusual in its predominant localisation to intracellular vesicles such as late endosomes and lysosomes, appears to partner with the H^+/K^+-ATPase, which is largely responsible for acidification of the stomach. CD63 knockdown, over-expression and mutation studies suggest that this tetraspanin is responsible for maintaining the H^+/K^+-ATPase in an intracellular localisation in unstimulated cells, and promoting its trafficking to the plasma membrane upon stimulation (Codina et al. 2005; Duffield et al. 2003).

It is important to note that the surface expression of tetraspanin partner proteins is not always affected by the absence of the tetraspanin. For example, surface expression of integrin α3β1 is not affected by either over-expression or knockdown of CD151 (Charrin et al. 2003b; Winterwood et al. 2006). However, there is evidence for defects in α3β1 biosynthetic maturation and some aspects of trafficking in the absence of CD151, since glycosylation of the integrin is abnormal (Baldwin et al. 2008) and internalisation of the integrin in cells migrating on laminin is impaired (Winterwood et al. 2006).

3.4.4 Regulation of Proteolysis

Tetraspanins associate with a number of membrane-bound proteases and appear to regulate the function of several of them including ADAM proteases, γ-secretase and MT1-MMP. Three different studies have identified MT1-MMP as tetraspanin-associated (Lafleur et al. 2009; Takino et al. 2003; Yanez-Mo et al. 2008). MT1-MMP is a membrane protease with activity towards a variety of extracellular matrix proteins, and which is upregulated in many tumour cell types, promoting cell tumour invasion and malignancy. The first tetraspanin-related study showed that ectopically expressed MT1-MMP localises to lysosomes that contain CD63, and that over-expressed CD63 interacts with the protease and promotes its degradation (Takino et al. 2003). Two later studies have used tetraspanin knockdown to address the functional consequences of tetraspanin association (Lafleur et al. 2009; Yanez-Mo et al. 2008). On endothelial cells, MT1-MMP interacts with CD151 and its activity appears to be negatively regulated by this interaction (Yanez-Mo et al. 2008). In contrast, on tumour cells MT1-MMP associates with CD9, CD81 and Tspan12, and its activity is positively regulated by this interaction (Lafleur et al. 2009). The different conclusions drawn from these studies are likely due to the different cell types used, but nevertheless demonstrate a key role for tetraspanins in the regulation of this important protease. Indeed, since proteomic studies have identified other transmembrane proteases as tetraspanin associated, namely the serine protease matriptase/TADG-15 and the ectopeptidases CD13 and CD26/dipeptidyl peptidase IV (Andre et al. 2006; Lafleur et al. 2009; Le Naour et al. 2006), the regulation of transmembrane proteases is likely to emerge as a major general function for the tetraspanin web.

ADAM10 and ADAM17 are related ectodomain sheddases that cleave off the extracellular region of an ever increasing list of important targets, many of which are shared. These include Notch, for which ADAM10 cleavage is essential for development, amyloid precursor protein and cellular prion protein, via which ADAM10 and ADAM17 have neuroprotective roles, and several other targets with roles in inflammation (for example, IL-6, TNF-α, transmembrane chemokines CX3CL1 and CXCL16) and cancer (for example, EGF receptor ligands) (Blobel 2005; Reiss and Saftig 2009). ADAM10 is particularly strongly associated with several tetraspanins (Arduise et al. 2008) and several anti-tetraspanin mAbs were shown to stimulate the ADAM10-mediated shedding of EGF and TNFα, through redistributing ADAM10 at the cell surface (Arduise et al. 2008). ADAM10 was shown to interact in a relatively more stable manner with Tspan12, which increases maturation of ADAM10 and promotes its activity towards APP, the β-amyloid peptide precursor (Xu et al. 2009). ADAM10 was subsequently shown to interact directly with TspanC8 tetraspanins which positively regulate total and surface ADAM10 expression (Dornier et al. 2012; Haining et al. 2012), and the expression of Tspan15 was shown to promote various ADAM10-mediated cleavage events (Prox et al. 2012). In another study, ADAM17 has also been reported to interact with CD9 (Gutierrez-Lopez et al. 2011). However, the interaction of ADAM17 with tetraspanins is likely to be not as strong as that of ADAM10, because no interaction could be detected in the studies by Arduise et al. and Xu et al., and no ADAM17 peptides were obtained in mass-spectrometry analyses of CD9 and CD81 immunoprecipitates (Andre et al. 2006; Lafleur et al. 2009; Le Naour et al. 2006). Knockdown of CD9 resulted in an increase in ADAM17 activity towards two of its substrates, TNF-α and ICAM-1, suggesting negative regulation of ADAM17 by this tetraspanin (Gutierrez-Lopez et al. 2011).

ADAM10- or ADAM17-mediated cleavage is usually followed by a second proteolytic cleavage within the transmembrane regions of its target proteins mediated by γ-secretase. γ-secretase was recently shown to interact with tetraspanins, at least using the mild detergent CHAPS or CHAPSO (Wakabayashi et al. 2009).

Manipulating CD9 or CD81 (as well as that of their partner CD9P-1/EWI-F) modulated the γ-secretase-dependant cleavage of several substrates, including APP (Wakabayashi et al. 2009). In addition, a screen of tetraspanins, and initial work in *C. elegans*, led to the discovery that the TspanC8 tetraspanins Tspan5 and Tspan33 played a role in Notch signalling (Dunn et al. 2010), which relies on a sequential cleavage by ADAM10 and γ-secretase. The authors suggested that these tetraspanins influenced Notch signalling by regulating γ-secretase activity. However, the subsequent demonstration that TspanC8 tetraspanins regulate ADAM10 expression levels offered a different interpretation of these results (Dornier et al. 2012).

3.5 Concluding Remarks

When the first tetraspanins were identified just over 20 years ago, the four-transmembrane structure of these relatively small proteins offered few clues as to their biological function. Since then, a large number of tetraspanins have been discovered in animals, plants and certain fungi, and their fundamental property to associate with many other integral membrane proteins has been uncovered. While most transmembrane proteins do not appear to interact with tetraspanins, a subset clearly do, including proteins involved in adhesion, signalling, proteolysis and pathogen entry. For some, specific tetraspanins have been found to interact with these so-called partner proteins and their respective interacting regions characterised biochemically. Analyses of experimentally-generated tetraspanin-deficient organisms, and naturally occurring tetraspanin mutations in humans, have proved tetraspanins to play essential roles in processes such as cell adhesion, motility, signalling and cell-cell fusion. In select cases, the mechanisms responsible for these phenotypes have been pin-pointed to dysregulated biosynthesis, intracellular trafficking and/or adhesive function of the tetraspanin partner.

Despite these advances, we have yet to identify partners for most of the 33 human tetraspanins, in part due to the lack of monoclonal antibody reagents to most tetraspanins, and indeed half have not been studied at all. While accepting such clear gaps in our current knowledge, 20 years of biochemistry, and more recent high-resolution imaging experiments, have nevertheless allowed us to develop a model for tetraspanin function at the cell surface. The biochemistry yielded the idea that tetraspanins, through their multiple levels of interaction, played a role in membrane compartmentalization, which can be defined as non-random distribution of membrane proteins or molecular interactions. High-resolution imaging allowed a more precise scenario to be conceived. We propose that tetraspanins patrol the plasma membrane in small, transient complexes, held together by tetraspanin-partner interactions and tetraspanin-tetraspanin interactions. In certain scenarios, such as cell stimulation impacting on certain tetraspanin-associated proteins, these tetraspanin clusters would be induced to associate into larger microdomains that could function as platforms to facilitate processes such as cell adhesion, signalling and proteolysis.

This network of dynamic tetraspanin interactions with partner proteins has been termed the tetraspanin web. In summary, tetraspanins can be regarded as proteins that optimise, or fine-tune, the responses of their partner proteins by maintaining them in partially pre-assembled, 'ready-to-go' complexes. In future, fully testing this model will depend on the identification of the full complement of tetraspanin partner proteins, together with high resolution imaging and structural data on tetraspanins and partners. Only then can we determine the extent to which the tetraspanin web can be manipulated with drugs to treat human disease.

References

Abache T, Le Naour F, Planchon S, Harper F, Boucheix C, Rubinstein E (2007) The transferrin receptor and the tetraspanin web molecules CD9, CD81, and CD9P-1 are differentially sorted into exosomes after TPA treatment of K562 cells. J Cell Biochem 102:650–664

Andre M, Le Caer JP, Greco C, Planchon S, El Nemer W, Boucheix C, Rubinstein E, Chamot-Rooke J, Le Naour F (2006) Proteomic analysis of the tetraspanin web using LC-ESI-MS/MS and MALDI-FTICR-MS. Proteomics 6:1437–1449

Andre M, Chambrion C, Charrin S, Soave S, Chaker J, Boucheix C, Rubinstein E, Le Naour F (2009) In situ chemical cross-linking on living cells reveals CD9P-1 cis-oligomer at cell surface. J Proteomics 73:93–102

Angelisova P, Hilgert I, Horejsi V (1994) Association of four antigens of the tetraspans family (CD37, CD53, TAPA-1, and R2/C33) with MHC class II glycoproteins. Immunogenetics 39:249–256

Anzai N, Lee Y, Youn BS, Fukuda S, Kim YJ, Mantel C, Akashi M, Broxmeyer HE (2002) C-kit associated with the transmembrane 4 superfamily proteins constitutes a functionally distinct subunit in human hematopoietic progenitors. Blood 99:4413–4421

Arduise C, Abache T, Li L, Billard M, Chabanon A, Ludwig A, Mauduit P, Boucheix C, Rubinstein E, Le Naour F (2008) Tetraspanins regulate ADAM10-mediated cleavage of TNF-alpha and epidermal growth factor. J Immunol 181:7002–7013

Baldwin G, Novitskaya V, Sadej R, Pochec E, Litynska A, Hartmann C, Williams J, Ashman L, Eble JA, Berditchevski F (2008) Tetraspanin CD151 regulates glycosylation of (alpha)3(beta)1 integrin. J Biol Chem 283:35445–35454

Baleato RM, Guthrie PL, Gubler MC, Ashman LK, Roselli S (2008) Deletion of CD151 results in a strain-dependent glomerular disease due to severe alterations of the glomerular basement membrane. Am J Pathol 173:927–937

Bandyopadhyay S, Zhan R, Chaudhuri A, Watabe M, Pai SK, Hirota S, Hosobe S, Tsukada T, Miura K, Takano Y, Saito K, Pauza ME, Hayashi S, Wang Y, Mohinta S, Mashimo T, Iiizumi M, Furuta E, Watabe K (2006) Interaction of KAI1 on tumor cells with DARC on vascular endothelium leads to metastasis suppression. Nat Med 12:933–938

Barreiro O, Yanez-Mo M, Sala-Valdes M, Gutierrez-Lopez MD, Ovalle S, Higginbottom A, Monk PN, Cabanas C, Sanchez-Madrid F (2005) Endothelial tetraspanin microdomains regulate leukocyte firm adhesion during extravasation. Blood 105:2852–2861

Barreiro O, Zamai M, Yanez-Mo M, Tejera E, Lopez-Romero P, Monk PN, Gratton E, Caiolfa VR, Sanchez-Madrid F (2008) Endothelial adhesion receptors are recruited to adherent leukocytes by inclusion in preformed tetraspanin nanoplatforms. J Cell Biol 183:527–542

Barylko B, Mao YS, Wlodarski P, Jung G, Binns DD, Sun HQ, Yin HL, Albanesi JP (2009) Palmitoylation controls the catalytic activity and subcellular distribution of phosphatidylinositol 4-kinase II{alpha}. J Biol Chem 284:9994–10003

Bassani S, Cingolani LA, Valnegri P, Folci A, Zapata J, Gianfelice A, Sala C, Goda Y, Passafaro M (2012) The X-linked intellectual disability protein TSPAN7 regulates excitatory synapse development and AMPAR trafficking. Neuron 73:1143–1158

Baudoux B, Castanares-Zapatero D, Leclercq-Smekens M, Berna N, Poumay Y (2000) The tetraspanin CD9 associates with the integrin alpha6beta4 in cultured human epidermal keratinocytes and is involved in cell motility. Eur J Cell Biol 79:41–51

Bell GM, Seaman WE, Niemi EC, Imboden JB (1992) The OX-44 molecule couples to signaling pathways and is associated with CD2 on rat T lymphocytes and a natural killer cell line. J Exp Med 175:527–536

Berditchevski F, Bazzoni G, Hemler ME (1995) Specific association of CD63 with the VLA-3 and VLA-6 integrins. J Biol Chem 270:17784–17790

Berditchevski F, Zutter MM, Hemler ME (1996) Characterization of novel complexes on the cell surface between integrins and proteins with 4 transmembrane domains (TM4 proteins). Mol BiolCell 7:193–207

Berditchevski F, Tolias KF, Wong K, Carpenter CL, Hemler ME (1997) A novel link between integrins, transmembrane-4 superfamily proteins (CD63 and CD81), and phosphatidylinositol 4-kinase. J Biol Chem 272:2595–2598

Berditchevski F, Gilbert E, Griffiths MR, Fitter S, Ashman L, Jenner SJ (2001) Analysis of the CD151-alpha3beta1 integrin and CD151-tetraspanin interactions by mutagenesis. J Biol Chem 276:41165–41174

Berditchevski F, Odintsova E, Sawada S, Gilbert E (2002) Expression of the palmitoylation-deficient CD151 weakens the association of alpha 3 beta 1 integrin with the tetraspanin-enriched microdomains and affects integrin-dependent signaling. J Biol Chem 277: 36991–37000

Blobel CP (2005) ADAMs: key components in EGFR signalling and development. Nat Rev Mol Cell Biol 6:32–43

Boucheix C, Rubinstein E (2001) Tetraspanins. Cell Mol Life Sci 58:1189–1205

Bradbury LE, Kansas GS, Levy S, Evans RL, Tedder TF (1992) The CD19/CD21 signal transducing complex of human B lymphocytes includes the target of antiproliferative antibody-1 and Leu-13 molecules. J Immunol 149:2841–2850

Cailleteau L, Estrach S, Thyss R, Boyer L, Doye A, Domange B, Johnsson N, Rubinstein E, Boucheix C, Ebrahimian T, Silvestre JS, Lemichez E, Meneguzzi G, Mettouchi A (2010) Alpha2beta1 integrin controls association of Rac with the membrane and triggers quiescence of endothelial cells. J Cell Sci 123:2491–2501

Carter RH, Fearon DT (1992) CD19: lowering the treshold for antigen receptor stimulation of B lymphocytes. Science 256:105–107

Chang Y, Finnemann SC (2007) Tetraspanin CD81 is required for the alpha v beta5-integrin-dependent particle-binding step of RPE phagocytosis. J Cell Sci 120:3053–3063

Charrin S, Alcover A (2006) Role of ERM (Ezrin-Radixin-Moesin) proteins in T lymphocyte polarization, immune synapse formation and in T cell receptor-mediated signaling. Front Biosci 11:1987–1997

Charrin S, Le Naour F, Oualid M, Billard M, Faure G, Hanash SM, Boucheix C, Rubinstein E (2001) The major CD9 and CD81 molecular partner: identification and characterization of the complexes. J Biol Chem 276:14329–14337

Charrin S, Manie S, Oualid M, Billard M, Boucheix C, Rubinstein E (2002) Differential stability of tetraspanin/tetraspanin interactions: role of palmitoylation. FEBS Lett 516:139–144

Charrin S, Le Naour F, Labas V, Billard M, Le Caer JP, Emile JF, Petit MA, Boucheix C, Rubinstein E (2003a) EWI-2 is a new component of the tetraspanin web in hepatocytes and lymphoid cells. Biochem J 373:409–421

Charrin S, Manie S, Billard M, Ashman L, Gerlier D, Boucheix C, Rubinstein E (2003b) Multiple levels of interactions within the tetraspanin web. Biochem Biophys Res Commun 304:107–112

Charrin S, Manie S, Thiele C, Billard M, Gerlier D, Boucheix C, Rubinstein E (2003c) A physical and functional link between cholesterol and tetraspanins. Eur J Immunol 33:2479–2489

Charrin S, Le Naour F, Silvie O, Milhiet PE, Boucheix C, Rubinstein E (2009a) Lateral organization of membrane proteins: tetraspanins spin their web. Biochem J 420:133–154
Charrin S, Yalaoui S, Bartosch B, Cocquerel L, Franetich JF, Boucheix C, Mazier D, Rubinstein E, Silvie O (2009b) The Ig domain protein CD9P-1 down-regulates CD81 ability to support Plasmodium yoelii infection. J Biol Chem 284:31572–31578
Cherezov V, Rosenbaum DM, Hanson MA, Rasmussen SG, Thian FS, Kobilka TS, Choi HJ, Kuhn P, Weis WI, Kobilka BK, Stevens RC (2007) High-resolution crystal structure of an engineered human beta2-adrenergic G protein-coupled receptor. Science 318:1258–1265
Chometon G, Zhang ZG, Rubinstein E, Boucheix C, Mauch C, Aumailley M (2006) Dissociation of the complex between CD151 and laminin-binding integrins permits migration of epithelial cells. Exp Cell Res 312:983–995
Claas C, Stipp CS, Hemler ME (2001) Evaluation of prototype transmembrane 4 superfamily protein complexes and their relation to lipid rafts. J Biol Chem 276:7974–7984
Claas C, Wahl J, Orlicky DJ, Karaduman H, Schnolzer M, Kempf T, Zoller M (2005) The tetraspanin D6.1A and its molecular partners on rat carcinoma cells. Biochem J 389:99–110
Clark KL, Zeng Z, Langford AL, Bowen SM, Todd SC (2001) Pgrl is a major cd81-associated protein on lymphocytes and distinguishes a new family of cell surface proteins. J Immunol 167:5115–5121
Clark KL, Oelke A, Johnson ME, Eilert KD, Simpson PC, Todd SC (2004) CD81 associates with 14-3-3 in a redox-regulated palmitoylation-dependent manner. J Biol Chem 279:19401–19406
Clergeot PH, Gourgues M, Cots J, Laurans F, Latorse MP, Pepin R, Tharreau D, Notteghem JL, Lebrun MH (2001) PLS1, a gene encoding a tetraspanin-like protein, is required for penetration of rice leaf by the fungal pathogen *Magnaporthe grisea*. Proc Natl Acad Sci USA 98:6963–6968
Cnops G, Neyt P, Raes J, Petrarulo M, Nelissen H, Malenica N, Luschnig C, Tietz O, Ditengou F, Palme K, Azmi A, Prinsen E, Van Lijsebettens M (2006) The TORNADO1 and TORNADO2 genes function in several patterning processes during early leaf development in *Arabidopsis thaliana*. Plant Cell 18:852–866
Codina J, Li J, DuBose TD Jr (2005) CD63 interacts with the carboxy terminus of the colonic H + -K + -ATPase to decrease [corrected] plasma membrane localization and 86Rb + uptake. Am J Physiol Cell Physiol 288:C1279–C1286
Coffey GP, Rajapaksa R, Liu R, Sharpe O, Kuo CC, Krauss SW, Sagi Y, Davis RE, Staudt LM, Sharman JP, Robinson WH, Levy S (2009) Engagement of CD81 induces ezrin tyrosine phosphorylation and its cellular redistribution with filamentous actin. J Cell Sci 122:3137–3144
Cowin AJ, Adams D, Geary SM, Wright MD, Jones JC, Ashman LK (2006) Wound healing is defective in mice lacking tetraspanin CD151. J Invest Dermatol 126:680–689
Danglot L, Chaineau M, Dahan M, Gendron MC, Boggetto N, Perez F, Galli T (2010) Role of TI-VAMP and CD82 in EGFR cell-surface dynamics and signaling. J Cell Sci 123:723–735
Delaguillaumie A, Harriague J, Kohanna S, Bismuth G, Rubinstein E, Seigneuret M, Conjeaud H (2004) Tetraspanin CD82 controls the association of cholesterol-dependent microdomains with the actin cytoskeleton in T lymphocytes: relevance to co-stimulation. J Cell Sci 117: 5269–5282
Dornier E, Coumailleau F, Ottavi JF, Moretti J, Boucheix C, Mauduit P, Schweisguth F, Rubinstein E (2012) TspanC8 tetraspanins regulate ADAM10/Kuzbanian trafficking and promote Notch activation in flies and mammals. J Cell Biol 199:481–496
Doyle EL, Ridger V, Ferraro F, Turmaine M, Saftig P, Cutler DF (2011) CD63 is an essential cofactor to leukocyte recruitment by endothelial P-selectin. Blood 118:4265–4273
Draber P, Vonkova I, Stepanek O, Hrdinka M, Kucova M, Skopcova T, Otahal P, Angelisova P, Horejsi V, Yeung M, Weiss A, Brdicka T (2011) SCIMP, a transmembrane adaptor protein involved in major histocompatibility complex class II signaling. Mol Cell Biol 31:4550–4562
Duffield A, Kamsteeg EJ, Brown AN, Pagel P, Caplan MJ (2003) The tetraspanin CD63 enhances the internalization of the H, K-ATPase beta-subunit. Proc Natl Acad Sci USA 100:15560–15565

Dunn CD, Sulis ML, Ferrando AA, Greenwald I (2010) A conserved tetraspanin subfamily promotes Notch signaling in Caenorhabditis elegans and in human cells. Proc Natl Acad Sci USA 107:5907–5912
Ellerman DA, Ha C, Primakoff P, Myles DG, Dveksler GS (2003) Direct binding of the ligand PSG17 to CD9 requires a CD9 site essential for sperm-egg fusion. Mol Biol Cell 14:5098–5103
Espenel C, Margeat E, Dosset P, Arduise C, Le Grimellec C, Royer CA, Boucheix C, Rubinstein E, Milhiet PE (2008) Single-molecule analysis of CD9 dynamics and partitioning reveals multiple modes of interaction in the tetraspanin web. J Cell Biol 182:765–776
Fearon DT, Carter RH (1995) The CD19/CR2/TAPA-1 complex of B lymphocytes: linking natural to acquired immunity. Annu Rev Immunol 13:127–149
Feigelson SW, Grabovsky V, Shamri R, Levy S, Alon R (2003) The CD81 tetraspanin facilitates instantaneous leukocyte VLA-4 adhesion strengthening to Vascular Cell Adhesion Molecule 1 (VCAM-1) under shear flow. J Biol Chem 278:51203–51212
Fitter S, Sincock PM, Jolliffe CN, Ashman LK (1999) Transmembrane 4 superfamily protein CD151 (PETA-3) associates with beta 1 and alpha IIb beta 3 integrins in haemopoietic cell lines and modulates cell-cell adhesion. Biochem J 338:61–70
Flannery AR, Czibener C, Andrews NW (2010) Palmitoylation-dependent association with CD63 targets the Ca2+ sensor synaptotagmin VII to lysosomes. J Cell Biol 191:599–613
Force T, Woulfe K, Koch WJ, Kerkela R (2007) Molecular scaffolds regulate bidirectional cross-talk between Wnt and classical seven-transmembrane-domain receptor signaling pathways. Sci STKE 2007:e41
Gambin Y, Lopez-Esparza R, Reffay M, Sierecki E, Gov NS, Genest M, Hodges RS, Urbach W (2006) Lateral mobility of proteins in liquid membranes revisited. Proc Natl Acad Sci USA 103:2098–2102
Garcia-Espana A, Chung PJ, Sarkar IN, Stiner E, Sun TT, Desalle R (2008) Appearance of new tetraspanin genes during vertebrate evolution. Genomics 91:326–334
Goldberg AF (2006) Role of peripherin/rds in vertebrate photoreceptor architecture and inherited retinal degenerations. Int Rev Cytol 253:131–175
Goschnick MW, Lau LM, Wee JL, Liu YS, Hogarth PM, Robb LM, Hickey MJ, Wright MD, Jackson DE (2006) Impaired "outside-in" integrin alphaIIbbeta3 signaling and thrombus stability in TSSC6-deficient mice. Blood 108:1911–1918
Greco C, Bralet MP, Ailane N, Dubart-Kupperschmitt A, Rubinstein E, Le Naour F, Boucheix C (2010) E-cadherin/p120-catenin and tetraspanin Co-029 cooperate for cell motility control in human colon carcinoma. Cancer Res 70:7674–7683
Gutierrez-Lopez MD, Gilsanz A, Yanez-Mo M, Ovalle S, Lafuente EM, Dominguez C, Monk PN, Gonzalez-Alvaro I, Sanchez-Madrid F, Cabanas C (2011) The sheddase activity of ADAM17/TACE is regulated by the tetraspanin CD9. Cell Mol Life Sci 68:3275–3292
Ha CT, Waterhouse R, Wessells J, Wu JA, Dveksler GS (2005) Binding of pregnancy-specific glycoprotein 17 to CD9 on macrophages induces secretion of IL-10, IL-6, PGE2, and TGF-beta1. J Leukoc Biol 77:948–957
Haining EJ, Yang J, Bailey RL, Khan K, Collier R, Tsai S, Watson SP, Frampton J, Garcia P, Tomlinson MG (2012) The TspanC8 subgroup of tetraspanins interacts with a disintegrin and metalloprotease 10 (ADAM10) and regulates its maturation and cell surface expression. J Biol Chem 287:39753–39765
Harris HJ, Farquhar MJ, Mee CJ, Davis C, Reynolds GM, Jennings A, Hu K, Yuan F, Deng H, Hubscher SG, Han JH, Balfe P, McKeating JA (2008) CD81 and claudin 1 coreceptor association: role in hepatitis C virus entry. J Virol 82:5007–5020
Hemler ME (2003) Tetraspanin proteins mediate cellular penetration, invasion, and fusion events and define a novel type of membrane microdomain. Annu Rev Cell Dev Biol 19:397–422
Hemler ME (2005) Tetraspanin functions and associated microdomains. Nat Rev Mol Cell Biol 6:801–811
Ho SH, Martin F, Higginbottom A, Partridge LJ, Parthasarathy V, Moseley GW, Lopez P, Cheng-Mayer C, Monk PN (2006) Recombinant extracellular domains of tetraspanin proteins are

potent inhibitors of the infection of macrophages by human immunodeficiency virus type 1. J Virol 80:6487–6496
Horváth G, Serru V, Clay D, Billard M, Boucheix C, Rubinstein E (1998) CD19 is linked to the integrin-associated tetraspans CD9, CD81, and CD82. J Biol Chem 273:30537–30543
Huang S, Yuan S, Dong M, Su J, Yu C, Shen Y, Xie X, Yu Y, Yu X, Chen S, Zhang S, Pontarotti P, Xu A (2005) The phylogenetic analysis of tetraspanins projects the evolution of cell-cell interactions from unicellular to multicellular organisms. Genomics 86:674–684
Imai T, Yoshie O (1993) C33 antigen and M38 antigen recognized by monoclonal antibodies inhibitory to syncytium formation by human T cell leukemia virus type 1 are both members of the transmembrane 4 superfamily and associate with each other and with CD4 or CD8 in T cells. J Immunol 151:6470–6481
Imai T, Kakizaki M, Nishimura M, Yoshie O (1995) Molecular analyses of the association of CD4 with two members of the transmembrane 4 superfamily CD81 and CD82. J Immunol 155:1229–1239
Indig FE, Diaz-Gonzalez F, Ginsberg MH (1997) Analysis of the tetraspanin CD9-integrin à IIb á3 (GPIIb-IIIa) complex in platelet membranes and transfected cells. Biochem J 327:291–298
Ishibashi T, Ding L, Ikenaka K, Inoue Y, Miyado K, Mekada E, Baba H (2004) Tetraspanin protein CD9 is a novel paranodal component regulating paranodal junctional formation. J Neurosci 24:96–102
Israels SJ, McMillan-Ward EM, Easton J, Robertson C, McNicol A (2001) CD63 associates with the alphaIIb beta3 integrin-CD9 complex on the surface of activated platelets. Thromb Haemost 85:134–141
Iwamoto R, Higashiyama S, Mitamura T, Taniguchi N, Klagsbrun M, Mekada E (1994) Heparin-binding EGF-like growth factor, which acts as the diphtheria toxin receptor, forms a complex with membrane protein DRAP27/CD9, which up-regulates functional receptors and diphtheria toxin sensitivity. EMBO J 13:2322–2330
Jones EY, Davis SJ, Williams AF, Harlos K, Stuart DI (1992) Crystal structure at 2.8 a resolution of a soluble form of the cell adhesion molecule CD2. Nature 360:232–239
Jones PH, Bishop LA, Watt FM (1996) Functional significance of CD9 association with á1 integrins in human epidermal keratinocytes. Cell Adhes Commun 4:297–305
Jung KK, Liu XW, Chirco R, Fridman R, Kim HR (2006) Identification of CD63 as a tissue inhibitor of metalloproteinase-1 interacting cell surface protein. EMBO J 25:3934–3942
Junge HJ, Yang S, Burton JB, Paes K, Shu X, French DM, Costa M, Rice DS, Ye W (2009) TSPAN12 regulates retinal vascular development by promoting Norrin- but not Wnt-induced FZD4/beta-catenin signaling. Cell 139:299–311
Kaji K, Oda S, Shikano T, Ohnuki T, Uematsu Y, Sakagami J, Tada N, Miyazaki S, Kudo A (2000) The gamete fusion process is defective in eggs of Cd9-deficient mice. Nat Genet 24:279–282
Kaji K, Oda S, Miyazaki S, Kudo A (2002) Infertility of CD9-deficient mouse eggs is reversed by mouse CD9, human CD9, or mouse CD81; polyadenylated mRNA injection developed for molecular analysis of sperm-egg fusion. Dev Biol 247:327–334
Karamatic CV, Burton N, Kagan A, Green CA, Levene C, Flinter F, Brady RL, Daniels G, Anstee DJ (2004) CD151, the first member of the tetraspanin (TM4) superfamily detected on erythrocytes, is essential for the correct assembly of human basement membranes in kidney and skin. Blood 104:2217–2223
Kazarov AR, Yang X, Stipp CS, Sehgal B, Hemler ME (2002) An extracellular site on tetraspanin CD151 determines alpha 3 and alpha 6 integrin-dependent cellular morphology. J Cell Biol 158:1299–1309
Kitadokoro K, Bordo D, Galli G, Petracca R, Falugi F, Abrignani S, Grandi G, Bolognesi M (2001) CD81 extracellular domain 3D structure: insight into the tetraspanin superfamily structural motifs. EMBO J 20:12–18
Klosek SK, Nakashiro K, Hara S, Shintani S, Hasegawa H, Hamakawa H (2005) CD151 forms a functional complex with c-Met in human salivary gland cancer cells. Biochem Biophys Res Commun 336:408–416
Kolesnikova TV, Stipp CS, Rao RM, Lane WS, Luscinskas FW, Hemler ME (2004) EWI-2 modulates lymphocyte integrin {alpha}4{beta}1 functions. Blood 103:3013–3019

Kong F, Garcia AJ, Mould AP, Humphries MJ, Zhu C (2009) Demonstration of catch bonds between an integrin and its ligand. J Cell Biol 185:1275–1284

Kopczynski CC, Davis GW, Goodman CS (1996) A neural tetraspanin, encoded by late bloomer, that facilitates synapse formation. Science 271:1867–1870

Kovalenko OV, Yang XH, Hemler ME (2007) A novel cysteine cross-linking method reveals a direct association between claudin-1 and tetraspanin CD9. Mol Cell Proteomics 6:1855–1867

Krementsov DN, Rassam P, Margeat E, Roy NH, Schneider-Schaulies J, Milhiet PE, Thali M (2010) HIV-1 assembly differentially alters dynamics and partitioning of tetraspanins and raft components. Traffic 11:1401–1414

Lafleur MA, Xu D, Hemler ME (2009) Tetraspanin proteins regulate membrane type-1 matrix metalloproteinase-dependent pericellular proteolysis. Mol Biol Cell 20:2030–2040

Lagaudriere-Gesbert C, Le Naour F, Lebel-Binay S, Billard M, Lemichez E, Boucheix C, Conjeaud H, Rubinstein E (1997a) Functional analysis of four tetraspans, CD9, CD53, CD81, and CD82, suggests a common role in costimulation, cell adhesion, and migration: only CD9 upregulates HB-EGF activity. Cell Immunol 182:105–112

Lagaudriere-Gesbert C, Lebel-Binay S, Wiertz E, Ploegh HL, Fradelizi D, Conjeaud H (1997b) The tetraspanin protein CD82 associates with both free HLA class I heavy chain and heterodimeric beta2-microglobulin complexes. J Immunol 158:2790–2797

Lagaudriere-Gesbert C, Lebel-Binay S, Hubeau C, Fradelizi D, Conjeaud H (1998) Signaling through the tetraspanin CD82 triggers its association with the cytoskeleton leading to sustained morphological changes and T cell activation. Eur J Immunol 28:4332–4344

Lammerding J, Kazarov AR, Huang H, Lee RT, Hemler ME (2003) Tetraspanin CD151 regulates {alpha}6{beta}1 integrin adhesion strengthening. Proc Natl Acad Sci USA 100:7616–7621

Latysheva N, Muratov G, Rajesh S, Padgett M, Hotchin NA, Overduin M, Berditchevski F (2006) Syntenin-1 is a new component of tetraspanin-enriched microdomains: mechanisms and consequences of the interaction of syntenin-1 with CD63. Mol Cell Biol 26:7707–7718

Lau LM, Wee JL, Wright MD, Moseley GW, Hogarth PM, Ashman LK, Jackson DE (2004) The tetraspanin superfamily member CD151 regulates outside-in integrin alphaIIbbeta3 signaling and platelet function. Blood 104:2368–2375

Le Naour F, Rubinstein E, Jasmin C, Prenant M, Boucheix C (2000) Severely reduced female fertility in CD9-deficient mice. Science 287:319–321

Le Naour F, Andre M, Greco C, Billard M, Sordat B, Emile JF, Lanza F, Boucheix C, Rubinstein E (2006) Profiling of the tetraspanin web of human colon cancer cells. Mol Cell Proteomics 5:845–857

Lebel-Binay S, Lagaudrière C, Fradelizi D, Conjeaud H (1995) CD82, member of the tetra-span-transmembrane protein family, is a costimulatory protein for T cell activation. J Immunol 155:101–110

Lekishvili T, Fromm E, Mujoomdar M, Berditchevski F (2008) The tumour-associated antigen L6 (L6-Ag) is recruited to the tetraspanin-enriched microdomains: implication for tumour cell motility. J Cell Sci 121:685–694

Levy S, Shoham T (2005) The tetraspanin web modulates immune-signalling complexes. Nat Rev Immunol 5:136–148

Levy S, Nguyen VQ, Andria ML, Takahashi S (1991) Structure and membrane topology of TAPA-1. J Biol Chem 266:14597–14602

Lineberry N, Su L, Soares L, Fathman CG (2008) The single subunit transmembrane E3 ligase Gene Related to Anergy In Lymphocytes (GRAIL) captures and then ubiquitinates transmembrane proteins across the cell membrane. J Biol Chem 283:28497–28505

Lingwood D, Simons K (2010) Lipid rafts as a membrane-organizing principle. Science 327:46–50

Little KD, Hemler ME, Stipp CS (2004) Dynamic regulation of a GPCR-tetraspanin-G protein complex on intact cells: central role of CD81 in facilitating GPR56-Galpha q/11 association. Mol Biol Cell 15:2375–2387

Longhurst CM, White MM, Wilkinson DA, Jennings LK (1999) A CD9, IIb3, integrin-associated protein, and GPIb/V/IX complex on the surface of human platelets is influenced by IIb3 conformational states. Eur J Biochem 263:104–111

Lozahic S, Christiansen D, Manie S, Gerlier D, Billard M, Boucheix C, Rubinstein E (2000) CD46 (membrane cofactor protein) associates with multiple beta1 integrins and tetraspans. Eur J Immunol 30:900–907

Mangin PH, Kleitz L, Boucheix C, Gachet C, Lanza F (2009) CD9 negatively regulates integrin alpha(llb)beta(3) activation and could thus prevent excessive platelet recruitment at sites of vascular injury. J Thromb Haemost 7:900–902

Mannion BA, Berditchevski F, Kraeft SK, Chen LB, Hemler ME (1996) Transmembrane-4 superfamily proteins CD81 (TAPA-1), CD82, CD63, and CD53 specifically associated with integrin alpha 4 beta 1 (CD49d/CD29). J Immunol 157:2039–2047

Marakalala MJ, Graham LM, Brown GD (2010) The role of Syk/CARD9-coupled C-type lectin receptors in immunity to *Mycobacterium* tuberculosis infections. Clin Dev Immunol 2010:567571

Marsh D (2008) Protein modulation of lipids, and vice-versa, in membranes. Biochim Biophys Acta 1778:1545–1575

Matsumoto AK, Martin DR, Carter RH, Klickstein LB, Ahearn JM, Fearon DT (1993) Functional dissection of the CD21/CD19/TAPA-1/leu-13 complex of B lymphocytes. J Exp Med 178:1407–1417

Meyer-Wentrup F, Figdor CG, Ansems M, Brossart P, Wright MD, Adema GJ, van Spriel AB (2007) Dectin-1 interaction with tetraspanin CD37 inhibits IL-6 production. J Immunol 178:154–162

Miao WM, Vasile E, Lane WS, Lawler J (2001) CD36 associates with CD9 and integrins on human blood platelets. Blood 97:1689–1696

Min G, Wang H, Sun TT, Kong XP (2006) Structural basis for tetraspanin functions as revealed by the cryo-EM structure of uroplakin complexes at 6-A resolution. J Cell Biol 173:975–983

Mittelbrunn M, Yanez-Mo M, Sancho D, Ursa A, Sanchez-Madrid F (2002) Cutting edge: dynamic redistribution of tetraspanin CD81 at the central zone of the immune synapse in both T lymphocytes and APC. J Immunol 169:6691–6695

Miyado K, Yamada G, Yamada S, Hasuwa H, Nakamura Y, Ryu F, Suzuki K, Kosai K, Inoue K, Ogura A, Okabe M, Mekada E (2000) Requirement of CD9 on the egg plasma membrane for fertilization. Science 287:321–324

Montpellier C, Tews BA, Poitrimole J, Rocha-Perugini V, D'Arienzo V, Potel J, Zhang XA, Rubinstein E, Dubuisson J, Cocquerel L (2011) Interacting regions of CD81 and two of its partners, EWI-2 and EWI-2wint and their effect on hepatitis C virus infection. J Biol Chem 286:13954–13965

Moribe H, Yochem J, Yamada H, Tabuse Y, Fujimoto T, Mekada E (2004) Tetraspanin protein (TSP-15) is required for epidermal integrity in *Caenorhabditis elegans*. J Cell Sci 117:5209–5220

Murayama Y, Shinomura Y, Oritani K, Miyagawa J, Yoshida H, Nishida M, Katsube F, Shiraga M, Miyazaki T, Nakamoto T, Tsutsui S, Tamura S, Higashiyama S, Shimomura I, Hayashi N (2008) The tetraspanin CD9 modulates epidermal growth factor receptor signaling in cancer cells. J Cell Physiol 216:135–143

Nakamura K, Mitamura T, Takahashi T, Kobayashi T, Mekada E (2000) Importance of the major extracellular domain of CD9 and the Epidermal Growth Factor (EGF)-like domain of heparin-binding EGF-like growth factor for Up-regulation of binding and activity. J Biol Chem 275:18284–18290

Nichols TC, Guthridge JM, Karp DR, Molina H, Fletcher DR, Holers VM (1998) Gamma-glutamyl transpeptidase, an ecto-enzyme regulator of intracellular redox potential, is a component of TM4 signal transduction complexes. Eur J Immunol 28:4123–4129

Nikopoulos K, Gilissen C, Hoischen A, van Nouhuys CE, Boonstra FN, Blokland EA, Arts P, Wieskamp N, Strom TM, Ayuso C, Tilanus MA, Bouwhuis S, Mukhopadhyay A, Scheffer H, Hoefsloot LH, Veltman JA, Cremers FP, Collin RW (2010) Next-generation sequencing of a 40 Mb linkage interval reveals TSPAN12 mutations in patients with familial exudative vitreoretinopathy. Am J Hum Genet 86:240–247

Nishiuchi R, Sanzen N, Nada S, Sumida Y, Wada Y, Okada M, Takagi J, Hasegawa H, Sekiguchi K (2005) Potentiation of the ligand-binding activity of integrin alpha3beta1 via association with tetraspanin CD151. Proc Natl Acad Sci USA 102:1939–1944

Novitskaya V, Romanska H, Dawoud M, Jones JL, Berditchevski F (2010) Tetraspanin CD151 regulates growth of mammary epithelial cells in three-dimensional extracellular matrix: implication for mammary ductal carcinoma in situ. Cancer Res 70:4698–4708

Odintsova E, Sugiura T, Berditchevski F (2000) Attenuation of EGF receptor signaling by a metastasis suppressor, the tetraspanin CD82/KAI-1. Curr Biol 10:1009–1012

Odintsova E, Voortman J, Gilbert E, Berditchevski F (2003) Tetraspanin CD82 regulates compartmentalisation and ligand-induced dimerization of EGFR. J Cell Sci 116:4557–4566

Odintsova E, Butters TD, Monti E, Sprong H, van Meer G, Berditchevski F (2006) Gangliosides play an important role in the organization of CD82-enriched microdomains. Biochem J 400:315–325

Olmos E, Reiss B, Dekker K (2003) The ekeko mutant demonstrates a role for tetraspanin-like protein in plant development. Biochem Biophys Res Commun 310:1054–1061

Ono M, Handa K, Sonnino S, Withers DA, Nagai H, Hakomori S (2001) GM3 ganglioside inhibits CD9-facilitated haptotactic cell motility: coexpression of GM3 and CD9 is essential in the downregulation of tumor cell motility and malignancy. Biochemistry 40:6414–6421

Orlowski E, Chand R, Yip J, Wong C, Goschnick MW, Wright MD, Ashman LK, Jackson DE (2009) A platelet tetraspanin superfamily member, CD151, is required for regulation of thrombus growth and stability in vivo. J Thromb Haemost 7:2074–2084

Pan Y, Brown C, Wang X, Geisert EE (2007) The developmental regulation of CD81 in the rat retina. Mol Vis 13:181–189

Parthasarathy V, Martin F, Higginbottom A, Murray H, Moseley GW, Read RC, Mal G, Hulme R, Monk PN, Partridge LJ (2009) Distinct roles for tetraspanins CD9, CD63 and CD81 in the formation of multinucleated giant cells. Immunology 127:237–248

Poulter JA, Ali M, Gilmour DF, Rice A, Kondo H, Hayashi K, Mackey DA, Kearns LS, Ruddle JB, Craig JE, Pierce EA, Downey LM, Mohamed MD, Markham AF, Inglehearn CF, Toomes C (2010) Mutations in TSPAN12 cause autosomal-dominant familial exudative vitreoretinopathy. Am J Hum Genet 86:248–253

Protty MB, Watkins NA, Colombo D, Thomas SG, Heath VL, Herbert JM, Bicknell R, Senis YA, Ashman LK, Berditchevski F, Ouwehand WH, Watson SP, Tomlinson MG (2009) Identification of Tspan9 as a novel platelet tetraspanin and the collagen receptor GPVI as a component of tetraspanin microdomains. Biochem J 417:391–400

Prox J, Willenbrock M, Weber S, Lehmann T, Schmidt-Arras D, Schwanbeck R, Saftig P, Schwake M (2012) Tetraspanin15 regulates cellular trafficking and activity of the ectodomain sheddase ADAM10. Cell Mol Life Sci 69:2919–2932

Puklin-Faucher E, Sheetz MP (2009) The mechanical integrin cycle. J Cell Sci 122:179–186

Qi JC, Wang J, Mandadi S, Tanaka K, Roufogalis BD, Madigan MC, Lai K, Yan F, Chong BH, Stevens RL, Krilis SA (2006) Human and mouse mast cells use the tetraspanin CD9 as an alternate interleukin-16 receptor. Blood 107:135–142

Rao TP, Kuhl M (2010) An updated overview on Wnt signaling pathways: a prelude for more. Circ Res 106:1798–1806

Reiss K, Saftig P (2009) The "A Disintegrin And Metalloprotease" (ADAM) family of sheddases: physiological and cellular functions. Semin Cell Dev Biol 20:126–137

Romanska HM, Berditchevski F (2011) Tetraspanins in human epithelial malignancies. J Pathol 223:4–14

Rubinstein E, Le Naour F, Billard M, Prenant M, Boucheix C (1994) CD9 antigen is an accessory subunit of the VLA integrin complexes. Eur J Immunol 24:3005–3013

Rubinstein E, Le Naour F, Lagaudrière C, Billard M, Conjeaud H, Boucheix C (1996) CD9, CD63, CD81 and CD82 are components of a surface tetraspan network connected to HLA-DR and VLA integrins. Eur J Immunol 26:2657–2665

Rubinstein E, Ziyyat A, Prenant M, Wrobel E, Wolf JP, Levy S, Le Naour F, Boucheix C (2006) Reduced fertility of female mice lacking CD81. Dev Biol 290:351–358

Sachs N, Kreft M, van den Bergh Weerman MA, Beynon AJ, Peters TA, Weening JJ, Sonnenberg A (2006) Kidney failure in mice lacking the tetraspanin CD151. J Cell Biol 175:33–39

Sadej R, Romanska H, Baldwin G, Gkirtzimanaki K, Novitskaya V, Filer AD, Krcova Z, Kusinska R, Ehrmann J, Buckley CD, Kordek R, Potemski P, Eliopoulos AG, Lalani E,

Berditchevski F (2009) CD151 regulates tumorigenesis by modulating the communication between tumor cells and endothelium. Mol Cancer Res 7:787–798
Sadej R, Romanska H, Kavanagh D, Baldwin G, Takahashi T, Kalia N, Berditchevski F (2010) Tetraspanin CD151 regulates transforming growth factor beta signaling: implication in tumor metastasis. Cancer Res 70:6059–6070
Sala-Valdes M, Ursa A, Charrin S, Rubinstein E, Hemler ME, Sanchez-Madrid F, Yanez-Mo M (2006) EWI-2 and EWI-F link the tetraspanin web to the actin cytoskeleton through their direct association with ezrin-radixin-moesin proteins. J Biol Chem 281:19665–19675
Schick MR, Levy S (1993) The TAPA-1 molecule is associated on the surface of B cells with HLA-DR molecules. J Immunol 151:4090–4097
Schmidt C, Kunemund V, Wintergerst ES, Schmitz B, Schachner M (1996) CD9 of mouse brain is implicated in neurite outgrowth and cell migration in vitro and is associated with the alpha 6/beta 1 integrin and the neural adhesion molecule L1. J Neurosci Res 43:12–31
Seehafer JG, Shou Ching T, Slupsky JR, Shaw ARE (1988) The functional glycoprotein CD9 is variably acylated: localization of the variably acylated region to a membrane-associated peptide containing the binding site for the agonistic monoclonal antibody 50H.19. Biochim Biophys Acta 957:399–410
Seigneuret M (2006) Complete predicted three-dimensional structure of the facilitator transmembrane protein and hepatitis C virus receptor CD81: conserved and variable structural domains in the tetraspanin superfamily. Biophys J 90:212–227
Serru V, Le Naour F, Billard M, Azorsa DO, Lanza F, Boucheix C, Rubinstein E (1999) Selective Tetraspan/integrin complexes (CD81/à4á1, CD151/à3á1, CD151/à6á1) under conditions disrupting tetraspan interactions. Biochem J 340:103–111
Sharma C, Yang XH, Hemler ME (2008) DHHC2 affects palmitoylation, stability, and functions of tetraspanins CD9 and CD151. Mol Biol Cell 19:3415–3425
Sheng KC, van Spriel AB, Gartlan KH, Sofi M, Apostolopoulos V, Ashman L, Wright MD (2009) Tetraspanins CD37 and CD151 differentially regulate Ag presentation and T-cell co-stimulation by DC. Eur J Immunol 39:50–55
Shi W, Fan H, Shum L, Derynck R (2000) The tetraspanin CD9 associates with transmembrane TGF-alpha and regulates TGF-alpha-induced EGF receptor activation and cell proliferation. J Cell Biol 148:591–602
Shibagaki N, Hanada K, Yamashita H, Shimada S, Hamada H (1999) Overexpression of CD82 on human T cells enhances LFA-1/ICAM-1-mediated cell-cell adhesion: functional association between CD82 and LFA-1 in T cell activation. Eur J Immunol 29:4081–4091
Shigeta M, Sanzen N, Ozawa M, Gu J, Hasegawa H, Sekiguchi K (2003) CD151 regulates epithelial cell-cell adhesion through PKC- and Cdc42-dependent actin cytoskeletal reorganization. J Cell Biol 163:165–176
Shiomi T, Inoki I, Kataoka F, Ohtsuka T, Hashimoto G, Nemori R, Okada Y (2005) Pericellular activation of proMMP-7 (promatrilysin-1) through interaction with CD151. Lab Invest 85:1489–1506
Shoham T, Rajapaksa R, Boucheix C, Rubinstein E, Poe JC, Tedder TF, Levy S (2003) The tetraspanin CD81 regulates the expression of CD19 during B cell development in a postendoplasmic reticulum compartment. J Immunol 171:4062–4072
Shoham T, Rajapaksa R, Kuo CC, Haimovich J, Levy S (2006) Building of the tetraspanin web: distinct structural domains of CD81 function in different cellular compartments. Mol Cell Biol 26:1373–1385
Silvie O, Charrin S, Billard M, Franetich JF, Clark KL, van Gemert GJ, Sauerwein RW, Dautry F, Boucheix C, Mazier D, Rubinstein E (2006) Cholesterol contributes to the organization of tetraspanin-enriched microdomains and to CD81-dependent infection by malaria sporozoites. J Cell Sci 119:1992–2002
Simons K, Toomre D (2000) Lipid rafts and signal transduction. Nat Rev Mol Cell Biol 1:31–39
Sincock PM, Fitter S, Parton RG, Berndt MC, Gamble JR, Ashman LK (1999) PETA-3/CD151, a member of the transmembrane 4 superfamily, is localised to the plasma membrane and endocytic

system of endothelial cells, associates with multiple integrins and modulates cell function. J Cell Sci 112:833–844
Skubitz KM, Campbell KD, Iida J, Skubitz APN (1996) CD63 associates with tyorsine kinase activity and CD11/CD18; and transmits an activation signal in neutrophils. J Immunol 157:3617–3626
Slupsky JR, Seehafer JG, Tang SC, Masellis-Smith A, Shaw AR (1989) Evidence that monoclonal antibodies against CD9 antigen induce specific association between CD9 and the platelet glycoprotein IIb-IIIa complex. J Biol Chem 264:12289–12293
Slupsky JR, Kamiguti AS, Rhodes NP, Cawley JC, Shaw AR, Zuzel M (1997) The platelet antigens CD9, CD42 and integrin alpha IIb beta IIIa can be topographically associated and transduce functionally similar signals. Eur J Biochem 244:168–175
Sridhar SC, Miranti CK (2006) Tetraspanin KAI1/CD82 suppresses invasion by inhibiting integrin-dependent crosstalk with c-Met receptor and Src kinases. Oncogene 25:2367–2378
Sterk LM, Geuijen CA, Oomen LC, Calafat J, Janssen H, Sonnenberg A (2000) The tetraspan molecule CD151, a novel constituent of hemidesmosomes, associates with the integrin alpha6beta4 and may regulate the spatial organization of hemidesmosomes. J Cell Biol 149:969–982
Sterk LM, Geuijen CA, van den Berg JG, Claessen N, Weening JJ, Sonnenberg A (2002) Association of the tetraspanin CD151 with the laminin-binding integrins alpha3beta1, alpha6beta1, alpha6beta4 and alpha7beta1 in cells in culture and in vivo. J Cell Sci 115:1161–1173
Stipp CS, Kolesnikova TV, Hemler ME (2001a) EWI-2 is a major CD9 and CD81 partner and member of a novel Ig protein subfamily. J Biol Chem 276:40545–40554
Stipp CS, Orlicky D, Hemler ME (2001b) FPRP, a major, highly stoichiometric, highly specific CD81- and CD9-associated protein. J Biol Chem 276:4853–4862
Stipp CS, Kolesnikova TV, Hemler ME (2003) EWI-2 regulates alpha3beta1 integrin-dependent cell functions on laminin-5. J Cell Biol 163:1167–1177
Sun TT (2006) Altered phenotype of cultured urothelial and other stratified epithelial cells: implications for wound healing. Am J Physiol Renal Physiol 291:F9–F21
Szöllösi J, Horejsi V, Bene L, Angelisova P, Damjanovich S (1996) Supramolecular complexes of MHC class I, MHC class II, CD20, and tetraspan molecules (CD53, CD81, and CD82) at the surface of a B cell line JY. J Immunol 157:2939–2946
Tachibana I, Bodorova J, Berditchevski F, Zutter MM, Hemler ME (1997) NAG-2 a novel Transmembrane-4 Superfamily (TM4SF) protein that complexes with integrins and other TM4SF proteins. J Biol Chem 272:29181–29189
Tai XG, Toyooka K, Yashiro Y, Abe R, Park CS, Hamaoka T, Kobayashi M, Neben S, Fujiwara H (1997) CD9-mediated costimulation of TCR-triggered naive T cells leads to activation followed by apoptosis. J Immunol 159:3799–3807
Takahashi S, Doss C, Levy S, Levy R (1990) TAPA-1, the target of an antiproliferative antibody, is associated on the cell surface with the Leu-13 antigen. J Immunol 145:2207–2213
Takahashi M, Sugiura T, Abe M, Ishii K, Shirasuna K (2007) Regulation of c-Met signaling by the tetraspanin KAI-1/CD82 affects cancer cell migration. Int J Cancer 121:1919–1929
Takeda Y, Tachibana I, Miyado K, Kobayashi M, Miyazaki T, Funakoshi T, Kimura H, Yamane H, Saito Y, Goto H, Yoneda T, Yoshida M, Kumagai T, Osaki T, Hayashi S, Kawase I, Mekada E (2003) Tetraspanins CD9 and CD81 function to prevent the fusion of mononuclear phagocytes. J Cell Biol 161:945–956
Takeda Y, Kazarov AR, Butterfield CE, Hopkins BD, Benjamin LE, Kaipainen A, Hemler ME (2007) Deletion of tetraspanin Cd151 results in decreased pathologic angiogenesis in vivo and in vitro. Blood 109:1524–1532
Takeda Y, He P, Tachibana I, Zhou B, Miyado K, Kaneko H, Suzuki M, Minami S, Iwasaki T, Goya S, Kijima T, Kumagai T, Yoshida M, Osaki T, Komori T, Mekada E, Kawase I (2008) Double deficiency of tetraspanins CD9 and CD81 alters cell motility and protease production of macrophages and causes chronic obstructive pulmonary disease-like phenotype in mice. J Biol Chem 283:26089–26097
Takino T, Miyamori H, Kawaguchi N, Uekita T, Seiki M, Sato H (2003) Tetraspanin CD63 promotes targeting and lysosomal proteolysis of membrane-type 1 matrix metalloproteinase. Biochem Biophys Res Commun 304:160–166

Tarrant JM, Groom J, Metcalf D, Li R, Borobokas B, Wright MD, Tarlinton D, Robb L (2002) The absence of Tssc6, a member of the tetraspanin superfamily, does not affect lymphoid development but enhances in vitro T-cell proliferative responses. Mol Cell Biol 22:5006–5018

Tedder TF, Zhou LJ, Engel P (1994) The CD19/CD21 signal transduction complex of B lymphocytes. Immunol Today 15:437–442

Thomas W (2008) Catch bonds in adhesion. Annu Rev Biomed Eng 10:39–57

Todeschini AR, Dos Santos JN, Handa K, Hakomori SI (2007) Ganglioside GM2-tetraspanin CD82 complex inhibits met and its cross-talk with integrins, providing a basis for control of cell motility through glycosynapse. J Biol Chem 282:8123–8133

Toyo-Oka K, Yashiro-Ohtani Y, Park CS, Tai XG, Miyake K, Hamaoka T, Fujiwara H (1999) Association of a tetraspanin CD9 with CD5 on the T cell surface: role of particular transmembrane domains in the association. Int Immunol 11:2043–2052

Tsai YC, Mendoza A, Mariano JM, Zhou M, Kostova Z, Chen B, Veenstra T, Hewitt SM, Helman LJ, Khanna C, Weissman AM (2007) The ubiquitin ligase gp78 promotes sarcoma metastasis by targeting KAI1 for degradation. Nat Med 13:1504–1509

Tu L, Sun TT, Kreibich G (2002) Specific heterodimer formation is a prerequisite for uroplakins to exit from the endoplasmic reticulum. Mol Biol Cell 13:4221–4230

Ubarretxena-Belandia I, Stokes DL (2010) Present and future of membrane protein structure determination by electron crystallography. Adv Protein Chem Struct Biol 81:33–60

Unternaehrer JJ, Chow A, Pypaert M, Inaba K, Mellman I (2007) The tetraspanin CD9 mediates lateral association of MHC class II molecules on the dendritic cell surface. Proc Natl Acad Sci USA 104:234–239

van Spriel AB, Puls KL, Sofi M, Pouniotis D, Hochrein H, Orinska Z, Knobeloch KP, Plebanski M, Wright MD (2004) A regulatory role for CD37 in T cell proliferation. J Immunol 172:2953–2961

van Zelm MC, Smet J, Adams B, Mascart F, Schandene L, Janssen F, Ferster A, Kuo CC, Levy S, van Dongen JJ, van der Burg M (2010) CD81 gene defect in humans disrupts CD19 complex formation and leads to antibody deficiency. J Clin Invest 120:1265–1274

Voisset C, Lavie M, Helle F, Op DB, Bilheu A, Bertrand-Michel J, Terce F, Cocquerel L, Wychowski C, Vu-Dac N, Dubuisson J (2008) Ceramide enrichment of the plasma membrane induces CD81 internalization and inhibits hepatitis C virus entry. Cell Microbiol 10:606–617

Wakabayashi T, Craessaerts K, Bammens L, Bentahir M, Borgions F, Herdewijn P, Staes A, Timmerman E, Vandekerckhove J, Rubinstein E, Boucheix C, Gevaert K, de Strooper B (2009) Analysis of the gamma-secretase interactome and validation of its association with tetraspanin-enriched microdomains. Nat Cell Biol 11:1340–1346

Wang XQ, Yan Q, Sun P, Liu JW, Go L, McDaniel SM, Paller AS (2007) Suppression of epidermal growth factor receptor signaling by protein kinase C-alpha activation requires CD82, caveolin-1, and ganglioside. Cancer Res 67:9986–9995

Waterhouse R, Ha C, Dveksler GS (2002) Murine CD9 is the receptor for pregnancy-specific glycoprotein 17. J Exp Med 195:277–282

Winterwood NE, Varzavand A, Meland MN, Ashman LK, Stipp CS (2006) A critical role for tetraspanin CD151 in alpha3beta1 and alpha6beta4 integrin-dependent tumor cell functions on laminin-5. Mol Biol Cell 17:2707–2721

Wright MD, Geary SM, Fitter S, Moseley GW, Lau LM, Sheng KC, Apostolopoulos V, Stanley EG, Jackson DE, Ashman LK (2004a) Characterization of mice lacking the tetraspanin superfamily member CD151. Mol Cell Biol 24:5978–5988

Wright MD, Moseley GW, van Spriel AB (2004b) Tetraspanin microdomains in immune cell signalling and malignant disease. Tissue Antigens 64:533–542

Wu XR, Kong XP, Pellicer A, Kreibich G, Sun TT (2009) Uroplakins in urothelial biology, function, and disease. Kidney Int 75:1153–1165

Xu H, Lee SJ, Suzuki E, Dugan KD, Stoddard A, Li HS, Chodosh LA, Montell C (2004) A lysosomal tetraspanin associated with retinal degeneration identified via a genome-wide screen. EMBO J 23:811–822

Xu D, Sharma C, Hemler ME (2009) Tetraspanin12 regulates ADAM10-dependent cleavage of amyloid precursor protein. FASEB J 23:3674–3681

Yamada M, Sumida Y, Fujibayashi A, Fukaguchi K, Sanzen N, Nishiuchi R, Sekiguchi K (2008a) The tetraspanin CD151 regulates cell morphology and intracellular signaling on laminin-511. FEBS J 275:3335–3351

Yamada M, Tamura Y, Sanzen N, Sato-Nishiuchi R, Hasegawa H, Ashman LK, Rubinstein E, Yanez-Mo M, Sanchez-Madrid F, Sekiguchi K (2008b) Probing the interaction of tetraspanin CD151 with integrin alpha 3 beta 1 using a panel of monoclonal antibodies with distinct reactivities toward the CD151-integrin alpha 3 beta 1 complex. Biochem J 415:417–427

Yanez-Mo M, Barreiro O, Gonzalo P, Batista A, Megias D, Genis L, Sachs N, Sala-Valdes M, Alonso MA, Montoya MC, Sonnenberg A, Arroyo AG, Sanchez-Madrid F (2008) MT1-MMP collagenolytic activity is regulated through association with tetraspanin CD151 in primary endothelial cells. Blood 112:3217–3226

Yang X, Claas C, Kraeft SK, Chen LB, Wang Z, Kreidberg JA, Hemler ME (2002) Palmitoylation of tetraspanin proteins: modulation of CD151 lateral interactions, subcellular distribution, and integrin-dependent cell morphology. Mol Biol Cell 13:767–781

Yang XH, Richardson AL, Torres-Arzayus MI, Zhou P, Sharma C, Kazarov AR, Andzelm MM, Strominger JL, Brown M, Hemler ME (2008) CD151 accelerates breast cancer by regulating alpha 6 integrin function, signaling, and molecular organization. Cancer Res 68:3204–3213

Yauch RL, Berditchevski F, Harler MB, Reichner J, Hemler ME (1998) Highly stoichiometric, stable, and specific association of integrin alpha3beta1 with CD151 provides a major link to phosphatidylinositol 4-kinase, and may regulate cell migration. Mol Biol Cell 9:2751–2765

Yauch RL, Kazarov AR, Desai B, Lee RT, Hemler ME (2000) Direct extracellular contact between integrin alpha(3)beta(1) and TM4SF protein CD151. J Biol Chem 275:9230–9238

Yoshida T, Kawano Y, Sato K, Ando Y, Aoki J, Miura Y, Komano J, Tanaka Y, Koyanagi Y (2008) A CD63 mutant inhibits T-cell tropic human immunodeficiency virus type 1 entry by disrupting CXCR4 trafficking to the plasma membrane. Traffic 9:540–558

Zevian S, Winterwood NE, Stipp CS (2011) Structure-function analysis of tetraspanin CD151 reveals distinct requirements for tumor cell behaviors mediated by α3{beta}1 versus α6{beta}4 integrin. J Biol Chem 286:7496–7506

Zhang XA, Bontrager AL, Hemler ME (2001) Transmembrane-4 superfamily proteins associate with activated Protein Kinase C (PKC) and link PKC to specific beta(1) integrins. J Biol Chem 276:25005–25013

Zhang XA, Kazarov AR, Yang X, Bontrager AL, Stipp CS, Hemler ME (2002) Function of the tetraspanin CD151-alpha6beta1 integrin complex during cellular morphogenesis. Mol Biol Cell 13:1–11

Zhang XA, Lane WS, Charrin S, Rubinstein E, Liu L (2003) EWI2/PGRL associates with the metastasis suppressor KAI1/CD82 and inhibits the migration of prostate cancer cells. Cancer Res 63:2665–2674

Zhou B, Liu L, Reddivari M, Zhang XA (2004) The palmitoylation of metastasis suppressor KAI1/CD82 is important for its motility- and invasiveness-inhibitory activity. Cancer Res 64:7455–7463

Zilber MT, Setterblad N, Vasselon T, Doliger C, Charron D, Mooney N, Gelin C (2005) MHC class II/CD38/CD9: a lipid-raft-dependent signaling complex in human monocytes. Blood 106:3074–3081

Zoller M (2009) Tetraspanins: push and pull in suppressing and promoting metastasis. Nat Rev Cancer 9:40–55

Zuo HJ, Lin JY, Liu ZY, Liu WF, Liu T, Yang J, Liu Y, Wang DW, Liu ZX (2010) Activation of the ERK signaling pathway is involved in CD151-induced angiogenic effects on the formation of CD151-integrin complexes. Acta Pharmacol Sin 31:805–812

Chapter 4
Dynamic Partitioning of Tetraspanins Within Plasma Membranes

Olga Barreiro, Francisco Sanchez-Madrid, Cédric Espenel, and Pierre-Emmanuel Milhiet

Abstract The study of the organization and dynamics of biological membranes has greatly benefited from the considerable technical advances achieved in the photonics field. Breaking the diffraction limit in optical microscopy to further increase spatial resolution has allowed to define the ultra-structure of the different proteolipid complexes within cellular membranes. Furthermore, the improvements in fluorescence sensitivity have prompted to the analysis of molecular dynamics up to single-molecule level with high temporal resolution. Thanks to all these advances, the concept of tetraspanin-enriched microdomains has been revisited in recent studies that shed new light on tetraspanin dynamics and interactions.

Abbreviations

Chl	Cholesterol
FCCS	Fluorescence Cross-Correlation Spectroscopy
FCS	Fluorescence Correlation Spectroscopy
FRAP	Fluorescent Recovery After Photobleaching

O. Barreiro (✉) • F. Sanchez-Madrid
Centro Nacional de Investigaciones Cardiovasculares, Madrid, Spain

Servicio de Inmunología, Hospital Universitario de la Princesa, Instituto de Investigación Sanitaria Princesa (IP), Madrid, Spain
e-mail: obarreiro@cnic.es; fsanchez.hlpr@salud.madrid.org

C. Espenel
Department of Cell and Developmental Biology, Weill Medical College, Cornell University, New York, USA

P.-E. Milhiet (✉)
Centre de Biochimie Structurale, CNRS UMR 5048, INSERM U1054, UM1, UM2, Montpellier, France
e-mail: pem@cbs.cnrs.fr

F. Berditchevski and E. Rubinstein (eds.), *Tetraspanins*, Proteins and Cell Regulation 9, DOI 10.1007/978-94-007-6070-7_4,

FRET-FLIM	Fluorescence Resonance Energy Transfer-Fluorescence Lifetime Imaging Microscopy
PALM	Photo-Activated Localization Microscopy
SMT	Single Molecule Tracking
STED	Stimulated Emission Depletion
STORM	Stochastic Optical Reconstruction Microscopy
TEM	Tetraspanin-Enriched Microdomain

4.1 Introduction

Molecular partitioning on cellular membranes and, in particular, the plasma membrane constitutes an essential requirement for the optimal function of an enormous variety of biological processes (Jacobson et al. 2007). Lipids, proteins and carbohydrates spatially coexist in diverse specialized microdomains or nanodomains of the membrane that are characterized by a specific molecular repertoire either in the outer or in the inner leaflets although coupling of the two leaflets is a matter of debate. Thus, receptors and related structural and signaling adaptors are confined within organized membrane units which allow cellular communication with the environment and signal transduction (Dehmelt and Bastiaens 2010). In this regard, the membrane polarity found in some tissues constitutes a further specialized cellular mechanism for compartmentalization of membrane bioactivity such as controlled endocytosis, secretion or proteolysis (Yanez-Mo et al. 2009).

The organizational diversity of various membrane microdomains is supported by a multitude of specific lipid-lipid, lipid-protein and protein-protein interactions. For example, both coalescence of glycosphingolipids and cholesterol (Chl) and their interaction with a number of membrane associated structural and signalling proteins (such as caveolin, flotillins, glycosylphosphatidylinositol (GPI)-linked proteins and the others) are necessary for the assembly of lipid rafts (Lingwood and Simons 2010). These microdomains have been proposed to function as platforms for sorting of GPI-anchored proteins (Mayor and Riezman 2004), the assembly of cytoplasmic signaling complexes (Lasserre et al. 2008; Parton and Hancock 2004) and caveolae (caveolin, cavin, flotillin)-mediated endocytosis (Hansen and Nichols 2010). Microdomains based on tetraspanin-mediated interactions (the so called tetraspanin-enriched microdomains TERMs (Berditchevski et al. 2002) or TEMs (Hemler 2003)) or tetraspanin web (Boucheix and Rubinstein 2001)) relies on the ability of tetraspanin proteins to simultaneously interact with each other and with molecular partners in *cis*. In this manner, a plethora of transmembrane receptors can be found embedded in a tetraspanin microenvironment which confers them an adequate molecular density to efficiently exert their physiological functions. Furthermore, TEMs spatially confine function-related receptors, which cooperate in certain cellular processes such as adhesion or migration (Barreiro et al. 2008). At the same time, tetraspanins promote local enrichment of actin-anchoring proteins and mediators which act downstream of tetraspanin partners and facilitate amplification of signal transduction (Charrin et al. 2009). TEMs exhibit a specific composition and stoichiometry depending on the cell type and, although a certain degree of redundancy can be found within the tetraspanin family, full restoration of a specific tetraspanin-dependent function cannot normally be achieved with related members of the family (Barreiro et al. 2005).

The initial evidence for existence of tetraspanin microdomains came from biochemical and proteomic approaches involving solubilisation of membranes with a wide range of detergents (Le Naour et al. 2006; Hemler 2005), from structural analysis of tetraspanin dimerization (Drummer et al. 2005) and from images of fixed cells and tissues obtained with conventional fluorescence and electron microscopy equipments (Berditchevski and Odintsova 1999; Abitorabi et.al. 1997). However, all these approaches do not reflect the dynamic nature of these membrane structures. The need for a formal demonstration of TEM existence in living cells led us to study in depth the complex and specific interactions among different tetraspanins and their partners as well as their biophysical properties in native conditions of the plasma membrane even at a single-molecule level (Barreiro et al. 2008; Espenel et al. 2008). Thus, analytical microscopy and spectroscopy techniques are able to "break" the diffraction limit allowed us to produce conclusive data on the differential and unique nature of TEMs in comparison with GPI protein-containing classical raft domains.

4.2 How to Probe Membrane Dynamics

During the last 10 years, important technological improvements in fluorescence microscopy have been done allowing analysis of biological molecules at the single molecule level. This includes the development of highly sensitive camera, high numerical aperture objectives and stable and bright probes. Two techniques in particular have been successfully used to probe the dynamics of biological membranes: fluorescence fluctuation spectroscopy (FFS) and its derivatives, and fluorescence single molecule tracking (SMT). Membrane dynamics can also be probed with FRAP (Fluorescence Recovery After Photobleaching), a technique that renders information about the average diffusion coefficient of a molecular ensemble as well as the percentage of mobile and immobile fractions out of the overall molecular population analyzed. However, the spatial resolution is poor because the size of the bleached area often exceeds the diffraction limit. In addition, it is impossible to determine different modes of motion of molecules within membranes.

First used to study molecules in solution, FCS (Fluorescence Correlation Spectroscopy) is a single-molecule sensitive technique that analyzes fluctuations in the fluorescence emission of small molecular ensembles in a sub-femtoliter volume. Fluorescence fluctuation analysis thus provides information about a multitude of parameters such as stoichiometry, concentration and molecular diffusion of fluorescently labeled molecules with a very high temporal resolution (reviewed in (Schwille et al. 1999)). In addition, FCS is less invasive than FRAP since it does not require

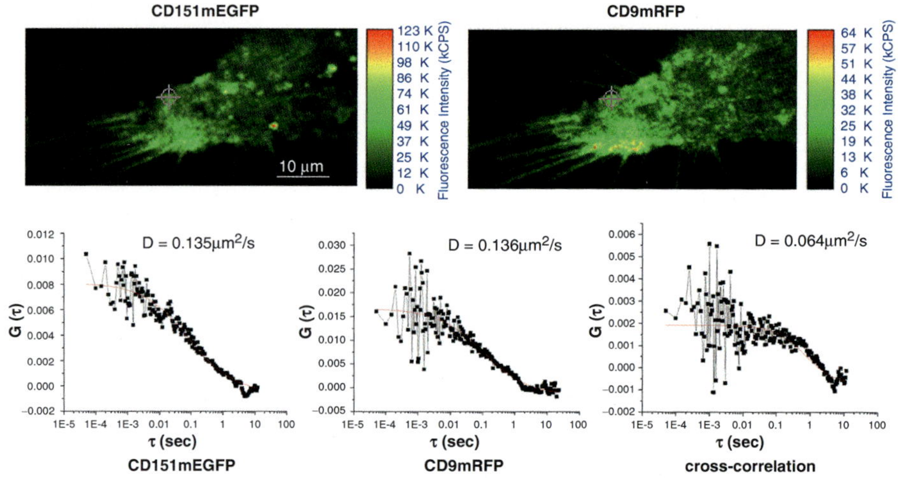

Fig. 4.1 Detection of CD9-CD151 complexes on the plasma membrane of living human primary endothelial cells by fluorescence cross-correlation spectroscopy (FCCS). Representative FCCS measurement at the plasma membrane of transiently transfected primary endothelial cells co-expressing very low levels of mRFP-tagged CD9 and mEGFP-tagged CD151. The figure shows the fluorescence intensity image (kCPS, kilo counts per second), the autocorrelation function (ACF) (*black line*) derived from the fluorescence intensity trace acquired at the point marked with a *grey cross*, the best-fitted curve using an anomalous diffusion model (*red line*), and the diffusion coefficient (D). In the FCS autocorrelation curves, the x axis (τ) represents the delay time in seconds, and the y axis (G(τ)) is the autocorrelation amplitude as a function of delay time. The detection of cross-correlation of *green* and *red* fluorescence fluctuations indicates that both fluorescently tagged molecular species co-diffuse in the same supramolecular complexes over time. In this particular example, the diffusion coefficient obtained in the cross-correlation seems to correspond more to EAP diffusion rather than to small tetraspanin complexes exchangeable among platforms which exhibit faster dynamics

heavy-loaded labeling and, for this reason, molecular species can be observed in a more preserved environment. In order to limit the observation volume, classical FCS setups combine high numerical aperture objective and pinholes (one can also use two-photon excitation as an alternative to pinholes). When compared to one-photon confocal FCS, two-Photon FCS requires an extremely high photon flux density but increases the signal to noise ratio and stability for long-term measurement (Chen et al. 2001), two parameters that are important for measuring diffusion in membranes. Moreover, two-photon FCS allows simultaneous excitation of two or more spectrally distinct dyes with a single line simplifying cross-correlation experiments and enables the detection of co-diffusion of two distinct molecular species (Fig. 4.1). The FCS approach has contributed significantly to the field of membrane organisation and dynamics, especially to studies of lipid rafts using giant unilamellar vesicles or model lipid bilayers (see the reviews (Garcia-Saez and Schwille 2008; Marguet et al. 2006)). A significant progress in the understanding of membrane organization has been achieved with the development by Marguet and colleagues of a new method which allows identification of the motion mode of membrane components by analyzing the FCS diffusion law of τ_d (i.e. the average time a fluorescent molecule stays within the illuminated area) versus the beam area (Wawrezinieck et al. 2005). More recently, a combination of STED (Stimulated Emission Depletion) microscopy with FCS has been used to probe the dynamics of membrane lipids in living cells (Eggeling et al. 2009). The advantage of this technique is the possibility to tune the diameter of the probed area up to 70-fold below the diffraction barrier. Different approaches facilitating FCS on membranes have also been developed focusing on the problem of photobleaching phenomenon occurring during FCS measurement. It mainly concerns scanning FCS (SFCS) in which the detection volume is scanned through the sample. This includes the continuous wave excitation SFCS (Ries and Schwille 2006) and raster image correlation spectroscopy (RICS), a version of FCS in which involves the spatio-temporal correlation analysis of fluorescence fluctuations in the intensity of a given pixel in an image obtained with a confocal laser-scanning microscope (see an example of diffusion analysis in cell membrane in Digman and Gratton 2009). The measurement of diffusion coefficient of molecules diffusing slowly in the membrane is possible with these two latter techniques (this is generally impossible to achieve with classical FCS).

SMT within membranes allows the direct visualization of the diffusion of a membrane component. It requires the labeling of only a few molecules in order to track them individually. Precursor studies have investigated the membrane behavior using latex or colloidal gold beads bound to the molecule of interest. The main drawback of this so-called single particle tracking (SPT) technique concerns the size of the bead, which is about two orders of magnitude larger that the size of the tracked membrane component, and can constrain its diffusion (reviewed in Marguet et al. 2006). With the development of very sensitive detectors and organic fluorescent probes that are very stable and bright, it is now possible to track membrane components for long time (within 1 min range). This technique, often referred to as single dye tracking (SDT), requires the nanometer-ranged determination of the localization of the fluorescence spot before linking the different position of the molecules to construct the trajectory (details in Fig. 4.2). This trajectory can give information about the behavior of a membrane component by calculating the mean square displacement (MSD) of the molecules (Saxton and Jacobson 1997). The most popular fluorescent probes to track molecules are the cyanin derivatives such as the cationic Atto647N (Eggeling et al. 2009; Espenel et al. 2008) and quantum dots (Qdots). These Qdots are very bright and the most stable probes available and can be considered as a tool of choice for tracking proteins within membranes, even though the control of the valency of labeling, an important parameter for the tracking of single molecule, still requires further improvement (see Pinaud et al. 2010).

It is worth mentioning that labeling with fluorescent proteins (e.g. Green Fluorescent Protein (GFP)) only allows tracking for up to a few seconds because of their weak stability. Therefore, labeling is generally performed with antibodies labeled with the probes mentioned above. Fab fragments should be preferred to full antibodies because of cross-linking ability of the latter reagents (this drawback of using a bivalent antibody was observed in SMT experiments which studied the dynamics of tetraspanin CD9 (Espenel et al. 2008)). On the other hand, both

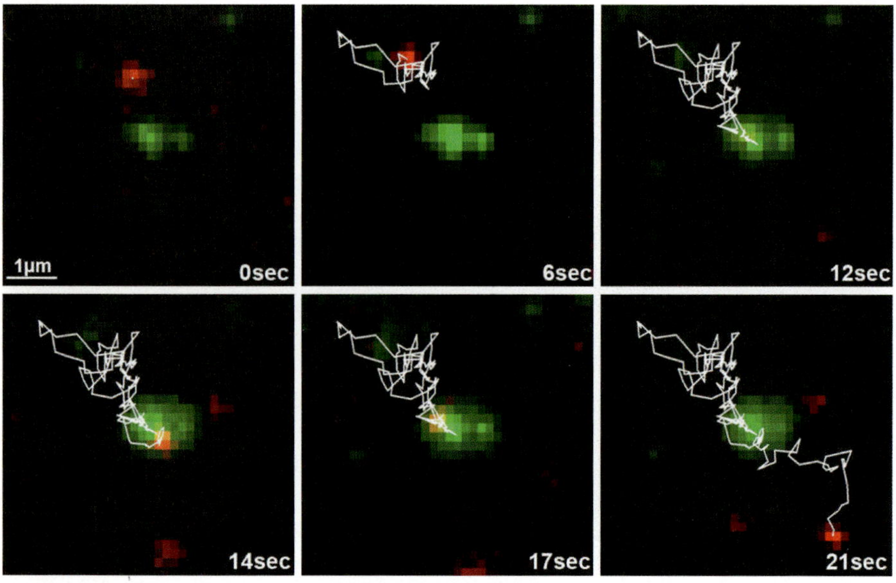

Fig. 4.2 Single molecule tracking of CD9 in PC3 cells. Time lapse of a mixed trajectory of CD9 obtained by superposition of ensemble CD9 labeling achieved with anti-CD9 Cy3B-labeled antibodies (*green*) and a fluorescent single CD9 molecule (*red spot*) labeled with Atto647N-conjugated Fab fragments of the anti-CD9 SYB-1 (trajectories of the single molecule is indicated by the *white thin line*). The single CD9 molecule enters and exits from a tetraspanin-enriched domain (*green*) and is transiently confined when its position overlaps with the ensemble labeling (Adapted from (Espenel et al. 2008))

approaches (cell labeling with antibodies and the use of GFP-derived tags) may occlude epitopes important for interactions of a target protein with its ligands and membrane or cytoplasmic partners.

4.3 Current View of the Dynamics of Tetraspanins

Characterization of the membrane dynamics of tetraspanins (TM4) remains an understudied area of research. Two papers, published almost simultaneously, have provided the first description of TEMs as physical entities in the plasma membrane of living cells and demonstrated that TM4, especially CD9, are very dynamic proteins (Barreiro et al. 2008; Espenel et al. 2008). Despite the fact that different biophysical approaches have been used in these studies, the authors have reached a similar general conclusion—although most of the TM4 molecules are dynamic, freely diffusing in the plasma membrane, they can also be confined in protein assemblies containing TM4 aggregates.

The work by Barreiro et al. have combined FRAP and FCS experiments to evaluate the dynamics of GFP-tagged CD9 and CD151 in living primary human endothelial cells in the context of cell adhesion. In FRAP experiments, CD9 diffusion coefficient was evaluated to be 0.47 μm²/s with a mobile fraction representing up to 88%; the diffusion coefficient of CD151 was ~0.41 μm²/s and its mobile fraction was 84%. The median values obtained with FCS although lower (0.17 μm²/s for CD9 and 0.14 μm²/s for CD151), were in the range similar to those obtained using FRAP. These results are in agreement with the differential size of the molecular samples measured by the two techniques (FRAP renders the average apparent diffusion of a huge molecular population versus a small number of molecules measured by FCS, whose average diffusion is much similar to single-molecule diffusion). FCS autocorrelation curves were describable by an anomalous diffusion model with an anomalous coefficient of 0.77 for CD9 and 0.71 for CD151. This parameter was interpreted as a transient interaction or confinement of TM4 diffusing molecules within endothelial adhesive platforms (EAPs) that also contain receptors such as ICAM-1 and VCAM-1. This interpretation is supported by an increase of the anomalous coefficient (and a corresponding decrease of the diffusion coefficient) when receptor-mediated docking structures were formed (i.e. increasing engagement of TM4 molecules in these structures). The difference found in the anomalous coefficients for these TM4 could suggest that CD151 is slightly more confined within EAP than CD9. This result can be related to the preferential interaction of CD151 with VCAM-1 that diffuses slower and is more confined than ICAM-1, a preferable partner for CD9 in endothelial cells. It underlines the specificity of interaction within TEMs and dependence of dynamics of individual TM4 on their specific partners. Furthermore, molecular complexes containing both CD9 and CD151 were also found by fluorescence cross-correlation analysis (Fig. 4.1; Barreiro and Zamai unpublished data). Interestingly, even in a different cell type and measuring by SMT at the basal instead of the apical membrane, the diffusion coefficient of CD9 molecules was in the same range (Espenel et al. 2008). Specifically, using SMT, the diffusion coefficient of CD9 in the prostate cancer cells PC3 was found to be 0.23 μm²/s. Furthermore, three different types of trajectories were identified from MSD analysis: Brownian, pure confined and mixed (a mixture of alternated confined and Brownian motion). Pure confined trajectories (23% of the total trajectories) can be paralleled with the percentage of immobile fraction in FRAP experiments in endothelial cells and, in both cell types, proteins displaying immobile and confined behavior represent the smaller fraction of the molecules. Using dual-view SMT setup, the authors were able to show that transient confinement of CD9 described above occurred in TM4 assemblies that behave like interactive platforms in permanent exchange with the rest of the membrane (Fig. 4.3). These platforms are somewhat similar to the specialized EAPs found in activated endothelial cells that contain adhesion receptors such as VCAM-1 and ICAM-1. In more recent experiments we extended these observations to other cell lines (such as HeLa and CHO cells) to show that the dynamic behaviour of CD9 (i.e. diffusion coefficient, type of motion and percentage of confinement in these cells) is similar to that observed in PC3 (Rassam and Milhiet, unpublished results). Altogether these results suggest that CD9 is a very dynamic membrane protein but a considerable fraction of CD9 molecules can be transiently trapped in TEMs. A tiny percentage of molecules also seem to be immobile in stable membrane platforms, at least for a few minutes. The relationship between the confinement of TM4 in membrane platforms and their

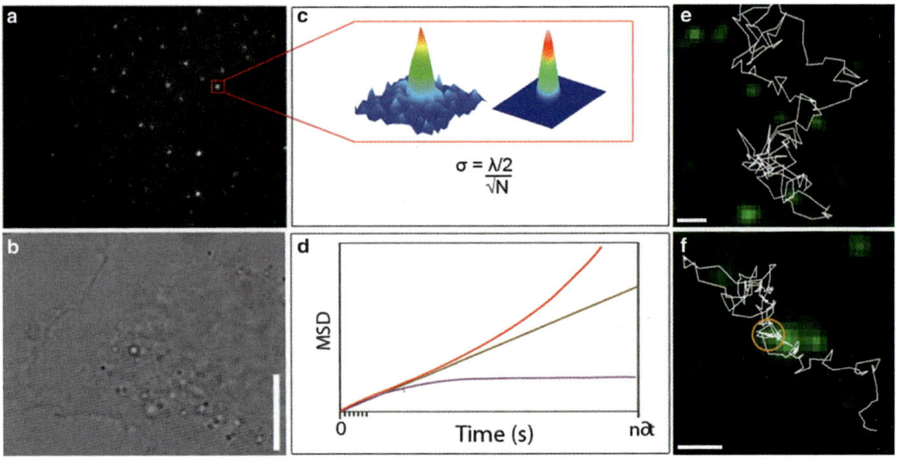

Fig. 4.3 Single molecule tracking of fluorescent molecules. *Setup:* In order to record single molecule fluorescence signal (**a**), a setup equipped with a high numerical aperture objective (~1.4) and an EM-CCD camera is necessary. Alternating-laser excitation (ALEX) prevents crosstalk in multicolor experiments (Margeat et al. 2006). Most commonly used probes are cyanin derivatives or quantum dots. (**b**) is the DIC image corresponding to the area shown in (**a**). *Analysis and processing of single molecule traces*: Trajectories are constructed using the individual diffraction limited signal of each molecule (**a**). The center of each fluorescence peak is determined with sub-pixel resolution by fitting a two-dimensional elliptical Gaussian function (**c**). The accuracy of the position measurement ∂ is dependent on the wavelength λ and the number of photons N collected. The two-dimensional trajectories of single molecules were constructed frame per frame. The mean square displacement (MSD) of each molecule is plotted as a function of the time interval ∂t and the shape of the curve determine the type of motion (**d**). For Brownian trajectories, the MSD varies linearly with ∂t and the value of the slope is 4D (*brown line* in **d**). If the MSD-∂t (or τ) plot shows positive or negative deviation from a straight line, the MSD is respectively adjusted with a quadratic curve $(4Dt+v^2t^2)$(directed diffusion, *red line* in **d**) or with an exponential curve $\frac{L^2}{3}[1-\exp(\frac{-12Dt}{L^2})]$ (confined diffusion, *purple line* in **d**) where L is the side of a square domain (Kusumi et al. 1993). The apparent diffusion coefficient values are determined from a linear fit to the MSD-τ plots between the second and the fourth points ($D_{2\text{-}4}$) according to the equation MSD(t) = 4Dt. Thanks to a dual view imaging, it is possible to superimpose trajectories in the context of an ensemble labeling (see a Brownian trajectory in **e** and a mixed trajectory in **f**; a mixed trajectory is a combination of Brownian and confined (*orange circle*) motion modes). All the movies were analyzed using a homemade software (named "PaTrack") implemented in visual C++ and motion modes within trajectories were automatically attributed using a neural network (manuscript in preparation). Scale bars are 10 μm in **b** and 1 μm in **e** and **f**

functions deserves further investigation. All these results obviously raises a question about the membrane behavior of other TM4. CD81 molecules have been shown to be part of TEMs in primary endothelial cells (EAPs) (Barreiro et al. 2005) and PC3 cells and is considered to be one of the preferable partners of CD9 in TM4 assemblies (Charrin et al. 2009). Study of its dynamics should further extend our understanding of TM4 membrane behavior.

In addition to protein-protein interactions that could affect TM4 membrane behavior, lipids also influence CD9 dynamics and partitioning. In this regard, it has been observed that Chl depletion led to a decrease in the association of CD9 with TEMs and, consequently, affected dynamic behaviour of the tetraspanin in the plasma membrane. In addition, the experiments using a palmitoylation-deficient mutant have shown that CD9 Brownian diffusion is slowed down by palmitoylation, a phenomenon that can be explained by the preferential interaction of palmitoylated CD9 with more ordered lipid phase (as has already been suggested for classical raft microdomains (Brown and London 2000)). In addition, palmitoylation is also involved in the interaction of CD9 with TEMs. In agreement with the effects on CD9 dynamics observed upon Chl depletion, saturated fatty acid chains such as palmitate could favor association of CD9 with TEMs that are enriched in Chl. However, because Chl depletion does not alter TEMs in terms of shape and position, it is tempting to speculate that different pools of Chl co-exist on the plasma membrane. The first pool will include molecules mostly involved in the general physical properties of membrane by interacting with fatty acid chains of lipids. Another pool could be more involved in organizing the architecture of protein assemblies. This second pool will be less accessible to the drugs used for Chl depletion and weakly exchanged with the rest of the membrane explaining the preservation of TEMs upon depletion of this lipid. The existence of the second pool of Chl is in agreement with the ability of this lipid to be chemically cross-linked with CD9 (Charrin et al. 2003b).

Because of the development of the dual-view SMT mentioned above, it has been possible to perform dynamic co-localization experiments using Fab fragments of anti-CD9 mAb (Charrin et al. 2003a), labeled with spectrally distinct fluorophores (for more details, see (Espenel et al. 2008)). We have been able to characterize a complex of two molecules of CD9 diffusing in the membrane, both exhibiting a Brownian motion, before and during their association. CD9-CD9 dynamic co-localization that concerned 15% of the Brownian trajectories was always transient. The diffusion coefficient of both molecules was not modified when interacting with each other, suggesting that CD9 molecules were more likely diffusing in a cluster of molecules than as individual transmembrane protein. This cluster or nanodomain could accommodate several proteins and lipids including TM4, their partners and Chl. This hypothesis is supported by FCS experiments performed on endothelial cells since photon-counting histograms revealed that more than one fluorescent protein was present in diffusing entities (Barreiro et al. 2008). Association of proteins within clusters appears to be dependent on Chl and CD9 palmitoylation.

Several properties of TEMs can be related to so-called lipid rafts, especially the involvement of Chl in their formation as well as their resistance to the extraction by detergent. In addition, our SMT experiments demonstrate that CD55, often used as a raft marker (Charrin et al. 2002), exhibits a membrane behavior similar to that of CD9 in terms of motion mode (Brownian, confined, and mixed diffusion). It is important to mention that a similar behavior has been described for other membrane proteins such as EGFR and the GPI-anchored proteins CD59 (Suzuki et al. 2007; Sergé et al. 2008). This behavior is likely more related to the general organization of the plasma membrane in eukaryotic cells (a mosaic of nano or microdomains) than to the properties of rafts or TEMs. However, our results provide compelling evidence that rafts and TEMs are different physical entities at the plasma membrane. The SMT approach demonstrates that CD55 was never confined in TEMs whereas most of the CD9 trajectories passed through these areas. Moreover, the dynamics of a raft marker (a GPI-anchored protein coupled to EGFP) in EAP of endothelial cells evaluated by FRAP was clearly different from those of the tetraspanins CD9 and CD151. In addition, classical detergent-based biochemical experiments with endothelial cells indicate that the two types of domains were not located in the same lipid environment (Barreiro et al. 2008).

4.4 Dynamics and Functions

Although there are no clear biochemical functions directly ascribed to tetraspanin proteins (except for their role as membrane organizers), it is well established that tetraspanins act as modulators of the function of their lateral partners regulating a multitude of processes such as cell-cell or cell-matrix adhesion and migration (Yanez-Mo et al. 2009), protein trafficking (Berditchevski and Odintsova 2007), fusion (Rubinstein et al. 2006), cancer invasion (Stipp 2010), pathogen infection (Thali 2009; Tham et al. 2010) and others. Thus, the insertion in TEMs bestows some transmembrane receptors with the local molecular density (clustering) thus resulting in fully functional proteins. Hence, the spatial organization of receptors at the plasma membrane is as critical as their appropriate expression level in order to accomplish their function. In most cases, receptor clustering is induced upon certain stimuli by dynamic coalescence of receptor-containing TEMs. This is the case for endothelial adhesion receptors (Barreiro et al. 2008). We have investigated in depth the spatial organization and the functional regulation of endothelial receptors which seem to be governed by their inclusion in TEMs. Inflammatory conditions induce the expression of a variety of adhesion receptors (e.g. VCAM-1, ICAM-1, E- and P-selectins) at the luminal surface of endothelium to allow leukocyte-endothelial cell contacts and promote leukocyte extravasation (Barreiro et al. 2007). These endothelial adhesion receptors are not evenly distributed in the plasma membrane, but show a patterned distribution in submicron-sized domains, as observed with scanning electron microscopy of immunogold-labelled samples (Barreiro et al. 2008). The treatment with the exogenous large extracellular loop of CD9 coupled to GST disturbs this spatial organization, decrease receptor clustering, alter the molecular dynamics of the clusters and, consequently, negatively affect the receptor adhesive functions (Barreiro et al. 2005, 2008). This argues in favour of TEMs acting as cell-cell adhesion modulators. Once a leukocyte contacts with the endothelium in order to extravasate towards the inflammatory focus, the tetraspanin nanoclusters containing endothelial adhesion receptors that we have coined as EAPs rapidly coalesce around the leukocyte forming the so-called endothelial docking structure (Barreiro et al. 2002). The insertion of a variety of adhesion receptors in this specialized kind of TEMs implies the co-recruitment of all these receptors with complementary adhesive functions towards the leukocyte-endothelial contact area. In this manner, endothelial adhesion receptors are exposed to be available in case of leukocyte transendothelial migration. Noteworthy, most of the endothelial receptors embedded within EAPs are driven to the vicinity of the adhered leukocytes without recognizing any ligand on the leukocyte side and even without the participation of the actin cytoskeleton, but only due to their interaction with the tetraspanins (Barreiro et al. 2008). In this regard, the complexity of the molecular inter-relationships within EAPs has been dissected by using FRET-FLIM (fluorescence resonance energy transfer-fluorescence lifetime imaging microscopy) and discovered the preferential binding of ICAM-1 to CD9, that of VCAM-1 to CD151, and the interaction of CD9 and CD151 with themselves and each other in living primary human endothelial cells. Thus, an array of receptors with related functions becomes spatially congregated to ensure the progression of the stepwise process of extravasation (Barreiro et al. 2008).

Integrins also constitute another representative example of molecules which can be regulated by avidity (clustering), although they also exhibit conformational changes to enhance their affinity for ligands (Carman and Springer 2003; Luo et al. 2007). Different integrin conformations have distinct diffusion profiles which control integrin clustering, vary according to the activation state of the cell and seem to be governed by the binding of the integrin to the actin cytoskeleton (Das et al. 2009). In this context, CD81 has been involved in the regulation of VLA-4 ($\alpha 4\beta 1$) and VLA-5 ($\alpha 5\beta 1$) adhesion strengthening to multivalent ligands, in particular during the interaction of VLA-4 to its endothelial ligand VCAM-1 under physiological flow conditions (Feigelson et al. 2003). Moreover, an anti-CD81 antibody has been shown to induce high avidity of the integrin LFA-1, the counter-receptor of the endothelial ICAM-1 (VanCompernolle et al. 2001). Therefore, the leukocyte integrins and their endothelial ligands participating in the leukocyte-endothelial adhesive interactions appear to be regulated in terms of avidity by their inclusion in TEMs. This regulatory mechanism could modulate the adhesion threshold of these transmembrane receptors to avoid undesired interactions at non-inflammatory sites of the vasculature but to promote a robust adhesion under stringent conditions as high shear flow stress at proper inflammatory scenarios. Same arguments could be applied to integrins functioning in cell-extracellular matrix interactions, a key process for cell adhesion, migration and invasion.

More recently, TM4 has been suggested to be involved in the sequestration of VAMP7, a membrane protein mediating exocytosis during neuritogenesis, phagocytosis and lysosomal secretion, within lipid microdomains (Danglot et al. 2010). Using Qdot-based single molecule tracking, it was shown that depletion in the tetraspanin CD82 in HeLa cells restrain EGFR diffusion. Effects upon CD82 depletion was explained by a modification of the membrane content in Chl and gangliosides, because this TM4 can regulate the expression of these lipids (Charrin et al. 2003b; Delaguillaumie et al. 2004; Le Naour et al. 2006; Odintsova et al. 2006; Regina Todeschini and Hakomori 2008), favouring the exit of EGFR molecules from lipid microdomains, and thus their dimerization and endocytosis. In this case, CD82 would be more involved in the homeostasis of the plasma membrane in terms of lipid composition than in the formation of a network of protein-protein interactions including EGFR.

The involvement of TEMs in receptor compartmentalization in other cellular contexts such as enzymatic activity (Yanez-Mo et al. 2008), antigen presentation at immune synapse (Mittelbrunn et al. 2002; Unternaehrer et al. 2007), viral and bacterial infection (Nydegger et al. 2006; Silvie et al. 2006; Tham et al. 2010), as well as gamete fusion (Ziyyat et al. 2006) has been also described. However, no detailed biophysical approaches have been applied on these topics to date, except in the case of HIV-1 assembly (Krementsov et al 2010). The employment of powerful analytical microscopy and spectroscopy techniques will help in solving the contribution of TEMs to these crucial physiopathological processes. The performance of biophysical studies will also favour the development of specific new therapeutic agents directed against TM4, which we hypothesize could have a general and more potent effect than the single targeting and inhibition of one of the receptors involved in cancer invasion, inflammation or pathogen infection. However, due to the ubiquitous nature of TM4, it will also be critical to develop new strategies for the administration, specific targeting and local delivery of TM4-based therapies (Barreiro et al. 2009).

4.5 Perspectives and Future Developments

The studies described in this chapter have demonstrated for the first time that TM4 are dynamic molecules that can exhibit a Brownian motion within plasma membrane or be transiently confined in TEMs (Fig. 4.4). Further works are now required to have a more precise view of the dynamic network of protein-protein interactions. Specifically, we need to uncover how TM4 interact with each other and with their membrane partners, how dynamic these TEMs are and what are precise molecular composition and stoichiometry of these membrane platforms. The relationship of these membrane platforms with the cytoskeleton needs also further clarification since some TM4 interact with proteins such as EWI-2, ICAM-1, VCAM-1 or CD44 which in turn bind to the actin cytoskeleton linkers ezrin–radixin–moesin (ERM) proteins. ERM could also act as a linker between TEMs and the cytoskeleton (Sala-Valdes et al. 2006). Whether TM4 could remain only transiently bound to the actin cytoskeleton (and this binding could be induced upon certain stimuli) is still a matter of investigation. It is noteworthy to mention that, even if TM4 are most of the time related to membrane assemblies, it is likely that isolated molecules have a key role in their cellular functions and single molecule analysis is a prerequisite for better understanding the molecular mechanisms underlying TM4 function.

Future studies should benefit from recent technological developments in optical fluorescence microscopy that circumvent the diffraction law (Patterson 2009). Among them, super resolution microscopy such PALM (Photo-Activated Localization Microscopy) and STORM (Stochastic Optical Reconstruction Microscopy) should

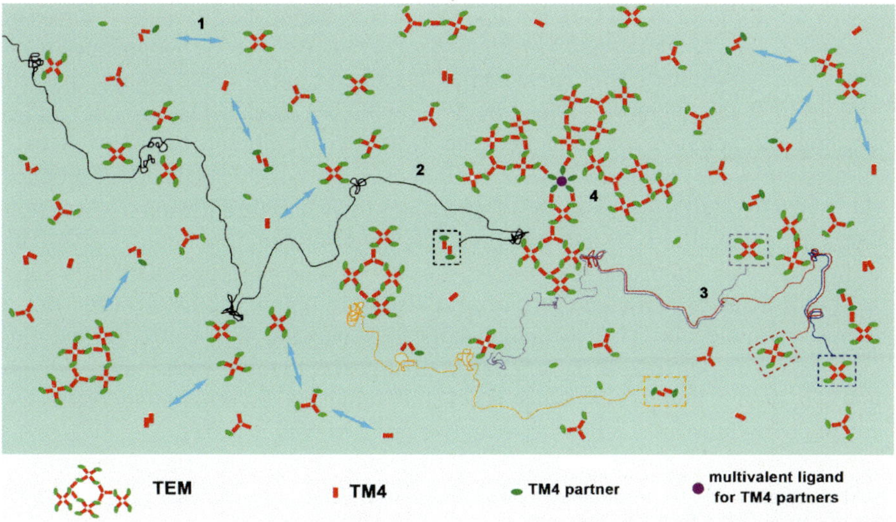

Fig. 4.4 Model of tetraspanin membrane organization. The scheme represents the dynamic concept of tetraspanin-enriched microdomains (*TEMs*) obtained in biophysical studies. Tetraspanins (*TM4*) interact with themselves and their partners at the membrane conforming TEMs, which are supramolecular platforms with variable composition and stoichiometry. These platforms are dynamic, slowly diffusing throughout the membrane and showing a continuous exchange of components (*1*, bidirectional *blue arrows*). There are also smaller tetraspanin ensembles which diffuse faster exhibiting a mixture of alternated Brownian and confined motion, i.e., free diffusion in membrane versus retention within platforms (*2*, coloured traces). Some of these ensembles can coalesce and co-diffuse over time and then diverge (*3*). Finally, it is also shown the clustering of TEMs into bigger immobilized platforms induced by the binding of tetraspanin partners to ligands and cytoskeleton constraints (*4*). A multivalent ligand able to bind to several TM4 partners simultaneously is exemplified

greatly improve our knowledge in the composition of TM4 assemblies. These two technologies that are relatively easy to implement will allow the estimation of the number of molecules per TEMs, their stoichiometry, and their spatial organization in these microdomains with a lateral resolution of 20 nm. Both methods are based on photoswitchable or photoactivable fluorophores. In each imaging cycle, only a fraction of the fluorophores is turned on, allowing their positions to be determined with nanometer accuracy. The fluorophore positions obtained from a series of imaging cycles are used to reconstruct the entire image (Betzig et al. 2006; Rust et al. 2006). With STORM, small fluorescent probes are used and it is easy to perform multicolor experiments that are necessary to make a mapping of several TM4 and partners on the same cell. Interestingly it can now be applied to conventional probes (Heilemann et al. 2008). PALM often required eosFP, a mutant of the Green Fluorescent Protein that can be photoactivated only once. Therefore, it could be used to quantitatively determine the number of a given TM4 within TEMs. Because PALM does not require the fixation of cells, it will be interesting in the future to combine PALM and tracking of activated TM4 proteins in living cells, the so-called sptPALM (Manley et al. 2008).

Further development in SMT should also help in deciphering the molecular mechanism underlying the dynamics and partitioning of TM4. Besides developments in the detection of single molecules (new EM-CCD cameras, more stable probes) that will improve time and lateral resolution, tracking in 3D is a crucial step for better understanding the behavior of TM4 because a distinct membrane topology of cells cultured under 3D conditions is likely to influence the dynamic behavior of proteins. More recently, Hell's group has developed a new optical setup allowing tracking with a very high spatial (20–40 nm) and time (<1 ms) resolution (Sahl et al. 2010). This setup is based on a confocal microscope, where the pinhole is replaced by three separate point detectors arranged in close proximity to obtain the localization of the fluorescent molecule. This technique appears especially well-suited for probing membrane components on a very large time scale. Finally, FRET at the single molecule level in living cells, which is not still available, represents another attractive way to probe direct interaction between two proteins within TEMs.

Acknowledgements We would like to thank Dr. Moreno Zamai for participating in data acquisition and analysis of Fig. 4.1. This work was supported by the Spanish Ministry of Science and Innovation (SAF2008-06235); and MEICA from Genoma España Foundation, INSINET from Comunidad Autónoma de Madrid [FONCICYT-C002-2009-1ALA/127249], and RECAVA [RD06/0014-0030] from Instituto de Salud Carlos III to F.S-M. This work was also supported by the Association Nationale pour la Recherche (ANR-06-BLAN-0378) to P.E.M. O.B. is a postdoctoral fellow from Spanish Ministry of Science and Innovation 'Juan de la Cierva Program' (JDC 08). Centro Nacional de Investigaciones Cardiovasculares is supported by the Ministry of Science and Innovation and the Pro CNIC Foundation.

References

Abitorabi MA, Pachynski RK, Ferrando RE, Tidswell M, Erle DJ (1997) Presentation of integrins on leukocyte microvilli: a role for the extracellular domain in determining membrane localization. J Cell Biol 139(2):563–571. doi:10.1083/jcb.139.2.563

Barreiro O, Yanez-Mo M, Serrador JM, Montoya MC, Vicente-Manzanares M, Tejedor R, Furthmayr H, Sanchez-Madrid F (2002) Dynamic interaction of VCAM-1 and ICAM-1 with moesin and ezrin in a novel endothelial docking structure for adherent leukocytes. J Cell Biol 157(7):1233–1245. doi:10.1083/jcb.200112126

Barreiro O, Yanez-Mo M, Sala-Valdes M, Gutierrez-Lopez MD, Ovalle S, Higginbottom A, Monk PN, Cabanas C, Sanchez-Madrid F (2005) Endothelial tetraspanin microdomains regulate leukocyte firm adhesion during extravasation. Blood 105(7):2852–2861. doi:10.1182/blood-2004-09-3606

Barreiro O, de la Fuente H, Mittelbrunn M, Sanchez-Madrid F (2007) Functional insights on the polarized redistribution of leukocyte integrins and their ligands during leukocyte migration and immune interactions. Immunol Rev 218:147–164. doi:10.1111/j.1600-065X.2007.00529

Barreiro O, Zamai M, Yanez-Mo M, Tejera E, Lopez-Romero P, Monk PN, Gratton E, Caiolfa VR, Sanchez-Madrid F (2008) Endothelial adhesion receptors are recruited to adherent leukocytes by inclusion in preformed tetraspanin nanoplatforms. J Cell Biol 183(3):527–542. doi:10.1083/jcb.200805076

Barreiro O, Aguilar RJ, Tejera E, Megias D, de Torres-Alba F, Evangelista A, Sanchez-Madrid F (2009) Specific targeting of human inflamed endothelium and in situ vascular tissue transfection by the use of ultrasound contrast agents. JACC Cardiovasc Imag 2(8):997–1005. doi:10.1016/j.jcmg.2009.04.012

Berditchevski F, Odintsova E (1999) Characterization of integrin-tetraspanin adhesion complexes: role of tetraspanins in integrin signaling. J Cell Biol 146(2):477–492. doi:10.1083/jcb.146.2.477
Berditchevski F, Odintsova E (2007) Tetraspanins as regulators of protein trafficking. Traffic 8(2):89–96. doi:10.1111/j.1600-0854.2006.00515.x
Berditchevski F, Odintsova E, Sawada S, Gilbert E (2002) Expression of the palmitoylation-deficient CD151 weakens the association of alpha 3 beta 1 integrin with the tetraspanin-enriched microdomains and affects integrin-dependent signaling. J Biol Chem 277(40):36991–37000
Betzig E, Patterson GH, Sougrat R, Lindwasser OW, Olenych S, Bonifacino JS, Davidson MW, Lippincott-Schwartz J, Hess HF (2006) Imaging intracellular fluorescent proteins at nanometer resolution. Science 313(5793):1642–1645. doi:10.1126/science.1127344
Boucheix C, Rubinstein E (2001) Tetraspanins. Cell Mol Life Sci 58:1189–1205
Brown DA, London E (2000) Structure and function of sphingolipid- and cholesterol-rich membrane rafts. J Biol Chem 275(23):17221–17224. doi:10.1074/jbc.R000005200
Carman CV, Springer TA (2003) Integrin avidity regulation: are changes in affinity and conformation underemphasized? Curr Opin Cell Biol 15(5):547–556. doi:S0955067403001121 [pii]
Charrin S, Manie S, Oualid M, Billard M, Boucheix C, Rubinstein E (2002) Differential stability of tetraspanin/tetraspanin interactions: role of palmitoylation. FEBS Lett 516:139–144. doi:S001457930202522X [pii]
Charrin S, Manie S, Billard M, Ashman L, Gerlier D, Boucheix C, Rubinstein E (2003a) Multiple levels of interactions within the tetraspanin web. Biochem Biophys Res Commun 304(1):107–112. doi:S0006291X0300545X [pii]
Charrin S, Manie S, Thiele C, Billard M, Gerlier D, Boucheix C, Rubinstein E (2003b) A physical and functional link between cholesterol and tetraspanins. Eur J Immunol 33:2479–2489. doi:10.1002/eji.200323884
Charrin S, le Naour F, Silvie O, Milhiet PE, Boucheix C, Rubinstein E (2009) Lateral organization of membrane proteins: tetraspanins spin their web. Biochem J 420(2):133–154. doi:10.1042/BJ20082422
Chen Y, Muller JD, Eid JS, Gratton E (2001) Two-photon fluorescence fluctuation spectroscopy. In: Valeur B, Brochon JC (eds) New trends in fluorescence spectroscopy. Springer, Berlin, pp 277–296
Danglot L, Chaineau M, Dahan M, Gendron MC, Boggetto N, Perez F, Galli T (2010) Role of TI-VAMP and CD82 in EGFR cell-surface dynamics and signaling. J Cell Sci 123(Pt 5):723–735. doi:10.1242/jcs.062497
Das R, Cairo CW, Coombs D (2009) A hidden Markov model for single particle tracks quantifies dynamic interactions between LFA-1 and the actin cytoskeleton. PLoS Comput Biol 5(11):e1000556. doi:10.1371/journal.pcbi.1000556
Dehmelt L, Bastiaens PI (2010) Spatial organization of intracellular communication: insights from imaging. Nat Rev Mol Cell Biol 11(6):440–452. doi:10.1038/nrm2903
Delaguillaumie A, Harriague J, Kohanna S, Bismuth G, Rubinstein E, Seigneuret M, Conjeaud H (2004) Tetraspanin CD82 controls the association of cholesterol-dependent microdomains with the actin cytoskeleton in T lymphocytes: relevance to co-stimulation. J Cell Sci 117(Pt 22):5269–5282. doi:10.1242/jcs.01380
Digman MA, Gratton E (2009) Analysis of diffusion and binding in cells using the RICS approach. Microsc Res Tech 72(4):323–332. doi:10.1002/jemt.20655
Drummer HE, Wilson KA, Poumbourios P (2005) Determinants of CD81 dimerization and interaction with hepatitis C virus glycoprotein E2. Biochem Biophys Res Commun 328(1):251–257. doi:10.1016/j.bbrc.2004.12.160
Eggeling C, Ringemann C, Medda R, Schwarzmann G, Sandhoff K, Polyakova S, Belov VN, Hein B, von Middendorff C, Schonle A, Hell SW (2009) Direct observation of the nanoscale dynamics of membrane lipids in a living cell. Nature 457(7233):1159–1162. doi:10.1038/nature07596
Espenel C, Margeat E, Dosset P, Arduise C, Le Grimellec C, Royer CA, Boucheix C, Rubinstein E, Milhiet PE (2008) Single-molecule analysis of CD9 dynamics and partitioning reveals multiple modes of interaction in the tetraspanin web. J Cell Biol 182(4):765–776. doi:10.1083/jcb.200803010

Feigelson SW, Grabovsky V, Shamri R, Levy S, Alon R (2003) The CD81 tetraspanin facilitates instantaneous leukocyte VLA-4 adhesion strengthening to vascular cell adhesion molecule 1 (VCAM-1) under shear flow. J Biol Chem 278(51):51203–51212. doi:10.1074/jbc.M303601200

Garcia-Saez AJ, Schwille P (2008) Fluorescence correlation spectroscopy for the study of membrane dynamics and protein/lipid interactions. Methods 46(2):116–122. doi:10.1016/j.ymeth.2008.06.011

Hansen CG, Nichols BJ (2010) Exploring the caves: cavins, caveolins and caveolae. Trends Cell Biol 20(4):177–186. doi:10.1016/j.tcb.2010.01.005

Heilemann M, Van De Linde S, Schüttpelz M, Kasper R, Seefeldt B, Mukherjee A, Tinnefeld P, Sauer M, 33 (2008) Subdiffraction-resolution fluorescence imaging with conventional fluorescent probes. Angew Chem Int Ed 47(33):6172–6176, doi:10.1002/anie.v47:33

Hemler ME (2003) Tetraspanin proteins mediate cellular penetration, invasion, and fusion events and define a novel type of membrane microdomain. Annu Rev Cell Dev Biol 19:397–422. doi:10.1146/annurev.cellbio.19.111301.153609

Hemler ME (2005) Tetraspanin functions and associated microdomains. Nat Rev Mol Cell Biol 6(10):801–811. doi:10.1038/nrm1736

Jacobson K, Mouritsen OG, Anderson RG (2007) Lipid rafts: at a crossroad between cell biology and physics. Nat Cell Biol 9(1):7–14. doi:10.1038/ncb0107-7

Krementsov DN, Rassam P, Margeat E, Roy NH, Schneider-Schaulies J, Milhiet PE, Thali M (2010) HIV-1 assembly differentially alters dynamics and partitioning of tetraspanins and raft components. Traffic 11(11):1401–1414. doi:10.1111/j.1600-0854.2010.01111.x

Kusumi A, Sako Y, Yamamoto N (1993) Confined lateral diffusion of membrane receptors as studied by single particle tracking (nanovid microscopy). Effects of calcium-induced differentiation in cultured epithelial cells. Biophys J 65(5):2021–2040. doi: 0006-3495/93/11/2021/20

Lasserre R, Guo XJ, Conchonaud F, Hamon Y, Hawchar O, Bernard AM, Soudja SM, Lenne PF, Rigneault H, Olive D, Bismuth G, Nunes JA, Payrastre B, Marguet D, He HT (2008) Raft nanodomains contribute to Akt/PKB plasma membrane recruitment and activation. Nat Chem Biol 4(9):538–547. doi:10.1038/nchembio.103

Le Naour F, Andre M, Boucheix C, Rubinstein E (2006) Membrane microdomains and proteomics: lessons from tetraspanin microdomains and comparison with lipid rafts. Proteomics 6(24):6447–6454. doi:10.1002/pmic.200600282

Lingwood D, Simons K (2010) Lipid rafts as a membrane-organizing principle. Science 327(5961):46–50. doi:10.1126/science.1174621

Luo BH, Carman CV, Springer TA (2007) Structural basis of integrin regulation and signaling. Annu Rev Immunol 25:619–647. doi:10.1146/annurev.immunol.25.022106.141618

Manley S, Gillette JM, Patterson GH, Shroff H, Hess HF, Betzig E, Lippincott-Schwartz J (2008) High-density mapping of single-molecule trajectories with photoactivated localization microscopy. Nat Meth 5(2):155–157. doi:10.1038/nmeth.1176

Margeat E, Kapanidis AN, Tinnefeld P, Wang Y, Mukhopadhyay J, Ebright RH, Weiss S (2006) Direct observation of abortive initiation and promoter escape within single immobilized transcription complexes. Biophys J 90(4):1419–1431

Marguet D, Lenne PF, Rigneault H, He HT (2006) Dynamics in the plasma membrane: how to combine fluidity and order. EMBO J 25(15):3446–3457. doi:10.1038/sj.emboj.7601204

Mayor S, Riezman H (2004) Sorting GPI-anchored proteins. Nat Rev Mol Cell Biol 5(2):110–120. doi:10.1038/nrm1309

Mittelbrunn M, Yanez-Mo M, Sancho D, Ursa A, Sanchez-Madrid F (2002) Cutting edge: dynamic redistribution of tetraspanin CD81 at the central zone of the immune synapse in both T lymphocytes and APC. J Immunol 169(12):6691–6695

Nydegger S, Khurana S, Krementsov DN, Foti M, Thali M (2006) Mapping of tetraspanin-enriched microdomains that can function as gateways for HIV-1. J Cell Biol 173(5):795–807. doi:10.1083/jcb.200508165

Odintsova E, Butters TD, Monti E, Sprong H, van Meer G, Berditchevski F (2006) Gangliosides play an important role in the organization of CD82-enriched microdomains. Biochem J 400(2):315–325

Parton RG, Hancock JF (2004) Lipid rafts and plasma membrane microorganization: insights from Ras. Trends Cell Biol 14(3):141–147. doi:10.1016/j.tcb.2004.02.001
Patterson G (2009) Fluorescence microscopy below the diffraction limit. Semin Cell Dev Biol. doi:10.1016/j.semcdb.2009.08.006
Pinaud F, Clarke S, Sittner A, Dahan M (2010) Probing cellular events, one quantum dot at a time. Nat Meth 7(4):275–285. doi:10.1038/nmeth.1444
Regina Todeschini A, Hakomori SI (2008) Functional role of glycosphingolipids and gangliosides in control of cell adhesion, motility, and growth, through glycosynaptic microdomains. Biochim Biophys Acta 1780(3):421–433
Ries J, Schwille P (2006) Studying slow membrane dynamics with continuous wave scanning fluorescence correlation spectroscopy. Biophys J 91(5):1915–1924. doi:10.1529/biophysj.106.082297
Rubinstein E, Ziyyat A, Wolf JP, Le Naour F, Boucheix C (2006) The molecular players of sperm-egg fusion in mammals. Semin Cell Dev Biol 17(2):254–263. doi:10.1016/j.semcdb.2006.02.012
Rust MJ, Bates M, Zhuang X (2006) Sub-diffraction-limit imaging by stochastic optical reconstruction microscopy (STORM). Nat Meth 3(10):793–796. doi:10.1038/nmeth929
Sahl SJ, Leutenegger M, Hilbert M, Hell SW, Eggeling C (2010) Fast molecular tracking maps nanoscale dynamics of plasma membrane lipids. Proc Natl Acad Sci USA 107(15):6829–6834. doi:10.1073/pnas.0912894107
Sala-Valdes M, Ursa A, Charrin S, Rubinstein E, Hemler ME, Sanchez-Madrid F, Yanez-Mo M (2006) EWI-2 and EWI-F link the tetraspanin web to the actin cytoskeleton through their direct association with ezrin-radixin-moesin proteins. J Biol Chem 281(28):19665–19675. doi:10.1074/jbc.M602116200
Saxton MJ, Jacobson K (1997) Single-particle tracking: applications to membrane dynamics. Annu Rev Biophys Biomol Struct 26:373–399. doi:10.1146/annurev.biophys.26.1.373
Schwille P, Korlach J, Webb WW (1999) Fluorescence correlation spectroscopy with single-molecule sensitivity on cell and model membranes. Cytometry 36(3):176–182. doi:10.1002/(SICI)1097-0320(19990701)36:3<176::AID-CYTO5>3.0.CO;2-F
Sergé A, Bertaux N, Rigneault H, Marguet D (2008) Dynamic multiple-target tracing to probe spatiotemporal cartography of cell membranes. Nat Meth 5(8):687–694. doi:10.1038/nmeth.1233
Silvie O, Charrin S, Billard M, Franetich JF, Clark KL, van Gemert GJ, Sauerwein RW, Dautry F, Boucheix C, Mazier D, Rubinstein E (2006) Cholesterol contributes to the organization of tetraspanin-enriched microdomains and to CD81-dependent infection by malaria sporozoites. J Cell Sci 119(Pt 10):1992–2002. doi:10.1242/jcs.02911
Stipp CS (2010) Laminin-binding integrins and their tetraspanin partners as potential antimetastatic targets. Expert Rev Mol Med 12:e3. doi:10.1017/S1462399409001355
Suzuki KG, Fujiwara TK, Edidin M, Kusumi A (2007) Dynamic recruitment of phospholipase C gamma at transiently immobilized GPI-anchored receptor clusters induces IP3-Ca2+ signaling: single-molecule tracking study 2. J Cell Biol 177(4):731–742. doi:10.1083/jcb.200609175
Thali M (2009) The roles of tetraspanins in HIV-1 replication. Curr Top Microbiol Immunol 339:85–102. doi:10.1007/978-3-642-02175-6_5
Tham TN, Gouin E, Rubinstein E, Boucheix C, Cossart P, Pizarro-Cerda J (2010) Tetraspanin CD81 is required for Listeria monocytogenes invasion. Infect Immun 78(1):204–209. doi:10.1128/IAI.00661-09
Unternaehrer JJ, Chow A, Pypaert M, Inaba K, Mellman I (2007) The tetraspanin CD9 mediates lateral association of MHC class II molecules on the dendritic cell surface. Proc Natl Acad Sci USA 104(1):234–239. doi:10.1073/pnas.0609665104
VanCompernolle SE, Levy S, Todd SC (2001) Anti-CD81 activates LFA-1 on T cells and promotes T cell-B cell collaboration. Eur J Immunol 31(3):823–831. doi:10.1002/1521-4141(200103)31:3<823::AID-IMMU823>3.0.CO;2-D
Wawrezinieck L, Rigneault H, Marguet D, Lenne PF (2005) Fluorescence correlation spectroscopy diffusion laws to probe the submicron cell membrane organization. Biophys J 89(6):4029–4042. doi:10.1529/biophysj.105.067959

Yanez-Mo M, Barreiro O, Gonzalo P, Batista A, Megias D, Genis L, Sachs N, Sala-Valdes M, Alonso MA, Montoya MC, Sonnenberg A, Arroyo AG, Sanchez-Madrid F (2008) MT1-MMP collagenolytic activity is regulated through association with tetraspanin CD151 in primary endothelial cells. Blood 112(8):3217–3226. doi:10.1182/blood-2008-02-139394

Yanez-Mo M, Barreiro O, Gordon-Alonso M, Sala-Valdes M, Sanchez-Madrid F (2009) Tetraspanin-enriched microdomains: a functional unit in cell plasma membranes. Trends Cell Biol 19(9):434–446. doi:10.1016/j.tcb.2009.06.004

Ziyyat A, Rubinstein E, Monier-Gavelle F, Barraud V, Kulski O, Prenant M, Boucheix C, Bomsel M, Wolf JP (2006) CD9 controls the formation of clusters that contain tetraspanins and the integrin alpha 6 beta 1, which are involved in human and mouse gamete fusion. J Cell Sci 119(Pt 3):416–424. doi:10.1242/jcs.02730

Chapter 5
Tetraspanins as Regulators of Protein Trafficking

Elena Odintsova and Fedor Berditchevski

Abstract The four transmembrane proteins of the tetraspanin superfamily are the main structural units of specialised membrane microdomains referred to as tetraspanin-enriched microdomains, (TERM or TEM). Variations in homotypic and heterotypic interactions within TERM result in the modulation of signalling pathways involving tetraspanin associated receptors including integrins and receptor tyrosine kinases. It has recently become apparent that, in addition to their purely structural function as organisers of TERM, tetraspanins also regulate various aspects of trafficking and biosynthetic processing of associated receptors. In this chapter we will specifically focus on this aspect of tetraspanin function.

The tetraspanin superfamily of small, four transmembrane domain proteins (up to 350 amino acids) consists of 33 members in humans and mice and includes proteins that are involved in physiological processes as diverse as egg-sperm fusion (Rubinstein et al. 2006), immunological responses (Levy and Shoham 2005) and tissue differentiation (Hemler 2005). There is emerging evidence that the involvement of tetraspanins in these processes may be due to their ability to regulate the biosynthetic maturation and trafficking of their associated partners. In this chapter, we will focus on this aspect of tetraspanin function and discuss what is known about the mechanisms which control distribution of tetraspanins in cells. Details on the structural organisation of tetraspanin microdomains and their various physiological functions can be found in other chapters of this volume.

Biosynthetic trafficking of tetraspanins. Similar to most other transmembrane proteins the tetraspanin trafficking route to the plasma membrane and various endocytic organelles involves a sequential passage through the endoplasmic reticulum (ER) and Golgi compartments. Although the molecular mechanisms that control

E. Odintsova (✉) • F. Berditchevski
School of Cancer Sciences, University of Birmingham, Edgbaston, Birmingham B15 2TT, UK
e-mail: e.odintsova@bham.ac.uk

F. Berditchevski and E. Rubinstein (eds.), *Tetraspanins*, Proteins and Cell Regulation 9, DOI 10.1007/978-94-007-6070-7_5,

these initial stages of trafficking have not been investigated for most tetraspanins, available data point to the importance of intra- and intermolecular interactions involving various parts of tetraspanin molecules, as a means of regulation.

5.1 Chaperones and Exit from the ER

The role of transmembrane domains in the ER exit. The cryo-EM structure of uroplakins (Uroplakin 1B/UP1B/TSPAN20 and Uroplakin 1A/UP1A/TSPAN21) suggested that the interactions of tightly packed transmembrane helices may define the folding of tetraspanin extracellular domains and, thereby, play a critical role in the ER-quality control processes (Min et al. 2006). Indeed, it has been reported that TM1, TM2 and TM4 of UPIB are all required for the ER exit (Tu et al. 2006). It has also been observed that conserved glutamic acid in TM3 is critical for allowing the protein to leave the ER. Another study reported that the deletion of the TM2-TM3 or TM4-cytoplasmic tail regions of CD9/TSPAN29 prevented the expression of these mutant proteins on the surface of thymoma cells (Toyo-oka et al. 1999). Similarly, mutants of CD151/TSPAN24 with deleted TM1-TM2, TM2-TM3 and TM1-TM2-TM3 regions were mostly trapped in the ER (Berditchevski et al. 2001). Finally, it has been demonstrated that disrupted trafficking to the cell surface of the TM2-4 mutant of CD82/TSPAN27 (missing the first transmembrane segment of the protein) could be restored in cells co-expressing CD82-TM1 as a separate polypeptide (Cannon and Cresswell 2001). Taken together these results indicate that the interactions between the transmembrane regions of tetraspanins are necessary for their biosynthetic trafficking.

Chaperones and tetraspanins. Folding and quality control of glycosylated transmembrane proteins in the ER is intimately tied to the function of calnexin, an ER-associated transmembrane chaperone. The association of tetraspanins with calnexin has been reported in two studies. Rubinstein and colleagues described co-immunoprecipitation of CD9 with calnexin from B-cells (Rubinstein et al. 1997). The authors also demonstrated that this interaction with calnexin is independent of the association of CD9 with the immature form of the β1 integrin subunit. In subsequent work calnexin (but not calreticulin) was also found to interact with CD82 (Cannon and Cresswell 2001). The interaction was transient and did not rely on the glycan-binding properties of the chaperone, suggesting the involvement of transmembrane domains in the interaction (see also below).

BAP31 is another ER-associated transmembrane chaperone which is involved in transport of transmembrane proteins (Schamel et al. 2003; Annaert et al. 1997; Paquet et al. 2004; Lambert et al. 2001). Interestingly, it was found that surface levels of CD9 and CD81/TSPAN28 were decreased in BAP31-null cells, thereby suggesting that BAP31 controls transport of these tetraspanins to the plasma membrane (Stojanovic et al. 2005). Although the interactions between the proteins were not specifically examined in this study, BAP31 is known to form complexes with calnexin (Zuppini et al. 2002), and, therefore, its role in trafficking of tetraspanins may be dependent on this association.

5.2 Post-Golgi Trafficking of Tetraspanins in Cells

Our knowledge about the precise steady-state distribution of tetraspanins in cells (i.e. electron microscopy) is mainly based on studies performed on cells of haematopoietic origin. Specifically, it has been shown that tetraspanins CD9, CD37/TSPAN26, CD53/TSPAN25, CD63/TSPAN30, CD81 and CD82 are abundant in various types of intracellular membranous organelles (see Table 5.1 for references). These include various endocytic organelles such as lysosomes and multivesicular bodies as well as different types of cytoplasmic granules. The trafficking routes and molecular signals that direct various tetraspanins to these compartments are largely unknown, as is the functional significance of such specific accumulation of tetraspanins in these compartments. Trafficking pathways have been most comprehensively studied for the tetraspanin CD63.

Distribution and trafficking of CD63. For almost two decades CD63 has been used as a marker for late endosomes and lysosomes (Pols and Klumperman 2009). The protein is also abundant on various specialized intracellular organelles including Weibel-Palade bodies in endothelial cells (Vischer and Wagner 1993), cytotoxic T lymphocytes (CTL) granules (Peters et al. 1991a), azurophil granules in neutrophils

Table 5.1 Distribution of tetraspanins on intracellular organelles

Tetraspanin	Localisation (cells)	Reference
CD9/TSPAN29	α-granules (platelets) MVB (dendritic cells)	Cramer et al. (1994), Brisson et al. (1997)
CD37/TSPAN26	MVB, multilamellae endosomes (B-cells) MVB, multilamellae endosomes (B-cells)	Escola et al. (1998)
CD53/TSPAN25	Dense granules (platelets), α-granules (platelets, megakaryocytes), cytotoxic T-cell granules (T-cells), azurophil granules (neutrophils), eosinophil secretory granules	Escola et al. (1998)
CD63/TSPAN30	MVB and multilamellae organelles (epithelial, endothelial cells, dendritic cells, B-cells, platelets, megakaryocytes), late endosomal tubules (epithelial cells), Weibel-Palade bodies (endothelial cells), lysosomes (dendritic cells) MVB, multilamellae endosomes (B-cells) MVB, multilamellae endosomes (B-cells, dendritic cells, macrophages)	Heijnen et al. (1998), Nishibori et al. (1993), Nieuwenhuis et al. (1987), Peters et al. (1991a, b), Kobayashi et al. (2000), van der Wel et al. (2003), Kobayashi et al. (2002), Escola et al. (1998)
CD81/TSPAN28		Escola et al. (1998)
CD82/TSPAN27		Escola et al. (1998), Garcia et al. (2005), Pelchen-Matthews et al. (2003), Hammond et al. (1998)

(Kuijpers et al. 1991) and platelet dense granules (Nishibori et al. 1993). It is widely accepted that the targeting of CD63 to late endocytic organelles and specialized granules is largely controlled by a classical tyrosine-based sorting motif (Y-X-X-Φ), which is found in the predicted C-terminal cytoplasmic domain of the protein (i.e. G-Y-E-V-M sequence). Indeed, the C-terminus of CD63 directly interacts with μ4, μ2, μ3A and μ3B subunits of AP adaptor complexes, which, in turn, interact with clathrin and thus connect CD63 to clathrin-dependent trafficking routes (Rous et al. 2002). The importance of the G-Y-E-V-M motif for late endocytic targeting is further emphasized by the observation that the mutation of tyrosine to alanine in this sequence completely abolishes trafficking of CD63 to lysosomes in rat fibroblasts (Rous et al. 2002).

It seems certain, however, that in addition to these conventional, clathrin-dependent trafficking pathways, CD63 may also use alternative routes/pathways for trafficking (Janvier and Bonifacino 2005). For example, in mouse fibroblasts localisation of CD63 to lysosomes was independent of the tyrosine-based sequence (Ryu et al. 2000). Furthermore, the presence of the G-Y-E-V-M sequence may not be sufficient for directing the protein to late endocytic organelles. This was demonstrated by using a CD9-CD63(GYEVM) chimeric protein which did not traffic to Cathepsin D–positive organelles in these cells (Ryu et al. 2000). Whilst AP-/clathrin-independent trafficking routes for CD63 remain largely unknown, recent data has identified two proteins, which may contribute to a diversification of the CD63 trafficking itinerary. Firstly, we established that the C-terminal cytoplasmic part of CD63 binds to syntenin-1, a PDZ domain containing cytoplasmic adaptor, which is known to regulate the trafficking of the associated transmembrane partners (Latysheva et al. 2006). Importantly, the sites of interaction with syntenin-1 and AP complexes in CD63 overlap and, therefore, binding will occur in a mutually exclusive manner. Thus, simplistically, syntenin-1 may function as a competitive inhibitor of the AP-/clathrin-dependent trafficking of CD63. In support of this notion we found that overexpression of syntenin-1 slows antibody-induced internalization of CD63 (Latysheva et al. 2006). This effect is accentuated in cells overexpressing the mutated variant of syntenin-1 that has a deletion of the N-terminal part of the protein. Thus, it appears that this region of syntenin-1 is critical for defining an alternative trafficking route for CD63. Whilst the relevant network of molecular interactions involving the N-terminus of syntenin-1 remains to be uncovered, it is noteworthy that the N-terminal 70 amino acids of syntenin-1 include three YPxL motifs. This sequence is also found in a number of viral proteins, including p9 of EIAV and p6 of HIV-1 (Strack et al. 2003; von Schwedler et al. 2003; Martin-Serrano et al. 2003) and has been shown to be essential for their interaction with Alix, a multi-domain cytoplasmic adaptor (Odorizzi 2006). Given that Alix itself forms complexes with ESCRT trafficking/sorting machinery (Martin-Serrano et al. 2003; von Schwedler et al. 2003), it is possible that the syntenin-1-Alix-ESCRT axis defines the alternative trafficking route for CD63. Finally, it has been reported that syntenin-1 interacts with a number of other proteins including the serine-threonine kinases Ulk1 and Ulk2 (Tomoda et al. 2004), small GTPases Rab5 (Tomoda et al. 2004) and Rab7 (Tomoda et al. 2004), and syntaxins (Ohno et al. 2004). Whilst the functional significance of each of syntenin-1—containing complexes remains to be examined, all these syntenin-1 partners are well-established regulators of vesicular movements in cells.

In more recent experiments we found that the surface expression of CD63 is increased in cells depleted of L6-Ag, an unrelated four transmembrane domain protein (Lekishvili et al. 2008). In spite a significant overlap in the steady-state distribution of L6-Ag and CD63, the kinetics and mechanisms controlling internalization of these proteins are clearly distinct: the relatively slow internalization of L6-Ag, which is regulated by protein ubiquitination, contrasts with the fast constitutive endocytosis of CD63. Thus, it would seem unlikely that L6-Ag influences the internalization of CD63. Rather, this observation suggests that L6-Ag regulates post-endocytic (or post-Golgi) trafficking of this tetraspanin.

Finally, it has been reported that in several cell types CD63 is localized to caveoli, the flask-shaped invaginations on the plasma membrane that are thought to control clathrin-independent endocytosis and transcytosis of membrane proteins (Pols and Klumperman 2009). Although this observation suggests that CD63 can be removed from the cell surface via caveoli, further experiments will be necessary to confirm the direct involvement of caveoli (or their molecular components) in endocytic trafficking of this tetraspanin.

Trafficking of CD82. The internalisation of tetraspanin CD82 has been studied by Xu and colleagues (Xu et al. 2009a). These authors found that in prostate cancer cell lines, the mAb-induced internalisation of this tetraspanin is a relatively slow process, and it is increased in cells treated with EGF. Most of the internalised CD82 is destined to late endocytic organelles (endosomes, lysosomes) and only a small proportion of the protein is found in the recycling compartments. Endocytosis of CD82 does not seem to involve clathrin or dynamin, nor does it require the activity of PI 3-kinase or Arf6, a small GTPase that is known to regulate one of the clathrin-independent endocytic and recycling pathways. On the other hand, whilst depletion of cholesterol did not affect steady-state levels of CD82 on the plasma membrane, it markedly inhibited the antibody-induced internalisation of the protein. Furthermore, expression of the dominant-negative mutant of cdc42 GTPase (but not Rac1 or RhoA) and disruption of the actin cytoskeleton also inhibited the uptake of anti-CD82 mAb. Finally, these authors reported that the mutation of a tyrosine based sorting sequence (Y-S-K-V) in the C-terminal cytoplasmic region did not affect internalisation of CD82. Our results using mammary epithelial cells have confirmed that the tyrosine-based sorting motif is not required for internalisation of CD82. However, we have shown that the mutation of tyrosine to alanine can prevent trafficking of the protein from sorting endosomes to lysosomes. We also found that deletion of the N-terminal cytoplasmic region of CD82 enhances antibody-induced internalisation and delivery of the tetraspanin to late endocytic compartments. Moreover, fraction of the internalised CD82 with deleted N-terminal was found in coated membrane invaginations and vesicles. These results suggest that the N-terminal part of the protein controls endocytic and recycling processes involving CD82. Thus, interactions of intracellular domains and, possibly, posttranslational modifications of CD82 play an essential role in controlling the protein's fate after internalisation.

Trafficking of CO-029/TSPAN8. CO-029 was initially described as a tumour-specific antigen (Szala et al. 1990). Subsequent studies have shown that this tetraspanin controls the growth of primary tumours by regulating tumour angiogenesis, through a process involving the production of exosomes (Nazarenko et al. 2010). Recently, endocytosis of CO-029 was investigated in cells treated with PMA, an activator of typical isoforms of the protein kinase C family (Rana et al. 2010). Using chimeric tetraspanin proteins (CO-029-CD9 and CO-029-CD151), it was shown that the N-terminal cytoplasmic domain plays an important role in PKC-dependent endocytic removal of CO-029 from the plasma membrane. Further examination of trafficking routes showed that CO-029/TSPAN8 co-localises with clathrin and intersectin-2, a clathrin-interacting adaptor protein (Rana et al. 2010). Accordingly, knock-down experiments have confirmed the role of intersectin-2 in endocytosis of this tetraspanin. Finally, in agreement with the proposed involvement of the clathrin-dependent pathway, internalisation of CO-029 was inhibited in cells pre-treated with Dynasore, a specific inhibitor of dynamins.

Trafficking of CD81. Like CO-029, sequence of CD81 does not reveal any obvious targeting/sorting signals. Therefore, it is likely that intracellular trafficking of this tetraspanin relies on its transmembrane partners. Initial studies have revealed that both cholesterol depletion and an increase in the amount of ceramide can lead to the down-regulation of surface levels of CD81 (Voisset et al. 2008; Kapadia et al. 2007). Whilst neither of these reports investigated the molecular mechanisms involved, the authors proposed that these lipids control the membrane compartmentalisation of CD81 and, specifically, their localisation to microdomains with various capacities for internalisation of their protein residents. More recently, Farquhar and colleagues studied internalisation pathways for CD81 following the binding of various anti-CD81 mAbs and incubation with soluble E2 (personal communication), an HCV-encoded glycoprotein which is critical for viral infection. Antibody-induced internalisation of CD81 was inhibited in cells expressing dominant-negative dynamin and Eps15 indicating the involvement of the clathrin-dependent pathway. Accordingly, viral infection was also blocked in cells that were pre-treated with an inhibitor of dynamin. Further analysis has shown that in addition to targeting to late endosomes and lysosomes, a significant proportion of internalised CD81 is localised to various recycling endosomes (i.e. Rab4 and Rab-11 positive compartments). These data suggest a complex trafficking itinerary for internalised CD81. Interestingly, the deletion of either N- or C-terminal cytoplasmic regions did not alter the internalisation rate of the tetraspanin. These observations led the authors to propose a model wherein antibody-induced endocytosis requires the interaction of CD81 with another transmembrane protein. In support of this model, the authors found that treatment with anti-CD81 mAb induced co-internalisation of CD81 with Claudin-1, an unrelated four transmembrane domain protein. With regard to this, it has been reported that the Claudin-1 interacts with and is ubiquitinated by the E3 ubiquitin ligase LNX1p80. It is, therefore, possible that the Claudin-1/CD81 complex traffics along ubiquitination/ESCRT-dependent endocytic routes (Takahashi et al. 2009).

Table 5.2 Sorting motifs in tetraspanins

A. Tyrosine-based motifs	
TSPAN1	TM4-*YCNL*Q
TSPAN3	TM4-CRRSRDPA*YELL*ITGGTYA
TSPAN4	TM4-TM*YCQV*VKADTYCA
TSPAN6	TM4-YCLSRAITNNQ*YEIV*
CD231/TSPAN7	TM4-CCLSRFITANQ*YEMV*
CO-029/TSPAN8	TM4-*YCQI*GNK
UP1A/TSPAN21[a]	TM4*(Y)-TML*
CD151/TSPAN24	TM4-CCL*YRSL*KLEHY
CD37/TSPAN26	TM4-LCRNLDHV*YNRL*ARYR
CD82/TSPAN27	TM4-LCRHVHSED*YSKV*PKY
CD63/TSPAN30	TM4-KSIRSG*YEVM*
NET-7/TSPAN15	MPRGDSEQVRYCARFS*YLWL*KF-TM1
TM4-B/TSPAN16	MAEIHTP*YSSL*KK-TM1
B. Dileucine-based motifs	
CD231/TSPAN7	MASRRM*ETKPVI*TCLKT-TM1
OCUL/TSPAN10	MEEG*ERSPLL*SQ-66-TM1
TSSC-6/TSPAN32	TM4-22-QPQ*EPSLL*RC-22
C. Acidic clusters	
RDS/TSPAN22	TM4-RYQTSLDGVSNP*EESESESE*GW-40

[a]Tyr[255] in UP1A/TSPAN21 is predicted (http://www.ch.embnet.org/software/TMPRED_form.html) to be the last amino acid in the TM4

Trafficking of CD151. The C-terminal cytoplasmic region of CD151 has a putative tyrosine-based sorting motif (Y-R-S-L), which may potentially recruit AP adaptor complexes and, thus, facilitate its trafficking via clathrin-dependent pathways. The importance of this sequence in the cellular distribution and trafficking of CD151 was analysed by Liu and colleagues (Liu et al. 2007). It was demonstrated that the mutation of critical Tyr and Leu to alanines (Yala mutant) leads to a moderate increase in the surface level of the protein and precludes its accumulation in EEA-1—and Rab5-positive early endosomes. Accordingly, antibody-induced internalisation of the Yala mutant was severely compromised. However, further analysis suggested that the internalisation (and, perhaps, cellular distribution) of CD151 is unlikely to involve AP complexes. Indeed, the authors found that CD151 is endocytosed via a dynamin-independent mechanism, which is obligatory for AP-/clathrin-dependent internalisation. In addition, it was found that CD151 internalisation requires the activity of phosphatidylinositol 3-kinase and an intact actin cytoskeleton, but is independent of Arf6.

Sorting/targeting motifs in other tetraspanins. Post-Golgi trafficking of transmembrane proteins is controlled by a number of short linear sequences—known as trafficking/sorting motifs (Braulke and Bonifacino 2009; Pandey 2009; Seaman 2008), some of which are embedded into the predicted cytoplasmic domains of tetraspanins. For example, in addition to being found in CD63, CD82 and CD151 (see above), the tyrosine-based motif (Y-X-X-Φ) is present in ten other tetraspanins (Table 5.2).

Table 5.3 Potential PDZ-binding motifs in tetraspanins

TSPAN2	TM4-CCAIRNSR***DVI***
TSPAN3[a]	TM4-LCRRSRDPAYELLITGG***TYA***
TSPAN6[a]	TM4-SRAITNNQY***EIV***
CD231/TSPAN7	TM4-SRFITANQY***EMV***
NET-2/TSPAN12	TM4-50-MANSFNTHFEM***EEL***
UP1B/TSPAN20	TM4-WSR***IEY***
UP1A/TSPAN21	TM4-MWTLPVMLIAMYFY***TML***
CD53/TSPAN25	TM4-NCQIDKTSQT***IGL***
CD81/TSPAN28[b]	TM4-CCGIRNS***SVY***
CD9/TSPAN29	TM4-CCAIRRNR***EMV***
CD63/TSPAN30[b]	TM4-VKSIRSGY***EVM***

[a]Interact with syntenin-1 (F. Berditchevski, unpublished data)
[b]Interaction with PDZ domain containing proteins has been reported (Latysheva et al. 2006; Pan et al. 2007)

Whilst this suggests possible binding of AP adaptors and the involvement of clathrin-dependent endocytosis, it is important to note that in more than half of Y-X-X-Φ—containing tetraspanins, critical tyrosine or bulky hydrophobic residues are located close to predicted membrane spanning regions (0–3 amino acids away, Table 5.2). This may interfere with productive interactions of the proteins with the μ subunits of AP adaptor complexes. Furthermore, tyrosine-based motifs in three of the tetraspanins (CO-029, TSPAN1 AND TSPAN4) also include cysteine, which, when palmitoylated (palmitoylation of tetraspanins is discussed in Chap. 3 of this volume), is likely to affect their binding to AP complexes.

Dileucine motifs in the N-terminal cytoplasmic domains of Oculospanin/TSPAN10 and CD231/A15/TSPAN7 and the C-terminal cytoplasmic region of TSSC6/TSPAN32 may also represent suitable targets for AP adaptors and clathrin-dependent trafficking. A cluster of glutamic acid residues with intermittent serines (a potential site for binding of PACS proteins, a family of multifunctional membrane traffic regulators (Youker et al. 2009)) and the SES motif (a sequence which is critical for ligand-independent internalisation of CXCR4 (Futahashi et al. 2007)) are found in the C-terminal cytoplasmic region of the tetraspanin RDS. A GxxY sequence (shown to be important for the internalisation of insulin receptor (Rajagopalan et al. 1991)) is present in the C-terminal cytoplasmic regions of TSPAN3, TSPAN9, TSPAN10 and TSPAN17 and in the N-terminus of TSPAN5. A SLL sequence, which plays a critical role in the internalisation of glucose transporter 4 (GLUT4) (Capilla et al. 2007), is found in the C-terminal cytoplasmic region of TSSC6. Furthermore, the C-terminal cytoplasmic regions of 11 tetraspanins have the potential capacity to bind PDZ domain containing proteins (Table 5.3), which may further diversify their trafficking routes. The precise functionality of all of these motifs/sequences remains currently unknown. Finally, it is important to note that predicted cytoplasmic regions of all tetraspanins have variable numbers of lysines and ubiquitination has been reported for CD82, CD151, CD81 (Lineberry et al. 2008; Tsai et al. 2007). Whilst the physiological signals that regulate tetraspanin ubiquitination have yet to be identified, ubiquitinated tetraspanins may certainly represent suitable cargos for ESCRT sorting/trafficking machinery (Raiborg and Stenmark 2009).

5.3 Tetraspanins and Transport of Associated Proteins to the Plasma Membrane

CD81-CD19-CD21-Leu13 complex. The tetraspanin CD81 has been described as a component of CD19-CD21-Leu13 complex in B-cells (Bradbury et al. 1992). In addition to its role in surface compartmentalisation and signalling function of the complex (Cherukuri et al. 2004), CD81 controls surface expression of CD19 (Maecker and Levy 1997; Miyazaki et al. 1997). Detailed analysis has revealed that CD81 regulates the transit of CD19 from the ER to the Golgi compartments (Shoham et al. 2003). The specific role of CD81 in trafficking of CD19 is further emphasised by the fact that CD9, a tetraspanin that shows the most similarity to CD81, could not rescue the CD19^{low} phenotype of the CD81-negative B-cells (Shoham et al. 2006). Similarly, CD81-deficiency in human patients results in the absence of CD19 from the surface of B lymphocytes (van Zelm et al. 2010). Importantly, the surface expression of both CD21 and Leu13/CD225 was not affected which stresses further the specificity of CD81-CD19 trafficking link. The "slow trafficking" phenotype can be rescued by the presence of CD81-TM1 in the CD81-CD9 chimera (Shoham et al. 2006). The authors in this study proposed a mechanism whereby CD81-TM1 facilitates the release of CD19 from its complex with calnexin. In this regard, TM1 of CD81 (but not CD9) contains the Gly-X-X-X-Gly motif, which has previously been shown to drive dimerization of membrane spanning proteins through transmembrane helix-helix interactions (Russ and Engelman 2000;Russ and Engelman 1999). Thus, dimerization of CD81 may be a critical factor responsible for the functional specificity of this tetraspanin towards CD19.

UPIA and trafficking of Uroplakin II. The tetraspanin UPIA directly interacts with uroplakin II/UPII, a single pass transmembrane protein (Sun et al. 1999). When expressed alone, these proteins cannot exit the ER and are rapidly directed to proteosomal-dependent degradation (Tu et al. 2002). Co-expression of UPIA and UPII results in the efficient transport of these proteins from the ER to the Golgi and in subsequent furin-dependent proteolytic processing of pro-UPII. Similarly, expression of tetraspanin UPIB rescues Golgi-directed trafficking of its molecular partner, uroplakin III (Tu et al. 2002). It has been subsequently suggested that specific UPIA-UPII and UPIB-UPIII interactions change conformations of both proteins within each of the pairs to allow their transit along the biosynthetic pathway (Hu et al. 2004, 2005).

TspanC8 tetraspanins and trafficking of ADAM10. A subgroup of tetraspanins which have eight Cysteines in their LEL (named as TspanC8 tetraspanins) include six proteins—Tspan5, Tspan10, Tspan14, Tspan15, Tspan17 and Tspan33. These proteins were recently shown to interact directly with ADAM10, a member of the ADAM superfamily of proteases which target a number transmembrane proteins (e.g. TNFα, FAS ligand) and play a critical role in various physiological processes (Prox et al. 2012, Rubinstein and Tomlinson, personal communications). Biochemical and immunofluorescence experiments have shown that Tspan5, Tspan14 and Tspan15 control trafficking of ADAM10 from the ER to the plasma membrane. Ectopic expression of Tspan33 also facilitated maturation of ADAM10 and resulted in higher surface levels of the protease. Conversely, the surface level of ADAM10 in the Tspan33-deficient erythrocytes was decreased by ~75 % as compared to the wild-type erythrocytes (M.Tomlinson, personal communication). It has been proposed that TspanC8 proteins regulate exit from the ER by shielding the arginine-rich ER-retention motif in the cytoplasmic tail of ADAM10 (Prox et al. 2012).

5.4 Tetraspanins and Internalisation of Associated Proteins

Integrins and tetraspanins. Tetraspanins form complexes with various integrins and regulate their ligand-binding and signalling properties (Berditchevski 2001; Hemler 2005). Recent data has shown that tetraspanins may also control trafficking of integrin complexes. An initial study by Winterwood and colleagues, has shown that down-regulation of CD151 expression slows the internalisation rate of α3β1 in squamous carcinoma cells plated on laminin-5, one of the main ligands for this integrin (Winterwood et al. 2006). In a subsequent study, it was found that CD151 is co-internalised with α3β1, α5β1 and α6β1 integrins in mouse fibroblasts (Liu et al. 2007). Importantly, the endocytic rate of integrins associated with the CD151 Yala mutant (see above) was reduced when compared to that associated with the wild-type CD151. The authors also found that expression of the CD151 Yala mutant did not affect endocytosis of CD9, another partner for CD151 in these cells.

Overexpression of the tetraspanin CD82 in prostate cancer cells results in elevated surface levels of α6β1 and α6β4 integrins (He et al. 2005). Whilst these CD82-dependent changes are likely to be cell-type specific (we observed no such correlation in breast epithelial or fibrosarcoma cells), the authors also found that CD82 potentiates ligand-induced internalisation of integrins (He et al. 2005). In dendritic cells, the surface expression of β1- and β2-integrins decreases when cells are incubated with anti-CD63 and anti-CD82 mAbs (Mantegazza et al. 2004). Importantly, surface levels of β3 integrins or the MHC class II complex, both of which are established partners for tetraspanins, are not affected. Consistent with the role of tetraspanins in endocytosis of integrins, diminished expression of CD151 or antibody-induced internalisation of CD63 have both been shown to affect integrin-mediated cell migration (Mantegazza et al. 2004; Winterwood et al. 2006).

Forced overexpression of CD81 (but not CD9) in retinal pigment epithelial cells results in a specific increase of surface levels of the αvβ5 integrin (Chang and Finnemann 2007). Biochemical experiments have shown that CD81 and αvβ5 are associated in these cells. Thus, it is likely that CD81 has a direct effect on integrin trafficking.

CD63 and trafficking of H^+-K^+ ATPases. Two reports have described the association of CD63 with two members of the H^+,K^+-ATPase family of heterodimeric proton pumps (Codina et al. 2005; Duffield et al. 2003). CD63 can be co-immunoprecipitated with the β subunit of the gastric H^+,K^+-ATPase (HKβ). Furthermore, elevated expression of CD63 enhances the internalisation of HKβ and its redistribution from the plasma membrane to intracellular vesicles (Duffield et al. 2003). Interestingly, CD63 predominantly interacts with the high-mannose species of HKβ, which are thought to accumulate in the ER and/or pre-Golgi compartments (Vagin et al. 2004). Thus, it is conceivable that CD63, which directly interacts with the μ subunits of adaptor protein (AP) complexes (see above), links H^+,K^+-ATPase to trafficking pathways that target the enzyme to late endocytic organelles for degradation (Duffield et al. 2003). Consistent with a linker function for CD63, the mutation of tyrosine to alanine in the sorting motif completely abolishes the activity of the tetraspanin towards HKβ. It has also been demonstrated that CD63 interacts with the α2 subunit of colonic H^+,K^+-ATPase and that depletion of the tetraspanin increases the surface expression of the ATPase (Codina et al. 2005). Taken together, these data support the idea that CD63 functions as an accessory modulator in the trafficking of associated proteins.

CD63 and trafficking of CXCR4. Ectopic overexpression of CD63 in different cell types has been shown to decrease the surface level of CXCR4, a G-protein coupled receptor for chemokine SDF-1 (Yoshida et al. 2009). Conversely, depletion of CD63 from Hela cells expressing CXCR4 led to an increase in the surface expression of the receptor. The suppressive effect of CD63 was even more pronounced when cells expressed a mutant with a deletion of the N-terminal cytoplasmic part of the protein (CD63ΔN). Further analysis revealed that CD63 can be co-immunoprecipitated with CXCR4 and that in CD63ΔN-expressing Hela cells, CXCR4 can be found predominantly in cis-Golgi and late endocytic organelles. The authors also demonstrated partial co-localisation of CXCR4 with calnexin and ERGIC-53, a mannose-specific membrane lectin operating as a cargo receptor for the transport of glycoproteins from the ER to the ER-Golgi intermediate compartment. Thus, it has been concluded that CD63 controls various aspects of intracellular trafficking of CXCR4 (Yoshida et al. 2009).

CD63 and surface expression of P-selectin. Depletion of CD63 from endothelial cells (EC) using siRNA resulted in the decreased steady-state surface levels of P-selectin, an adhesion receptor, which plays a critical role in the interaction of leukocytes with EC and their recruitment to the sites of inflammation (Doyle et al. 2011). Furthermore, PKC-induced increase in the surface expression of P-selectin was also suppressed in the CD63-depleted cells. This correlated with the decrease in clustering of P-selectin on the plasma membrane. Whilst the molecular mechanisms underlying CD63-dependent changes in surface presentation of P-selectin has not been investigated by the authors, they proposed that decreased clustering facilitated re-internalisation of the adhesion protein.

CD63 and trafficking of synaptotagmin VII. Synaptotagmin VII (Syt-VII) is a Ca^{2+} sensor that regulates lysosome exocytosis and plasma membrane repair

(Flannery et al. 2010). Under steady-state conditions Syt-VII is predominantly found in CD63/Lamp-I—positive late endocytic organelles (LEO). Flannery and colleagues have recently established that the trafficking of Syt-VII to LEO is dependent on palmitoylation of the protein, and that palmitoylated Syt-VII forms complexes with CD63 (Flannery et al. 2010). Furthermore, depletion of CD63 suppresses trafficking of Syt-VII (both the wild-type and palmitoylation-deficient mutant) from Golgi to LEO, thereby suggesting that the function of CD63 as a trafficking chaperone cannot be compensated by other tetraspanins.

CD82 and trafficking of EGF receptor. CD82 is associated with EGF receptor, and we have previously found that elevated expression of this tetraspanin facilitates ligand-induced internalisation of EGFR (Odintsova et al. 2000, 2003). Interestingly, under conditions of a saturating concentration of EGF, its binding to the receptor destabilizes the EGFR-CD82 interaction, suggesting that CD82 may indirectly influence endocytosis of EGFR. Indeed, subsequent experiments have shown that expression of CD82 leads to changes in the glycosphingolipid composition of the plasma membrane (Odintsova et al. 2003, 2006) and also affects the compartmentalisation of EGFR. Thus, it is conceivable that the CD82-induced relocalisation of the receptor within the plasma membrane places it in close proximity to the endocytic machinery and facilitates its endocytosis following ligand stimulation. In a more recent study, Danglot and colleagues have shown that depletion of CD82 from Hela cells resulted in a reduction of both surface and total levels of EGFR (Danglot et al. 2010). It was also found that ligand-induced internalization of EGFR was increased in CD82-depleted cells and that CD82 contributes to the dynamics of the ligand-bound receptor on the cell surface. These results led the authors to propose that in Hela cells, CD82 prevents the effective recruitment of activated EGFR to clathrin-coated pits and, consequently, its endocytosis. Thus, it appears that the role of CD82 in EGFR endocytosis is dependent on cellular context. More recently, we found that CD82 may affect the sorting of ligand-bound EGFR. CD82 overexpression in mammary epithelial cells diverts HB-EGF-bound EGFR (but not EGF-bound receptor) to the recycling route. Interestingly, deletion CD82ΔC mutant did not have this effect indicating that the C-terminal of CD82 may still be important for bridging the associated receptors to the endocytic machinery. Finally, it has recently been shown that internalised EGFR is delivered to a distinct population of multivesicular endosomes that are positive for CD63 (White et al. 2006) and CD82 (FB and EO, unpublished results). Taken together these results show that tetraspanins are also involved in post-endocytic trafficking of EGFR.

Tetraspanins and dectin-1. Two reports have described the association of tetraspanins with dectin-1, a lectin-like transmembrane receptor responsible for the internalisation of β-glucan-bearing ligands (Brown 2006). Initially, it was observed that CD63 is rapidly internalised when dendritic cells are induced to phagocytose yeast (Mantegazza et al. 2004). It was then found that CD63 is associated with dectin-1, a primary receptor for yeast particles on dendritic cells. This observation led the authors to suggest that CD63 may be involved in dectin-1—dependent uptake and processing of foreign antigens by dendritic cells. More recently, Meyer-Wentrup and colleagues found that dectin-1 also interacts with the tetraspanin CD37 (Meyer-Wentrup et al. 2007).

Notably, the expression level of dectin-1 on the surface of CD37-deficient macrophages was significantly decreased. This was due to a higher rate of endocytosis of the receptor. It has recently been established that endocytosis of dectin-1, which is known to localize to lipid rafts (Xu et al. 2009b), is controlled by the actin cytoskeleton and dynamin (Hernanz-Falcon et al. 2009). Given the established role of tetraspanins in the surface compartmentalization of their associated receptors, one can speculate that the effect of CD37 on endocytosis of dectin-1 is linked to the tetraspanin-dependent re-distribution of the receptor on the cell surface leading to changes in its internalisation itinerary.

CD231/Talla-1/TSPAN7 and trafficking of the AMPA receptor. TSPAN7 is abundantly expressed in neuronal cells, muscles and spleen (Takagi et al. 1995). Two mutations in *TSPAN7* are associated with X-linked mental retardation (MR) (Zemni et al. 2000). Furthermore, decreased expression of TSPAN7 expression is associated in a patient with balanced translocation [46, X, t(X;2)(p11.4; p21.3)], and 2 bp deletion in the LEL TSPAN7 detected in one of the X-linked MR families results in a frameshift mutation and premature termination of the protein. More recently, one of the X-linked MR mutations in the LEL TSPAN7 (P172H) was found in one patient diagnosed with autism spectrum disorder (Noor et al. 2009). At the cellular level TSPAN7 was shown to regulate excitatory synapse development and Bassani with colleagues established that TSPAN7 plays an important role in stabilization receptor for α-amino-3-hydroxy-5-methyl-4-isoxazolepropionic acid (AMPAR) on the surface on neurons (Bassani et al. 2012). Indeed, siRNA-induced depletion of TSPAN7 resulted in increased endocytosis of the GluA2 subunit of AMPAR. Further experiments have shown that like GluA2, TSPAN7 is associated with PICK1, a PDZ-domain containing adaptor protein which is known to control trafficking of various synaptic transmembrane proteins. The authors concluded that in neurons TSPAN7 competes with GluA2 for PICK1 binding and thus regulates the dynamic equilibrium between the endocytic and surface pools of AMPAR.

The tetraspanin Sunglasses and degradation of Rhodopsin in Drosophila. The tetraspanin Sunglasses (*Sun*/42E1) was identified in a genome-wide screen for eye-enriched genes in *Drosophila* (Xu et al. 2004). *Sun*-deficient flies exhibited light-dependent degeneration of photoreceptor cells due to a defect in the degradation of major rhodopsin (Rh1), a photoactivated G-protein—coupled receptor which is known to be endocytosed via the arrestin-mediated clathrin-dependent route. This initial study established that *Sun* is abundant in lysosomes, and it was proposed that the protein participates in the targeting of Rh1 for lysosomal degradation. Further analysis has revealed that *Sun* is also found on the plasma membrane, where it controls Gq-dependent endocytosis of Rh1 (Han et al. 2007). Although the molecular details of its regulatory role are currently unknown, the authors found that *Sun* can form complexes with Gq proteins and, therefore, may potentially influence the Gq-dependent signaling pathways that lead to receptor internalization. Given a recently established link between tetraspanins and arrestin in endocytosis (see above for CO-029), it is tempting to speculate that *Sun* may also be involved in the regulation of interactions within the arrestin-dependent endocytic network. Finally, one can't exclude the possibility that *Sun* influences the autophagic-dependent degradation of Rho1. The potential link between tetraspanins and autophagy has recently been suggested by a report describing autophagic cell death in multiple myeloma cells which overexpress tetraspanins CD82 and CD81 (Zismanov et al. 2009).

5.5 Tetraspanins, Exosomes and Release of Intracellular Granules

As mentioned above, many tetraspanins are abundant on various types of endocytic organelles and intracellular granules. The role played by tetraspanins in the biogenesis of these organelles and their intracellular transport remains largely unknown.

Tetraspanins, multivesicular bodies and exosomes. Multivesicular endosomes or multivesicular bodies (MVBs) can be broadly defined as endocytic organelles containing internal vesicles (Woodman and Futter 2008). Recent studies have revealed that MVBs serve as depots for the sorting of transmembrane proteins along various post-endocytic trafficking pathways (e.g. recycling or degradation). A number of reports have described abundant localisation of tetraspanins to the internal/intraluminal membranes of MVBs (Table 5.1), thereby raising the possibility that tetraspanins may be involved in the biogenesis of MVBs. The effect of tetraspanin removal on the morphology of endocytic organelles was analysed by Ruiz-Mateos and colleagues, who found that CD63 is not required for the maintenance of MVBs and lysosomes in monocyte-derived macrophages (Ruiz-Mateos et al. 2008). In agreement with these data, no obvious late endosomal/lysosomal abnormalities were observed for cells derived from CD63-knockout animals (Schroder et al. 2009). The role of other tetraspanins in the biogenesis of MVBs has not been reported.

Exosomes are 30–100 nm membranous particles with characteristic cup-shape morphology, which are thought to derive from the intraluminal vesicles of MVBs when they fuse with the plasma membrane (Simons and Raposo 2009). Whilst exosomal secretion was initially thought to represent an alternative pathway for the release of redundant (or overexpressed) proteins, it is now well established that exosomes can function as mediators of communication between various cell types. For example, exosomes derived from tumour cells can activate the endothelium and induce specific MHC-dependent T-cell responses. Tetraspanins are abundant on exosomes and two recent reports have shown that elevated expression of tetraspanin proteins can change the protein composition of these organelles. For example, overexpression of CO-029/TSPAN8 in a pancreatic cancer cell line facilitates exosomal recruitment of other tetraspanins and integrins (Gesierich et al. 2006; Nazarenko et al. 2010). More recent experiments have shown that CD82 and CD9 also influence the composition of exosomes by facilitating exosomal recruitment of β-catenin, an important signalling component of the Wnt- and E-cadherin signalling pathways (Chairoungdua et al. 2010). Interaction with E-cadherin appears to be particularly critical for tetraspanin-dependent redistribution of β-catenin in cells. This study also demonstrated that dendritic cells (DC) from CD9 knockout animals produced less exosomes when compared to wild-type DC (Chairoungdua et al. 2010). Although, this phenomenon has not been investigated further, the tetraspanin-dependent regulation of exosome secretion by DC may represent a previously unsuspected mode of action of tetraspanins in antigen presentation.

The role of CD63 and biogenesis of melanosomes. In pigmented melanoma cell line CD63 is associated and partially colocalised with PMEL, amyloidogenic pigment cell-specific type I integral membrane protein, on the intraluminal vesicles (ILVs) in pre-melanosomes (van Niel et al. 2011). Depletion of CD63 from the cells significantly decreased the number of ILVs in pre-melanosomes and, as a consequence, decreased recruitment of PMEL to these organelles. These authors also established that depletion of CD63 has a more general effect on endocytic organelles in these cells: the number of stage II premelanosomes and stage III pigmented melanosomes decreased, while pigmented round melanosomal structures with unstructured melanin deposits accumulated. There was also 4-fold increase in the number of compartments with lysosomal morphology. Importantly, the number of melanosomes in retinal pigment epithelium of CD63 knockout animals was also significantly decreased. Thus, CD63 appears to play a critical role in the biogenesis of these organelles. However, it is likely that the effect of CD63 on biogenesis of endocytic organelles is cell type/tissue specific. Indeed, detailed analysis of endocytic compartments in other cells from CD63 knockout animals did not reveal any obvious abnormalities (Schroder et al. 2009). Furthermore, siRNA-based depletion of CD63 from human macrophages did not affect the morphology of multivesicular compartments (Ruiz-Mateos et al. 2008).

Tetraspanins and intracellular granules. Neutrophil elastase (NE) is a proteolytic enzyme which plays an important role in innate immunity (Heutinck et al. 2010). It is stored as an active enzyme within the primary granules of neutrophils. These granules are released when cells reach the extravascular space to support the antimicrobial host response. In promyelocytic cells NE is synthesised as an inactive pro-NE, which associates with CD63 and is then targeted to secretory granules via a CD63-dependent pathway (Kallquist et al. 2008). Depletion of wild-type CD63 or overexpression of a CD63 variant with a mutation in the C-terminal sorting motif of the protein (i.e. YEVM → AEVM, see above) decrease the number of dense granules in cells. These observations strongly suggest that CD63 plays a critical role in the biogenesis of NE-positive granules.

It was demonstrated that alveolar macrophages isolated from $CD81^{-/-}CD9^{-/-}$ mice develop many lysosomes and vacuoles. Some of these cells contain intracytoplasmic needle-shaped inclusions which appear to be phagocytosed collagen. These structures are not observed in macrophages isolated from wild-type or single knockout animals. Additionally, $CD81^{-/-}CD9^{-/-}$ type II epithelial cells have been shown to contain many lamellar bodies (Takeda et al. 2008). Not only do these results suggest a certain degree of redundancy in the trafficking functions of CD9 and CD81 in these cells, but they also point to a previously unsuspected function of these tetraspanins in the biogenesis of post-endocytic organelles.

5.6 Concluding Remarks and Future Directions

The future will undoubtedly provide further evidence for the various roles of tetraspanins in trafficking of transmembrane proteins. The identification of specific targets and critical trafficking steps affected by tetraspanins in various cell types will be important areas for forthcoming investigations. Perhaps the main question that remains is how tetraspanins regulate trafficking of their respective targets/cargos. Is this intricately linked to the trafficking of tetraspanins themselves? In this regard, functionality of classical sorting motifs, found in several tetraspanins, has, thus far, only been established for CD63 and, to a certain extent, CD82. However, even in these cases, there is the possibility of diversion from the predicted AP-dependent trafficking routes. Not all tetraspanins that are routinely found in late endocytic organelles or exosomes possess recognizable sorting motifs. What are the signals that control the trafficking of these tetraspanins? Does ubiquitylation play a role in their sorting to various intracellular compartments? Is there a role for palmitoylation?

Finally, the propensity of tetraspanins to form complex networks of interacting molecular aggregates ("tetraspanin webs"; TERM) adds an additional level of complexity to the issue of tetraspanin-dependent trafficking. Indeed, there seem to be intricate links between various tetraspanins so that modulation in the expression level of one affects the distribution of the others (Barreiro et al. 2005). Looking to the future, it seems certain that new integrated approaches (e.g. combinatorial knock-outs and knock-downs in conjunction with the detailed proteomic analysis of the depleted TERM; simultaneous real-time imaging of various tetraspanins and their partner receptors) will pave the way to answering all the current "How?"s, "What?"s and "Do?"s.

Acknowledgements We would like to thank Dr. M. Zöller, Dr. E. Rubinstein, Dr. M.Tomlinson and Dr. M. Farquhar for shering their results prior to publication and A. Berditchevskaia for editorial help. We are grateful to all members of our laboratories for their help during preparation of this article. The work in the authors' laboratory described in this article was supported by Cancer Research UK grants C1322/A2945 and C1322/A5705 and Breast Cancer Campaign grant 2003:712.

References

Annaert WG, Becker B, Kistner U, Reth M, Jahn R (1997) Export of cellubrevin from the endoplasmic reticulum is controlled by BAP31. J Cell Biol 139:1397–1410

Barreiro O, Yanez-Mo M, Sala-Valdes M, Gutierrez-Lopez MD, Ovalle S, Higginbottom A, Monk PN, Cabanas C, Sanchez-Madrid F (2005) Endothelial tetraspanin microdomains regulate leukocyte firm adhesion during extravasation. Blood 105:2852–2861

Bassani S, Cingolani LA, Valnegri P, Folci A, Zapata J, Gianfelice A, Sala C, Goda Y, Passafaro M (2012) The X-linked intellectual disability protein TSPAN7 regulates excitatory synapse development and AMPAR trafficking. Neuron 73:1143–1158

Berditchevski F (2001) Complexes of tetraspanins with integrins: more than meets the eye. J Cell Sci 115:4143–4151

Berditchevski F, Gilbert E, Griffiths MR, Fitter S, Ashman LK, Jenner SJ (2001) Analysis of the CD151-alpha3beta1 integrin and CD151-tetraspanin interactions by mutagenesis. J Biol Chem 276:41165–41174

Bradbury LE, Kansas GS, Levy S, Evans RL, Tedder TF (1992) The CD19/CD21 signal transducing complex of human B lymphocytes includes the target of antiproliferative antibody-1 and Leu-13 molecules. J Immunol 149:2841–2850

Braulke T, Bonifacino JS (2009) Sorting of lysosomal proteins. Biochim Biophys Acta 1793:605–614

Brisson C, Azorsa DO, Jennings LK, Moog S, Cazenave JP, Lanza F (1997) Co-localization of CD9 and GPIIb-IIIa (alpha IIb beta 3 integrin) on activated platelet pseudopods and alpha-granule membranes. Histochem J 29:153–165

Brown GD (2006) Dectin-1: a signalling non-TLR pattern-recognition receptor. Nat Rev Immunol 6:33–43

Cannon KS, Cresswell P (2001) Quality control of transmembrane domain assembly in the tetraspanin CD82. EMBO J 20:2443–2453

Capilla E, Suzuki N, Pessin JE, Hou JC (2007) The glucose transporter 4 FQQI motif is necessary for Akt substrate of 160-kilodalton-dependent plasma membrane translocation but not Golgi-localized (gamma)-ear-containing Arf-binding protein-dependent entry into the insulin-responsive storage compartment. Mol Endocrinol 21:3087–3099

Chairoungdua A, Smith DL, Pochard P, Hull M, Caplan MJ (2010) Exosome release of beta-catenin: a novel mechanism that antagonizes Wnt signaling. J Cell Biol 190:1079–1091

Chang Y, Finnemann SC (2007) Tetraspanin CD81 is required for the alpha v beta5-integrin-dependent particle-binding step of RPE phagocytosis. J Cell Sci 120:3053–3063

Cherukuri A, Shoham T, Sohn HW, Levy S, Brooks S, Carter R, Pierce SK (2004) The tetraspanin CD81 is necessary for partitioning of coligated CD19/CD21-B cell antigen receptor complexes into signaling-active lipid rafts. J Immunol 172:370–380

Codina J, Li J, Dubose TD Jr (2005) CD63 interacts with the carboxy terminus of the colonic H + -K + -ATPase to increase plasma membrane localization and 86Rb + uptake. Am J Physiol Cell Physiol 288:C1279–C1286

Cramer EM, Berger G, Berndt M (1994) Platelet a-granule and plasma membrane share two new components: CD9 and PECAM-1. Blood 84:1722–1730

Danglot L, Chaineau M, Dahan M, Gendron MC, Boggetto N, Perez F, Galli T (2010) Role of TI-VAMP and CD82 in EGFR cell-surface dynamics and signaling. J Cell Sci 123:723–735

Doyle EL, Ridger V, Ferraro F, Turmaine M, Saftig P, Cutler DF (2011) CD63 is an essential cofactor to leukocyte recruitment by endothelial P-selectin. Blood 118:4265–4273

Duffield A, Kamsteeg EJ, Brown AN, Pagel P, Caplan MJ (2003) The tetraspanin CD63 enhances the internalization of the H, K-ATPase beta-subunit. Proc Natl Acad Sci USA 100:15560–15565

Escola JM, Kleijmeer MJ, Stoorvogel W, Griffith J, Yoshie O, Geuze HJ (1998) Selective enrichment of tetraspan proteins on the internal vesicles of multivesicular endosomes and exosomes secreted by human B-lymphocytes. J Biol Chem 273:20121–20127

Flannery AR, Czibener C, Andrews NW (2010) Palmitoylation-dependent association with CD63 targets the Ca^{2+} sensor synaptotagmin VII to lysosomes. J Cell Biol 191:599–613

Futahashi Y, Komano J, Urano E, Aoki T, Hamatake M, Miyauchi K, Yoshida T, Koyanagi Y, Matsuda Z, Yamamoto N (2007) Separate elements are required for ligand-dependent and -independent internalization of metastatic potentiator CXCR4. Cancer Sci 98:373–379

Garcia E, Pion M, Pelchen-Matthews A, Collinson L, Arrighi JF, Blot G, Leuba F, Escola JM, Demaurex N, Marsh M, Piguet V (2005) HIV-1 trafficking to the dendritic cell-T-cell infectious synapse uses a pathway of tetraspanin sorting to the immunological synapse. Traffic 6:488–501

Gesierich S, Berezovskiy I, Ryschich E, Zoller M (2006) Systemic induction of the angiogenesis switch by the tetraspanin D6.1A/CO-029. Cancer Res 66:7083–7094

Hammond C, Denzin LK, Pan M, Griffith JM, Geuze HJ, Cresswell P (1998) The tetraspan protein CD82 is a resident of MHC class II compartments where it associates with HLA-DR, -DM, and -DO molecules. J Immunol 161:3282–3291

Han J, Reddig K, Li HS (2007) Prolonged G(q) activity triggers fly rhodopsin endocytosis and degradation, and reduces photoreceptor sensitivity. EMBO J 26:4966–4973

He B, Liu L, Cook GA, Grgurevich S, Jennings LK, Zhang XA (2005) Tetraspanin CD82 attenuates cellular morphogenesis through down-regulating integrin alpha6-mediated cell adhesion. J Biol Chem 280:3346–3354

Heijnen HF, Debili N, Vainchencker W, Breton-Gorius J, Geuze HJ, Sixma JJ (1998) Multivesicular bodies are an intermediate stage in the formation of platelet alpha-granules. Blood 91: 2313–2325

Hemler ME (2005) Tetraspanin functions and associated microdomains. Nat Rev Mol Cell Biol 6:801–811

Hernanz-Falcon P, Joffre O, Williams DL, Reis e Sousa (2009) Internalization of dectin-1 terminates induction of inflammatory responses. Eur J Immunol 39:507–513

Heutinck KM, ten Berge IJ, Hack CE, Hamann J, Rowshani AT (2010) Serine proteases of the human immune system in health and disease. Mol Immunol 47:1943–1955

Hu CA, Liang F, Zhou G, Tu L, Kreibich G, Sun T (2004) Tetraspanin uroplakins Ia and Ib interact and stabilize their partner uroplakins II and III. Mol Biol Cell 15:311A–312A

Hu CCA, Liang FX, Zhou G, Tu LY, Tang CHA, Zhou J, Kreibich G, Sun TT (2005) Assembly of urothelial plaques: tetraspanin function in membrane protein trafficking. Mol Biol Cell 16:3937–3950

Janvier K, Bonifacino JS (2005) Role of the endocytic machinery in the sorting of lysosome-associated membrane proteins. Mol Biol Cell 16:4231–4242

Kallquist L, Hansson M, Persson AM, Janssen H, Calafat J, Tapper H, Olsson I (2008) The tetraspanin CD63 is involved in granule targeting of neutrophil elastase. Blood 112:3444–3454

Kapadia SB, Barth H, Baumert T, McKeating JA, Chisari FV (2007) Initiation of hepatitis C virus infection is dependent on cholesterol and cooperativity between CD81 and scavenger receptor B type I. J Virol 81:374–383

Kobayashi T, Vischer UM, Rosnoblet C, Lebrand C, Lindsay M, Parton RG, Kruithof EK, Gruenberg J (2000) The tetraspanin CD63/lamp3 cycles between endocytic and secretory compartments in human endothelial cells. Mol Biol Cell 11:1829–1843

Kobayashi T, Beuchat MH, Chevallier J, Makino A, Mayran N, Escola JM, Lebrand C, Cosson P, Kobayashi T, Gruenberg J (2002) Separation and characterization of late endosomal membrane domains. J Biol Chem 277:32157–32164

Kuijpers TW, Tool AT, van der Schoot CE, Ginsel LA, Onderwater JJ, Roos D, Verhoeven AJ (1991) Membrane surface antigen expression on neutrophils: a reappraisal of the use of surface markers for neutrophil activation. Blood 78:1105–1111

Lambert G, Becker B, Schreiber R, Boucherot A, Reth M, Kunzelmann K (2001) Control of cystic fibrosis transmembrane conductance regulator expression by BAP31. J Biol Chem 276:20340–20345

Latysheva N, Muratov G, Rajesh S, Padgett M, Hotchin NA, Overduin M, Berditchevski F (2006) Syntenin-1 is a new component of tetraspanin-enriched microdomains: mechanisms and consequences of the interaction of syntenin-1 with CD63. Mol Cell Biol 26:7707–7718

Lekishvili T, Fromm E, Mujoomdar M, Berditchevski F (2008) The tumour-associated antigen L6 (L6-Ag) is recruited to the tetraspanin-enriched microdomains: implication for tumour cell motility. J Cell Sci 121:685–694

Levy S, Shoham T (2005) The tetraspanin web modulates immune-signalling complexes. Nat Rev Immunol 5:136–148

Lineberry N, Su L, Soares L, Fathman CG (2008) The single subunit transmembrane E3 ligase gene related to anergy in lymphocytes (GRAIL) captures and then ubiquitinates transmembrane proteins across the cell membrane. J Biol Chem 283:28497–28505

Liu L, He B, Liu WM, Zhou D, Cox JV, Zhang XA (2007) Tetraspanin CD151 promotes cell migration by regulating integrin trafficking. J Biol Chem 282:31631–31642

Maecker HT, Levy S (1997) Normal lymphocyte development but delayed humoral immune response in CD81-null mice. J Exp Med 185:1505–1510

Mantegazza AR, Barrio MM, Moutel S, Bover L, Weck M, Brossart P, Teillaud JL, Mordoh J (2004) CD63 tetraspanin slows down cell migration and translocates to the endosomal-lysosomal-MIICs route after extracellular stimuli in human immature dendritic cells. Blood 104:1183–1190

Martin-Serrano J, Yarovoy A, Perez-Caballero D, Bieniasz PD (2003) Divergent retroviral late-budding domains recruit vacuolar protein sorting factors by using alternative adaptor proteins. Proc Natl Acad Sci USA 100:12414–12419

Meyer-Wentrup F, Figdor CG, Ansems M, Brossart P, Wright MD, Adema GJ, van Spriel AB (2007) Dectin-1 interaction with tetraspanin CD37 inhibits IL-6 production. J Immunol 178:154–162

Min G, Wang H, Sun TT, Kong XP (2006) Structural basis for tetraspanin functions as revealed by the cryo-EM structure of uroplakin complexes at 6-a resolution. J Cell Biol 173:975–983

Miyazaki T, Muller U, Campbell KS (1997) Normal development but differentially altered proliferative responses of lymphocytes in mice lacking CD81. EMBO J 16:4217–4225

Nazarenko I, Rana S, Baumann A, McAlear J, Hellwig A, Trendelenburg M, Lochnit G, Preissner KT, Zoller M (2010) Cell surface tetraspanin Tspan8 contributes to molecular pathways of exosome-induced endothelial cell activation. Cancer Res 70:1668–1678

Nieuwenhuis HK, van Oosterhout JJG, Rozemuller E, van Iwaarden F, Sixma JJ (1987) Studies with a monoclonal antibody against activated platelets: evidence that a secreted 53,000-molecular weight lysosome-like granule protein is exposed on the surface of activated platelets in the circulation. Blood 70:838–845

Nishibori M, Cham B, McNicol A, Shalev A, Jain N, Gerrard JM (1993) The protein CD63 is in platelet dense granules, is deficient in a patient with hermansky-pudlak syndrome, and appears identical to granulophysin. J Clin Invest 91:1775–1782

Noor A, Gianakopoulos PJ, Fernandez B, Marshall CR, Szatmari P, Roberts W, Scherer SW, Vincent JB (2009) Copy number variation analysis and sequencing of the X-linked mental retardation gene TSPAN7/TM4SF2 in patients with autism spectrum disorder. Psychiatr Genet 19:154–155

Odintsova E, Sugiura T, Berditchevski F (2000) Attenuation of EGF receptor signaling by a metastasis suppressor tetraspanin KAI-1/CD82. Curr Biol 10:1009–1012

Odintsova E, Voortman J, Gilbert E, Berditchevski F (2003) Tetraspanin CD82 regulates compartmentalisation and ligand-induced dimerization of EGFR. J Cell Sci 116:4557–4566

Odintsova E, Butters TD, Monti E, Sprong H, van Meer G, Berditchevski F (2006) Gangliosides play an important role in the organisation of CD82-enriched microdomains. Biochem J 400:315–325

Odorizzi G (2006) The multiple personalities of alix. J Cell Sci 119:3025–3032

Ohno K, Koroll M, El Far O, Scholze P, Gomeza J, Betz H (2004) The neuronal glycine transporter 2 interacts with the PDZ domain protein syntenin-1. Mol Cell Neurosci 26:518–529

Pan Y, Brown C, Wang X, Geisert EE (2007) The developmental regulation of CD81 in the rat retina. Mol Vis 13:181–189

Pandey KN (2009) Functional roles of short sequence motifs in the endocytosis of membrane receptors. Front Biosci 14:5339–5360

Paquet ME, Cohen-Doyle M, Shore GC, Williams DB (2004) Bap29/31 influences the intracellular traffic of MHC class I molecules. J Immunol 172:7548–7555

Pelchen-Matthews A, Kramer B, Marsh M (2003) Infectious HIV-1 assembles in late endosomes in primary macrophages. J Cell Biol 162:443–455

Peters PJ, Borst J, Oorschot V, Fukuda M, Krahenbuhl O, Tschopp J, Slot JW, Geuze HJ (1991a) Cytotoxic T lymphocyte granules are secretory lysosomes, comtaining both perforin and granzymes. J Exp Med 173:1099–1109

Peters PJ, Neefjes JJ, Oorschot V, Ploegh HL, Geuze HJ (1991b) Segregation of MHC class II molecules from MHC class I molecules in the Golgi complex for transport to lysosomal compartments. Nature 349:669–676

Pols MS, Klumperman J (2009) Trafficking and function of the tetraspanin CD63. Exp Cell Res 315:1584–1592

Prox J, Willenbrock M, Weber S, Lehmann T, Schmidt-Arras D, Schwanbeck R, Saftig P, Schwake M (2012) Tetraspanin15 regulates cellular trafficking and activity of the ectodomain sheddase ADAM10. Cell Mol Life Sci 69(17):2919–2932
Raiborg C, Stenmark H (2009) The ESCRT machinery in endosomal sorting of ubiquitylated membrane proteins. Nature 458:445–452
Rajagopalan M, Neidigh JL, McClain DA (1991) Amino acid sequences Gly-Pro-Leu-Tyr and Asn-Pro-Glu-Tyr in the submembranous domain of the insulin receptor are required for normal endocytosis. J Biol Chem 266:23068–23073
Rana S, Claas C, Kretz CC, Zoeller M (2010) Activation-induced internalization differs for the tetraspanins CD9 and Tspan8: impact on tumor cell motility. Int J Biochem Cell Biol 43(1):106–119
Rous BA, Reaves BJ, Ihrke G, Briggs JA, Gray SR, Stephens DJ, Banting G, Luzio JP (2002) Role of adaptor complex AP-3 in targeting wild-type and mutated CD63 to lysosomes. Mol Biol Cell 13:1071–1082
Rubinstein E, Poindessous-Jazat V, Le Naour F, Billard M, Boucheix C (1997) CD9, but not other tetraspans, associates with the beta1 integrin precursor. Eur J Immunol 27:1919–1927
Rubinstein E, Ziyyat A, Wolf JP, Le Naour F, Boucheix C (2006) The molecular players of sperm-egg fusion in mammals. Semin Cell Dev Biol 17:254–263
Ruiz-Mateos E, Pelchen-Matthews A, Deneka M, Marsh M (2008) CD63 is not required for production of infectious human immunodeficiency virus type 1 in human macrophages. J Virol 82:4751–4761
Russ WP, Engelman DM (1999) TOXCAT: a measure of transmembrane helix association in a biological membrane. Proc Natl Acad Sci USA 96:863–868
Russ WP, Engelman DM (2000) The GxxxG motif: a framework for transmembrane helix-helix association. J Mol Biol 296:911–919
Ryu F, Takahashi T, Nakamura K, Takahashi Y, Kobayashi T, Shida S, Kameyama T, Mekada E (2000) Domain analysis of the tetraspanins: studies of CD9/CD63 chimeric molecules on subcellular localization and upregulation activity for diphtheria toxin binding. Cell Struct Funct 25:317–327
Schamel WW, Kuppig S, Becker B, Gimborn K, Hauri HP, Reth M (2003) A high-molecular-weight complex of membrane proteins BAP29/BAP31 is involved in the retention of membrane-bound IgD in the endoplasmic reticulum. Proc Natl Acad Sci USA 100:9861–9866
Schroder J, Lullmann-Rauch R, Himmerkus N, Pleines I, Nieswandt B, Orinska Z, Koch-Nolte F, Schroder B, Bleich M, Saftig P (2009) Deficiency of the tetraspanin CD63 associated with kidney pathology but normal lysosomal function. Mol Cell Biol 29:1083–1094
Seaman MN (2008) Endosome protein sorting: motifs and machinery. Cell Mol Life Sci 65:2842–2858
Shoham T, Rajapaksa R, Boucheix C, Rubinstein E, Poe JC, Tedder TF, Levy S (2003) The tetraspanin CD81 regulates the expression of CD19 during B cell development in a postendoplasmic reticulum compartment. J Immunol 171:4062–4072
Shoham T, Rajapaksa R, Kuo CC, Haimovich J, Levy S (2006) Building of the tetraspanin web: distinct structural domains of CD81 function in different cellular compartments. Mol Cell Biol 26:1373–1385
Simons M, Raposo G (2009) Exosomes—vesicular carriers for intercellular communication. Curr Opin Cell Biol 21:575–581
Stojanovic M, Germain M, Nguyen M, Shore GC (2005) BAP31 and its caspase cleavage product regulate cell surface expression of tetraspanins and integrin-mediated cell survival. J Biol Chem 280:30018–30024
Strack B, Calistri A, Craig S, Popova E, Gottlinger HG (2003) AIP1/ALIX is a binding partner for HIV-1 p6 and EIAV p9 functioning in virus budding. Cell 114:689–699
Sun TT, Liang FX, Wu XR (1999) Uroplakins as markers of urothelial differentiation. Adv Exp Med Biol 462:7–18
Szala S, Kasai Y, Steplewski Z, Rodeck U, Koprowski H, Linnenbach AJ (1990) Molecular cloning of cDNA for the human tumor-associated antigen CO-029 and identification of related transmembrane antigens. Proc Natl Acad Sci USA 87:6833–6837

Takagi S, Fujikawa K, Imai T, Fukuhara N, Fukudome K, Minegishi M, Tsuchiya S, Konno T, Hinuma Y, Yoshie O (1995) Identification of highly specific surface marker of T-cell acute lymphoblastic leukemia and neuroblastoma as a new member of the transmembrane four superfamily. Int J Cancer 61:706–715
Takahashi S, Iwamoto N, Sasaki H, Ohashi M, Oda Y, Tsukita S, Furuse M (2009) The E3 ubiquitin ligase LNX1p80 promotes the removal of claudins from tight junctions in MDCK cells. J Cell Sci 122:985–994
Takeda Y, He P, Tachibana I, Zhou B, Miyado K, Kaneko H, Suzuki M, Minami S, Iwasaki T, Goya S, Kijima T, Kumagai T, Yoshida M, Osaki T, Komori T, Mekada E, Kawase I (2008) Double deficiency of tetraspanins CD9 and CD81 alters cell motility and protease production of macrophages and causes chronic obstructive pulmonary disease-like phenotype in mice. J Biol Chem 283:26089–26097
Tomoda T, Kim JH, Zhan C, Hatten ME (2004) Role of Unc51.1 and its binding partners in CNS axon outgrowth. Genes Dev 18:541–558
Toyo-oka K, Yashiro-Ohtani Y, Park CS, Tai XG, Miyake K, Hamaoka T, Fujiwara H (1999) Association of a tetraspanin CD9 with CD5 on the T cell surface: role of particular transmembrane domains in the association. Int Immunol 11:2043–2052
Tsai YC, Mendoza A, Mariano JM, Zhou M, Kostova Z, Chen B, Veenstra T, Hewitt SM, Helman LJ, Khanna C, Weissman AM (2007) The ubiquitin ligase gp78 promotes sarcoma metastasis by targeting KAI1 for degradation. Nat Med 13:1504–1509
Tu L, Sun TT, Kreibich G (2002) Specific heterodimer formation is a prerequisite for uroplakins to exit from the endoplasmic reticulum. Mol Biol Cell 13:4221–4230
Tu L, Kong XP, Sun TT, Kreibich G (2006) Integrity of all four transmembrane domains of the tetraspanin uroplakin Ib is required for its exit from the ER. J Cell Sci 119:5077–5086
Vagin O, Turdikulova S, Sachs G (2004) The H, K-ATPase beta subunit as a model to study the role of N-glycosylation in membrane trafficking and apical sorting. J Biol Chem 279:39026–39034
van der Wel NN, Sugita M, Fluitsma DM, Cao X, Schreibelt G, Brenner MB, Peters PJ (2003) CD1 and major histocompatibility complex II molecules follow a different course during dendritic cell maturation. Mol Biol Cell 14:3378–3388
van Niel G, Charrin S, Simoes S, Romao M, Rochin L, Saftig P, Marks MS, Rubinstein E, Raposo G (2011) The tetraspanin CD63 regulates ESCRT-independent and -dependent endosomal sorting during melanogenesis. Dev Cell 21:708–721
van Zelm MC, Smet J, Adams B, Mascart F, Schandene L, Janssen F, Ferster A, Kuo CC, Levy S, van Dongen JJ, van der Burg M (2010) CD81 gene defect in humans disrupts CD19 complex formation and leads to antibody deficiency. J Clin Invest 120:1265–1274
Vischer UM, Wagner DD (1993) CD63 is a component of Weibel-Palade bodies of human endothelial cells. Blood 82:1184–1191
Voisset C, Lavie M, Helle F, De Op BA, Bilheu A, Bertrand-Michel J, Terce F, Cocquerel L, Wychowski C, Vu-Dac N, Dubuisson J (2008) Ceramide enrichment of the plasma membrane induces CD81 internalization and inhibits hepatitis C virus entry. Cell Microbiol 10:606–617
von Schwedler UK, Stuchell M, Muller B, Ward DM, Chung HY, Morita E, Wang HE, Davis T, He GP, Cimbora DM, Scott A, Krausslich HG, Kaplan J, Morham SG, Sundquist WI (2003) The protein network of HIV budding. Cell 114:701–713
White IJ, Bailey LM, Aghakhani MR, Moss SE, Futter CE (2006) EGF stimulates annexin 1-dependent inward vesiculation in a multivesicular endosome subpopulation. EMBO J 25:1–12
Winterwood NE, Varzavand A, Meland MN, Ashman LK, Stipp CS (2006) A critical role for tetraspanin CD151 in alpha3beta1 and alpha6beta4 integrin-dependent tumor cell functions on laminin-5. Mol Biol Cell 17:2707–2721
Woodman PG, Futter CE (2008) Multivesicular bodies: co-ordinated progression to maturity. Curr Opin Cell Biol 20:408–414
Xu H, Lee SJ, Suzuki E, Dugan KD, Stoddard A, Li HS, Chodosh LA, Montell C (2004) A lysosomal tetraspanin associated with retinal degeneration identified via a genome-wide screen. EMBO J 23:811–822

Xu C, Zhang YH, Thangavel M, Richardson MM, Liu L, Zhou B, Zheng Y, Ostrom RS, Zhang XA (2009a) CD82 endocytosis and cholesterol-dependent reorganization of tetraspanin webs and lipid rafts. FASEB J 23:3273–3288

Xu S, Huo J, Gunawan M, Su IH, Lam KP (2009b) Activated dectin-1 localizes to lipid raft microdomains for signaling and activation of phagocytosis and cytokine production in dendritic cells. J Biol Chem 284:22005–22011

Yoshida T, Ebina H, Koyanagi Y (2009) N-linked glycan-dependent interaction of CD63 with CXCR4 at the Golgi apparatus induces downregulation of CXCR4. Microbiol Immunol 53:629–635

Youker RT, Shinde U, Day R, Thomas G (2009) At the crossroads of homoeostasis and disease: Roles of the PACS proteins in membrane traffic and apoptosis. Biochem J 421:1–15

Zemni R, Bienvenu T, Vinet MC, Sefiani A, Carrie A, Billuart P, McDonell N, Couvert P, Francis F, Chafey P, Fauchereau F, Friocourt G, des Portes V, Cardona A, Frints S, Meindl A, Brandau O, Ronce N, Moraine C, van Bokhoven H, Ropers HH, Sudbrak R, Kahn A, Fryns JP, Beldjord C, Chelly J (2000) A new gene involved in X-linked mental retardation identified by analysis of an X;2 balanced translocation. Nat Genet 24:167–170

Zismanov V, Lishner M, Tartakover-Matalon S, Radnay J, Shapiro H, Drucker L (2009) Tetraspanin-induced death of myeloma cell lines is autophagic and involves increased UPR signalling. Br J Cancer 101:1402–1409

Zuppini A, Groenendyk J, Cormack LA, Shore G, Opas M, Bleackley RC, Michalak M (2002) Calnexin deficiency and endoplasmic reticulum stress-induced apoptosis. Biochemistry 41:2850–2858

Chapter 6
The Role of Tetraspanins in Cell Migration and Intercellular Adhesion

Emilio Tejera*, Mónica Sala-Valdés*, Vera Rocha-Perugini, and María Yáñez-Mó

Abstract Tetraspanin-enriched microdomains (TEMs) are specialized platforms in the plasma membrane that include certain adhesion receptors, mainly integrins and receptors of the Ig superfamily. Insertion into TEMs increases the local concentration of these adhesion receptors, facilitating their function as avidity regulators. TEMs also regulate interaction and crosstalk between different receptors at the plasma membrane, as well as their internalization rate. Moreover, certain signaling pathways are regulated by association with tetraspanins. Thus, tetraspanins emerge as critical regulators of biological phenomena involving adhesion to the extracellular matrix or homotypic or heterotypic intercellular interactions. These proteins are implicated in different steps of cancer progression, the regulation of intercellular adhesion between polarized epithelial cells, angiogenesis, antigen presentation and extravasation of leukocytes or tumor cells. In addition, several pathogens hijack these tetraspanin-adhesion platforms to increase their infectivity.

6.1 Introduction

Some tetraspanins were initially described as molecules induced in tumors that regulate the capacity of tumor cells to migrate and metastasize. Much of the evidence for this came from survival studies in cancer patients and the ability of

---

*These authors contributed equally.

E. Tejera • M. Yáñez-Mó (✉)
Unidad de Investigación, Hospital Santa Cristina, Instituto de Investigación Sanitaria Princesa (IIS-IP), C/Maestro Amadeo Vives 2, Madrid 28009, Spain
e-mail: myanez.hlpr@salud.madrid.org

M. Sala-Valdés • V. Rocha-Perugini
Servicio de Inmunología, Hospital de la Princesa, Instituto de Investigación Sanitaria Princesa (IIS-IP), C/Diego de León 62, Madrid 28006

F. Berditchevski and E. Rubinstein (eds.), *Tetraspanins*, Proteins and Cell Regulation 9, DOI 10.1007/978-94-007-6070-7_6, © Springer Science+Business Media Dordrecht 2013

tetraspanin-specific antibodies to reduce the capacity of animal tumors to induce metastasis (Hemler 2008; Charrin et al. 2009a; Lazo 2007; Wright et al. 2004a; Boucheix et al. 2001; Testa et al. 1999; Romanska and Berditchevski 2011; Richardson et al. 2011). These data are described in detail in the chapter on tetraspanins in cancer. Tetraspanins also regulate cell motility during wound healing. Anti-tetraspanin antibodies inhibit the wound repair capacity of monolayers of human endothelial cells, keratinocytes, melanocytes and hepatic stellate cells (Mazzocca et al. 2002; Klein-Soyer et al. 2000; Deissler et al. 2007; Penas et al. 2000; Yanez-Mo et al. 1998; Garcia-Lopez et al. 2005). Smooth muscle cells treated with anti-CD9 or derived from CD9 null-mice show impaired cell migration (Scherberich et al. 2002; Kotha et al. 2009), although this effect is not seen in all circumstances (Lijnen et al. 2000). CD151-null mice present defects in wound healing and migration of skin cells (Cowin et al. 2006; Wright et al. 2004b; Geary et al. 2008). Moreover, microarray studies have shown up-regulated tetraspanin expression in pathologies associated with wound healing (Vazquez-Chona et al. 2004; Peters et al. 2001).

The first studies into tetraspanin function were mainly based on the use of anti-tetraspanin antibodies (Yanez-Mo et al. 1998; Jones et al. 1996; Sincock et al. 1999; Mantegazza et al. 2004; Domanico et al. 1997; Anton et al. 1995), which did not provide understanding of underlying mechanisms. More recently, genetic approaches such as overexpression, gene knock-down and the use of cells from knock-out mice, confirmed roles for tetraspanins in physiological and pathological processes associated with cell migration. Knock-down of CD9 and CD151 enhances primary melanocyte motility (Garcia-Lopez et al. 2005). In contrast, CD151-deficiency impairs the angiogenesis associated with certain diseases (Takeda et al. 2007; Shi et al. 2010), while its overexpression promotes neovascularization (Shi et al. 2010; Zheng and Liu 2007a, b; Zuo et al. 2009a). CD151 overexpression enhances melanoma (Hong et al. 2006) or prostate cancer cell motility (Ang et al. 2010). CD82 overexpression suppresses the migration of oligodendrocyte precursors (Mela and Goldman 2009). TSPAN8 silencing reduces melanoma cell migration and invasivity (Berthier-Vergnes et al. 2011). In the immune system, tetraspanins regulate the migration and trafficking of natural killer (NK) (Kramer et al. 2009) and dendritic cells (Nattermann et al. 2006; Quast et al. 2011), and CD9/CD81 double knockout mice show defects in macrophage cell motility (Takeda et al. 2008).

Several studies also revealed a role for tetraspanins in the regulation of adhesion to other cells (Yanez-Mo et al. 2001a). Tetraspanins are involved in homotypic adhesion, like the one that occurs between neighboring skin cells, and in heterotypic adhesions characteristic of such diverse phenomena as sperm-egg fusion (Sutovsky 2009), leukocyte extravasation (Barreiro et al. 2005), embryonic implantation (Dominguez et al. 2010), and infection with viruses (van Spriel and Figdor 2010; Yanez-Mo et al. 2009), bacteria (Tham et al. 2010; Green et al. 2011) and parasites (Silvie et al. 2003). In some circumstances adhesion leads to membrane fusion, as is the case with viral induced syncytia formation (Gordon-Alonso et al. 2006), myotube formation and fertilization (Hemler 2003). The implication of tetraspanins in such a diverse array of biological processes suggests that the tools developed for their study might have significant therapeutic potential (Hemler 2008; Stipp 2010). Antibodies tested for this purpose include anti-CD81 in multiple sclerosis (Dijkstra et al. 2008), anti-CD9 in gastric cancer (Nakamoto et al. 2009) and anti-CD37 in B cell malignancies (Robak et al. 2009).

While some of the effects of tetraspanins on cell migration and adhesion can be attributed to their interaction with the transmembrane form of growth factors (pro-growth factors (Shi et al. 2000; Lagaudriere-Gesbert et al. 1997)), growth factor receptors (Odintsova et al. 2000) and metalloproteases (Yanez-Mo et al. 2011), tetraspanins also affect cell motility directly via association with integrins (Berditchevski 2001), their ligands (Barreiro et al. 2005) and other adhesion receptors.

6.2 Tetraspanins and Their Relation with Integrins

Integrins were among the first molecules shown, by a variety of biochemical approaches, to associate with tetraspanins (Table 6.1). Most integrin-tetraspanin interactions described involve beta1 integrins. Laminin-binding integrins associate with CD151 in a strong, direct and stoichiometric manner (Yauch et al. 1998, 2000; Sterk et al. 2002). The association between CD151 and alpha3beta1 integrin regulates the normal migration of various cell types such as neutrophils, endothelial and epithelial cells (Penas et al. 2000; Yanez-Mo et al. 1998; Geary et al. 2008; Yauch et al. 1998), neurite outgrowth on laminin-5 (Stipp and Hemler 2000) and the pathological migration of carcinoma cells (Hasegawa et al. 2007; Winterwood et al. 2006). This molecular association has been studied in depth and shown to be dependent on an extracellular site on the integrin alpha chain and certain residues on the second large extracellular loop of CD151 (Yauch et al. 2000; Kazarov et al. 2002; Yamada et al. 2008a), although recent data could not confirm these observations (Zevian et al. 2011). In many of these scenarios, there is also evidence for a role for the association between CD151 and alpha6beta1 (Sterk et al. 2002; Zhang et al. 2002; Lammerding et al. 2003). Other tetraspanins such as TSPAN1 and TSSC6 (Herlevsen et al. 2003), CD81 (Domanico et al. 1997), CD9 (Schmidt et al. 1996), CO-029 (Gesierich et al. 2005) and CD82 (He et al. 2005) also regulate alpha6beta1-dependent motility and invasion (Table 6.1).

Association with other beta1 integrins is not so robust. CD81 was reported to regulate the avidity of both alpha4beta1 and alpha5beta1 during leukocyte extravasation (Feigelson et al. 2003). CD9 regulates alpha5beta1-dependent adhesion, spreading and migration (Kotha et al. 2009; Cook et al. 1999), although direct binding of the integrin ligand fibronectin to CD9 was also reported (Cook et al. 2002;

Table 6.1 Summary of known tetraspanin-integrin interactions, highlighting the biochemical methods used to demonstrate complex formation, as well as some of the main functions of these complexes related to motility and adhesion

Integrin heterodimer	Associated with	Methods	Functions related to motility or adhesion
Alpha1beta1	CD9	Bri97 (Lozahic et al. 2000)	
Alpha2beta1	CD9, CD151	CHAPS coIP (Sincock et al. 1999; Scherberich et al. 1998)	In VSMC overexpression of CD9 and use of antibodies increased the cell-mediated collagen gel contraction (Scherberich et al. 1998)
			CD9, through alpha2beta1, interferes with membrane anchorage of Rac and subsequently on assembly of mature focal adhesions in endothelial cells (Cailleteau et al. 2010). This interaction also induces cell-cycle arrest
Alpha3beta1	CD151, CD9, CD81, CD63, NAG-2/TSPAN4, CD82	Brij 96 coIP (Stipp and Hemler v2000; Berditchevski et al. 1995; Berditchevski et al. 1996), crosslinking, TX-100 (for CD151) coIP (Jones et al. 1996; Sincock et al. 1999; Thorne et al. 2000)	CD151 controls alpha3beta1-dependent cell adhesion to laminin (Zevian et al. 2011), and subsequent signaling (Yamada et al. 2008b), cell spreading and matrix remodeling (Kazarov et al. 2002)
			CD151-alpha3beta1 complexes regulate migration of neutrophils (Yauch et al. 1998), keratinocytes and fibroblasts (Geary et al. 2008), endothelial and epithelial cells (Penas et al. 2000; Yanez-Mo et al. 1998; Zevian et al. 2011), and carcinoma cells (Hasegawa et al. 2007; Winterwood et al. 2006)
			CD151-alpha3beta1 complexes are important for angiogenesis *in vitro* (Yanez-Mo et al. 1998; Sincock et al. 1999; Yanez-Mo et al. 2008) and *in vivo* (Takeda et al. 2007; Liu et al. 2011). They also regulate angiotensin-II elicited angiogenic responses (Dominguez-Jimenez et al. 2001)

			CD151 interaction with integrin is needed for correct assembly of basement membrane in kidney (Karamatic Crew et al. 2004; Baleato et al. 2008). CD151 favors alpha3beta1-dependent strengthening of adhesion in podocytes (Sachs et al. 2006)
			CD151 and CD81-alpha3beta1 complexes are involved in neurite outgrowth on laminin-5 (Stipp and Hemler 2000)
			CD151-alpha3beta1 complexes are essential for epithelial intercellular adhesion (Penas et al. 2000; Jones et al. 1996; Yanez-Mo et al. 2001b; Nakamura et al. 2001), forming part of a cadherin-catenin multimolecular complex with signaling effects on PKC and Rho GTPases (Chattopadhyay et al. 2003; Johnson et al. 2009; Shigeta et al. 2003)
			CD151-alpha3beta1 complexes are functionally relevant in skin wound healing (Penas et al. 2000; Wright et al. 2004b; Geary et al. 2008)
			CD151-alpha3beta1 complexes are important for tumoral-endothelial cell communication (Sadej et al. 2009)
Alpha4beta1	CD81 CD9, CD82	Brij 58 coIP (Feigelson et al. 2003) Brij 97 and digitonin CoIP (Serru et al. 1999), CHAPS (Rubinstein et al. 1994, 1996)	CD81-alpha4beta1 complexes participate in leukocyte extravasation (Feigelson et al. 2003)
Alpha5beta1	CD9, CD151, CD81, CD82	CHAPS (Sincock et al. 1999; Scherberich et al. 1998; Rubinstein et al. 1994) Brij 58 (Feigelson et al. 2003) and Brij 98 coIP (Ono et al. 2000)	CD81 facilitates alpha5beta1 avidity on fibronectin (Feigelson et al. 2003) CD9 favors high affinity conformation of beta1 integrin (Kotha et al. 2008), increasing cell adhesion, spreading (Cook et al. 1999) and motility on fibronectin

(continued)

Table 6.1 (continued)

Integrin heterodimer	Associated with	Methods	Functions related to motility or adhesion
			CD9 collaborates with alpha5beta1 integrin in the invasion of trophoblasts during implantation and placentation (Hirano et al. 1999b)
			Disruption of CD82-alpha5beta1 complexes increases cell motility on fibronectin (Ono et al. 2000)
			CD151 increases cell adhesion to fibronectin (Hasegawa et al. 1998) and favors cell motility (Liu et al. 2007)
Alpha6beta1	CD151, CD9, CO-029, TSPAN1, CD82, NAG-2/TSPAN4	Proteomics of Brij 97 coIP (Le Naour et al. 2006), Brij 96, 98 and Brij 58 coIP (Gesierich et al. 2005; Thorne et al. 2000; Huang et al. 2008), CHAPS coIP (Herlevsen et al. 2003; He et al. 2005; Hadjiargyrou et al. 1996), TX-100 coIP (Sincock et al. 1999)	CD151 interacts with alpha6beta1 in cell migration (Lammerding et al. 2003), integrin trafficking(Liu et al. 2007), and also in intercellular adhesion (Sincock et al. 1999; Zhang et al. 2002), cell spreading and matrix remodeling (Kazarov et al. 2002)
			CD151 interaction with this integrin is needed for correct assembly of the basement membrane in the kidney (Karamatic Crew et al. 2004)
			CD151- (Yang et al. 2008a), CO-029/Tspan8- (Gesierich et al. 2005; Claas et al. 1998), TSPAN1- (Huang et al. 2008) alpha6beta1 complexes promote motility and cancer invasion
			CD9 antibodies affect alpha6beta1-dependant neuron migration (Schmidt et al. 1996)
			CD9-alpha6beta1 complexes participate in parietal endoderm cell migration (Mulrooney et al. 2002)
			CD81-alpha6beta1 association promotes cell migration driven by other integrins (Domanico et al. 1997)
			CD82 diminishes integrin expression and adhesion to substrate (He et al. 2005)
Alpha7beta1	CD151,CD9	CHAPS coIP (Sterk et al. 2002; Tachibana and Hemler 1999)	

Alpha6beta4	CD151, CD9, CO-029	CHAPS (Herlevsen et al. 2003), Nonidet P-40 coIP (Jones et al. 1996; Sincock et al. 1999)	CD151 and alpha6 participate in the formation of hemidesmosomes (Sterk et al. 2000) important for skin organization (Karamatic Crew et al. 2004)
			CD151-alpha6beta4 complexes are important for tumor-endothelial cell communication (Sadej et al. 2009)
			PKC increases association of CD151 and CO-029 with alpha6beta4 integrin, leading to higher internalization and favoring relocalization to the leading edge and increased cell motility (Herlevsen et al. 2003; Gesierich et al. 2005)
			CD9 interacts with alpha6beta4 in keratinocyte motility (Baudoux et al. 2000)
AlphaLbeta2	CD82	Brij 96 coIP (Shibagaki et al. 1999)	CD82 is present in the pSMAC associated with LFA-1. CD82 increases LFA-1-induced cell adhesion (Shibagaki et al. 1999), and acts as a costimulatory molecule (Lebel-Binay et al. 1995; Shibagaki et al. 1998), inducing changes in actin cytoskeleton through Rho GTPases (Delaguillaumie et al. 2004; Delaguillaumie et al. 2002; Lagaudriere-Gesbert et al. 1998)
			CD81 does not physically associate with LFA-1, but influences LFA-1 dependent adhesion (Volkov et al. 2006)
AlphaMbeta2	CD63	Brij 58 (Skubitz et al. 1996) and CHAPS coIP (Mantegazza et al. 2004)	CD63 associates with alphaMbeta2 and increases the adhesion of neutrophils to HUVEC monolayers (Skubitz et al. 1996)
			CD63 participates in alphaMbeta2 internalization and slows down DC migration (Mantegazza et al. 2004)

(continued)

Table 6.1 (continued)

Integrin heterodimer	Associated with	Methods	Functions related to motility or adhesion
AlphaIIbbeta3	CD9, CD151, TSSC6, CD63	Brij 35 and Brij 96 coIP (Lau et al. 2004; Indig et al. 1997; Israels and McMillan-Ward 2010), CHAPS coIP (Fitter et al. 1999), T-X-100 coIP (Goschnick et al. 2006; Israels and McMillan-Ward 2010), Crosslinking (Slupsky et al. 1989)	In platelets TSSC6 and CD151 are necessary for alphaIIbbeta3 outside-in signaling, so that platelets deficient in these tetraspanins show defective aggregation (Goschnick et al. 2006; Lau et al. 2004) CD9 seems to have an opposite function (Mangin et al. 2009), although anti-CD9 antibodies induce activation of integrin-dependent signaling (Wu et al. 1999)
AlphaVbeta3	CD82	Brij 96 coIP (Ruseva et al. 2009)	CD82 inhibits ovarian cancer cell migration by increasing alphavbeta3 adhesion to vitronectin (Ruseva et al. 2009)
AlphaVbeta5	CD81	Brij 97 coIP (Chang and Finnemann 2007)	

coIP co-immunoprecipitation

Longhurst et al. 2002). CD9 can also associate with alpha3 and alpha5 integrins in human extravillous trophoblasts (Hirano et al. 1999a). In vascular smooth muscle cells (VSMC), CD9 associates with alpha5beta1, alpha2beta1 and alpha3beta1 integrins, and overexpression or crosslinking of this tetraspanin with specific antibodies increases cell-mediated collagen gel contraction (Scherberich et al. 1998).

Regarding non-beta1 integrins, CD151 associates with alpha6beta4 in skin hemidesmosomes (Sterk et al. 2000) and CD151 ablation changes the subcellular distribution of this integrin, reducing tumor cell migration (Yang et al. 2008a). CD9 also associates with alpha6beta4 in keratinocytes *in vitro* (Baudoux et al. 2000). Association of tetraspanins with beta2 integrins appears to be in many cases indirect, but is nonetheless important for cellular adhesion and signaling in the immune system (Volkov et al. 2006; Shibagaki et al. 1999; VanCompernolle et al. 2001) (see below). In platelets, TSSC6 and CD151 are necessary for alphaIIbbeta3 outside-in signaling, so that platelets deficient for these tetraspanins show defective aggregation (Goschnick et al. 2006; Lau et al. 2004). In contrast, CD9 seems to have an opposite function (Mangin et al. 2009), although alpha2bbeta3 can also be activated by anti-CD9 antibodies (Wu et al. 1999). Finally, CD82 was shown to inhibit migration of ovarian cancer cells by increasing the adhesion of alphavbeta3 to vitronectin (Ruseva et al. 2009).

6.3 Non-Integrin Tetraspanin Adhesion Partners

Tetraspanins form complexes with numerous non-integrin transmembrane proteins, some of which are related to adhesion or migration (Table 6.2). For CD9 and CD81 tetraspanins, the most robust protein partners found are the type I transmembrane proteins EWI-F (also called CD9P-1, CD315 or FPRP) and EWI-2 (PGRL, CD316 or IgSF8). EWI-2 and EWI-F, together with EWI-3 (IgSF3) and EWI-101 (CD101 or V7), constitute a novel subfamily of structurally related immunoglobulin superfamily proteins that share a specific Glu-Trp-Ile (EWI) motif in their second Ig domain (Stipp et al. 2001; Charrin et al. 2001; Clark et al. 2001). EWI-2 also associates with tetraspanin KAI1/CD82 (Zhang et al. 2003a), CD53 (Charrin et al. 2003a) and indirectly, through CD9 and CD81, with CD151 (Stipp et al. 2001).

Overexpression of EWI-2 impairs cell migration and inhibits invasion in several cellular contexts (Zhang et al. 2003a; Kolesnikova et al. 2009; Stipp et al. 2003), while its silencing enhances lymphocyte migration and polarization (Sala-Valdes et al. 2006). EWI-2 overexpression affects integrin function by regulating TEM stoichiometries (Stipp et al. 2003; Kolesnikova et al. 2004) (see below). In addition, EWI-2 might also function as a receptor itself. Its expression is induced in dendritic cells upon maturation (Sala-Valdes et al. 2006; Kettner et al. 2007) where it can bind HSPA8 (Kettner et al. 2007). Anti-EWI-2 antibodies or expression of recombinant HSPA8 enhance DC chemotaxis while suppressing their antigen presenting capacity, suggesting an important regulatory role for this interaction (Kettner et al. 2007). EWI proteins can also regulate the infection of some pathogens. A truncated version of EWI-2 blocks the binding of HCV to CD81 (Rocha-Perugini et al. 2008), thus

Table 6.2 Summary of tetraspanin partners with described function in adhesion or migration

Partner	Associated with
CD44	CD9, CO-029/TSPAN8 (Le Naour et al. 2006; Kuhn et al. 2007; Schmidt et al. 2004)
Syndecan	CD9 (Jones et al. 1996)
EWI-2/PGRL/CD316/IgSF8	CD9, CD81, CD82 (Stipp et al. 2001; Clark et al. 2001; Zhang et al. 2003a)
EWI-F/CD9P-1/FPRP/CD315	CD9, CD81, CO-029/Tspan8 (Stipp et al. 2001; Charrin et al. 2001; Claas et al. 2005)
ICAM-1/CD54	CD9 (Barreiro et al. 2005; Barreiro et al. 2008)
VCAM-1/CD106	CD151 (Barreiro et al. 2005; Barreiro et al. 2008)
EpCAM/GA733-2	CD9, D6.1a/CO-029/TSPAN8 (Le Naour et al. 2006)
Claudin-1	CD9 (Kovalenko et al. 2007), CD81 (Kovalenko et al. 2007; Harris et al. 2008; Yang et al. 2008b)
Claudin-7	CO-029/TSPAN8 (Kuhn et al. 2007)
Dectin-1	CD63, CD37 (Mantegazza et al. 2004; Meyer-Wentrup et al. 2007)
E-Cadherin	CD151 (Chattopadhyay et al. 2003), CO-029/TSPAN8 (Greco et al. 2010)
Norrin	TSPAN12 (Junge et al. 2009)

influencing viral tropism, and EWI-F also inhibits the ability of CD81 to support *Plasmodium yoelii* infection (Charrin et al. 2009b). Overexpression of EWI-F/CD9P1 leads to enhanced motility on collagen while slowing down the migration on fibronectin (Chambrion and Le Naour 2010).

Other non-integrin adhesion partners identified in TEMs in various tumor cell lines include CD44, claudins-1 and -7 and EpCAM (Kuhn et al. 2007; Schmidt et al. 2004; Kovalenko et al. 2007). Claudin-7 recruits EpCAM into TEMs (Nubel et al. 2009), and the complex composed of CD44, claudin-7, EpCAM and CO-029/Tspan8 promotes cancer progression while the individual molecules do not have this effect (Kuhn et al. 2007). Crosslinking experiments suggest that claudin-1 interacts directly with CD9 (Kovalenko et al. 2007). In keratinocytes CD9 has also been detected in immunoprecipitates of CD44 and syndecan (Jones et al. 1996). Claudin-1 also associates with CD81 (Kovalenko et al. 2007; Harris et al. 2008; Yang et al. 2008b) in a cholesterol-dependent manner (Harris et al. 2010).

The endothelial adhesion receptors ICAM-1 and VCAM-1 associate directly with CD9 and CD151, respectively (Barreiro et al. 2008), with functional consequences for the interaction with leukocytes during inflammation-induced extravasation (Barreiro et al. 2005). In retinal endothelium Tspan12 associates with and regulates signaling downstream of Norrin/FZ-4 during vasculogenesis (Junge et al. 2009). CD151 also associates with E-cadherin in polarized epithelia (Chattopadhyay et al. 2003).

The C-type lectin dectin-1 has been shown to associate with both CD37 (Meyer-Wentrup et al. 2007) and CD63 (Mantegazza et al. 2004). In dendritic cells, interaction with CD37 stabilizes dectin-1 at the cell surface. Moreover, CD37 deficiency results in an elevated dectin-1 mediated IL-6 secretion (Meyer-Wentrup et al. 2007) that contributes to anti-fungal immunity (van Spriel et al. 2009).

6.4 Regulation of Adhesion Receptor Function by Tetraspanins

6.4.1 Avidity Versus Affinity

Avidity regulation involves an increase in the local concentration of a given receptor in the plasma membrane (Fig. 6.1a). Recently, avidity regulation by TEM was demonstrated *in situ* for the endothelial adhesion receptors ICAM-1 and VCAM-1 (Barreiro et al. 2008). In contrast, CD82 prevents ligand-induced EGFR dimerization at the plasma membrane (Odintsova et al. 2003).

Regarding integrins, tetraspanins seem to have little effect on the initial binding of integrins to their ligands (Berditchevski 2001; Hemler 2005), but rather regulate the avidity of these interactions (Ovalle et al. 2007) and the post-binding strengthening of adhesion (Zhang et al. 2002; Lammerding et al. 2003; Feigelson et al. 2003; Yamada et al. 2008b). Recent evidence suggest that insertion into TEM regulates the mode of diffusion of integrins at the plasma membrane (Yang et al. 2012). As a consequence, cells become progressively more resistant to detachment. For integrins, avidity regulation imposed by tetraspanins has been described in tumor cells (Ovalle et al. 2007) and leukocytes (Feigelson et al. 2003).

Integrins also undergo conformational changes that increase ligand affinity, and which are commonly detected by specific, conformation-dependent antibodies called LIBS (ligand-induced binding site) (Luo et al. 2007). Beta 1 integrins colocalized with tetraspanins in intercellular adhesions are not detected by LIBS antibodies (Yanez-Mo et al. 2001b), indicating that they are in a low affinity conformation or in a non-ligand-bound state. At cell-matrix contact sites, colocalization of tetraspanins with LIBS antibodies has been demonstrated (Kotha et al. 2008), while in other cases LIBS antibodies did not detect tetraspanin-enriched adhesions (Penas et al. 2000). This discrepancy might derive from the different LIBS antibodies, extracellular matrix or cell type used in each study. Nonetheless, association of tetraspanins with ligand-bound integrins has been demonstrated biochemically (Yanez-Mo et al. 2001b), and antibodies able to disrupt CD151-alpha3 integrin complexes reduce laminin binding and LIBS reactivity (Nishiuchi et al. 2005). Tetraspanins thus appear to modulate integrin-ligand affinity by inducing or stabilizing their conformational changes (Fig. 6.1a). Recently, CD9 was found in a RNAi screen to regulate beta1 activation (Pellinen et al. 2012). However, with the exception of CD151 included in hemidesmosomes, tetraspanins are usually found in nascent adhesion complexes, more relevant for migration, and are excluded from well-developed focal adhesions (Berditchevski and Odintsova 1999), although CD9 deficiency leads to impaired localization of talin to focal adhesions (Powner et al. 2011). Tetraspanin-integrin interaction can also induce conformational changes in tetraspanins, with certain epitopes in tetraspanin extracellular regions becoming exposed upon activation of the associated integrin. Such is the case with CD9, whose conformation-dependent epitope, detected by mAb PAINS-13, is only exposed when the associated beta1 integrin has been previously activated (Gutierrez-Lopez et al. 2003).

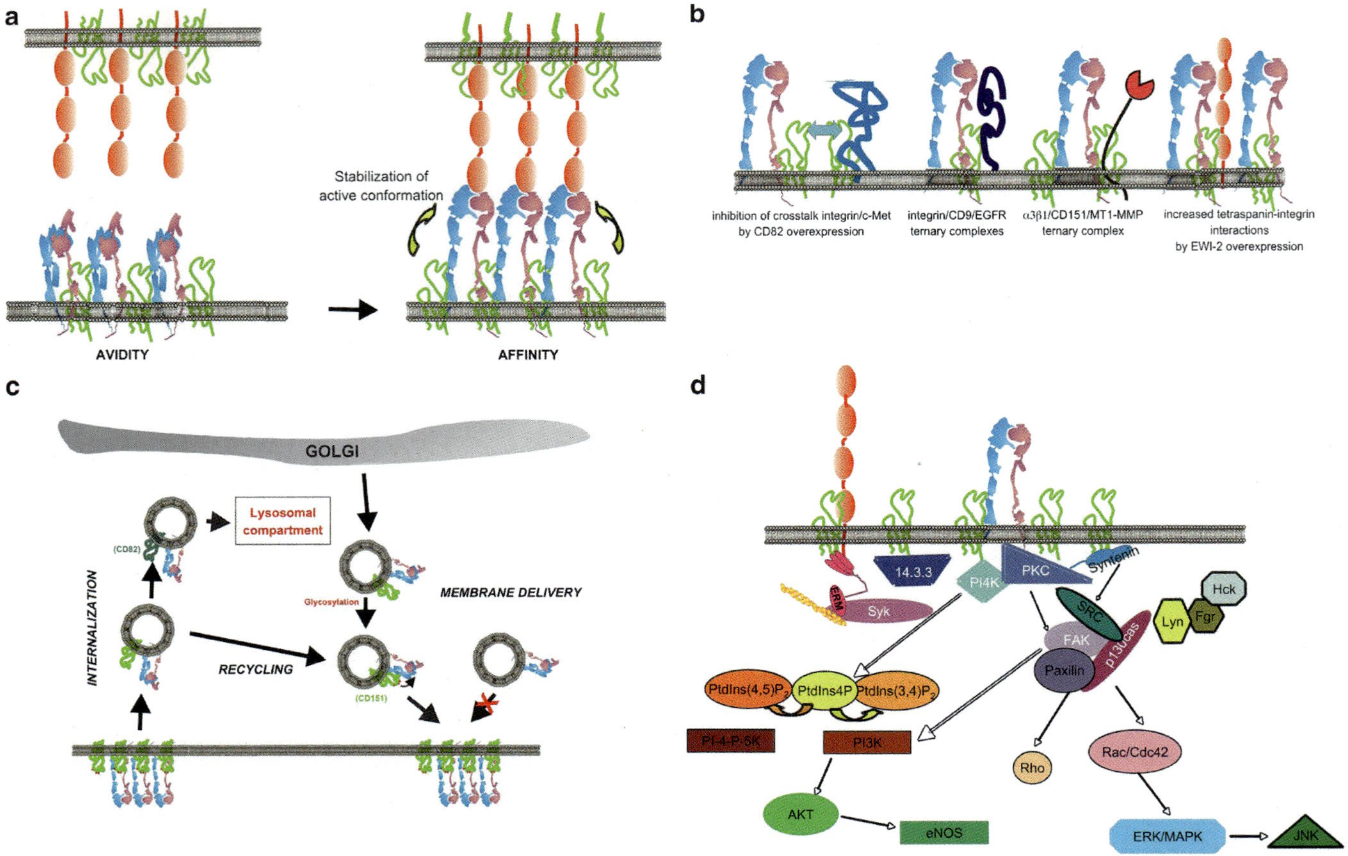

Fig. 6.1 Tetraspanins regulate cell adhesion and migration via the organization of associated receptors in specialized membrane microdomains. This kind of compartmentalization is involved in avidity regulation (**a**), crosstalk with other adhesion or signaling receptors (**b**), regulates internalization/recycling rates (**c**) and provides links to certain intracellular signaling pathways and the cortical cytosketeton (**d**)

6.4.2 Heterologous Associations on the Plasma Membrane

Insertion into tetraspanin-enriched microdomains may not only increase avidity of a given receptor at the plasma membrane, but also facilitate its interaction with other TEM components. For example, CD81, CD82 or EWI-2 overexpression reduces TEM-associated CD9 multimers, while CD81 or CD151 knock-down enhances CD9 multimerization (Yang et al. 2006). Genetic deletion or overexpression of tetraspanins has also been found to disrupt the lateral associations of integrins with other membrane proteins and may be responsible for their functional crosstalk. CD82 overexpression inhibits integrin-dependent crosstalk with the receptor tyrosine kinase c-Met (Sridhar and Miranti 2006) (Fig. 6.1b) possibly via a direct association of CD82 with c-Met (Takahashi et al. 2007). Functional complexes of CD9, beta1 integrins and EGFR have also been detected in tumor cells (Murayama et al. 2008) (Fig. 6.1b). CD151 acts as a molecular linker between laminin-binding integrins and c-Met (Klosek et al. 2005; Franco et al. 2010). CD151 knockdown reverses the laminin-5 induced resistance of breast cancer cells to anti-ErbB2 antitumor treatments inhibiting ErbB2 activation (Yang et al. 2010). CD151 deficiency also attenuates TGFβ signaling (Sadej et al. 2010), although in these reports, the formation of a ternary complex integrin/tetraspanin/growth factor receptor was not assessed.

In endothelial cells, CD151 gene deletion abolishes the molecular association of alpha3beta1 integrin with several undefined molecules (Takeda et al. 2007). CD151 and alpha3beta1 integrin are reported to form a ternary complex with the membrane metalloproteinase MT1-MMP, thus spatiotemporally directing pericellular proteolysis during endothelial cell migration and angiogenesis (Yanez-Mo et al. 2008) (Fig. 6.1b). CD9, CD81 and TSPAN12 regulate MT1-MMP proteolytic function in tumor cells (Lafleur et al. 2009). CD82, by regulating the subcellular localization of alpha5beta1 integrin, impairs uPAR dependent plasminogen activation and proteolysis (Bass et al. 2005). CD9 associates with the metalloproteases ADAM10 (Arduise et al. 2008; Yan et al. 2002) and ADAM-17 (Gutierrez-Lopez et al. 2011). Tetraspanin15 directly associates with ADAM10 and regulates its trafficking and maturation (Prox et al. 2012), while CD63 interacts with the soluble factor TIMP-1, a metalloprotease inhibitor involved in the regulation of proliferation, cell survival, differentiation, and renewal of extracellular matrix proteins (Jung et al. 2006). In other cases, induction of soluble metalloproteinase gene expression by tetraspanins has also been reported. For example, anti-tetraspanin antibodies induce PI3-kinase-dependent production of matrix metalloproteinase 2 (MMP-2) (Sugiura and Berditchevski 1999) and crosslinking of CD81 by HCV E2 glycoprotein upregulates MMP2 in hepatoma cell lines (Mazzocca et al. 2005). CD151 homophilic interactions stimulate integrin-dependent MMP-9 expression in human melanoma cells (Hong et al. 2006), and its expression associates with that of the metalloproteinase in hepatocellular carcinoma (Shi et al. 2010). CD9 expression induces MMP-2 in melanoma (Hong et al. 2005), while CD9 depletion augments it in small cell lung cancer cells (Saito et al. 2006) and mouse blastocysts (Liu et al. 2006). Overexpression of CD81 or CD82 in several myeloma cells reduces MMP2 expression (Tohami et al. 2007), and TSPAN8 overexpression has been reported to induce mRNA expression of the protease ADAM12, which mediates tetraspanin-induced esophageal cancer invasion (Zhou et al. 2008). These data indicate that tetraspanins might also influence not only invasion but also matrix deposition (Cook et al. 1999) and degradation. Supporting this, some phenotypes of tetraspanin-deficient mice, such as the disorganized basement membrane in the renal glomeruli (Sachs et al. 2006) and skin (Cowin et al. 2006) of CD151-KO mice, are consistent with a defect in extracellular matrix organization and podocyte adhesion (Blumenthal et al. 2012; Sachs et al. 2012).

EWI-2 overexpression alters the associative balance in TEMs such that interactions between tetraspanins and between tetraspanins and integrins appear to be increased (Stipp et al. 2003; Kolesnikova et al. 2004) (Fig. 6.1b). This results in decreased alpha-4beta1 dependent spreading and ruffling on VCAM-1 coated surfaces (Kolesnikova et al. 2004) and alpha3beta1 dependent aggregation and motility of epidermoid carcinoma cells on laminin-5 (Stipp et al. 2003). In contrast, EWI-2 expression decreases the association of tetraspanins with MT1-MMP (Kolesnikova et al. 2009).

Tetraspanin enriched microdomains can also include signaling-related molecules such as γ-glutamyl transpeptidase (Nichols et al. 1998), a transmembrane protein involved in the regulation of intracellular redox potential, or G-protein coupled receptors such as GPCR56 (Little et al. 2004). However, functional consequences of interactions of these molecules with integrins or other tetraspanin partners has not been investigated.

The interaction of tetraspanins with their partners, both integrins and non-integrins, is regulated by a variety of mechanisms. Tetraspanins interact with cholesterol (Charrin et al. 2003b) and cholesterol extraction from the membrane by methyl-beta-cyclodextrin treatment diminishes total CD81 expression in human hepatocytes (Kapadia et al. 2007; Rocha-Perugini et al. 2009) and TEM-associated CD81 in mouse hepatocytes (Silvie et al. 2006). Membrane enrichment in ceramides induced by sphingomyelinase treatment reduces CD81 total expression while it enhances TEM-associated CD81 (Rocha-Perugini et al. 2009; Voisset et al. 2008). The gangliosides GM2, GM3, GM1 and GD1a enhance the interaction of CD82 with c-Met (Todeschini et al. 2007, 2008) and EGFR (Odintsova et al. 2006). These lipids also enhance the interaction between CD9 and the integrins alpha3beta1 (Kawakami et al. 2002) and alpha5beta1 (Miura et al. 2004), decreasing cell motility. Tetraspanin-integrin complexes can also be regulated by posttranslational modifications of tetraspanins such as glycosylation (Ono et al. 2000) and palmitoylation (Israels and McMillan-Ward 2010; Yang et al. 2002; Berditchevski et al. 2002; Yang et al. 2004; Charrin et al. 2002). Extracellular stimuli such as cytokines (Huang et al. 2008) have also been reported to regulate tetraspanin-integrin associations.

6.4.3 Biosynthesis and Trafficking

Tetraspanins have been found to associate with integrins and other partners during the early stages of biosynthesis (Kazarov et al. 2002; Berditchevski et al. 2001;

Shoham et al. 2006). The interaction with integrins can be modulated by the tetraspanin glycosylation state –such is the case of CD82 and alpha5, which interaction is diminished by CD82 glycosylation, leading to enhanced FN-induced motility (Ono et al. 2000). Association with tetraspanins can also regulate the glycosylation of the integrin itself—as it happens with CD151 and alpha3beta1. In this case, changes in integrin glycosylation by CD151 silencing impaired cell migration towards laminin (Baldwin et al. 2008) (Fig. 6.1c).

For other partners, such as CD19, association with tetraspanin CD81 regulates not only glycosylation but also exit from the ER (Shoham et al. 2006). As a consequence, CD81 deficiency in mice (Shoham et al. 2003) or humans (van Zelm et al. 2010) results in reduced levels of CD19 plasma membrane expression on B cells.

Although no changes in integrin expression have been detected in CD151 deficient mouse (Wright et al. 2004b; Takeda et al. 2007), they have been in some instances reported *in vitro* (Mantegazza et al. 2004; He et al. 2005; Furuya et al. 2005). Overexpression or silencing of CD151 regulates alpha3 integrin trafficking (Winterwood et al. 2006; Liu et al. 2007) and this effect depends on an intracellular internalization motif present in the CD151 C-terminal domain (Liu et al. 2007) (Fig. 6.1c). Suppression of CD151 expression by RNAi-mediated silencing (Winterwood et al. 2006) or exposure to anti-CD151 antibody (Zijlstra et al. 2008) impairs cell migration by inducing defective retraction of the trailing edge of the cell, for which proper integrin recycling is crucial.

CD82 expression enhances alpha6beta1 internalization, leading to reduced integrin membrane expression, and consequently to a diminished adhesiveness and alpha6beta1-mediated morphogenesis on laminin (He et al. 2005). Moreover, internalization of CD82 ameliorates its inhibition of cell migration (Fig. 6.1c) (Xu et al. 2009), suggesting that dynamic regulation of tetraspanin expression levels at the plasma membrane can rapidly regulate their functions. Activation of PKC increases the association of alpha6beta4 with CO-029/TSPAN8 and CD151 and their subsequent internalization (Herlevsen et al. 2003; Gesierich et al. 2005). In keratinocytes, alpha6beta4 is also internalized with CD9 (Baudoux et al. 2000). Internalization rates are different for different tetraspanins, as demonstrated for PMA-induced internalization of CD9 or TSPAN8, so that tetraspanin-enriched microdomains differ in composition along the recycling pathway (Rana et al. 2011). Regulation of intracellular traffic by tetraspanins is addressed in detail in another chapter of this book.

6.4.4 Intracellular Signaling and Cytoskeletal Connections

Tetraspanin regulation of post-adhesion events explains most of the effects of tetraspanins on cell morphology, spreading and migration. However, the mechanisms by which tetraspanins might support integrin-dependent adhesion strengthening are not completely understood. The most feasible possibility is that tetraspanins regulate actin cytoskeleton reorganization by modulating signaling pathways downstream their partners.

The first evidence that tetraspanins influence integrin-initiated signaling was the observation that anti-tetraspanin antibodies alter the tyrosine phosphorylation of focal adhesion kinase (FAK), either positively or negatively depending on the experimental conditions: plating serum-starved cells on anti-tetraspanin mAbs induces dephosphorylation of FAK, while anti-tetraspanin mAbs potentiated collagen-induced FAK phosphorylation (Berditchevski and Odintsova 1999). Moreover, increased invasion and cell motility induced by overexpression of CD151 only occurs in FAK (+/+) fibroblasts (Kohno et al. 2002). In several independent studies, RNAi-mediated silencing or genetic deletion of CD151 resulted in impaired signaling via laminin-binding integrins (Takeda et al. 2007; Yang et al. 2008a, b). Signaling molecules whose activities are reduced in CD151-deficient cells include the tyrosine kinases FAK, Src (although not in other reports (Takeda et al. 2007)), $p130^{cas}$ and paxilin (Yamada et al. 2008b); protein kinase B (AKT); endothelial nitric-oxide synthase (eNOS) and the small GTPases Rac1 and Cdc42 (Takeda et al. 2007). Overexpression of CD151 activates Jun N-teminal kinase (JNK), phosphatidylinositol-3 kinase (PI3K), AKT and eNOS in neovascularization processes (Zuo et al. 2009a; Zheng and Liu 2006). In another study, signaling through CD151 reduced signaling via the Ras-ERK/MAPK1/2 pathway, without apparent changes in FAK (Sawada et al. 2003). In contrast, FAK phosphorylation is reduced in CD9 deficient VSMCs (Scherberich et al. 2002). In these same cells, CD9 overexpression increased PI3K-dependent AKT activation, while the opposite effect was observed in small cell lung cancer cells (Saito et al. 2006). CD82 expression reduces the expression level of p130CAS and thus impairs p130CAS-dependent signaling (Zhang et al. 2003b).

This regulation of intracellular signaling downstream of integrins might originate from the direct association of tetraspanins with determined signaling molecules. Although most tetraspanins have short intracellular domains, comprising just a few aminoacids, their functional role has been demonstrated with mutant proteins lacking their C-term intracellular sequence (Lammerding et al. 2003; Wang et al. 2011) and some specific interactions with intracellular molecules have been described (Fig. 6.1d). At least five tetraspanins—CD9, CD63, CD81, CD151 and A15/TALLA/Tspan7—can associate with type II PI 4-kinase (Berditchevski et al. 1997; Yauch and Hemler 2000). Tetraspanin-dependent recruitment of PI4K is expected to result in the localized conversion of PtdIns into PtdIns4P, which is then available for further conversion to PtdIns(3,4)P_2 by PI3K or to PtdIns(4,5)P_2 by PI-4-P 5-kinase. This interaction may play a role in cell migration (Mazzocca et al. 2008), bacterial infection (Tham et al. 2010) and tumor cell proliferation (Carloni et al. 2004).

Stimulation with PKC-activating stimuli in several cell lines induces association of the kinase with CD9, CD53, CD81, CD82 and CD151 (Zhang et al. 2001a) (Fig. 6.1d). These tetraspanins provide a key linker function, ensuring that only $\alpha3$ integrin forms able to associate with tetraspanins are phosphorylated by PKC (Zhang et al. 2001a). PKC-mediated phosphorylation was further shown to be necessary for $\alpha3$–dependent signaling to FAK, $p130^{cas}$ and paxillin, inducing spreading and migration (Zhang et al. 2001b). Association of CD82 with PKC-alpha requires the presence of caveolin-1, and the interaction of caveolin-1 or PKC-alpha with EGFR requires the presence of CD82 and ganglioside GM3 (Wang et al. 2007). CD151 induces PKC-dependent activation of Rac and Cdc42 but not Rho (Shigeta et al. 2003), while its loss causes excessive RhoA activation (Johnson et al. 2009). Rho GTPases were also shown to mediate the morphological changes induced by CD82 crosslinking (Delaguillaumie et al. 2002).

The C-terminal cytoplasmic domain of CD63 binds to the PDZ adaptor protein syntenin-1 (Latysheva et al. 2006), which plays a potential role in cell migration via the activation of the Src-Rac pathway (Boukerche et al. 2008; Sala-Valdes et al. 2012). TSPAN7, whose deficiency is associated with the X-linked mental retardation (Zemni et al. 2000), binds to PICK1, another PDZ-containing protein, and regulates AMPA-R trafficking and spine formation (Bassani et al. 2012). Two Src family tyrosine-kinases, Lyn and Hck, and unspecified serine/threonine kinase activities are associated with the β2-integrin-CD63 complex and might play an important role in CD63-induced upregulation and activation of β2-integrins in human neutrophils (Skubitz et al. 1996). Hck and Fgr tyrosine kinases are tyrosine phosphorylated in response to CD9 ligation in human monocytes (Zilber et al. 2005) and, in B cells, CD9 overexpression induces the activation of tyrosine kinases and enhances beta-1 integrin-dependent migration (Shaw et al. 1995).

CD81 and its direct partners EWI-2 and EWI-F bind directly to ERM proteins (Sala-Valdes et al. 2006) (Fig. 6.1d). EWI-2 and EWI-F association with CD9 and ERMs at the membrane of oocytes could be important for the morphology and dynamics of microvilli and might have a role in microvilli function during sperm-oocyte fusion (Runge et al. 2007; Jegou et al. 2011). Other tetraspanin partners, such as VCAM-1 (Barreiro et al. 2002), CD44 and ICAM-1 bind to the actin cytoskeleton via ERM adaptor proteins. This connection with the actin cystoskeleton dictates TEM subcellular localization. The redundancy in this interaction, with both CD81 and its partners binding to ERM proteins, ensures that even cytoplasmic truncated mutants of CD44, VCAM-1 or EWI-2 maintain a normal subcellular localization (Sala-Valdes et al. 2006; Barreiro et al. 2008; Legg and Isacke 1998). Binding to ERMs connects TEMs not only to the actin cytoskeleton, but also to the tyrosine kinase Syk (Urzainqui et al. 2002), and engagement of CD81 induces Syk dependent ezrin phosphorylation in B cells (Coffey et al. 2009) and NK cells (Kramer et al. 2009). Finally, a redox dependent association of the CD81 C-terminal domain with 14-3-3 proteins was demonstrated in T cells (Clark et al. 2004).

6.5 Tetraspanin Functions in Migration and Intercellular Adhesion

The regulation of single cell motility by tetraspanins has mostly been studied in tumor cells. In this context, how tetraspanins regulate integrin-dependent adhesion and its coordination with matrix degradation or growth factor signaling may be of utmost importance. This area has been commented on in earlier sections here and is the main topic of another chapter of this book. TEMs organize adhesion receptors in specialized adhesive platforms that are involved in both homotypic and heterotypic intercellular adhesion, being crucial for such important physiological processes as sperm-egg binding and fusion, tissue architecture and repair, and immune cell communication and migration. Several pathogens use TEMs as gateways for infection, exploiting their intrinsic augmentation of adhesion receptor avidity and connections with intracellular trafficking pathways. Again, this area is covered in other chapters. The remaining sections here provide a brief overview of some of the physiological adhesion and migration phenomena in which tetraspanins play a major regulatory role.

6.5.1 Epithelial Cell-Cell Adhesion and Wound Healing

In skin, CD151 may contribute to the organization of hemidesmosomes through its interaction with the laminin binding integrins alpha6beta4 (preferentially) and alpha3beta1 (Sterk et al. 2000). It was also shown that tetraspanin-integrin complexes in epithelial cells show a similar localization to cadherins at cell-cell contact sites (Jones et al. 1996; Yanez-Mo et al. 2001b; Nakamura et al. 2001), although cadherin-mediated intercellular adhesion is not necessary for tetraspanin localization (Yanez-Mo et al. 2001b). In addition, complexes of alpha3beta1 integrin and CD151 are found associated with the cadherin-catenin multimolecular complex, which also includes PKCbetaII and PTPmu (Chattopadhyay et al. 2003). Tetraspanins and their partners thus appear to contribute to the establishment of both cell-matrix and cell-cell adhesions, and to the signaling processes necessary for their maintenance. Tetraspanin-partner associations regulate PKC signals and Rho GTPases balance (Johnson et al. 2009; Shigeta et al. 2003) and even alter gene expression to reinforce intercellular adhesion (Chattopadhyay et al. 2003). Tetraspanin-regulated cell-cell adhesion also modulates cell migration, so that CD151 deletion impairs migration of dispersed cells while enhancing migration of attached cells via destabilization of intercellular junctions after RhoA activation (Johnson et al. 2009). The staining pattern of an anti-CD151 antibody that recognizes only non-integrin-associated CD151 correlates with cells presenting increased migratory capacity (Chometon et al. 2006). In breast epithelium deletion of CD151 restored polarity and suppressed tumor growth (Novitskaya et al. 2010).

Some of the tumor suppressor properties of CD82 have been assigned to its capacity to stimulate signals that stabilize cadherin-catenin complexes (Abe et al. 2008). Moreover, the expression of some tetraspanins was found to change during epithelial to mesenchymal transition (Yanez-Mo et al. 2003). These studies provide molecular explanations of the phenotype of a CD151 mutation in humans that disrupts epidermal integrity (Karamatic Crew et al. 2004) and of the wound healing defect of CD151 deficient mice (Cowin et al. 2006). Deletion of TSP-15 in *C.elegans* also results in loss of epidermal integrity (Moribe et al. 2004). Gastrulation movements were shown to be disturbed by injection of Tspan-1 mRNA in Xenopus embryos, indicating the involvement of tetraspanins in early migratory and cell fate decisions (Yamamoto et al. 2007). Finally, a tetraspanin protein was found to be component of a cell-cell junction structure in metazoan demosponge epithelium (Adell et al. 2004).

6.5.2 Angiogenesis

The role of tetraspanins in cellular migration and intercellular adhesion is also important in endothelial cells during angiogenesis. Antibodies against CD9, CD81 or CD151 affect alpha3beta1 or alpha6beta1 dependent cell migration during wound healing and tube formation (Klein-Soyer et al. 2000; Deissler et al. 2007; Yanez-Mo et al. 1998; Sincock et al. 1999; Zhang et al. 2002). CD9, through alpha2beta1, interferes with membrane anchorage of Rac and subsequently on assembly of mature focal adhesions in endothelial cells (Cailleteau et al. 2010). CD151-alpha3beta1 complexes are key regulators of the angiotensin-II induced angiogenic response (Dominguez-Jimenez et al. 2001). Moreover, the proposed use of anti-CD9 antibodies in gastric cancer therapy is partially based on their inhibitory effect on tumor angiogenesis (Nakamoto et al. 2009). A role in retinal vascular development in mice has recently been ascribed to the association between Tspan12 and the Norrin receptor (Junge et al. 2009). Endothelial cells derived from CD151 deficient mice show defects in migration, spreading, invasion, matrigel contraction, tube formation and spheroid sprouting (Takeda et al. 2007). They also present selective signaling defects when grown on laminin substrates (Takeda et al. 2007) and impaired collagenolysis as a result of disruption of the complex formed by MT1-MMP with CD151 and alpha3beta1 integrin (Yanez-Mo et al. 2008). CD151 is also required to maintain capillary integrity by regulating cell-cell adhesions via a Rho/Rac balance (Zhang et al. 2011). CD151 stimulates the PI3K/AKT pathway (Zheng and Liu 2007a) and its overexpression increases angiogenesis and blood reperfusion after a myocardial infarction (Zuo et al. 2009b). A direct involvement of the association of CD151 with alpha3 integrin in angiogenesis was demonstrated by mutation of the QRD site of CD151 LEL (Liu et al. 2011).

Regarding tumoral angiogenesis, Tspan8/CO-029 is detected in exosomes released by tumor cells, and these exosomes increase endothelial angiogenesis (Nazarenko et al. 2010; Gesierich et al. 2006). Since exosomes are delivered in the bloodstream, this angiogenesis can be produced at sites distant from the tumor, and can be inhibited by anti-CO-029 antibodies. In a breast cancer model, CD151 expression by tumor cells induces tumor associated vascularization, so that CD151 depletion reduces tumorigenicity (Sadej et al. 2009).

6.5.3 Extravasation of Leukocytes and Tumor Cells

To exert their functions, leukocytes in circulating blood must enter inflamed tissues by extravasation of the blood vessel wall (Ley et al. 2007). Other cells, such as malignant tumor cells, can also extravasate and induce metastasis by settling in distant tissues, a process in which adhesion receptors are also involved (Konstantopoulos and Thomas 2009).

6.5.3.1 Leukocyte-Endothelium Interaction

Leukocyte extravasation proceeds via successive, well-defined steps: initial attachment (tethering), rolling, strong adhesion and finally transendothelial migration (Ley et al. 2007). During the firm adhesion phase, leukocyte integrins VLA-4 (alpha4beta1) and LFA-1 become activated and interact with their endothelial ligands, VCAM-1 and ICAM-1. On the leukocyte side, CD81 was found to increase the avidity of integrin VLA-4 for VCAM-1, thereby facilitating the rolling and arrest of leukocytes under shear flow (Feigelson et al. 2003). CD9 and CD82 have been recently reported to regulate Hematopoietic Stem Cell homing (Larochelle et al. 2012; Leung et al. 2011).

On the endothelial side, tetraspanins regulate apical surface presentation of VCAM-1 and ICAM-1, facilitating strong adhesion that prevents leukocyte detachment (Barreiro et al. 2005). Tetraspanins CD9 and CD151 have been shown to be responsible for organizing actin-associated (Barreiro et al. 2002) endothelial adherent platforms for leukocytes (Barreiro et al. 2008), and the dynamics of VCAM-1 and ICAM-1 are dependent on TEMs. These endothelial TEMs include other molecules involved in rolling (such as E-selectin) or transmigration (such as CD31 or JAM-A). The facilitatory role of tetraspanins in leukocyte extravasation is relevant in some disease processes. Endothelial CD81 is upregulated in the initial steps of atherosclerotic plaque formation (Rohlena et al. 2009), facilitating monocyte adhesion via VCAM-1 and ICAM-1. CD9 is also detected in atherosclerotic plaques (Nishida et al. 2000), and antibodies to CD81 impair monocyte transmigration across brain endothelial cell monolayers, with a therapeutic effect on autoimmune encephalomyelitis (Dijkstra et al. 2008).

6.5.3.2 Malignant Tumor Cell Extravasation

After intravasation, malignant cells can produce metastasis by traversing endothelial cell-cell junctions to exit the blood vessels and colonize new tissues. During this process a mosaic between tumor and endothelial cells is formed (Chang et al. 2000; Longo et al. 2001). CD9 decorates heterotypic intercellular contacts and anti-CD9 antibodies are able to prevent transendothelial migration (Longo et al. 2001). Moreover, CD9 expression was shown to be upregulated in tumor cells by the interaction with the endothelium (De Bruyne et al. 2006) and during extravasation (Sauer et al. 2003). Direct interaction of tumor expressed CD82 with the endothelial membrane protein DARC inhibits tumor cell proliferation and leads to metastasis suppression (Bandyopadhyay et al. 2006). CD151 KO mice present reduced metastasis because of impaired tumor-endothelium interactions (Takeda et al. 2011).

6.5.4 *Immune Synapse*

Although the importance of tetraspanins in the immune system is discussed in another chapter, we will briefly describe their role in the intercellular adhesion (the immune synapse) between T lymphocytes and antigen-presenting cells (APC) during antigen recognition.

Most of the available data in this area relate to tetraspanins CD81 and CD82 (Levy et al. 1998). CD81 has been found to localize, both in T cells and APC, in the cSMAC, the central region of the immune synapse where the antigen recognition molecules reside (Mittelbrunn et al. 2002). In T cells, CD81 associates with CD3 and coreceptors CD4 or CD8 (Levy and Shoham 2005; Imai et al. 1995; Todd et al. 1996). Anti-CD81 antibodies or direct ligands, such as HCV E2 glycoprotein, have a co-stimulatory effect on TCR-dependent signaling that is independent of the classic CD28 stimulatory pathway (Todd et al. 1996; Wack et al. 2001; Soldaini et al. 2003; Witherden et al. 2000). In contrast, these agents have the opposite effect on the activation of NK cells (Crotta et al. 2002, 2006; Tseng and Klimpel 2002). CD81 preferentially increases Th2 differentiation (Maecker 2003; Maecker and Levy 1997), and the absence of CD81 in T cells has been found to diminish signaling and cytokine production dependant on cognate interaction between T and B cells (Deng et al. 2002). Moreover, antibodies directed to CD81 are able to activate LFA-1 integrin and facilitate T-B cell interactions (VanCompernolle et al. 2001)

T cell expressed CD82 associates with LFA-1 (Shibagaki et al. 1999), a beta2-integrin present in the peripheral SMAC of the immune synapse, increasing its adhesion to ICAM-1 on the APC. CD82 has also been described as a co-stimulatory molecule that increases T-APC adhesion and adhesion-dependent signaling (Lebel-Binay et al. 1995; Shibagaki et al. 1998). This co-stimulation is able to induce changes in actin cytoskeleton through Rho-GTPase signaling (Delaguillaumie et al. 2004, 2002; Lagaudriere-Gesbert et al. 1998). The costimulatory action of CD9 facilitates the insertion of CD3 into lipid rafts (Yashiro-Ohtani et al. 2000), while other tetraspanins, such as CD37, TSSC6 and CD151, have been described as negative regulators of TCR-dependent T cell proliferation (Wright et al. 2004b; van Spriel et al. 2004; Tarrant et al. 2002).

In APCs (either dendritic or B cells), several tetraspanins (CD81, CD82, CD37, CD9 and CD53) associate with integrins (Rubinstein et al. 1996) and MHC-II (Levy and Shoham 2005; Angelisova et al. 1994; Szöllósi et al. 1996; Engering and Pieters 2001; Unternaehrer et al. 2007) and favor the presentation of MHC-II-peptide complexes to T cells (Vogt et al. 2002; Hoorn et al. 2012).

6.6 Conclusions

Tetraspanin-enriched microdomains, by regulating the functions of integrins and their relationship with other adhesion receptors, or by generating specialized adhesive platforms in the plasma membrane, regulate both migration and intercellular interactions throughout the body, behaving as sensors of the extracellular environment.

Acknowledgements We thank Monica Gordon-Alonso and Francisco Sánchez-Madrid for a critical reading of the manuscript, S. Bartlett for English editing and Miguel Vicente-Manzanares for his masterly cartoon drawing. This work was supported by grants PI080794 and PI11/01645 from the Instituto de Salud Carlos III to M.Y.-M.; MS-V is funded by a Carlos III Ayuda Predoctoral de Formación en Investigación fellowship (FI05/00238), E.T. by a FPU fellowship from Ministerio de Ciencia e Innovacion and V.R-P by RECAVA—RD06/0014/0030.

References

Abe M, Sugiura T, Takahashi M, Ishii K, Shimoda M, Shirasuna K (2008) A novel function of CD82/KAI-1 on E-cadherin-mediated homophilic cellular adhesion of cancer cells. Cancer Lett 266(2):163–170

Adell T, Gamulin V, Perovic-Ottstadt S, Wiens M, Korzhev M, Muller IM, Muller WE (2004) Evolution of metazoan cell junction proteins: the scaffold protein MAGI and the transmembrane receptor tetraspanin in the demosponge suberites domuncula. J Mol Evol 59(1):41–50

Ang J, Fang BL, Ashman LK, Frauman AG (2010) The migration and invasion of human prostate cancer cell lines involves CD151 expression. Oncol Rep 24(6):1593–1597

Angelisova P, Hilgert I, Horejsi V (1994) Association of four antigens of the tetraspans family (CD37, CD53, TAPA-1, and R2/C33) with MHC class II glycoproteins. Immunogenetics 39(4):249–256

Anton ES, Hadjiargyrou M, Patterson PH, Matthew WD (1995) CD9 plays a role in schwann cell migration in vitro. J Neurosci 15(1 Pt 2):584–595

Arduise C, Abache T, Li L, Billard M, Chabanon A, Ludwig A, Mauduit P, Boucheix C, Rubinstein E, Le Naour F (2008) Tetraspanins regulate ADAM10-mediated cleavage of TNF-alpha and epidermal growth factor. J Immunol 181(10):7002–7013

Baldwin G, Novitskaya V, Sadej R, Pochec E, Litynska A, Hartmann C, Williams J, Ashman L, Eble JA, Berditchevski F (2008) Tetraspanin CD151 regulates glycosylation of (alpha)3(beta)1 integrin. J Biol Chem 283(51):35445–35454

Baleato RM, Guthrie PL, Gubler MC, Ashman LK, Roselli S (2008) Deletion of CD151 results in a strain-dependent glomerular disease due to severe alterations of the glomerular basement membrane. Am J Pathol 173(4):927–937

Bandyopadhyay S, Zhan R, Chaudhuri A, Watabe M, Pai SK, Hirota S, Hosobe S, Tsukada T, Miura K, Takano Y, Saito K, Pauza ME, Hayashi S, Wang Y, Mohinta S, Mashimo T, Iiizumi M, Furuta E, Watabe K (2006) Interaction of KAI1 on tumor cells with DARC on vascular endothelium leads to metastasis suppression. Nat Med 12(8):933–938

Barreiro O, Yanez-Mo M, Serrador JM, Montoya MC, Vicente-Manzanares M, Tejedor R, Furthmayr H, Sanchez-Madrid F (2002) Dynamic interaction of VCAM-1 and ICAM-1 with moesin and ezrin in a novel endothelial docking structure for adherent leukocytes. J Cell Biol 157(7):1233–1245

Barreiro O, Yanez-Mo M, Sala-Valdes M, Gutierrez-Lopez MD, Ovalle S, Higginbottom A, Monk PN, Cabañas C, Sanchez-Madrid F (2005) Endothelial tetraspanin microdomains regulate leukocyte firm adhesion during extravasation. Blood 105(7):2852–2861

Barreiro O, Zamai M, Yanez-Mo M, Tejera E, Lopez-Romero P, Monk PN, Gratton E, Caiolfa VR, Sanchez-Madrid F (2008) Endothelial adhesion receptors are recruited to adherent leukocytes by inclusion in preformed tetraspanin nanoplatforms. J Cell Biol 183(3):527–542

Bass R, Werner F, Odintsova E, Sugiura T, Berditchevski F, Ellis V (2005) Regulation of urokinase receptor proteolytic function by the tetraspanin CD82. J Biol Chem 280(15):14811–14818

Bassani S, Cingolani LA, Valnegri P, Folci A, Zapata J, Gianfelice A, Sala C, Goda Y, Passafaro M (2012) The X-linked intellectual disability protein TSPAN7 regulates excitatory synapse development and AMPAR trafficking. Neuron 73(6):1143–1158

Baudoux B, Castanares-Zapatero D, Leclercq-Smekens M, Berna N, Poumay Y (2000) The tetraspanin CD9 associates with the integrin alpha6beta4 in cultured human epidermal keratinocytes and is involved in cell motility. Eur J Cell Biol 79(1):41–51

Berditchevski F (2001) Complexes of tetraspanins with integrins: more than meets the eye. J Cell Sci 114(Pt 23):4143–4151

Berditchevski F, Odintsova E (1999) Characterization of integrin-tetraspanin adhesion complexes: role of tetraspanins in integrin signaling. J Cell Biol 146(2):477–492

Berditchevski F, Bazzoni G, Hemler ME (1995) Specific association of CD63 with the VLA-3 and VLA-6 integrins. J Biol Chem 270(30):17784–17790

Berditchevski F, Zutter MM, Hemler ME (1996) Characterization of novel complexes on the cell surface between integrins and proteins with 4 transmembrane domains (TM4 proteins). Mol Biol Cell 7(2):193–207

Berditchevski F, Tolias KF, Wong K, Carpenter CL, Hemler ME (1997) A novel link between integrins, transmembrane-4 superfamily proteins (CD63 and CD81), and phosphatidylinositol 4-kinase. J Biol Chem 272(5):2595–2598

Berditchevski F, Gilbert E, Griffiths MR, Fitter S, Ashman L, Jenner SJ (2001) Analysis of the CD151-alpha3beta1 integrin and CD151-tetraspanin interactions by mutagenesis. J Biol Chem 276(44):41165–41174

Berditchevski F, Odintsova E, Sawada S, Gilbert E (2002) Expression of the palmitoylation-deficient CD151 weakens the association of alpha 3 beta 1 integrin with the tetraspanin-enriched microdomains and affects integrin-dependent signaling. J Biol Chem 277(40): 36991–37000

Berthier-Vergnes O, Kharbili ME, de la Fouchardiere A, Pointecouteau T, Verrando P, Wierinckx A, Lachuer J, Le Naour F, Lamartine J (2011) Gene expression profiles of human melanoma cells with different invasive potential reveal TSPAN8 as a novel mediator of invasion. Br J Cancer 104(1):155–165

Blumenthal A, Giebel J, Ummanni R, Schluter R, Endlich K, Endlich N (2012) Morphology and migration of podocytes are affected by CD151 levels. Am J Physiol Ren Physiol 302(10):F1265–F1277

Boucheix C, Duc GH, Jasmin C, Rubinstein E (2001) Tetraspanins and malignancy. Expert Rev Mol Med 2001:1–17

Boukerche H, Su ZZ, Prevot C, Sarkar D, Fisher PB (2008) Mda-9/Syntenin promotes metastasis in human melanoma cells by activating c-Src. Proc Natl Acad Sci USA 105(41):15914–15919

Cailleteau L, Estrach S, Thyss R, Boyer L, Doye A, Domange B, Johnsson N, Rubinstein E, Boucheix C, Ebrahimian T, Silvestre JS, Lemichez E, Meneguzzi G, Mettouchi A (2010) alpha2beta1 integrin controls association of Rac with the membrane and triggers quiescence of endothelial cells. J Cell Sci 123(Pt 14):2491–2501

Carloni V, Mazzocca A, Ravichandran KS (2004) Tetraspanin CD81 is linked to ERK/MAPKinase signaling by Shc in liver tumor cells. Oncogene 23(8):1566–1574

Chambrion C, Le Naour F (2010) The tetraspanins CD9 and CD81 regulate CD9P1-induced effects on cell migration. PLoS One 5(6):e11219

Chang Y, Finnemann SC (2007) Tetraspanin CD81 is required for the alpha v beta5-integrin-dependent particle-binding step of RPE phagocytosis. J Cell Sci 120(Pt 17):3053–3063

Chang YS, di Tomaso E, McDonald DM, Jones R, Jain RK, Munn LL (2000) Mosaic blood vessels in tumors: frequency of cancer cells in contact with flowing blood. Proc Natl Acad Sci USA 97(26):14608–14613

Charrin S, Le Naour F, Oualid M, Billard M, Faure G, Hanash SM, Boucheix C, Rubinstein E (2001) The major CD9 and CD81 molecular partner. Identification and characterization of the complexes. J Biol Chem 276(17):14329–14337
Charrin S, Manie S, Oualid M, Billard M, Boucheix C, Rubinstein E (2002) Differential stability of tetraspanin/tetraspanin interactions: role of palmitoylation. FEBS Lett 516(1–3):139–144
Charrin S, Le Naour F, Labas V, Billard M, Le Caer JP, Emile JF, Petit MA, Boucheix C, Rubinstein E (2003a) EWI-2 is a new component of the tetraspanin web in hepatocytes and lymphoid cells. Biochem J 373(Pt 2):409–421
Charrin S, Manie S, Thiele C, Billard M, Gerlier D, Boucheix C, Rubinstein E (2003b) A physical and functional link between cholesterol and tetraspanins. Eur J Immunol 33(9):2479–2489
Charrin S, le Naour F, Silvie O, Milhiet PE, Boucheix C, Rubinstein E (2009a) Lateral organization of membrane proteins: tetraspanins spin their web. Biochem J 420(2):133–154
Charrin S, Yalaoui S, Bartosch B, Cocquerel L, Franetich JF, Boucheix C, Mazier D, Rubinstein E, Silvie O (2009b) The Ig domain protein CD9P-1 down-regulates CD81 ability to support Plasmodium yoelii infection. J Biol Chem 284(46):31572–31578
Chattopadhyay N, Wang Z, Ashman LK, Brady-Kalnay SM, Kreidberg JA (2003) alpha3beta1 integrin-CD151, a component of the cadherin-catenin complex, regulates PTPmu expression and cell-cell adhesion. J Cell Biol 163(6):1351–1362
Chometon G, Zhang ZG, Rubinstein E, Boucheix C, Mauch C, Aumailley M (2006) Dissociation of the complex between CD151 and laminin-binding integrins permits migration of epithelial cells. Exp Cell Res 312(7):983–995
Claas C, Seiter S, Claas A, Savelyeva L, Schwab M, Zoller M (1998) Association between the rat homologue of CO-029, a metastasis-associated tetraspanin molecule and consumption coagulopathy. J Cell Biol 141(1):267–280
Claas C, Wahl J, Orlicky DJ, Karaduman H, Schnolzer M, Kempf T, Zoller M (2005) The tetraspanin D6.1A and its molecular partners on rat carcinoma cells. Biochem J 389(Pt 1):99–110
Clark KL, Zeng Z, Langford AL, Bowen SM, Todd SC (2001) PGRL is a major CD81-associated protein on lymphocytes and distinguishes a new family of cell surface proteins. J Immunol 167(9):5115–5121
Clark KL, Oelke A, Johnson ME, Eilert KD, Simpson PC, Todd SC (2004) CD81 associates with 14-3-3 in a redox-regulated palmitoylation-dependent manner. J Biol Chem 279(19):19401–19406
Coffey GP, Rajapaksa R, Liu R, Sharpe O, Kuo CC, Krauss SW, Sagi Y, Davis RE, Staudt LM, Sharman JP, Robinson WH, Levy S (2009) Engagement of CD81 induces ezrin tyrosine phosphorylation and its cellular redistribution with filamentous actin. J Cell Sci 122(Pt 17):3137–3144
Cook GA, Wilkinson DA, Crossno JT Jr, Raghow R, Jennings LK (1999) The tetraspanin CD9 influences the adhesion, spreading, and pericellular fibronectin matrix assembly of Chinese hamster ovary cells on human plasma fibronectin. Exp Cell Res 251(2):356–371
Cook GA, Longhurst CM, Grgurevich S, Cholera S, Crossno JT Jr, Jennings LK (2002) Identification of CD9 extracellular domains important in regulation of CHO cell adhesion to fibronectin and fibronectin pericellular matrix assembly. Blood 100(13):4502–4511
Cowin AJ, Adams D, Geary SM, Wright MD, Jones JC, Ashman LK (2006) Wound healing is defective in mice lacking tetraspanin CD151. J Invest Dermatol 126(3):680–689
Crotta S, Stilla A, Wack A, D'Andrea A, Nuti S, D'Oro U, Mosca M, Filliponi F, Brunetto RM, Bonino F, Abrignani S, Valiante NM (2002) Inhibition of natural killer cells through engagement of CD81 by the major hepatitis C virus envelope protein. J Exp Med 195(1):35–41
Crotta S, Ronconi V, Ulivieri C, Baldari CT, Valiante NM, Abrignani S, Wack A (2006) Cytoskeleton rearrangement induced by tetraspanin engagement modulates the activation of T and NK cells. Eur J Immunol 36(4):919–929
De Bruyne E, Andersen TL, De Raeve H, Van Valckenborgh E, Caers J, Van Camp B, Delaisse JM, Van Riet I, Vanderkerken K (2006) Endothelial cell-driven regulation of CD9 or motility-related protein-1 expression in multiple myeloma cells within the murine 5T33MM model and myeloma patients. Leukemia 20(10):1870–1879
Deissler H, Kuhn EM, Lang GE (2007) Tetraspanin CD9 is involved in the migration of retinal microvascular endothelial cells. Int J Mol Med 20(5):643–652

Delaguillaumie A, Lagaudriere-Gesbert C, Popoff MR, Conjeaud H (2002) Rho GTPases link cytoskeletal rearrangements and activation processes induced via the tetraspanin CD82 in T lymphocytes. J Cell Sci 115(Pt 2):433–443

Delaguillaumie A, Harriague J, Kohanna S, Bismuth G, Rubinstein E, Seigneuret M, Conjeaud H (2004) Tetraspanin CD82 controls the association of cholesterol-dependent microdomains with the actin cytoskeleton in T lymphocytes: relevance to co-stimulation. J Cell Sci 117(Pt 22):5269–5282

Deng J, Dekruyff RH, Freeman GJ, Umetsu DT, Levy S (2002) Critical role of CD81 in cognate T-B cell interactions leading to Th2 responses. Int Immunol 14(5):513–523

Dijkstra S, Kooij G, Verbeek R, van der Pol SM, Amor S, Geisert EE Jr, Dijkstra CD, van Noort JM, Vries HE (2008) Targeting the tetraspanin CD81 blocks monocyte transmigration and ameliorates EAE. Neurobiol Dis 31(3):413–421

Domanico SZ, Pelletier AJ, Havran WL, Quaranta V (1997) Integrin alpha 6A beta 1 induces CD81-dependent cell motility without engaging the extracellular matrix migration substrate. Mol Biol Cell 8(11):2253–2265

Dominguez F, Simon C, Quinonero A, Ramirez MA, Gonzalez-Munoz E, Burghardt H, Cervero A, Martinez S, Pellicer A, Palacin M, Sanchez-Madrid F, Yanez-Mo M (2010) Human endometrial CD98 is essential for blastocyst adhesion. PLoS One 5(10):e13380

Dominguez-Jimenez C, Yanez-Mo M, Carreira A, Tejedor R, Gonzalez-Amaro R, Alvarez V, Sanchez-Madrid F (2001) Involvement of alpha3 integrin/tetraspanin complexes in the angiogenic response induced by angiotensin II. FASEB J 15(8):1457–1459

Engering A, Pieters J (2001) Association of distinct tetraspanins with MHC class II molecules at different subcellular locations in human immature dendritic cells. Int Immunol 13(2):127–134

Feigelson SW, Grabovsky V, Shamri R, Levy S, Alon R (2003) The CD81 tetraspanin facilitates instantaneous leukocyte VLA-4 adhesion strengthening to vascular cell adhesion molecule 1 (VCAM-1) under shear flow. J Biol Chem 278(51):51203–51212

Fitter S, Sincock PM, Jolliffe CN, Ashman LK (1999) Transmembrane 4 superfamily protein CD151 (PETA-3) associates with beta 1 and alpha IIb beta 3 integrins in haemopoietic cell lines and modulates cell-cell adhesion. Biochem J 338:61–70

Franco M, Muratori C, Corso S, Tenaglia E, Bertotti A, Capparuccia L, Trusolino L, Comoglio PM, Tamagnone L (2010) The tetraspanin CD151 is required for Met-dependent signaling and tumor cell growth. J Biol Chem 285(50):38756–38764

Furuya M, Kato H, Nishimura N, Ishiwata I, Ikeda H, Ito R, Yoshiki T, Ishikura H (2005) Down-regulation of CD9 in human ovarian carcinoma cell might contribute to peritoneal dissemination: morphologic alteration and reduced expression of beta1 integrin subsets. Cancer Res 65(7):2617–2625

Garcia-Lopez MA, Barreiro O, Garcia-Diez A, Sanchez-Madrid F, Penas PF (2005) Role of tetraspanins CD9 and CD151 in primary melanocyte motility. J Invest Dermatol 125(5):1001–1009

Geary SM, Cowin AJ, Copeland B, Baleato RM, Miyazaki K, Ashman LK (2008) The role of the tetraspanin CD151 in primary keratinocyte and fibroblast functions: implications for wound healing. Exp Cell Res 314(11–12):2165–2175

Gesierich S, Paret C, Hildebrand D, Weitz J, Zgraggen K, Schmitz-Winnenthal FH, Horejsi V, Yoshie O, Herlyn D, Ashman LK, Zoller M (2005) Colocalization of the tetraspanins, CO-029 and CD151, with integrins in human pancreatic adenocarcinoma: impact on cell motility. Clin Cancer Res 11(8):2840–2852

Gesierich S, Berezovskiy I, Ryschich E, Zoller M (2006) Systemic induction of the angiogenesis switch by the tetraspanin D6.1A/CO-029. Cancer Res 66(14):7083–7094

Gordon-Alonso M, Yanez-Mo M, Barreiro O, Alvarez S, Munoz-Fernandez MA, Valenzuela-Fernandez A, Sanchez-Madrid F (2006) Tetraspanins CD9 and CD81 modulate HIV-1-induced membrane fusion. J Immunol 177(8):5129–5137

Goschnick MW, Lau LM, Wee JL, Liu YS, Hogarth PM, Robb LM, Hickey MJ, Wright MD, Jackson DE (2006) Impaired "outside-in" integrin alphaIIbbeta3 signaling and thrombus stability in TSSC6-deficient mice. Blood 108(6):1911–1918

Greco C, Bralet MP, Ailane N, Dubart-Kupperschmitt A, Rubinstein E, Le Naour F, Boucheix C (2010) E-cadherin/p120-catenin and tetraspanin Co-029 cooperate for cell motility control in human colon carcinoma. Cancer Res 70(19):7674–7683

Green LR, Monk PN, Partridge LJ, Morris P, Gorringe AR, Read RC (2011) Cooperative role for tetraspanins in adhesin-mediated attachment of bacterial species to human epithelial cells. Infect Immun 79(6):2241–2249

Gutierrez-Lopez MD, Ovalle S, Yanez-Mo M, Sanchez-Sanchez N, Rubinstein E, Olmo N, Lizarbe MA, Sanchez-Madrid F, Cabanas C (2003) A functionally relevant conformational epitope on the CD9 tetraspanin depends on the association with activated beta1 integrin. J Biol Chem 278(1):208–218

Gutierrez-Lopez MD, Gilsanz A, Yanez-Mo M, Ovalle S, Lafuente EM, Dominguez C, Monk PN, Gonzalez-Alvaro I, Sanchez-Madrid F, Cabanas C (2011) The sheddase activity of ADAM17/TACE is regulated by the tetraspanin CD9. Cell Mol Life Sci 68(19):3275–3292

Hadjiargyrou M, Kaprielian Z, Kato N, Patterson PH (1996) Association of the tetraspan protein CD9 with integrins on the surface of S-16 schwann cells. J Neurochem 67(6):2505–2513

Harris HJ, Farquhar MJ, Mee CJ, Davis C, Reynolds GM, Jennings A, Hu K, Yuan F, Deng H, Hubscher SG, Han JH, Balfe P, McKeating JA (2008) CD81 and claudin 1 coreceptor association: role in hepatitis C virus entry. J Virol 82(10):5007–5020

Harris HJ, Davis C, Mullins JG, Hu K, Goodall M, Farquhar MJ, Mee CJ, McCaffrey K, Young S, Drummer H, Balfe P, McKeating JA (2010) Claudin association with CD81 defines hepatitis C virus entry. J Biol Chem 285(27):21092–21102

Hasegawa H, Nomura T, Kishimoto K, Yanagisawa K, Fujita S (1998) SFA-1/PETA-3 (CD151), a member of the transmembrane 4 superfamily, associates preferentially with alpha 5 beta 1 integrin and regulates adhesion of human T cell leukemia virus type 1-infected T cells to fibronectin. J Immunol 161(6):3087–3095

Hasegawa M, Furuya M, Kasuya Y, Nishiyama M, Sugiura T, Nikaido T, Momota Y, Ichinose M, Kimura S (2007) CD151 dynamics in carcinoma-stroma interaction: integrin expression, adhesion strength and proteolytic activity. Lab Invest 87(9):882–892

He B, Liu L, Cook GA, Grgurevich S, Jennings LK, Zhang XA (2005) Tetraspanin CD82 attenuates cellular morphogenesis through down-regulating integrin alpha6-mediated cell adhesion. J Biol Chem 280(5):3346–3354

Hemler ME (2003) Tetraspanin proteins mediate cellular penetration, invasion, and fusion events and define a novel type of membrane microdomain. Annu Rev Cell Dev Biol 19:397–422

Hemler ME (2005) Tetraspanin functions and associated microdomains. Nat Rev Mol Cell Biol 6(10):801–811

Hemler ME (2008) Targeting of tetraspanin proteins-potential benefits and strategies. Nat Rev Drug Discov 7(9):747–758

Herlevsen M, Schmidt DS, Miyazaki K, Zoller M (2003) The association of the tetraspanin D6.1A with the alpha6beta4 integrin supports cell motility and liver metastasis formation. J Cell Sci 116(Pt 21):4373–4390

Hirano T, Higuchi T, Ueda M, Inoue T, Kataoka N, Maeda M, Fujiwara H, Fujii S (1999a) CD9 is expressed in extravillous trophoblasts in association with integrin alpha3 and integrin alpha5. Mol Hum Reprod 5(2):162–167

Hirano T, Higuchi T, Katsuragawa H, Inoue T, Kataoka N, Park KR, Ueda M, Maeda M, Fujiwara H, Fujii S (1999b) CD9 is involved in invasion of human trophoblast-like choriocarcinoma cell line, BeWo cells. Mol Hum Reprod 5(2):168–174

Hong IK, Kim YM, Jeoung DI, Kim KC, Lee H (2005) Tetraspanin CD9 induces MMP-2 expression by activating p38 MAPK, JNK and c-Jun pathways in human melanoma cells. Exp Mol Med 37(3):230–239

Hong IK, Jin YJ, Byun HJ, Jeoung DI, Kim YM, Lee H (2006) Homophilic interactions of Tetraspanin CD151 up-regulate motility and matrix metalloproteinase-9 expression of human melanoma cells through adhesion-dependent c-Jun activation signaling pathways. J Biol Chem 281(34):24279–24292

Hoorn T, Paul P, Janssen L, Janssen H, Neefjes J (2012) Dynamics within tetraspanin pairs affect MHC class II expression. J Cell Sci 125(Pt 2):328–339

Huang Y, Sook-Kim M, Ratovitski E (2008) Midkine promotes tetraspanin-integrin interaction and induces FAK-Stat1alpha pathway contributing to migration/invasiveness of human head and neck squamous cell carcinoma cells. Biochem Biophys Res Commun 377(2):474–478

Imai T, Kakizaki M, Nishimura M, Yoshie O (1995) Molecular analyses of the association of CD4 with two members of the transmembrane 4 superfamily, CD81 and CD82. J Immunol 155(3):1229–1239

Indig FE, Diaz-Gonzalez F, Ginsberg MH (1997) Analysis of the tetraspanin CD9-integrin alphaIIb-beta3 (GPIIb-IIIa) complex in platelet membranes and transfected cells. Biochem J 327:291–298

Israels SJ, McMillan-Ward EM (2010) Palmitoylation supports the association of tetraspanin CD63 with CD9 and integrin alphaIIbbeta3 in activated platelets. Thromb Res 125(2):152–158

Jegou A, Ziyyat A, Barraud-Lange V, Perez E, Wolf JP, Pincet F, Gourier C (2011) CD9 tetraspanin generates fusion competent sites on the egg membrane for mammalian fertilization. Proc Natl Acad Sci USA 108(27):10946–10951

Johnson JL, Winterwood N, DeMali KA, Stipp CS (2009) Tetraspanin CD151 regulates RhoA activation and the dynamic stability of carcinoma cell-cell contacts. J Cell Sci 122(Pt 13):2263–2273

Jones PH, Bishop LA, Watt FM (1996) Functional significance of CD9 association with beta 1 integrins in human epidermal keratinocytes. Cell Adhes Commun 4(4–5):297–305

Jung K, Liu X, Chirco R, Fridman R, Kim H (2006) Identification of CD63 as a tissue inhibitor of metalloproteinase-1 interacting cell surface protein. EMBO J 25(17):3934–3942

Junge HJ, Yang S, Burton JB, Paes K, Shu X, French DM, Costa M, Rice DS, Ye W (2009) TSPAN12 regulates retinal vascular development by promoting Norrin- but not Wnt-induced FZD4/beta-catenin signaling. Cell 139(2):299–311

Kapadia SB, Barth H, Baumert T, McKeating JA, Chisari FV (2007) Initiation of hepatitis C virus infection is dependent on cholesterol and cooperativity between CD81 and scavenger receptor B type I. J Virol 81(1):374–383

Karamatic Crew V, Burton N, Kagan A, Green CA, Levene C, Flinter F, Brady RL, Daniels G, Anstee DJ (2004) CD151, the first member of the tetraspanin (TM4) superfamily detected on erythrocytes, is essential for the correct assembly of human basement membranes in kidney and skin. Blood 104(8):2217–2223

Kawakami Y, Kawakami K, Steelant WF, Ono M, Baek RC, Handa K, Withers DA, Hakomori S (2002) Tetraspanin CD9 is a "proteolipid," and its interaction with alpha 3 integrin in microdomain is promoted by GM3 ganglioside, leading to inhibition of laminin-5-dependent cell motility. J Biol Chem 277(37):34349–34358

Kazarov AR, Yang X, Stipp CS, Sehgal B, Hemler ME (2002) An extracellular site on tetraspanin CD151 determines alpha 3 and alpha 6 integrin-dependent cellular morphology. J Cell Biol 158(7):1299–1309

Kettner S, Kalthoff F, Graf P, Priller E, Kricek F, Lindley I, Schweighoffer T (2007) EWI-2/CD316 is an inducible receptor of HSPA8 on human dendritic cells. Mol Cell Biol 27(21):7718–7726

Klein-Soyer C, Azorsa DO, Cazenave JP, Lanza F (2000) CD9 participates in endothelial cell migration during in vitro wound repair. Arterioscler Thromb Vasc Biol 20(2):360–369

Klosek SK, Nakashiro K, Hara S, Shintani S, Hasegawa H, Hamakawa H (2005) CD151 forms a functional complex with c-Met in human salivary gland cancer cells. Biochem Biophys Res Commun 336(2):408–416

Kohno M, Hasegawa H, Miyake M, Yamamoto T, Fujita S (2002) CD151 enhances cell motility and metastasis of cancer cells in the presence of focal adhesion kinase. Int J Cancer 97(3):336–343

Kolesnikova TV, Stipp CS, Rao RM, Lane WS, Luscinskas FW, Hemler ME (2004) EWI-2 modulates lymphocyte integrin alpha4beta1 functions. Blood 103(8):3013–3019

Kolesnikova TV, Kazarov AR, Lemieux ME, Lafleur MA, Kesari S, Kung AL, Hemler ME (2009) Glioblastoma inhibition by cell surface immunoglobulin protein EWI-2, in vitro and in vivo. Neoplasia 11(1):77–86

Konstantopoulos K, Thomas SN (2009) Cancer cells in transit: the vascular interactions of tumor cells. Annu Rev Biomed Eng 11:177–202

Kotha J, Longhurst C, Appling W, Jennings LK (2008) Tetraspanin CD9 regulates beta 1 integrin activation and enhances cell motility to fibronectin via a PI-3 kinase-dependent pathway. Exp Cell Res 314(8):1811–1822

Kotha J, Zhang C, Longhurst CM, Lu Y, Jacobs J, Cheng Y, Jennings LK (2009) Functional relevance of tetraspanin CD9 in vascular smooth muscle cell injury phenotypes: a novel target for the prevention of neointimal hyperplasia. Atherosclerosis 203(2):377–386

Kovalenko OV, Yang XH, Hemler ME (2007) A novel cysteine cross-linking method reveals a direct association between claudin-1 and tetraspanin CD9. Mol Cell Proteomics 6(11):1855–1867

Kramer B, Schulte D, Korner C, Zwank C, Hartmann A, Michalk M, Sohne J, Langhans B, Nischalke HD, Coenen M, Mohl C, Vogt A, Hennenberg M, Sauerbruch T, Spengler U, Nattermann J (2009) Regulation of NK cell trafficking by CD81. Eur J Immunol 39(12):3447–3458

Kuhn S, Koch M, Nubel T, Ladwein M, Antolovic D, Klingbeil P, Hildebrand D, Moldenhauer G, Langbein L, Franke WW, Weitz J, Zoller M (2007) A complex of EpCAM, claudin-7, CD44 variant isoforms, and tetraspanins promotes colorectal cancer progression. Mol Cancer Res 5(6):553–567

Lafleur MA, Xu D, Hemler ME (2009) Tetraspanin proteins regulate membrane type-1 matrix metalloproteinase-dependent pericellular proteolysis. Mol Biol Cell 20(7):2030–2040

Lagaudriere-Gesbert C, Le Naour F, Lebel-Binay S, Billard M, Lemichez E, Boquet P, Boucheix C, Conjeaud H, Rubinstein E (1997) Functional analysis of four tetraspans, CD9, CD53, CD81, and CD82, suggests a common role in costimulation, cell adhesion, and migration: only CD9 upregulates HB-EGF activity. Cell Immunol 182(2):105–112

Lagaudriere-Gesbert C, Lebel-Binay S, Hubeau C, Fradelizi D, Conjeaud H (1998) Signaling through the tetraspanin CD82 triggers its association with the cytoskeleton leading to sustained morphological changes and T cell activation. Eur J Immunol 28(12):4332–4344

Lammerding J, Kazarov AR, Huang H, Lee RT, Hemler ME (2003) Tetraspanin CD151 regulates alpha6beta1 integrin adhesion strengthening. Proc Natl Acad Sci USA 100(13):7616–7621

Larochelle A, Gillette JM, Desmond R, Ichwan B, Cantilena A, Cerf A, Barrett AJ, Wayne AS, Lippincott-Schwartz J, Dunbar CE (2012) Bone marrow homing and engraftment of human hematopoietic stem and progenitor cells is mediated by a polarized membrane domain. Blood 119(8):1848–1855

Latysheva N, Muratov G, Rajesh S, Padgett M, Hotchin NA, Overduin M, Berditchevski F (2006) Syntenin-1 is a new component of tetraspanin-enriched microdomains: mechanisms and consequences of the interaction of syntenin-1 with CD63. Mol Cell Biol 26(20):7707–7718

Lau LM, Wee JL, Wright MD, Moseley GW, Hogarth PM, Ashman LK, Jackson DE (2004) The tetraspanin superfamily member CD151 regulates outside-in integrin alphaIIbbeta3 signaling and platelet function. Blood 104(8):2368–2375

Lazo PA (2007) Functional implications of tetraspanin proteins in cancer biology. Cancer Sci 98(11):1666–1677

Le Naour F, Andre M, Boucheix C, Rubinstein E (2006) Membrane microdomains and proteomics: lessons from tetraspanin microdomains and comparison with lipid rafts. Proteomics 6(24):6447–6454

Lebel-Binay S, Lagaudrière C, Fradelizi D, Conjeaud H (1995) CD82, member of the tetra-span-transmembrane protein family, is a costimulatory protein for T cell activation. J Immunol 155:101–110

Legg JW, Isacke CM (1998) Identification and functional analysis of the ezrin-binding site in the hyaluronan receptor, CD44. Curr Biol 8(12):705–708

Leung KT, Chan KY, Ng PC, Lau TK, Chiu WM, Tsang KS, Li CK, Kong CK, Li K (2011) The tetraspanin CD9 regulates migration, adhesion, and homing of human cord blood CD34+ hematopoietic stem and progenitor cells. Blood 117(6):1840–1850

Levy S, Shoham T (2005) The tetraspanin web modulates immune-signalling complexes. Nat Rev Immunol 5(2):136–148

Levy S, Todd SC, Maecker HT (1998) CD81 (TAPA-1): a molecule involved in signal transduction and cell adhesion in the immune system. Annu Rev Immunol 16:89–109

Ley K, Laudanna C, Cybulsky MI, Nourshargh S (2007) Getting to the site of inflammation: the leukocyte adhesion cascade updated. Nat Rev Immunol 7(9):678–689

Lijnen HR, Lupu F, Collen D, Le Naour F, Boucheix C (2000) CD9 gene deficiency does not affect smooth muscle cell migration and neointima formation after vascular injury in mice. Thromb Haemost 83(6):956–961

Little KD, Hemler ME, Stipp CS (2004) Dynamic regulation of a GPCR-tetraspanin-G protein complex on intact cells: central role of CD81 in facilitating GPR56-Galpha q/11 association. Mol Biol Cell 15(5):2375–2387

Liu WM, Cao YJ, Yang YJ, Li J, Hu Z, Duan EK (2006) Tetraspanin CD9 regulates invasion during mouse embryo implantation. J Mol Endocrinol 36(1):121–130

Liu L, He B, Liu WM, Zhou D, Cox JV, Zhang XA (2007) Tetraspanin CD151 promotes cell migration by regulating integrin trafficking. J Biol Chem 282(43):31631–31642

Liu WF, Zuo HJ, Chai BL, Peng D, Fei YJ, Lin JY, Yu XH, Wang DW, Liu ZX (2011) Role of tetraspanin CD151-alpha3/alpha6 integrin complex: implication in angiogenesis CD151-integrin complex in angiogenesis. Int J Biochem Cell Biol 43(4):642–650

Longhurst CM, Jacobs JD, White MM, Crossno JT Jr, Fitzgerald DA, Bao J, Fitzgerald TJ, Raghow R, Jennings LK (2002) Chinese hamster ovary cell motility to fibronectin is modulated by the second extracellular loop of CD9. Identification of a putative fibronectin binding site. J Biol Chem 277(36):32445–32452

Longo N, Yanez-Mo M, Mittelbrunn M, de la Rosa G, Munoz ML, Sanchez-Madrid F, Sanchez-Mateos P (2001) Regulatory role of tetraspanin CD9 in tumor-endothelial cell interaction during transendothelial invasion of melanoma cells. Blood 98(13):3717–3726

Lozahic S, Christiansen D, Manie S, Gerlier D, Billard M, Boucheix C, Rubinstein E (2000) CD46 (membrane cofactor protein) associates with multiple beta1 integrins and tetraspans. Eur J Immunol 30(3):900–907

Luo BH, Carman CV, Springer TA (2007) Structural basis of integrin regulation and signaling. Annu Rev Immunol 25:619–647

Maecker HT (2003) Human CD81 directly enhances Th1 and Th2 cell activation, but preferentially induces proliferation of Th2 cells upon long-term stimulation. BMC Immunol 4:1

Maecker HT, Levy S (1997) Normal lymphocyte development but delayed humoral immune response in CD81-null mice. J Exp Med 185(8):1505–1510

Mangin PH, Kleitz L, Boucheix C, Gachet C, Lanza F (2009) CD9 negatively regulates integrin alphaIIbbeta3 activation and could thus prevent excessive platelet recruitment at sites of vascular injury. J Thromb Haemost 7(5):900–902

Mantegazza AR, Barrio MM, Moutel S, Bover L, Weck M, Brossart P, Teillaud JL, Mordoh J (2004) CD63 tetraspanin slows down cell migration and translocates to the endosomal-lysosomal-MIICs route after extracellular stimuli in human immature dendritic cells. Blood 104(4):1183–1190

Mazzocca A, Carloni V, Sciammetta S, Cordella C, Pantaleo P, Caldini A, Gentilini P, Pinzani M (2002) Expression of transmembrane 4 superfamily (TM4SF) proteins and their role in hepatic stellate cell motility and wound healing migration. J Hepatol 37(3):322–330

Mazzocca A, Sciammetta SC, Carloni V, Cosmi L, Annunziato F, Harada T, Abrignani S, Pinzani M (2005) Binding of hepatitis C virus envelope protein E2 to CD81 up-regulates matrix metalloproteinase-2 in human hepatic stellate cells. J Biol Chem 280(12):11329–11339

Mazzocca A, Liotta F, Carloni V (2008) Tetraspanin CD81-regulated cell motility plays a critical role in intrahepatic metastasis of hepatocellular carcinoma. Gastroenterology 135(1):244–256, e241

Mela A, Goldman JE (2009) The tetraspanin KAI1/CD82 is expressed by late-lineage oligodendrocyte precursors and may function to restrict precursor migration and promote oligodendrocyte differentiation and myelination. J Neurosci 29(36):11172–11181

Meyer-Wentrup F, Figdor CG, Ansems M, Brossart P, Wright MD, Adema GJ, van Spriel AB (2007) Dectin-1 interaction with tetraspanin CD37 inhibits IL-6 production. J Immunol 178(1):154–162

Mittelbrunn M, Yanez-Mo M, Sancho D, Ursa A, Sanchez-Madrid F (2002) Cutting edge: dynamic redistribution of tetraspanin CD81 at the central zone of the immune synapse in both T lymphocytes and APC. J Immunol 169(12):6691–6695

Miura Y, Kainuma M, Jiang H, Velasco H, Vogt PK, Hakomori S (2004) Reversion of the Jun-induced oncogenic phenotype by enhanced synthesis of sialosyllactosylceramide (GM3 ganglioside). Proc Natl Acad Sci USA 101(46):16204–16209

Moribe H, Yochem J, Yamada H, Tabuse Y, Fujimoto T, Mekada E (2004) Tetraspanin protein (TSP-15) is required for epidermal integrity in caenorhabditis elegans. J Cell Sci 117(Pt 22):5209–5220

Mulrooney JP, Allen J, Bickelhaupt E, Grabel LB (2002) CD9-alpha6beta1 interactions in migratory parietal endoderm cells. Cell Commun Adhes 9(5–6):249–258

Murayama Y, Shinomura Y, Oritani K, Miyagawa J, Yoshida H, Nishida M, Katsube F, Shiraga M, Miyazaki T, Nakamoto T, Tsutsui S, Tamura S, Higashiyama S, Shimomura I, Hayashi N (2008) The tetraspanin CD9 modulates epidermal growth factor receptor signaling in cancer cells. J Cell Physiol 216(1):135–143

Nakamoto T, Murayama Y, Oritani K, Boucheix C, Rubinstein E, Nishida M, Katsube F, Watabe K, Kiso S, Tsutsui S, Tamura S, Shinomura Y, Hayashi N (2009) A novel therapeutic strategy with anti-CD9 antibody in gastric cancers. J Gastroenterol 44(9):889–896

Nakamura Y, Handa K, Iwamoto R, Tsukamoto T, Takahasi M, Mekada E (2001) Immunohistochemical distribution of CD9, heparin binding epidermal growth factor-like growth factor, and integrin alpha3beta1 in normal human tissues. J Histochem Cytochem 49(4):439–444

Nattermann J, Zimmermann H, Iwan A, von Lilienfeld-Toal M, Leifeld L, Nischalke HD, Langhans B, Sauerbruch T, Spengler U (2006) Hepatitis C virus E2 and CD81 interaction may be associated with altered trafficking of dendritic cells in chronic hepatitis C. Hepatology 44(4):945–954

Nazarenko I, Rana S, Baumann A, McAlear J, Hellwig A, Trendelenburg M, Lochnit G, Preissner KT, Zoller M (2010) Cell surface tetraspanin Tspan8 contributes to molecular pathways of exosome-induced endothelial cell activation. Cancer Res 70(4):1668–1678

Nichols TC, Guthridge JM, Karp DR, Molina H, Fletcher DR, Holers VM (1998) Gamma-glutamyl transpeptidase, an ecto-enzyme regulator of intracellular redox potential, is a component of TM4 signal transduction complexes. Eur J Immunol 28(12):4123–4129

Nishida M, Miyagawa J, Yamashita S, Higashiyama S, Nakata A, Ouchi N, Tamura R, Yamamori K, Kihara S, Taniguchi N, Matsuzawa Y (2000) Localization of CD9, an enhancer protein for proheparin-binding epidermal growth factor-like growth factor, in human atherosclerotic plaques: possible involvement of juxtacrine growth mechanism on smooth muscle cell proliferation. Arterioscler Thromb Vasc Biol 20(5):1236–1243

Nishiuchi R, Sanzen N, Nada S, Sumida Y, Wada Y, Okada M, Takagi J, Hasegawa H, Sekiguchi K (2005) Potentiation of the ligand-binding activity of integrin alpha3beta1 via association with tetraspanin CD151. Proc Natl Acad Sci USA 102(6):1939–1944

Novitskaya V, Romanska H, Dawoud M, Jones JL, Berditchevski F (2010) Tetraspanin CD151 regulates growth of mammary epithelial cells in three-dimensional extracellular matrix: implication for mammary ductal carcinoma in situ. Cancer Res 70(11):4698–4708

Nubel T, Preobraschenski J, Tuncay H, Weiss T, Kuhn S, Ladwein M, Langbein L, Zoller M (2009) Claudin-7 regulates EpCAM-mediated functions in tumor progression. Mol Cancer Res 7(3):285–299

Odintsova E, Sugiura T, Berditchevski F (2000) Attenuation of EGF receptor signaling by a metastasis suppressor, the tetraspanin CD82/KAI-1. Curr Biol 10(16):1009–1012

Odintsova E, Voortman J, Gilbert E, Berditchevski F (2003) Tetraspanin CD82 regulates compartmentalisation and ligand-induced dimerization of EGFR. J Cell Sci 116(Pt 22):4557–4566

Odintsova E, Butters TD, Monti E, Sprong H, van Meer G, Berditchevski F (2006) Gangliosides play an important role in the organization of CD82-enriched microdomains. Biochem J 400(2):315–325

Ono M, Handa K, Withers DA, Hakomori S (2000) Glycosylation effect on membrane domain (GEM) involved in cell adhesion and motility: a preliminary note on functional alpha3, alpha5-CD82 glycosylation complex in ldlD 14 cells. Biochem Biophys Res Commun 279(3):744–750

Ovalle S, Gutierrez-Lopez MD, Olmo N, Turnay J, Lizarbe MA, Majano P, Molina-Jimenez F, Lopez-Cabrera M, Yanez-Mo M, Sanchez-Madrid F, Cabanas C (2007) The tetraspanin CD9 inhibits the proliferation and tumorigenicity of human colon carcinoma cells. Int J Cancer 121(10):2140–2152

Pellinen T, Rantala JK, Arjonen A, Mpindi JP, Kallioniemi O, Ivaska J (2012) A functional genetic screen reveals new regulators of beta1-integrin activity. J Cell Sci 125(Pt 3):649–661

Penas PF, Garcia-Diez A, Sanchez-Madrid F, Yanez-Mo M (2000) Tetraspanins are localized at motility-related structures and involved in normal human keratinocyte wound healing migration. J Invest Dermatol 114(6):1126–1135

Peters DG, Kassam AB, Feingold E, Heidrich-O'Hare E, Yonas H, Ferrell RE, Brufsky A (2001) Molecular anatomy of an intracranial aneurysm: coordinated expression of genes involved in wound healing and tissue remodeling. Stroke 32(4):1036–1042

Powner D, Kopp PM, Monkley SJ, Critchley DR, Berditchevski F (2011) Tetraspanin CD9 in cell migration. Biochem Soc Trans 39(2):563–567

Prox J, Willenbrock M, Weber S, Lehmann T, Schmidt-Arras D, Schwanbeck R, Saftig P, Schwake M (2012) Tetraspanin15 regulates cellular trafficking and activity of the ectodomain sheddase ADAM10. Cell Mol Life Sci 69(17):2919–2932

Quast T, Eppler F, Semmling V, Schild C, Homsi Y, Levy S, Lang T, Kurts C, Kolanus W (2011) CD81 is essential for the formation of membrane protrusions and regulates Rac1-activation in adhesion-dependent immune cell migration. Blood 118(7):1818–1827

Rana S, Claas C, Kretz CC, Nazarenko I, Zoeller M (2011) Activation-induced internalization differs for the tetraspanins CD9 and Tspan8: impact on tumor cell motility. Int J Biochem Cell Biol 43(1):106–119

Richardson MM, Jennings LK, Zhang XA (2011) Tetraspanins and tumor progression. Clin Exp Metastasis 28(3):261–270

Robak T, Robak P, Smolewski P (2009) TRU-016, a humanized anti-CD37 IgG fusion protein for the potential treatment of B-cell malignancies. Curr Opin Investig Drugs 10(12):1383–1390

Rocha-Perugini V, Montpellier C, Delgrange D, Wychowski C, Helle F, Pillez A, Drobecq H, Le Naour F, Charrin S, Levy S, Rubinstein E, Dubuisson J, Cocquerel L (2008) The CD81 partner EWI-2wint inhibits hepatitis C virus entry. PLoS One 3(4):e1866

Rocha-Perugini V, Lavie M, Delgrange D, Canton J, Pillez A, Potel J, Lecoeur C, Rubinstein E, Dubuisson J, Wychowski C, Cocquerel L (2009) The association of CD81 with tetraspanin-enriched microdomains is not essential for Hepatitis C Virus entry. BMC Microbiol 9(1):111

Rohlena J, Volger OL, van Buul JD, Hekking LH, van Gils JM, Bonta PI, Fontijn RD, Post JA, Hordijk PL, Horrevoets AJ (2009) Endothelial CD81 is a marker of early human atherosclerotic plaques and facilitates monocyte adhesion. Cardiovasc Res 81(1):187–196

Romanska HM, Berditchevski F (2011) Tetraspanins in human epithelial malignancies. J Pathol 223(1):4–14

Rubinstein E, Le Naour F, Billard M, Prenant M, Boucheix C (1994) CD9 antigen is an accessory subunit of the VLA integrin complexes. Eur J Immunol 24(12):3005–3013

Rubinstein E, Le Naour F, Lagaudriere-Gesbert C, Billard M, Conjeaud H, Boucheix C (1996) CD9, CD63, CD81, and CD82 are components of a surface tetraspan network connected to HLA-DR and VLA integrins. Eur J Immunol 26(11):2657–2665

Runge KE, Evans JE, He ZY, Gupta S, McDonald KL, Stahlberg H, Primakoff P, Myles DG (2007) Oocyte CD9 is enriched on the microvillar membrane and required for normal microvillar shape and distribution. Dev Biol 304(1):317–325

Ruseva Z, Geiger PX, Hutzler P, Kotzsch M, Luber B, Schmitt M, Gross E, Reuning U (2009) Tumor suppressor KAI1 affects integrin alphavbeta3-mediated ovarian cancer cell adhesion, motility, and proliferation. Exp Cell Res 315(10):1759–1771

Sachs N, Kreft M, van den Bergh Weerman MA, Beynon AJ, Peters TA, Weening JJ, Sonnenberg A (2006) Kidney failure in mice lacking the tetraspanin CD151. J Cell Biol 175(1):33–39

Sachs N, Claessen N, Aten J, Kreft M, Teske GJ, Koeman A, Zuurbier CJ, Janssen H, Sonnenberg A (2012) Blood pressure influences end-stage renal disease of Cd151 knockout mice. J Clin Invest 122(1):348–358

Sadej R, Romanska H, Baldwin G, Gkirtzimanaki K, Novitskaya V, Filer AD, Krcova Z, Kusinska R, Ehrmann J, Buckley CD, Kordek R, Potemski P, Eliopoulos AG, el Lalani N, Berditchevski F (2009) CD151 regulates tumorigenesis by modulating the communication between tumor cells and endothelium. Mol Cancer Res 7(6):787–798

Sadej R, Romanska H, Kavanagh D, Baldwin G, Takahashi T, Kalia N, Berditchevski F (2010) Tetraspanin CD151 regulates transforming growth factor beta signaling: implication in tumor metastasis. Cancer Res 70(14):6059–6070

Saito Y, Tachibana I, Takeda Y, Yamane H, He P, Suzuki M, Minami S, Kijima T, Yoshida M, Kumagai T, Osaki T, Kawase I (2006) Absence of CD9 enhances adhesion-dependent morphologic differentiation, survival, and matrix metalloproteinase-2 production in small cell lung cancer cells. Cancer Res 66(19):9557–9565

Sala-Valdes M, Ursa A, Charrin S, Rubinstein E, Hemler ME, Sanchez-Madrid F, Yanez-Mo M (2006) EWI-2 and EWI-F link the tetraspanin web to the actin cytoskeleton through their direct association with ezrin-radixin-moesin proteins. J Biol Chem 281(28):19665–19675

Sala-Valdes M, Gordon-Alonso M, Tejera E, Ibanez A, Cabrero JR, Ursa A, Mittelbrunn M, Lozano F, Sanchez-Madrid F, Yanez-Mo M (2012) Association of syntenin-1 with M-RIP polarizes Rac-1 activation during chemotaxis and immune interactions. J Cell Sci 125(5): 1235–1246

Sauer G, Windisch J, Kurzeder C, Heilmann V, Kreienberg R, Deissler H (2003) Progression of cervical carcinomas is associated with down-regulation of CD9 but strong local re-expression at sites of transendothelial invasion. Clin Cancer Res 9(17):6426–6431

Sawada S, Yoshimoto M, Odintsova E, Hotchin NA, Berditchevski F (2003) The tetraspanin CD151 functions as a negative regulator in the adhesion-dependent activation of Ras. J Biol Chem 278(29):26323–26326

Scherberich A, Moog S, Haan-Archipoff G, Azorsa DO, Lanza F, Beretz A (1998) Tetraspanin CD9 is associated with very late-acting integrins in human vascular smooth muscle cells and modulates collagen matrix reorganization. Arterioscler Thromb Vasc Biol 18(11):1691–1697

Scherberich A, Giannone G, Perennou E, Takeda K, Boucheix C, Rubinstein E, Lanza F, Beretz A (2002) FAK-mediated inhibition of vascular smooth muscle cell migration by the tetraspanin CD9. Thromb Haemost 87(6):1043–1050

Schmidt C, Kunemund V, Wintergerst ES, Schmitz B, Schachner M (1996) CD9 of mouse brain is implicated in neurite outgrowth and cell migration in vitro and is associated with the alpha 6/beta 1 integrin and the neural adhesion molecule L1. J Neurosci Res 43(1):12–31

Schmidt DS, Klingbeil P, Schnolzer M, Zoller M (2004) CD44 variant isoforms associate with tetraspanins and EpCAM. Exp Cell Res 297(2):329–347

Serru V, Le Naour F, Billard M, Azorsa DO, Lanza F, Boucheix C, Rubinstein E (1999) Selective tetraspan-integrin complexes (CD81/alpha4beta1, CD151/alpha3beta1, CD151/alpha6beta1) under conditions disrupting tetraspan interactions. Biochem J 340(Pt 1):103–111

Shaw AR, Domanska A, Mak A, Gilchrist A, Dobler K, Visser L, Poppema S, Fliegel L, Letarte M, Willett BJ (1995) Ectopic expression of human and feline CD9 in a human B cell line confers beta 1 integrin-dependent motility on fibronectin and laminin substrates and enhanced tyrosine phosphorylation. J Biol Chem 270(41):24092–24099

Shi W, Fan H, Shum L, Derynck R (2000) The tetraspanin CD9 associates with transmembrane TGF-alpha and regulates TGF-alpha-induced EGF receptor activation and cell proliferation. J Cell Biol 148(3):591–602

Shi GM, Ke AW, Zhou J, Wang XY, Xu Y, Ding ZB, Devbhandari RP, Huang XY, Qiu SJ, Shi YH, Dai Z, Yang XR, Yang GH, Fan J (2010) CD151 modulates expression of matrix metalloproteinase 9 and promotes neoangiogenesis and progression of hepatocellular carcinoma. Hepatology 52(1):183–196

Shibagaki N, Hanada K, Yamaguchi S, Yamashita H, Shimada S, Hamada H (1998) Functional analysis of CD82 in the early phase of T cell activation: roles in cell adhesion and signal transduction. Eur J Immunol 28(4):1125–1133

Shibagaki N, Hanada K, Yamashita H, Shimada S, Hamada H (1999) Overexpression of CD82 on human T cells enhances LFA-1/ICAM-1-mediated cell-cell adhesion: functional association between CD82 and LFA-1 in T cell activation. Eur J Immunol 29(12):4081–4091

Shigeta M, Sanzen N, Ozawa M, Gu J, Hasegawa H, Sekiguchi K (2003) CD151 regulates epithelial cell-cell adhesion through PKC- and Cdc42-dependent actin cytoskeletal reorganization. J Cell Biol 163(1):165–176

Shoham T, Rajapaksa R, Boucheix C, Rubinstein E, Poe JC, Tedder TF, Levy S (2003) The tetraspanin CD81 regulates the expression of CD19 during B cell development in a postendoplasmic reticulum compartment. J Immunol 171(8):4062–4072

Shoham T, Rajapaksa R, Kuo CC, Haimovich J, Levy S (2006) Building of the tetraspanin web: distinct structural domains of CD81 function in different cellular compartments. Mol Cell Biol 26(4):1373–1385

Silvie O, Rubinstein E, Franetich JF, Prenant M, Belnoue E, Renia L, Hannoun L, Eling W, Levy S, Boucheix C, Mazier D (2003) Hepatocyte CD81 is required for Plasmodium falciparum and Plasmodium yoelii sporozoite infectivity. Nat Med 9(1):93–96

Silvie O, Greco C, Franetich JF, Dubart-Kupperschmitt A, Hannoun L, van Gemert GJ, Sauerwein RW, Levy S, Boucheix C, Rubinstein E, Mazier D (2006) Expression of human CD81 differently affects host cell susceptibility to malaria sporozoites depending on the Plasmodium species. Cell Microbiol 8(7):1134–1146

Sincock PM, Fitter S, Parton RG, Berndt MC, Gamble JR, Ashman LK (1999) PETA-3/CD151, a member of the transmembrane 4 superfamily, is localised to the plasma membrane and endocytic system of endothelial cells, associates with multiple integrins and modulates cell function. J Cell Sci 112:833–844

Skubitz KM, Campbell KD, Iida J, Skubitz AP (1996) CD63 associates with tyrosine kinase activity and CD11/CD18, and transmits an activation signal in neutrophils. J Immunol 157(8):3617–3626

Slupsky JR, Seehafer JG, Tang SC, Masellis-Smith A, Shaw AR (1989) Evidence that monoclonal antibodies against CD9 antigen induce specific association between CD9 and the platelet glycoprotein IIb-IIIa complex. J Biol Chem 264(21):12289–12293

Soldaini E, Wack A, D'Oro U, Nuti S, Ulivieri C, Baldari CT, Abrignani S (2003) T cell costimulation by the hepatitis C virus envelope protein E2 binding to CD81 is mediated by Lck. Eur J Immunol 33(2):455–464

Sridhar SC, Miranti CK (2006) Tetraspanin KAI1/CD82 suppresses invasion by inhibiting integrin-dependent crosstalk with c-Met receptor and Src kinases. Oncogene 25(16):2367–2378

Sterk LM, Geuijen CA, Oomen LC, Calafat J, Janssen H, Sonnenberg A (2000) The tetraspan molecule CD151, a novel constituent of hemidesmosomes, associates with the integrin alpha6beta4 and may regulate the spatial organization of hemidesmosomes. J Cell Biol 149(4):969–982

Sterk LM, Geuijen CA, van den Berg JG, Claessen N, Weening JJ, Sonnenberg A (2002) Association of the tetraspanin CD151 with the laminin-binding integrins alpha3beta1, alpha6beta1, alpha6beta4 and alpha7beta1 in cells in culture and in vivo. J Cell Sci 115(Pt 6):1161–1173

Stipp CS (2010) Laminin-binding integrins and their tetraspanin partners as potential antimetastatic targets. Expert Rev Mol Med 12:e3

Stipp CS, Hemler ME (2000) Transmembrane-4-superfamily proteins CD151 and CD81 associate with alpha 3 beta 1 integrin, and selectively contribute to alpha 3 beta 1-dependent neurite outgrowth. J Cell Sci 113:1871–1882

Stipp CS, Kolesnikova TV, Hemler ME (2001) EWI-2 is a major CD9 and CD81 partner and member of a novel Ig protein subfamily. J Biol Chem 276(44):40545–40554

Stipp CS, Kolesnikova TV, Hemler ME (2003) EWI-2 regulates alpha3beta1 integrin-dependent cell functions on laminin-5. J Cell Biol 163(5):1167–1177

Sugiura T, Berditchevski F (1999) Function of alpha3beta1-tetraspanin protein complexes in tumor cell invasion. Evidence for the role of the complexes in production of matrix metalloproteinase 2 (MMP-2). J Cell Biol 146(6):1375–1389

Sutovsky P (2009) Sperm-egg adhesion and fusion in mammals. Expert Rev Mol Med 11:e11

Szöllósi J, Horejsí V, Bene L, Angelisová P, Damjanovich S (1996) Supramolecular complexes of MHC class I, MHC class II, CD20, and tetraspan molecules (CD53, CD81, and CD82) at the surface of a B cell line JY. J Immunol 157:2939–2946

Tachibana I, Hemler ME (1999) Role of transmembrane 4 superfamily (TM4SF) proteins CD9 and CD81 in muscle cell fusion and myotube maintenance. J Cell Biol 146(4):893–904

Takahashi M, Sugiura T, Abe M, Ishii K, Shirasuna K (2007) Regulation of c-Met signaling by the tetraspanin KAI-1/CD82 affects cancer cell migration. Int J Cancer 121(9):1919–1929

Takeda Y, Kazarov AR, Butterfield CE, Hopkins BD, Benjamin LE, Kaipainen A, Hemler ME (2007) Deletion of tetraspanin Cd151 results in decreased pathologic angiogenesis in vivo and in vitro. Blood 109(4):1524–1532

Takeda Y, He P, Tachibana I, Zhou B, Miyado K, Kaneko H, Suzuki M, Minami S, Iwasaki T, Goya S, Kijima T, Kumagai T, Yoshida M, Osaki T, Komori T, Mekada E, Kawase I (2008) Double deficiency of tetraspanins CD9 and CD81 alters cell motility and protease production of macrophages and causes chronic obstructive pulmonary disease-like phenotype in mice. J Biol Chem 283(38):26089–26097

Takeda Y, Li Q, Kazarov AR, Epardaud M, Elpek K, Turley SJ, Hemler ME (2011) Diminished metastasis in tetraspanin CD151-knockout mice. Blood 118(2):464–472

Tarrant JM, Groom J, Metcalf D, Li R, Borobokas B, Wright MD, Tarlinton D, Robb L (2002) The absence of Tssc6, a member of the tetraspanin superfamily, does not affect lymphoid development but enhances in vitro T-cell proliferative responses. Mol Cell Biol 22(14):5006–5018

Testa JE, Brooks PC, Lin JM, Quigley JP (1999) Eukaryotic expression cloning with an antimetastatic monoclonal antibody identifies a tetraspanin (PETA-3/CD151) as an effector of human tumor cell migration and metastasis. Cancer Res 59(15):3812–3820

Tham TN, Gouin E, Rubinstein E, Boucheix C, Cossart P, Pizarro-Cerda J (2010) Tetraspanin CD81 is required for Listeria monocytogenes invasion. Infect Immun 78(1):204–209

Thorne RF, Marshall JF, Shafren DR, Gibson PG, Hart IR, Burns GF (2000) The integrins alpha3beta1 and alpha6beta1 physically and functionally associate with CD36 in human melanoma cells. Requirement for the extracellular domain OF CD36. J Biol Chem 275(45):35264–35275

Todd SC, Lipps SG, Crisa L, Salomon DR, Tsoukas CD (1996) CD81 expressed on human thymocytes mediates integrin activation and interleukin 2-dependent proliferation. J Exp Med 184(5):2055–2060

Todeschini AR, Dos Santos JN, Handa K, Hakomori SI (2007) Ganglioside GM2-tetraspanin CD82 complex inhibits met and its cross-talk with integrins, providing a basis for control of cell motility through glycosynapse. J Biol Chem 282(11):8123–8133

Todeschini AR, Dos Santos JN, Handa K, Hakomori SI (2008) Ganglioside GM2/GM3 complex affixed on silica nanospheres strongly inhibits cell motility through CD82/cMet-mediated pathway. Proc Natl Acad Sci USA 105(6):1925–1930

Tohami T, Drucker L, Shapiro H, Radnay J, Lishner M (2007) Overexpression of tetraspanins affects multiple myeloma cell survival and invasive potential. FASEB J 21(3):691–699

Tseng CT, Klimpel GR (2002) Binding of the hepatitis C virus envelope protein E2 to CD81 inhibits natural killer cell functions. J Exp Med 195(1):43–49

Unternaehrer JJ, Chow A, Pypaert M, Inaba K, Mellman I (2007) The tetraspanin CD9 mediates lateral association of MHC class II molecules on the dendritic cell surface. Proc Natl Acad Sci USA 104(1):234–239

Urzainqui A, Serrador JM, Viedma F, Yanez-Mo M, Rodriguez A, Corbi AL, Alonso-Lebrero JL, Luque A, Deckert M, Vazquez J, Sanchez-Madrid F (2002) ITAM-based interaction of ERM proteins with Syk mediates signaling by the leukocyte adhesion receptor PSGL-1. Immunity 17(4):401–412

van Spriel AB, Figdor CG (2010) The role of tetraspanins in the pathogenesis of infectious diseases. Microbes Infect 12(2):106–112

van Spriel AB, Puls KL, Sofi M, Pouniotis D, Hochrein H, Orinska Z, Knobeloch KP, Plebanski M, Wright MD (2004) A regulatory role for CD37 in T cell proliferation. J Immunol 172(5):2953–2961 van Spriel AB, Sofi M, Gartlan KH, van der Schaaf A, Verschueren I, Torensma R, Raymakers RA, Loveland BE, Netea MG, Adema GJ, Wright MD, Figdor CG (2009) The tetraspanin protein CD37 regulates IgA responses and anti-fungal immunity. PLoS Pathog 5(3):e1000338

van Zelm MC, Smet J, Adams B, Mascart F, Schandene L, Janssen F, Ferster A, Kuo CC, Levy S, van Dongen JJ, van der Burg M (2010) CD81 gene defect in humans disrupts CD19 complex formation and leads to antibody deficiency. J Clin Invest 120(4):1265–1274

VanCompernolle SE, Levy S, Todd SC (2001) Anti-CD81 activates LFA-1 on T cells and promotes T cell-B cell collaboration. Eur J Immunol 31(3):823–831

Vazquez-Chona F, Song BK, Geisert EE Jr (2004) Temporal changes in gene expression after injury in the rat retina. Invest Ophthalmol Vis Sci 45(8):2737–2746

Vogt AB, Spindeldreher S, Kropshofer H (2002) Clustering of MHC-peptide complexes prior to their engagement in the immunological synapse: lipid raft and tetraspan microdomains. Immunol Rev 189:136–151

Voisset C, Lavie M, Helle F, Op De Beeck A, Bilheu A, Bertrand-Michel J, Terce F, Cocquerel L, Wychowski C, Vu-Dac N, Dubuisson J (2008) Ceramide enrichment of the plasma membrane induces CD81 internalization and inhibits hepatitis C virus entry. Cell Microbiol 10(3): 606–617

Volkov Y, Long A, Freeley M, Golden-Mason L, O'Farrelly C, Murphy A, Kelleher D (2006) The hepatitis C envelope 2 protein inhibits LFA-1-transduced protein kinase C signaling for T-lymphocyte migration. Gastroenterology 130(2):482–492

Wack A, Soldaini E, Tseng C, Nuti S, Klimpel G, Abrignani S (2001) Binding of the hepatitis C virus envelope protein E2 to CD81 provides a co-stimulatory signal for human T cells. Eur J Immunol 31(1):166–175

Wang XQ, Yan Q, Sun P, Liu JW, Go L, McDaniel SM, Paller AS (2007) Suppression of epidermal growth factor receptor signaling by protein kinase C-alpha activation requires CD82, caveolin-1, and ganglioside. Cancer Res 67(20):9986–9995

Wang HX, Li Q, Sharma C, Knoblich K, Hemler ME (2011) Tetraspanin protein contributions to cancer. Biochem Soc Trans 39(2):547–552

Winterwood NE, Varzavand A, Meland MN, Ashman LK, Stipp CS (2006) A critical role for tetraspanin CD151 in alpha3beta1 and alpha6beta4 integrin-dependent tumor cell functions on laminin-5. Mol Biol Cell 17(6):2707–2721

Witherden DA, Boismenu R, Havran WL (2000) CD81 and CD28 costimulate T cells through distinct pathways. J Immunol 165(4):1902–1909

Wright MD, Moseley GW, van Spriel AB (2004a) Tetraspanin microdomains in immune cell signalling and malignant disease. Tissue Antigens 64(5):533–542

Wright MD, Geary SM, Fitter S, Moseley GW, Lau LM, Sheng KC, Apostolopoulos V, Stanley EG, Jackson DE, Ashman LK (2004b) Characterization of mice lacking the tetraspanin superfamily member CD151. Mol Cell Biol 24(13):5978–5988

Wu H, Peng L, Li J, Wu W, Hou Q, Liu H, Wang M (1999) Reversible exposure of human platelet fibrinogen receptors by antiplatelet tetraspanin monoclonal antibodies via induction of a conformational change in membrane glycoprotein IIb/IIIa complex. Thromb Res 95(5):245–253

Xu C, Zhang YH, Thangavel M, Richardson MM, Liu L, Zhou B, Zheng Y, Ostrom RS, Zhang XA (2009) CD82 endocytosis and cholesterol-dependent reorganization of tetraspanin webs and lipid rafts. FASEB J 23(10):3273–3288

Yamada M, Tamura Y, Sanzen N, Sato-Nishiuchi R, Hasegawa H, Ashman LK, Rubinstein E, Yanez-Mo M, Sanchez-Madrid F, Sekiguchi K (2008a) Probing the interaction of tetraspanin CD151 with integrin alpha 3 beta 1 using a panel of monoclonal antibodies with distinct reactivities toward the CD151-integrin alpha 3 beta 1 complex. Biochem J 415(3):417–427

Yamada M, Sumida Y, Fujibayashi A, Fukaguchi K, Sanzen N, Nishiuchi R, Sekiguchi K (2008b) The tetraspanin CD151 regulates cell morphology and intracellular signaling on laminin-511. FEBS J 275(13):3335–3351

Yamamoto Y, Grubisic K, Oelgeschlager M (2007) Xenopus tetraspanin-1 regulates gastrulation movements and neural differentiation in the early Xenopus embryo. Differentiation 75(3):235–245

Yan Y, Shirakabe K, Werb Z (2002) The metalloprotease Kuzbanian (ADAM10) mediates the transactivation of EGF receptor by G protein-coupled receptors. J Cell Biol 158(2):221–226

Yanez-Mo M, Alfranca A, Cabanas C, Marazuela M, Tejedor R, Ursa MA, Ashman LK, de Landazuri MO, Sanchez-Madrid F (1998) Regulation of endothelial cell motility by complexes of tetraspan molecules CD81/TAPA-1 and CD151/PETA-3 with alpha3 beta1 integrin loc.lized at endothelial lateral junctions. J Cell Biol 141(3):791–804
Yanez-Mo M, Mittelbrunn M, Sanchez-Madrid F (2001a) Tetraspanins and intercellular interactions. Microcirculation 8(3):153–168
Yanez-Mo M, Tejedor R, Rousselle P, Sanchez-Madrid F (2001b) Tetraspanins in intercellular adhesion of polarized epithelial cells: spatial and functional relationship to integrins and cadherins. J Cell Sci 114(Pt 3):577–587
Yanez-Mo M, Lara-Pezzi E, Selgas R, Ramirez-Huesca M, Dominguez-Jimenez C, Jimenez-Heffernan JA, Aguilera A, Sanchez-Tomero JA, Bajo MA, Alvarez V, Castro MA, del Peso G, Cirujeda A, Gamallo C, Sanchez-Madrid F, Lopez-Cabrera M (2003) Peritoneal dialysis and epithelial-to-mesenchymal transition of mesothelial cells. N Engl J Med 348(5):403–413
Yanez-Mo M, Barreiro O, Gonzalo P, Batista A, Megias D, Genis L, Sachs N, Sala-Valdes M, Alonso MA, Montoya MC, Sonnenberg A, Arroyo AG, Sanchez-Madrid F (2008) MT1-MMP collagenolytic activity is regulated through association with tetraspanin CD151 in primary endothelial cells. Blood 112(8):3217–3226
Yanez-Mo M, Barreiro O, Gordon-Alonso M, Sala-Valdes M, Sanchez-Madrid F (2009) Tetraspanin-enriched microdomains: a functional unit in cell plasma membranes. Trends Cell Biol 19(9):434–446
Yanez-Mo M, Gutierrez-Lopez MD, Cabanas C (2011) Functional interplay between tetraspanins and proteases. Cell Mol Life Sci 68(20):3323–3335
Yang X, Claas C, Kraeft SK, Chen LB, Wang Z, Kreidberg JA, Hemler ME (2002) Palmitoylation of tetraspanin proteins: modulation of CD151 lateral interactions, subcellular distribution, and integrin-dependent cell morphology. Mol Biol Cell 13(3):767–781
Yang X, Kovalenko OV, Tang W, Claas C, Stipp CS, Hemler ME (2004) Palmitoylation supports assembly and function of integrin-tetraspanin complexes. J Cell Biol 167(6):1231–1240
Yang XH, Kovalenko OV, Kolesnikova TV, Andzelm MM, Rubinstein E, Strominger JL, Hemler ME (2006) Contrasting effects of EWI proteins, integrins, and protein palmitoylation on cell surface CD9 organization. J Biol Chem 281(18):12976–12985
Yang XH, Richardson AL, Torres-Arzayus MI, Zhou P, Sharma C, Kazarov AR, Andzelm MM, Strominger JL, Brown M, Hemler ME (2008a) CD151 accelerates breast cancer by regulating alpha 6 integrin function, signaling, and molecular organization. Cancer Res 68(9):3204–3213
Yang W, Qiu C, Biswas N, Jin J, Watkins SC, Montelaro RC, Coyne CB, Wang T (2008b) Correlation of the tight junction-like distribution of Claudin-1 to the cellular tropism of hepatitis C virus. J Biol Chem 283(13):8643–8653
Yang XH, Flores LM, Li Q, Zhou P, Xu F, Krop IE, Hemler ME (2010) Disruption of laminin-integrin-CD151-focal adhesion kinase axis sensitizes breast cancer cells to ErbB2 antagonists. Cancer Res 70(6):2256–2263
Yang XH, Mirchev R, Deng X, Yacono P, Yang HL, Golan DE, Hemler ME (2012) CD151 restricts alpha6 integrin diffusion mode. J Cell Sci 125(Pt6):1478–1487
Yashiro-Ohtani Y, Zhou XY, Toyo-Oka K, Tai XG, Park CS, Hamaoka T, Abe R, Miyake K, Fujiwara H (2000) Non-CD28 costimulatory molecules present in T cell rafts induce T cell costimulation by enhancing the association of TCR with rafts. J Immunol 164(3):1251–1259
Yauch RL, Hemler ME (2000) Specific interactions among transmembrane 4 superfamily (TM4SF) proteins and phosphoinositide 4-kinase. Biochem J 351:629–637
Yauch RL, Berditchevski F, Harler MB, Reichner J, Hemler ME (1998) Highly stoichiometric, stable, and specific association of integrin alpha3beta1 with CD151 provides a major link to phosphatidylinositol 4-kinase, and may regulate cell migration. Mol Biol Cell 9(10):2751–2765
Yauch RL, Kazarov AR, Desai B, Lee RT, Hemler ME (2000) Direct extracellular contact between integrin alpha(3)beta(1) and TM4SF protein CD151. J Biol Chem 275(13):9230–9238
Zemni R, Bienvenu T, Vinet MC, Sefiani A, Carrie A, Billuart P, McDonell N, Couvert P, Francis F, Chafey P, Fauchereau F, Friocourt G, des Portes V, Cardona A, Frints S, Meindl A, Brandau O, Ronce N, Moraine C, van Bokhoven H, Ropers HH, Sudbrak R, Kahn A, Fryns JP, Beldjord

C, Chelly J (2000) A new gene involved in X-linked mental retardation identified by analysis of an X;2 balanced translocation. Nat Genet 24(2):167–170

Zevian S, Winterwood NE, Stipp CS (2011) Structure-function analysis of tetraspanin CD151 reveals distinct requirements for tumor cell behaviors mediated by alpha3beta1 versus alpha-6beta4 integrin. J Biol Chem 286(9):7496–7506

Zhang XA, Bontrager AL, Hemler ME (2001a) Transmembrane-4 superfamily proteins associate with activated protein kinase C (PKC) and link PKC to specific beta(1) integrins. J Biol Chem 276(27):25005–25013

Zhang XA, Bontrager AL, Stipp CS, Kraeft SK, Bazzoni G, Chen LB, Hemler ME (2001b) Phosphorylation of a conserved integrin alpha 3 QPSXXE motif regulates signaling, motility, and cytoskeletal engagement. Mol Biol Cell 12(2):351–365

Zhang XA, Kazarov AR, Yang X, Bontrager AL, Stipp CS, Hemler ME (2002) Function of the tetraspanin CD151-alpha6beta1 integrin complex during cellular morphogenesis. Mol Biol Cell 13(1):1–11

Zhang XA, Lane WS, Charrin S, Rubinstein E, Liu L (2003a) EWI2/PGRL associates with the metastasis suppressor KAI1/CD82 and inhibits the migration of prostate cancer cells. Cancer Res 63(10):2665–2674

Zhang XA, He B, Zhou B, Liu L (2003b) Requirement of the p130CAS-Crk coupling for metastasis suppressor KAI1/CD82-mediated inhibition of cell migration. J Biol Chem 278(29):27319–27328

Zhang F, Michaelson JE, Moshiach S, Sachs N, Zhao W, Sun Y, Sonnenberg A, Lahti JM, Huang H, Zhang XA (2011) Tetraspanin CD151 maintains vascular stability by balancing the forces of cell adhesion and cytoskeletal tension. Blood 118(15):4274–4284

Zheng Z, Liu Z (2006) CD151 gene delivery activates PI3K/Akt pathway and promotes neovascularization after myocardial infarction in rats. Mol Med 12(9–10):214–220

Zheng ZZ, Liu ZX (2007a) Activation of the phosphatidylinositol 3-kinase/protein kinase Akt pathway mediates CD151-induced endothelial cell proliferation and cell migration. Int J Biochem Cell Biol 39(2):340–348

Zheng ZZ, Liu ZX (2007b) CD151 gene delivery increases eNOS activity and induces ECV304 migration, proliferation and tube formation. Acta Pharmacol Sin 28(1):66–72

Zhou Z, Ran YL, Hu H, Pan J, Li ZF, Chen LZ, Sun LC, Peng L, Zhao XL, Yu L, Sun LX, Yang ZH (2008) TM4SF3 promotes esophageal carcinoma metastasis via upregulating ADAM12m expression. Clin Exp Metastasis 25(5):537–548

Zijlstra A, Lewis J, Degryse B, Stuhlmann H, Quigley JP (2008) The inhibition of tumor cell intravasation and subsequent metastasis via regulation of in vivo tumor cell motility by the tetraspanin CD151. Cancer Cell 13(3):221–234

Zilber MT, Setterblad N, Vasselon T, Doliger C, Charron D, Mooney N, Gelin C (2005) MHC class II/CD38/CD9: a lipid-raft-dependent signaling complex in human monocytes. Blood 106(9):3074–3081

Zuo H, Liu Z, Liu X, Yang J, Liu T, Wen S, Zhang XA, Cianflone K, Wang D (2009a) CD151 gene delivery after myocardial infarction promotes functional neovascularization and activates FAK signaling. Mol Med 15(9–10):307–315

Zuo HJ, Liu ZX, Liu XC, Yang J, Liu T, Wen S, Wang DW, Zhang X (2009b) Assessment of myocardial blood perfusion improved by CD151 in a pig myocardial infarction model. Acta Pharmacol Sin 30(1):70–77

Chapter 7
Genetic Evidence for Tetraspanin Functions

Martin E. Hemler

Abstract Tetraspanin proteins have been suggested to associate with many different cell surface partner proteins, and in vitro studies have suggested involvement in a wide variety of cellular functions. However, knockout phenotypes in mice have so far been relatively mild, and there have been few clear examples of human tetraspanin mutations having functional consequences. Nonetheless a range of sometimes subtle, but interesting phenotypes are emerging from studies of mouse knockouts, and from analysis of human mutations.

7.1 Introduction

This chapter is focused largely on tetraspanin functions, as revealed through genetic evidence from mice, humans, and a few other vertebrate species. Tetraspanin functions on tumor cells, immune cells, and oocytes are only mentioned briefly, since those topics are covered in other chapters. Likewise, the uroplakin, peripherin and RDS tetraspanins are not discussed here, since they are covered in other chapters.

7.2 Genetic Evidence from Both Humans and Mice

TSPAN12 (NET-2)—Mice deleted for *TSPAN12* are viable and fertile, with no overt abnormalities. However, more detailed inspection revealed disruption of the retinal vasculature (Junge et al. 2009). The specific deficiencies in the multilayer retinal vasculature seen in *TSPAN12* mice closely resemble those seen for *FZD4*, *NORRIN*, and/or *LRP5* mutant mice. Mice lacking one allele each of *TSPAN12* and

M.E. Hemler (✉)
Dana-Farber Cancer Institute, 44 Binney Street, Boston, MA 02115, USA
e-mail: martin_hemler@dfci.harvard.edu

F. Berditchevski and E. Rubinstein (eds.), *Tetraspanins*, Proteins and Cell Regulation 9, DOI 10.1007/978-94-007-6070-7_7, © Springer Science+Business Media Dordrecht 2013

NORRIN, or *TSPAN12* and *LRP5* closely replicated deficits seen for single gene homozygous mutants. By contrast, absence of only one allele of *TSPAN12* or *NORRIN* or *LRP5* had minimal effect on retinal vasculature (Junge et al. 2009). These compound mutant mouse results demonstrate genetic interactions between *TSPAN12* and *NORRIN*, and between *TSPAN12* and *LRP5* (Table 7.1).

The Norrin protein binds to its receptor Frizzled-4 (FZD4), with co-receptor assistance from Lrp5 (low-density lipoprotein receptor-related protein-5) to signal through β-catenin, which interacts with TCF/LEF transcription factors to turn on target genes (Ye et al. 2010). TSPAN12 strongly enhanced Norrin/β-catenin signaling, only when FZD4, LRP5, and Norrin were present (Junge et al. 2009). TSPAN12 did not bind directly to Norrin, and did not enhance Norrin binding to FZD4. Instead, TSPAN12 appears to act by promoting receptor (FZD4) multimerization, thereby contributing to β-catenin signaling (Junge et al. 2009). Remarkably, TSPAN12 does not affect signaling triggered by Wnt3a, another ligand which can bind to FZD4/Lrp5 and activates β-catenin (Junge et al. 2009). It remains to be seen why TSPAN12 is required for Norrin/FZD4/β-catenin signaling but not Wnt3a/FZD4/β-catenin signaling in the same cells. These results reinforce an emerging theme for tetraspanin functions. Tetraspanins typically do not directly affect ligand binding, but instead act as co-receptors to enhance receptor-ligand interactions for key partner proteins, for example as seen elsewhere for CD151 (Lammerding et al. 2003), CD81 (Feigelson et al. 2003), and CD9 (Iwamoto et al. 1994).

In humans, mutations in *NORRIN*, *FZD4* and *LRP5* genes all are associated with familial exudative vitreoretinopathy (FEVR), an inherited disorder characterized by incomplete development of the retinal vasculature (Criswick and Schepens 1969). Because only ~40 % of human FEVR patients carry mutations in *NORRIN*, *FZD4* and *LRP5* genes (Toomes et al. 2004), this prompted examination of the *TSPAN12* locus. FEVR patients independently distributed throughout the world were found to have nine different *TSPAN12* mutations (Poulter et al. 2010; Nikopoulos et al. 2010). These FEVR patients displayed a range of disease phenotypes, from mild to severe (i.e., resulting in blindness), which closely mimic those seen in patients with *NORRIN*, *FZD4* and *LRP5* mutations. Hence, *TSPAN12* mutation is clearly a cause of human FEVR. Among the nine *TSPAN12* mutations (Poulter et al. 2010; Nikopoulos et al. 2010), five are frameshift or missense mutations likely resulting in expression of substantially truncated TSPAN12 protein. Also there are four distinct missense mutations, resulting in radical amino acid substitutions to residues highly conserved throughout *TSPAN12* vertebrate evolution. A G188R mutation places a bulky positively charged residue next to a critical C189 residue within the TSPAN12 large extracellular loop (Nikopoulos et al. 2010). This could disrupt critical disulfide bridge formation, and/or TSPAN12 associations with partner proteins such as FZD4. Another mutation, M210R, lies within the large extracellular loop, 10 residues from the fourth transmembrane domain (Poulter et al. 2010). Uroplakin type tetraspanins contain an extended helical structure, connecting through the fourth transmembrane domain into the large extracellular loop (Min et al. 2006). Such a structure, if present also in TSPAN12, would likely be disrupted. Other TSPAN12 mutations are L101H in the middle of the third transmembrane domain

Table 7.1 Genetic evidence for tetraspanin functions

TSPAN	Genetic alteration	Phenotype	Reference
TSPAN12	Human mutations	Familial Exudative Vitreoretinopathy (FEVR)	Poulter et al. (2010), Nikopoulos et al. (2010)
	Mouse knockout	FEVR-like alterations in retina	Junge et al. (2009)
CD151	Human mutation	Pretibial bullous skin lesions	Karamatic et al. (2004), Kagan et al. (1988)
		Glomerular basement membrane disruption	
		Sensorineural deafness	
		Severe anemia	
	Mouse knockout	Glomerular basement membrane disruption	Sachs et al. (2006), Baleato et al. (2008)
		Delayed skin wound healing	Cowin et al. (2006)
		Impaired pathological angiogenesis	Takeda et al. (2007)
		Hyperstimulatory dendritic cells	Sheng et al. (2009)
		Impaired platelet thrombus formation	Orlowski et al. (2009)
TSPAN7	Human mutation	X-linked mental retardation	Zemni et al. (2000), Abidi et al. (2002), Maranduba et al. (2004)
CD53	Human deficiency	Recurring infectious diseases	Mollinedo et al. (1997)
TSPAN8	Mouse knockout	Decreased body weight, lower bone density and decreased circulating fatty acids/glycerol	Champy et al. (2010)
CD37	Mouse knockout	T cell hyperproliferation, impaired antibody responses, increased dendritic cell antigen presentation, elevated IgA, elevated IL-6	Sheng et al. (2009), van Spriel et al. (2004), Knobeloch et al. (2000), Meyer-Wentrup et al. (2007), van Spriel et al. (2009), Rops et al. (2010)
CD81	Mouse knockout	Impaired CD19 expression on B lymphocytes	Miyazaki et al. (1997), Maecker and Levy (1997), Tsitsikov et al. (1997)
		Diminished adhesion strengthening	Feigelson et al. (2003)
		Altered Th2 immune responses	Maecker et al. (1998), Deng et al. (2000), (2002)
		Increased brain size	Geisert et al. (2002)
		Reduced oocyte fertilization	Rubinstein et al. (2006)
		Enhanced cell fusion	Takeda et al. (2003)
		Lost susceptibility to Plasmodium yoelii	Silvie et al. (2006)
		Loss of HCV entry	Bertaux and Dragic (2006)

(continued)

Table 7.1 (continued)

TSPAN	Genetic alteration	Phenotype	Reference
CD9	Mouse knockout	Greatly reduced oocyte fertilization	Miyado et al. (2000), Le Naour et al. (2000), Kaji et al. (2000)
		Disruption of paranodal junctions	Ishibashi et al. (2004)
		Enhanced cell fusion	Takeda et al. (2003)
		Altered immune cell functions	Unternaehrer et al. (2007), Ha et al. (2006)
CD63	Mouse knockout	Disruption of kidney collecting ducts and altered water balance	Schroder et al. (2009)
TSSC6	Mouse knockout	Hyperproliferative T cell response	Tarrant et al. (2002)
		Diminished cellular immunity	Gartlan et al. (2010)
		Unstable hemostasis	Goschnick et al. (2006)
TSPAN33	Mouse knockout	Impaired erythropoiesis	Heikens et al. (2007)
TSPAN1	Xenopus knockdown	Impaired gastrulation and neural differentiation	Yamamoto et al. (2007)

(Poulter et al. 2010) and A237P in the middle of the fourth transmembrane domain (Nikopoulos et al. 2010). These non-conservative mutations are likely to disrupt helical packing within transmembrane regions (Kovalenko et al. 2005), resulting in loss of structural integrity and possibly also protease degradation.

Mutations in *TSPAN12*, *NORRIN*, *FZD4*, and *LRP5* genes still only account for a fraction of human FEVR cases, suggesting that more FEVR genes remain to be identified (Poulter et al. 2010). Among these, some could potentially interact functionally and/or structurally with TSPAN12. In another human eye disorder, retinopathy of prematurity (ROP), *NORRIN*, *FZD4* and *LRP5* mutations account for 10–12 % cases (Shastry 2010). Hence, TSPAN12 could also contribute to ROP, but this has not yet been demonstrated. TSPAN12 is expressed in a variety of tissues and cell types besides retina vascular cells, but its functions on those cells are unknown. A possible clue may come from in vitro studies, in which TSPAN12 was shown to enhance the maturation and function of ADAM10, a membrane metalloprotease, in its cleavage of amyloid precursor protein (Xu et al. 2009). No evidence is yet available regarding functions of the atypically long (~60 aa) cytoplasmic tail of TSPAN12.

CD151 (TSPAN24)—Tetraspanin CD151 associates closely with laminin-binding integrins ($\alpha3\beta1$, $\alpha6\beta1$, $\alpha6\beta4$, $\alpha7\beta1$)(Kazarov et al. 2002; Berditchevski et al. 2001) and regulates a variety of integrin-dependent events, including adhesion strengthening, cell spreading, cell motility, invasion, chemotaxis, and neurite outgrowth (Lammerding et al. 2003; Kazarov et al. 2002; Sincock et al. 1999; Yánez-Mó et al. 1998; Zhang et al. 2002; Yang et al. 2002; Yauch et al. 1998). CD151 (and other tetraspanins) do not typically affect initial integrin ligand binding and static cell adhesion, except perhaps at very low laminin levels (Lammerding et al. 2003; Winterwood et al. 2006). Considering the close association of CD151 with laminin-binding integrins, it was expected that a CD151 phenotype might resemble skin, kidney, and/or muscle phenotypes seen in integrin α6, β4, α3 and α7-null mice (Belkin and Stepp 2000). Indeed, three individuals with a CD151 mutation (an insertion resulting in a premature stop codon) display hereditary nephritic syndrome leading to end-stage kidney failure (Karamatic et al. 2004; Kagan et al. 1988). A kidney biopsy from one of the patients revealed thickening and splitting of the tubular basement membrane and thickening, reticulation and fragmentation of the lamina densa of the glomerular basement membrane, with inclusion of electron-dense particles (Karamatic et al. 2004). Two of these individuals (sibs) were further shown to have pretibialbullous skin lesions, in which the dermal-epidermal junction appeared to be compromised (Karamatic et al. 2004; Kagan et al. 1988). These skin and kidney alterations support the idea that CD151 functions mostly by modulating functions of laminin-binding integrins. The two sibs also displayed sensorineural deafness, bilateral lacrimal duct stenosis, and nail dystrophy. These defects likely also involve dysregulated integrin-laminin interactions (Karamatic et al. 2004). The combination of progressive familial kidney failure and sensorineural deafness is characteristic of Alport syndrome (Kashtan 2000). However, pretibial epidermolysis bullosa is not characteristic of Alport syndrome, suggesting that human *CD151* mutations are unlikely to be responsible for Alport syndrome. Severe anemia was also observed in CD151-deficient patients, likely due to defective erythropoeisis (Karamatic et al. 2004; Kagan et al. 1988). It is speculated that impaired bone marrow response to erythropoietin may result from dysregulated integrin-laminin effects on erythroid progenitors (Karamatic et al. 2004).

Unexpectedly, CD151-null mice in C57BL/6 and 129Sv genetic backgrounds were viable and fertile, with only a few minor defects (Wright et al. 2004; Takeda et al. 2007). Despite the prominence of CD151 in hemidesmosomes, its removal had no apparent effect on normal epithelial integrity in skin or elsewhere (Wright et al. 2004; Sachs et al. 2006). However, crossing CD151 knockout mice into the FVB/N strain caused significant renal pathology, including proteinuria, focal glomerulosclerosis, and disorganization of the glomerular basement membrane (Sachs et al. 2006; Baleato et al. 2008). It is speculated that the initial event causing glomerular basement membrane disorganization is defective laminin network maturation and assembly. The first defect observed, shortly after birth of CD151-null FVB mice, is proteinuria accompanied by thickening and splitting of the glomerular basement membrane. Loss of podocytes becomes apparent by 3 weeks, and glomerulosclerosis by 12 weeks (Baleato et al. 2008). These defects strongly resemble defects seen in human and mouse Alport syndrome (Kashtan 2000). Striking differences in CD151 deletion effects in FVB and C57Bl/6 mice indicates the existence of genetic modifier loci, possibly coding for a compensating tetraspanin in C57Bl/6.

Although skin blistering was associated with CD151 mutation in humans (Karamatic et al. 2004), and with deletion of α3, α6, and β4 integrin subunits in mice (Belkin and Stepp 2000), CD151-null mice showed no obvious abnormalities in normal skin in either C57Bl/6 or FVB strains. However, wound healing in the skin of CD151-null mice was significantly impaired, accompanied by defective organization and deposition of laminin-5 (LN332) in the basement membrane, diminished keratinocyte proliferation in the wound, and diminished upregulation of α6β4 in the wound (Cowin et al. 2006). While CD151 is abundant on endothelial cells, CD151-null mice showed no vascular defects in the C57Bl/6 strain (Takeda et al. 2007). However pathological angiogenesis was impaired, as seen in in vivo assays involving endothelial invasion into a Matrigel plug, neovascularization in a corneal micropellet, and growth of implanted tumors (Takeda et al. 2007). These defects appeared to be mostly due to impaired integrin-laminin interactions (Takeda et al. 2007).

In another study, dendritic cells from CD151-null mice were hyperstimulatory to T cells, due to an uncontrolled co-stimulation effect (Sheng et al. 2009). It is not yet clear whether specific laminin-binding integrins are involved. Also, platelets from CD151-deficient mice show defective aggregation, impaired spreading on fibrinogen, and delayed clot retraction in vitro. These deficiencies are attributed to impaired outside-in signaling through integrin αIIbβ3, a receptor for fibrinogen (Lau et al. 2004). In vivo models firmly establish that platelet CD151 regulates the size and stability of thrombus formation (Orlowski et al. 2009). CD151 regulation of αIIbβ3-dependent functions in platelets provides perhaps the only example of CD151 regulating a non-laminin-binding integrin. Finally, there have been many reports that CD151 contributes to tumor cell growth, migration, invasion, and metastasis. For example, in human breast cancer cells, the removal of CD151 impairs tumor growth in mouse xenograft models (Yang et al. 2008; Sadej et al. 2009). However, except for diminished tumor growth linked to an angiogenesis defect (Takeda et al. 2007), alterations of tumor behavior in CD151-null mice have not yet been reported.

7.3 Genetic Evidence from Humans

TSPAN7 (CD231, TALLA-1, A15, TM4SF2)—Three distinct mutations within the TSPAN7 large extracellular loop point to TSPAN7 being involved in human X-linked mental retardation (Zemni et al. 2000; Abidi et al. 2002). The pathogenic nature of these TSPAN7 gene mutations was later questioned (Gomot et al. 2002). However, subsequent identification of another mental retardation patient with a P172H mutation strengthens the case for TSPAN7 involvement (Maranduba et al. 2004). The non-conservative P172H point mutation occurs within a highly conserved 'PXSC' motif found in the large extracellular loops of many tetraspanin proteins. Presumably this mutation could disrupt the association of TSPAN7 with key partner proteins which remain to be identified. The 2 other TSPAN7 mutations are stop codon truncations resulting in loss of the fourth transmembrane domain and C-terminal cytoplasmic tail. These are likely to lead to loss of cell surface expression.

TSPAN7 is also under investigation for possible roles in autism spectrum disorder and schizophrenia (Noor et al. 2009; Piton et al. 2010), but the evidence is not yet definitive. RNA in situ hybridization results show high TSPAN7 expression in the mouse central nervous system. Expression of TSPAN7 in the cerebral cortex and hippocampus is consistent with a role in physiological processes underlying memory and learning (Zemni et al. 2000).

CD53 (TSPAN25)—Tetraspanin CD53 is expressed on cells of the lymphoid and myeloid lineages. Three individuals within the same family showed greatly diminished CD53 expression on neutrophils and T lymphocytes (Mollinedo et al. 1997). This deficiency was accompanied by recurring infectious diseases, caused by bacteria, fungi, and viruses (Mollinedo et al. 1997). Consequently, CD53 could play a major role in human infectious disease pathogenesis. Unfortunately, after the initial report in 1997, there has been little or no follow-up with these patients or related patients to determine the basis for CD53 deficiency. Mice ablated for CD53 gene expression have not yet been reported. Other studies have linked CD53 to lymphocyte and macrophage survival and activation (Yunta and Lazo 2003; Kim et al. 2004; Ardman et al. 1992).

7.4 Genetic Evidence from Mice

TSPAN8 (CO-029, TM4SF3)—The TSPAN8 gene was knocked out in mice, resulting in no gross abnormalities or other obvious developmental defects. However, male C57Bl/6 mice showed lower body weight, decreased bone density and lower levels of circulating free fatty acids and glycerol (Champy et al. 2010). A possible explanation for these results is that TSPAN8 deficiency could alter male sex hormones regulating these phenotypic parameters. Despite changes in body weight and fat deposition, there were no changes in insulin or glucose levels, glucose clearance, or insulin sensitivity in mice lacking TSPAN8 (Champy et al. 2010). These results were unexpected because multiple studies had shown that the TSPAN8 gene might play a role in the onset of type II diabetes in humans (Zeggini et al. 2008; Grarup et al. 2008). However, functions of TSPAN8 in human type II diabetes might not translate to mice, since TSPAN8 is abundant in human pancreas but nearly absent in mouse pancreas (Champy et al. 2010).

Other studies have identified human TSPAN8 as a candidate gene for involvement in bipolar disorder and schizophrenia (Sklar et al. 2008; Scholz et al. 2010). In addition, TSPAN8 may be involved in tumor invasion, survival and angiogenesis (Zoller 2009).

CD37 (TSPAN26)—CD37 expression is restricted to lymphoid and myeloid lineages. Mice lacking CD37 show T cell hyperproliferation, impaired antibody responses, and increased capacity of dendritic cells to present antigen (Sheng et al. 2009; van Spriel et al. 2004; Knobeloch et al. 2000). CD37 null macrophages show elevated IL-6 production, due to unregulated IL-6 induction by dectin-1, a C-type lectin (Meyer-Wentrup et al. 2007). In addition, CD37 null mice show highly elevated serum IgA (van Spriel et al. 2009), and renal pathology associated with glomerular IgA deposition (Rops et al. 2010).

CD81 (TSPAN28)—Although CD81 is widely expressed on nearly all cell and tissue types, mice lacking CD81 are viable and fertile. Nonetheless, closer inspection reveals several immune cell deficiencies. B lymphocytes display impaired adhesion strengthening under shear flow conditions (Feigelson et al. 2003), and markedly reduced levels of the CD19 (Miyazaki et al. 1997; Maecker and Levy 1997; Tsitsikov et al. 1997), a known CD81 partner protein. Also CD81-null mice show impaired T helper 2 (Th2) immune cell responses (Maecker et al. 1998; Deng et al. 2000; Deng et al. 2002).

CD81-null mice showed substantially increased brain size, likely due to increased numbers of astrocytes and microglia (Geisert et al. 2002). In this regard, all four major classes of glia express abundant CD81. A possible mechanism is that CD81 on astrocytes (a glial cell subtype) may play a key role in facilitating neuron-induced inhibition of astrocyte proliferation (Kelic et al. 2001). The extent of brain size increase was profoundly influenced by the mouse genetic background (Geisert et al. 2002). This result strongly suggests a role for modifier loci, but such genes have not yet been identified. Other studies, involving inducible CD81 expression in rat brain, suggest that CD81 contributes to behavioral changes associated with cocaine sensitization (Bahi et al. 2005).

A 40 % reduction in female fertilization was observed for CD81-null mice, due to impaired oocyte ability to fuse with sperm (Rubinstein et al. 2006). This CD81 deletion effect (40 %) is mild compared to the effect of CD9 gene ablation (95 %). Mice lacking both CD81 and CD9 are completely infertile (Rubinstein et al. 2006), strongly emphasizing the critical roles of these tetraspanins for oocyte-sperm fusion. However, CD81-null mice alveolar macrophages and bone marrow cells formed elevated numbers of multinucleated cells, indicating that CD81 can at least sometimes inhibit cell fusion events (Takeda et al. 2003). In other studies, hepatocytes from CD81-null mice showed loss of susceptibility to Plasmodium yoelii sporozoites (Silvie et al. 2006), and CD81 in hepatocytes is also required for hepatitis C virus entry (Bertaux and Dragic 2006). Interestingly, a major CD81 partner protein (a truncated form of EWI-2 called EWI-2wint) blocked HCV entry (Rocha-Perugini et al. 2008). Echoing this theme, another CD81 partner protein (EWI-F/CD9P-1) inhibited Plasmodium yoelii sporozoite infection of hepatocytes. These results establish a precedent, suggesting that other CD81-dependent (and CD9-dependent) functions may be affected by the partner proteins EWI-2 and EWI-F/CD9P-1.

CD9 (TSPAN29)—Knockout mouse results show that CD9 on mouse oocytes plays a major role in supporting sperm-egg fusion (Miyado et al. 2000; Le Naour et al. 2000; Kaji et al. 2000), by a mechanism that includes stabilization of oocyte microvilli (Runge et al. 2007). Conversely, in CD9-null mice, alveolar macrophages and bone marrow cells formed elevated numbers of multinucleated cells, indicating that CD9 can also inhibit cell fusion events (Takeda et al. 2003). In addition, CD9 null mice show severe disruption of paranodal junctions (specialized cell-cell junctions between neuron and glial cells). However, formation of compact myelin was not altered (Ishibashi et al. 2004). In another study of CD9-null mice, oligodendrocyte production and myelination appeared normal (Terada et al. 2002). In the immune system, CD9 mediates heterologous association among MHC II molecules (Unternaehrer et al. 2007), and regulates B1 cell emigration from the peritoneal cavity (Ha et al. 2006).

Going forward, it will be interesting to ascertain how many of CD9 dependent functions also involve participation of closely associated EWI-2 and EWI-F/CD9P-1 partner proteins (Stipp et al. 2001a, b; Charrin et al. 2001). In this regard, oocyte EWI-2 does appear to contribute to CD9-dependent fertilization (Glazar and Evans 2009). Although there is a large literature on the role of CD9 in the regulation of tumor cell behavior, this has still not been analyzed in a spontaneous mouse tumor model.

CD63 (TSPAN30)—Numerous in vitro studies have implicated CD63 as a regulator of a variety of cellular activities including lymphocyte activation (Pfistershammer et al. 2004), monocyte and neutrophil adhesion (Koyama et al. 1990; Toothill et al. 1990), platelet functions (Israels and McMillan-Ward 2005), HIV infection (von Lindern et al. 2003) and CD63 partner protein trafficking (Pols and Klumperman 2009). With respect to partner proteins, CD63 may associate with many different proteins including integrins, other tetraspanins, kinases, adaptor proteins, transport proteins, proteases, and various cell surface receptors (Pols and Klumperman 2009).

However, CD63 knockout mice have a relatively mild phenotype. They are viable and fertile, with no overt abnormalities (Schroder et al. 2009). No detectable differences were seen in immune cell numbers, distribution, cell surface markers, or activation in bone marrow, spleen or lymph node. Within mouse embryo fibroblasts, the late endosomal/lysosomal compartment was unaltered, in terms of lysosomal marker distribution, enzyme activities, and lysosomal maturation, acidification, and endocytic transport activities. Platelet numbers, marker protein expression, adhesion, aggregation, and thrombus formation were all indistinguishable between null and wild type cells. There was a small reduction in reversibility of platelet aggregation in CD63-null platelets. However, the significance of this is unclear, since thrombus formation was not altered either under flow conditions or in vivo (Schroder et al. 2009).

Detailed analysis of brain, liver, spleen and lung revealed no abnormalities in CD63-null mice. However, within CD63-null kidneys, there was an accumulation of lamellar inclusions in the principle cells of the collecting duct. In addition, the mice showed increased water loss, and decreased water uptake, indicative of altered water balance (Schroder et al. 2009). The molecular basis for altered water balance and the appearance of lamellar inclusion bodies is not yet evident. Kidney defects seen in CD63-null mice are quite distinct from those seen in CD151-null mice, suggesting distinct and non-compensating functions.

TSPAN32 (TSSC6, Phemx)—Expression of TSSC6 is primarily restricted to hematopoietic cells and organs (Nicholson et al. 2000; Robb et al. 2001). Mice lacking TSSC6 are healthy and fertile, with no obvious abnormalities (Tarrant et al. 2002). Likewise, no alterations appear in bone marrow, spleen, blood cell counts, or in hematopoiesis or lymphocyte development (Tarrant et al. 2002). However, T cells show a hyperproliferative response in vitro, ascribed to increased interleukin 2 production, following T cell receptor stimulation (Tarrant et al. 2002). In a later study, TSSC6-null mice showed reduced in vivo cellular immune responses to pathogens (Gartlan et al. 2010).

TSSC6 also associates with platelet integrin αIIBβ3, and regulates signaling and adhesion events (Goschnick et al. 2006). Consequently, TSSC6-null mice show unstable hemostasis, as manifested by increased blood loss and a tendency for re-bleeds in a tail bleeding model (Goschnick et al. 2006). Also, TSSC6 appears to stabilize platelet thrombus formation, suggesting an important role in hemostasis and clot retraction during wound healing (Goschnick et al. 2006).

TSPAN33 (Penumbra)—TSPAN33 is predominately expressed in erythroblasts, with little expression in non-erythroid cells or tissues. Mice lacking TSPAN33 are viable and fertile, but develop marked splenomegaly and severe anemia as they age (Heikens et al. 2007). In addition, they express an abundance of immature cells called 'basophilic macrocytes', suggestive of abnormal reticulocyte/erythrocyte differentiation (Heikens et al. 2007). A multipotent cell line, lacking TSPAN33, showed impaired differentiation of erythroid progenitors in response to erythropoietin. This defect was rescued by TSPAN33 expression, further supporting a positive function for TSPAN33 during erythropoiesis (Heikens et al. 2007). Little information is yet available regarding TSPAN33-associated partner proteins and molecular mechanisms.

Human TSPAN33 maps to a chromosomal region (7q32) sometimes deleted in myelodysplastic syndrome (MDS) and acute myelogenous leukemia (AML) (Chen et al. 2005). Hence, it is reasonable to consider that TSPAN33 deletion might contribute to anemia in some cases of human MDS and AML (Heikens et al. 2007; Chen et al. 2005). TSPAN33 also may contribute to γ-secretase cleavage of Notch, as seen in a cell culture system (Dunn et al. 2010).

7.5 Genetic Evidence from Other Species

TSPAN1 (NET-1)—Genetic evidence is not yet available regarding a functional role for TSPAN1 during normal mammalian development. However, tissue expression and knockdown data in the early Xenopus embryo suggest that TSPAN1 could link bone morphogenetic protein (BMP) signalling with gastrulation movements and neural differentiation (Yamamoto et al. 2007).

7.6 Genetic Evidence Not Yet Available

TSPAN2 (NET-3)—Gene expression for TSPAN2 is restricted to the nervous system (in rats) where it shows strong expression on cells of the oligodendrocyte lineage. Increasing levels of expression postnatally, from days 3–22, suggests a possible role both in early oligodendrocyte differentiation and in mature myelin-forming glia (Birling et al. 1999). Additional genetic evidence for TSPAN2 functions is not yet available.

TSPAN3 (TM4SF8)—TSPAN3 is expressed in cells throughout the central nervous system, including neurons and astrocytes (Tiwari-Woodruff et al. 2004). TSPAN3 may participate in oligodendrocyte proliferation and migration (Tiwari-Woodruff et al. 2001), but definitive genetic evidence is not yet available. TSPAN3 is also expressed on various immune cell types in mice, including dendritic cells (DCs). Downregulation of TSPAN3 during DC activation may suggest a regulatory role (Tokoro et al. 2001).

TSPAN4 (NAG-2, TM4SF7)—TSPAN4 is widely expressed in a variety of cell and tissue types, but not in brain (Tachibana et al. 1997). TSPAN4 is targeted by autoantibodies in patients with systemic sclerosis (Traggiai et al. 2010), but otherwise its functions are unknown.

TSPAN5 (NET-4, TM4SF9)—Developmental expression changes suggest that TSPAN5 may regulate maturation in mouse brain and other tissues (Juenger et al. 2005; Garcia-Frigola et al. 2001). Data from cell culture systems suggests that TSPAN5 may contribute to γ-secretase cleavage of Notch (Dunn et al. 2010), and may play a role during osteoclastogenesis (Iwai et al. 2007). Definitive genetic evidence for TSPAN5 functions remains to be obtained.

TSPAN9 (NET-5)—TSPAN9 is preferentially expressed on megakaryocytes and platelets in mice and humans (Protty et al. 2009). TSPAN9 functions are unknown.

TSPAN10 (Oculospanin)—TSPAN10 is expressed in several regions of the eye, but functions are unknown.

TSPAN13 (NET-6, TM4SF13)—In a model cell line, TSPAN13 may inhibit osteoclastogenesis (Iwai et al. 2007). Otherwise, its functions are unknown.

CD82 (KAI1, TSPAN27)—An abundance of evidence supports CD82 acting as a suppressor of tumor growth, migration, invasion and/or metastasis (Liu and Zhang 2006). Other reports have suggested roles in T cell activation (Lagaudriere-Gesbert et al. 1998), neural development (Mela and Goldman 2009), and maternal-fetal interface communication (Gellersen et al. 2007). As of the end of 2010, no published reports have described a CD82 mouse knockout.

TSPAN6 (T245, TM4SF6); **TSPAN11**; **TSPAN14**; **TSPAN15 (NET-7)**; **TSPAN16 (TM4-B, TM4SF16)**; **TSPAN17**; **TSPAN18**; **TSPAN19**; **TSPAN31 (SAS)**—Little is yet known regarding functions and/or expression patterns for these tetraspanin proteins.

7.7 Summary and Longterm Perspectives

Genetic information is not yet available regarding the functions of more than half of the tetraspanin family members 17/33. However from the other 16 tetraspanins, genetic analyses yield a central theme. In multiple cases, specific tetraspanins serve to enhance indirectly the functions of associated proteins, most likely by affecting the lateral organization of those proteins in the plasma membrane.

Because cells typically express several tetraspanins, with possibly overlapping functions, it appears that functional compensation may explain the failure to observe more dramatic genetic phenotypes. Consequently, it is predicted that double and triple knockouts may be necessary to achieve phenotypes that are more dramatic, and therefore more useful with respect to gaining mechanistic insights. In this regard, oocyte fertilization is only partially inhibited in CD9 and CD81 single knockout mice, but completely in double knockout mice (Rubinstein et al. 2006). Likewise, the extent of spontaneous fusion to form multinucleated giant cells in lung may be enhanced in CD9/CD81 double knockout mice compared to single knockout mice (Takeda et al. 2003). In another example, cellular immune functions were significantly more altered in cells from CD37/TSSC6 double knockout mice compared to single knockouts (Gartlan et al. 2010). With CD9/CD81 and CD37/TSSC6 double knockouts having been already produced, there are still 526 possible tetraspanin double knockouts remaining to be generated!

In several cases the functions of tetraspanins become more obvious during pathological situations, thus making them attractive targets for therapeutic intervention, e.g., during infectious disease, tumor progression, and other pathological situations (Hemler 2008). Precisely because tetraspanins are generally not required for the development and maintenance of normal cellular functions, therapeutic intervention might be achievable without undue side effects.

References

Abidi FE, Holinski-Feder E, Rittinger O, Kooy F, Lubs HA, Stevenson RE, Schwartz CE (2002) A novel 2 bp deletion in the TM4SF2 gene is associated with MRX58. J Med Genet 39:430–433

Ardman B, Sikorski MA, Staunton DE (1992) CD43 interferes with T-lymphocyte adhesion. Proc Natl Acad Sci USA 89:5001–5005

Bahi A, Boyer F, Kolira M, Dreyer JL (2005) In vivo gene silencing of CD81 by lentiviral expression of small interference RNAs suppresses cocaine-induced behaviour. J Neurochem 92:1243–1255

Baleato RM, Guthrie PL, Gubler MC, Ashman LK, Roselli S (2008) Deletion of CD151 results in a strain-dependent glomerular disease due to severe alterations of the glomerular basement membrane. Am J Pathol 173:927–937

Belkin AM, Stepp MA (2000) Integrins as receptors for laminins. Microsc Res Tech 51:280–301

Berditchevski F, Gilbert E, Griffiths MR, Fitter S, Ashman L, Jenner SJ (2001) Analysis of the CD151-alpha3beta1 integrin and CD151-tetraspanin interactions by mutagenesis. J Biol Chem 276:41165–41174

Bertaux C, Dragic T (2006) Different domains of CD81 mediate distinct stages of hepatitis C virus pseudoparticle entry. J Virol 80:4940–4948

Birling MC, Tait S, Hardy RJ, Brophy PJ (1999) A novel rat tetraspan protein in cells of the oligodendrocyte lineage. J Neurochem 73:2600–2608

Champy MF, Le Voci L, Selloum M, Peterson LB, Cumiskey AM, Blom D (2010) Reduced body weight in male Tspan8-deficient mice. Int J Obes (Lond) 35:605–617

Charrin S, Le Naour F, Oualid M, Billard M, Faure G, Hanash SM, Boucheix C, Rubinstein E (2001) The major CD9 and CD81 molecular partner. Identification and characterization of the complexes. J Biol Chem 276:14329–14337

Chen Z, Pasquini M, Hong B, DeHart S, Heikens M, Tsai S (2005) The human Penumbra gene is mapped to a region on chromosome 7 frequently deleted in myeloid malignancies. Cancer Genet Cytogenet 162:95–98

Cowin AJ, Adams D, Geary SM, Wright MD, Jones JC, Ashman LK (2006) Wound healing is defective in mice lacking tetraspanin CD151. J Invest Dermatol 126:680–689

Criswick VG, Schepens CL (1969) Familial exudative vitreoretinopathy. Am J Ophthalmol 68:578–594

Deng J, Yeung VP, Tsitoura D, DeKruyff RH, Umetsu DT, Levy S (2000) Allergen-induced airway hyperreactivity is diminished in CD81-deficient mice. J Immunol 165:5054–5061

Deng J, DeKruyff RH, Freeman GJ, Umetsu DT, Levy S (2002) Critical role of CD81 in cognate T-B cell interactions leading to Th2 responses. Int Immunol 14:513–523

Dunn CD, Sulis ML, Ferrando AA, Greenwald I (2010) A conserved tetraspanin subfamily promotes Notch signaling in Caenorhabditis elegans and in human cells. Proc Natl Acad Sci USA 107:5907–5912

Feigelson SW, Grabovsky V, Shamri R, Levy S, Alon R (2003) The CD81 tetraspanin facilitates instantaneous leukocyte VLA-4 adhesion strengthening to vascular cell adhesion molecule 1 (VCAM-1) under shear flow. J Biol Chem 278:51203–51212

Garcia-Frigola C, Burgaya F, de Lecea L, Soriano E (2001) Pattern of expression of the tetraspanin Tspan-5 during brain development in the mouse. Mech Dev 106:207–212

Gartlan KH, Belz GT, Tarrant JM, Minigo G, Katsara M, Sheng KC, Sofi M, van Spriel AB, Apostolopoulos V, Plebanski M, Robb L, Wright MD (2010) A complementary role for the tetraspanins CD37 and Tssc6 in cellular immunity. J Immunol 185:3158–3166

Geisert EE Jr, Williams RW, Geisert GR, Fan L, Asbury AM, Maecker HT, Deng J, Levy S (2002) Increased brain size and glial cell number in CD81-null mice. J Comp Neurol 453:22–32

Gellersen B, Briese J, Oberndorfer M, Redlin K, Samalecos A, Richter DU, Loning T, Schulte HM, Bamberger AM (2007) Expression of the metastasis suppressor KAI1 in decidual cells at the human maternal-fetal interface: regulation and functional implications. Am J Pathol 170:126–139

Glazar AI, Evans JP (2009) Immunoglobulin superfamily member IgSF8 (EWI-2) and CD9 in fertilisation: evidence of distinct functions for CD9 and a CD9-associated protein in mammalian sperm-egg interaction. Reprod Fertil Dev 21:293–303

Gomot M, Ronce N, Dessay S, Zemni R, Ayrault AD, Moizard MP, Nivelon A, Gilgenkrantz S, Dourlens J, des Portes V, Chelly J, Moraine C (2002) TM4SF2 gene involvement reconsidered in an XLMR family after neuropsychological assessment. Am J Med Genet 112:400–404

Goschnick MW, Lau LM, Wee JL, Liu YS, Hogarth PM, Robb LM, Hickey MJ, Wright MD, Jackson DE (2006) Impaired 'outside-in' integrin alphaIIbbeta3 signaling and thrombus stability in TSSC6-deficient mice. Blood 108:1911–1918

Grarup N, Andersen G, Krarup NT, Albrechtsen A, Schmitz O, Jorgensen T, Borch-Johnsen K, Hansen T, Pedersen O (2008) Association testing of novel type 2 diabetes risk alleles in the JAZF1, CDC123/CAMK1D, TSPAN8, THADA, ADAMTS9, and NOTCH2 loci with insulin release, insulin sensitivity, and obesity in a population-based sample of 4,516 glucose-tolerant middle-aged Danes. Diabetes 57:2534–2540

Ha SA, Tsuji M, Suzuki K, Meek B, Yasuda N, Kaisho T, Fagarasan S (2006) Regulation of B1 cell migration by signals through Toll-like receptors. J Exp Med 203:2541–2550

Heikens MJ, Cao TM, Morita C, Dehart SL, Tsai S (2007) Penumbra encodes a novel tetraspanin that is highly expressed in erythroid progenitors and promotes effective erythropoiesis. Blood 109:3244–3252

Hemler ME (2008) Targeting of tetraspanin proteins—potential benefits and strategies. Nat Rev Drug Discov 7:747–758

Ishibashi T, Ding L, Ikenaka K, Inoue Y, Miyado K, Mekada E, Baba H (2004) Tetraspanin protein CD9 is a novel paranodal component regulating paranodaljunctional formation. J Neurosci 24:96–102

Israels SJ, McMillan-Ward EM (2005) CD63 modulates spreading and tyrosine phosphorylation of platelets on immobilized fibrinogen. Thromb Haemost 93:311–318

Iwai K, Ishii M, Ohshima S, Miyatake K, Saeki Y (2007) Expression and function of transmembrane-4 superfamily (tetraspanin) proteins in osteoclasts: reciprocal roles of Tspan-5 and NET-6 during osteoclastogenesis. Allergol Int 56:457–463

Iwamoto R, Higashiyama S, Mitamura T, Taniguchi N, Klagsbrun M, Mekada E (1994) Heparin-binding EGF-like growth factor, which acts as a diphtheria toxin receptor, forms a complex with membrane protein DRAP27/CD9, which upregulates functional receptors and diphtheria toxin sensitivity. EMBO J 13:2322–2330

Juenger H, Holst MI, Duffe K, Jankowski J, Baader SL (2005) Tetraspanin-5 (Tm4sf9) mRNA expression parallels neuronal maturation in the cerebellum of normal and L7En-2 transgenic mice. J Comp Neurol 483:318–328

Junge HJ, Yang S, Burton JB, Paes K, Shu X, French DM, Costa M, Rice DS, Ye W (2009) TSPAN12 regulates retinal vascular development by promoting Norrin- but not Wnt-induced FZD4/beta-catenin signaling. Cell 139:299–311

Kagan A, Feld S, Chemke J, Bar-Khayim Y (1988) Occurrence of hereditary nephritis, pretibia-lepidermolysisbullosa and beta-thalassemia minor in two siblings with end-stage renal disease. Nephron 49:331–332

Kaji K, Oda S, Shikano T, Ohnuki T, Uematsu Y, Sakagami J, Tada N, Miyazaki S, Kudo A (2000) The gamete fusion process is defective in eggs of Cd9-deficient mice. Nat Genet 24:279–282

Karamatic CV, Burton N, Kagan A, Green CA, Levene C, Flinter F, Brady LR, Daniels G, Anstee DJ (2004) CD151, the first member of the tetraspanin (TM4) superfamily detected on erythrocytes, is essential for the correct assembly of human basement membranes in kidney and skin. Blood 104:2217–2223

Kashtan CE (2000) Alport syndromes: phenotypic heterogeneity of progressive hereditary nephritis. Pediatr Nephrol 14:502–512

Kazarov AR, Yang X, Stipp CS, Sehgal B, Hemler ME (2002) An extracellular site on tetraspanin CD151 determines α3 and α6 integrin-dependent cellular morphology. J Cell Biol 158:1299–1309

Kelic S, Levy S, Suarez C, Weinstein DE (2001) CD81 regulates neuron-induced astrocyte cell-cycle exit. Mol Cell Neurosci 17:551–560

Kim TR, Yoon JH, Kim YC, Yook YH, Kim IG, Kim YS, Lee H, Paik SG (2004) LPS-induced CD53 expression: a protection mechanism against oxidative and radiation stress. Mol Cells 17:125–131

Knobeloch KP, Wright MD, Ochsenbein AF, Liesenfeld O, Lohler J, Zinkernagel RM, Horak I, Orinska Z (2000) Targeted inactivation of the tetraspanin CD37 impairs T-cell-dependent B-cell response under suboptimal costimulatory conditions. Mol Cell Biol 20:5363–5369

Kovalenko OV, Metcalf DG, DeGrado WF, Hemler ME (2005) Structural organization and interactions of transmembrane domains in tetraspanin proteins. BMC Struct Biol 5:11

Koyama Y, Yamanoha B, Yoshida T (1990) A novel monoclonal antibody induces the differentiation of monocyte leukemic cells. Biochem Biophys Res Commun 168:898–904

Lagaudriere-Gesbert C, Lebel-Binay S, Hubeau C, Fradelizi D, Conjeaud H (1998) Signaling through the tetraspanin CD82 triggers its association with the cytoskeleton leading to sustained morphological changes and T cell activation. Eur J Immunol 28:4332–4344

Lammerding J, Kazarov AR, Huang H, Lee RT, Hemler ME (2003) Tetraspanin CD151 regulates alpha6beta1 integrin adhesion strengthening. Proc Natl Acad Sci USA 100:7616–7621

Lau LM, Wee JL, Wright MD, Moseley GW, Hogarth PM, Ashman LK, Jackson DE (2004) The tetraspanin superfamily member, CD151 regulates outside-in integrin {alpha}IIb{beta}3 signalling and platelet function. Blood 104:2368–2375

Le Naour F, Rubinstein E, Jasmin C, Prenant M, Boucheix C (2000) Severely reduced female fertility in CD9-deficient mice. Science 287:319–321

Liu WM, Zhang XA (2006) KAI1/CD82, a tumor metastasis suppressor. Cancer Lett 240:183–194

Maecker HT, Levy S (1997) Normal lymphocyte development but delayed humoral immune response in CD81-null mice. J Exp Med 185:1505–1510

Maecker HT, Do MS, Levy S (1998) CD81 on B cells promotes interleukin 4 secretion and antibody production during T helper type 2 immune responses. Proc Natl Acad Sci USA 95:2458–2462

Maranduba CM, Sa ME, Muller OG, Pavanello RC, Vianna-Morgante AM, Passos-Bueno MR (2004) Does the P172H mutation at the TM4SF2 gene cause X-linked mental retardation? Am J Med Genet A 124A:413–415

Mela A, Goldman JE (2009) The tetraspanin KAI1/CD82 is expressed by late-lineage oligodendrocyte precursors and may function to restrict precursor migration and promote oligodendrocyte differentiation and myelination. J Neurosci 29:11172–11181

Meyer-Wentrup F, Figdor CG, Ansems M, Brossart P, Wright MD, Adema GJ, van Spriel AB (2007) Dectin-1 interaction with tetraspanin CD37 inhibits IL-6 production. J Immunol 178:154–162

Min G, Wang H, Sun TT, Kong XP (2006) Structural basis for tetraspanin functions as revealed by the cryo-EM structure of uroplakin complexes at 6-A resolution. J Cell Biol 173:975–983

Miyado K, Yamada G, Yamada S, Hasuwa H, Nakamura Y, Ryu F, Suzuki K, Kosai K, Inoue K, Ogura A, Okabe M, Mekada E (2000) Requirement of CD9 on the egg plasma membrane for fertilization. Science 287:321–324

Miyazaki T, Muller U, Campbell KS (1997) Normal development but differentially altered proliferative responses of lymphocytes in mice lacking CD81. EMBO J 16:4217–4225

Mollinedo F, Fontan G, Barasoain I, Lazo PA (1997) Recurrent infectious diseases in human CD53 deficiency. Clin Diagn Lab Immunol 4:229–231

Nicholson RH, Pantano S, Eliason JF, Galy A, Weiler S, Kaplan J, Hughes MR, Ko MS (2000) Phemx, a novel mouse gene expressed in hematopoietic cells maps to the imprinted cluster on distal chromosome 7. Genomics 68:13–21

Nikopoulos K, Gilissen C, Hoischen A, van Nouhuys CE, Boonstra FN, Blokland EA, Arts P, Wieskamp N, Strom TM, Ayuso C, Tilanus MA, Bouwhuis S, Mukhopadhyay A, Scheffer H, Hoefsloot LH, Veltman JA, Cremers FP, Collin RW (2010) Next-generation sequencing of a 40 Mb linkage interval reveals TSPAN12 mutations in patients with familial exudative vitreoretinopathy. Am J Hum Genet 86:240–247

Noor A, Gianakopoulos PJ, Fernandez B, Marshall CR, Szatmari P, Roberts W, Scherer SW, Vincent JB (2009) Copy number variation analysis and sequencing of the X-linked mental retardation gene TSPAN7/TM4SF2 in patients with autism spectrum disorder. Psychiatr Genet 19:154–155

Orlowski E, Chand R, Yip J, Wong C, Goschnick MW, Wright MD, Ashman LK, Jackson DE (2009) A platelet tetraspanin superfamily member, CD151, is required for regulation of thrombus growth and stability in vivo. J Thromb Haemost 7:2074–2084

Pfistershammer K, Majdic O, Stockl J, Zlabinger G, Kirchberger S, Steinberger P, Knapp W (2004) CD63 as an activation-linked T cell costimulatory element. J Immunol 173:6000–6008

Piton A, Gauthier J, Hamdan FF, Lafreniere RG, Yang Y, Henrion E, Laurent S, Noreau A, Thibodeau P, Karemera L, Spiegelman D, Kuku F, Duguay J, Destroismaisons L, Jolivet P, Cote M, Lachapelle K, Diallo O, Raymond A, Marineau C, Champagne N, Xiong L, Gaspar C, Riviere JB, Tarabeux J, Cossette P, Krebs MO, Rapoport JL, Addington A, Delisi LE, Mottron L, Joober R, Fombonne E, Drapeau P, Rouleau GA (2010) Systematic resequencing of X-chromosome synaptic genes in autism spectrum disorder and schizophrenia. Mol Psychiatry 16:867–880

Pols MS, Klumperman J (2009) Trafficking and function of the tetraspanin CD63. Exp Cell Res 315:1584–1592

Poulter JA, Ali M, Gilmour DF, Rice A, Kondo H, Hayashi K, Mackey DA, Kearns LS, Ruddle JB, Craig JE, Pierce EA, Downey LM, Mohamed MD, Markham AF, Inglehearn CF, Toomes C (2010) Mutations in TSPAN12 cause autosomal-dominant familial exudative vitreoretinopathy. Am J Hum Genet 86:248–253

Protty MB, Watkins NA, Colombo D, Thomas SG, Heath VL, Herbert JM, Bicknell R, Senis YA, Ashman LK, Berditchevski F, Ouwehand WH, Watson SP, Tomlinson MG (2009) Identification

of Tspan9 as a novel platelet tetraspanin and the collagen receptor GPVI as a component of tetraspanin microdomains. Biochem J 417:391–400
Robb L, Tarrant J, Groom J, Ibrahim M, Li R, Borobakas B, Wright MD (2001) Molecular characterisation of mouse and human TSSC6: evidence that TSSC6 is a genuine member of the tetraspanin superfamily and is expressed specifically in haematopoietic organs. Biochim Biophys Acta 1522:31–41
Rocha-Perugini V, Montpellier C, Delgrange D, Wychowski C, Helle F, Pillez A, Drobecq H, Le Naour F, Charrin S, Levy S, Rubinstein E, Dubuisson J, Cocquerel L (2008) The CD81 partner EWI-2wint inhibits hepatitis C virus entry. PLoS One 3:e1866
Rops AL, Figdor CG, van der Schaaf A, Tamboer WP, Bakker MA, Berden JH, Dijkman HB, Steenbergen EJ, van der Vlag J, van der Spriel AB (2010) The tetraspanin CD37 protects against glomerular IgA deposition and renal pathology. Am J Pathol 176:2188–2197
Rubinstein E, Ziyyat A, Prenant M, Wrobel E, Wolf JP, Levy S, Le Naour F, Boucheix C (2006) Reduced fertility of female mice lacking CD81. Dev Biol 290:351–358
Runge KE, Evans JE, He ZY, Gupta S, McDonald KL, Stahlberg H, Primakoff P, Myles DG (2007) Oocyte CD9 is enriched on the microvillar membrane and required for normal microvillar shape and distribution. Dev Biol 304:317–325
Sachs N, Kreft M, van den Bergh Weerman MA, Beynon AJ, Peters TA, Weening JJ, Sonnenberg A (2006) Kidney failure in mice lacking the tetraspanin CD151. J Cell Biol 175:33–39
Sadej R, Romanska H, Baldwin G, Gkirtzimanaki K, Novitskaya V, Filer AD, Krcova Z, Kusinska R, Ehrmann J, Buckley CD, Kordek R, Potemski P, Eliopoulos AG, Lalani e-N, Berditchevski F (2009) CD151 regulates tumorigenesis by modulating the communication between tumor cells and endothelium. Mol Cancer Res 7:787–798
Scholz CJ, Jacob CP, Buttenschon HN, Kittel-Schneider S, Boreatti-Hummer A, Zimmer M, Walter U, Lesch KP, Mors O, Kneitz S, Deckert J, Reif A (2010) Functional variants of TSPAN8 are associated with bipolar disorder and schizophrenia. Am J Med Genet BNeuropsychiatr Genet 153B:967–972
Schroder J, Lullmann-Rauch R, Himmerkus N, Pleines I, Nieswandt B, Orinska Z, Koch-Nolte F, Schroder B, Bleich M, Saftig P (2009) Deficiency of the tetraspanin CD63 associated with kidney pathology but normal lysosomal function. Mol Cell Biol 29:1083–1094
Shastry BS (2010) Genetic susceptibility to advanced retinopathy of prematurity (ROP). J Biomed Sci 17:69
Sheng KC, van Spriel AB, Gartlan KH, Sofi M, Apostolopoulos V, Ashman L, Wright MD (2009) Tetraspanins CD37 and CD151 differentially regulate Ag presentation and T-cell co-stimulation by DC. Eur J Immunol 39:50–55
Silvie O, Greco C, Franetich JF, Dubart-Kupperschmitt A, Hannoun L, van Gemert GJ, Sauerwein RW, Levy S, Boucheix C, Rubinstein E, Mazier D (2006) Expression of human CD81 differently affects host cell susceptibility to malaria sporozoites depending on the Plasmodium species. Cell Microbiol 8:1134–1146
Sincock PM, Fitter S, Parton RG, Berndt MC, Gamble JR, Ashman LK (1999) PETA-3/CD151, a member of the transmembrane 4 superfamily, is localised to the plasma membrane and endocytic system of endothelial cells, associates with multiple integrins and modulates cell function. J Cell Sci 112:833–844
Sklar P, Smoller JW, Fan J, Ferreira MA, Perlis RH, Chambert K, Nimgaonkar VL, McQueen MB, Faraone SV, Kirby A, de Bakker PI, Ogdie MN, Thase ME, Sachs GS, Todd-Brown K, Gabriel SB, Sougnez C, Gates C, Blumenstiel B, Defelice M, Ardlie KG, Franklin J, Muir WJ, McGhee KA, MacIntyre DJ, McLean A, VanBeck M, McQuillin A, Bass NJ, Robinson M, Lawrence J, Anjorin A, Curtis D, Scolnick EM, Daly MJ, Blackwood DH, Gurling HM, Purcell SM (2008) Whole-genome association study of bipolar disorder. Mol Psychiatry 13:558–569
Stipp CS, Orlicky D, Hemler ME (2001a) FPRP: a major, highly stoichiometric, highly specific CD81 and CD9-associated protein. J Biol Chem 276:4853–4862
Stipp CS, Kolesnikova TV, Hemler ME (2001b) EWI-2 is a major CD9 and CD81 partner, and member of a novel Ig protein subfamily. J Biol Chem 276:40545–40554

Tachibana I, Bodorova J, Berditchevski F, Zutter MM, Hemler ME (1997) NAG-2, a novel transmembrane-4 superfamily (TM4SF) protein that complexes with integrins and other TM4SF proteins. J Biol Chem 272:29181–29189

Takeda Y, Tachibana I, Miyado K, Kobayashi M, Miyazaki T, Funakoshi T, Kimura H, Yamane H, Saito Y, Goto H, Yoneda T, Yoshida M, Kumagai T, Osaki T, Hayashi S, Kawase I, Mekada E (2003) Tetraspanins CD9 and CD81 function to prevent the fusion of mononuclear phagocytes. J Cell Biol 161:945–956

Takeda Y, Kazarov AR, Butterfield CE, Hopkins BD, Benjamin LE, Kaipainen A, Hemler ME (2007) Deletion of tetraspanin Cd151 results in decreased pathologic angiogenesis in vivo and in vitro. Blood 109:1524–1532

Tarrant JM, Groom J, Metcalf D, Li R, Borobokas B, Wright MD, Tarlinton D, Robb L (2002) The absence of Tssc6, a member of the tetraspanin superfamily, does not affect lymphoid development but enhances in vitro T-cell proliferative responses. Mol Cell Biol 22: 5006–5018

Terada N, Baracskay K, Kinter M, Melrose S, Brophy PJ, Boucheix C, Bjartmar C, Kidd G, Trapp BD (2002) The tetraspanin protein, CD9, is expressed by progenitor cells committed to oligodendrogenesis and is linked to beta1 integrin, CD81, and Tspan-2. Glia 40:350–359

Tiwari-Woodruff SK, Buznikov AG, Vu TQ, Micevych PE, Chen K, Kornblum HI, Bronstein JM (2001) OSP/claudin-11 forms a complex with a novel member of the tetraspanin super family and beta1 integrin and regulates proliferation and migration of oligodendrocytes. J Cell Biol 153:295–305

Tiwari-Woodruff SK, Kaplan R, Kornblum HI, Bronstein JM (2004) Developmental expression of OAP-1/Tspan-3, a member of the tetraspanin superfamily. J Neurosci Res 77:166–173

Tokoro Y, Shibuya K, Osawa M, Tahara-Hanaoka S, Iwama A, Kitamura T, Nakauchi H, Shibuya A (2001) Molecular cloning and characterization of mouse Tspan-3, a novel member of the tetraspanin superfamily, expressed on resting dendritic cells. Biochem Biophys Res Commun 288:178–183

Toomes C, Bottomley HM, Scott S, Mackey DA, Craig JE, Appukuttan B, Stout JT, Flaxel CJ, Zhang K, Black GC, Fryer A, Downey LM, Inglehearn CF (2004) Spectrum and frequency of FZD4 mutations in familial exudativevitreoretinopathy. Invest Ophthalmol Vis Sci 45:2083–2090

Toothill VJ, van Mourik JA, Nieuwenhuis HK, Metzelaar MJ, Pearson JD (1990) Characterization of the enhanced adhesion of neutrophil leukocytes to thrombin-stimulated endothelial cells. J Immunol 145:283–291

Traggiai E, Lunardi C, Bason C, Dolcino M, Tinazzi E, Corrocher R, Puccetti A (2010) Generation of anti-NAG-2 mAb from patients' memory B cells: implications for a novel therapeutic strategy in systemic sclerosis. Int Immunol 22:367–374

Tsitsikov EN, Gutierrez-Ramos JC, Geha RS (1997) Impaired CD19 expression and signaling, enhanced antibody response to type II T independent antigen and reduction of B-1 cells in CD81-deficient mice. Proc Natl Acad Sci USA 94:10844–10849

Unternaehrer JJ, Chow A, Pypaert M, Inaba K, Mellman I (2007) The tetraspanin CD9 mediates lateral association of MHC class II molecules on the dendritic cell surface. Proc Natl Acad Sci USA 104:234–239

van Spriel AB, Sofi M, Gartlan KH, van der Schaaf A, Verschueren I, Torensma R, Raymakers RA, Loveland BE, Netea MG, Adema GJ, Wright MD, Figdor CG (2009) The tetraspanin protein CD37 regulates IgA responses and anti-fungal immunity. PLoS Pathog 5:e1000338

van Spriel AB, Puls KL, Sofi M, Pouniotis D, Hochrein H, Orinska Z, Knobeloch KP, Plebanski M, Wright MD (2004) A regulatory role for CD37 in T cell proliferation. J Immunol 172:2953–2961

von Lindern JJ, Rojo D, Grovit-Ferbas K, Yeramian C, Deng C, Herbein G, Ferguson MR, Pappas TC, Decker JM, Singh A, Collman RG, O'Brien WA (2003) Potential role for CD63 in CCR5-mediated human immunodeficiency virus type 1 infection of macrophages. J Virol 77:3624–3633

Winterwood NE, Varzavand A, Meland MN, Ashman LK, Stipp CS (2006) A critical role for tetraspanin CD151 in {alpha}3{beta}1 and {alpha}6{beta}4 integrin-dependent tumor cell functions on laminin-5. Mol Biol Cell 17:2707–2721

Wright MD, Geary SM, Fitter S, Moseley GW, Lau LM, Sheng KC, Apostolopoulos V, Stanley EG, Jackson DE, Ashman LK (2004) Characterization of mice lacking the tetraspanin superfamily member CD151. Mol Cell Biol 24:5978–5988

Xu D, Sharma C, Hemler ME (2009) Tetraspanin12 regulates ADAM10-dependent cleavage of amyloid precursor protein. FASEB J 23:3674–3681

Yamamoto Y, Grubisic K, Oelgeschlager M (2007) Xenopus Tetraspanin-1 regulates gastrulation movements and neural differentiation in the early Xenopus embryo. Differentiation 75:235–245

Yánez-Mó M, Alfranca A, Cabañas C, Marazuela M, Tejedor R, Ursa MA, Ashman LK, De Landázuri MO, Sánchez-Madrid F (1998) Regulation of endothelial cell motility by complexes of tetraspan molecules CD81/TAPA-1 and CD151/PETA-3 with α3β1 integrin localized at endothelial lateral junctions. J Cell Biol 141:791–804

Yang X, Claas C, Kraeft SK, Chen LB, Wang Z, Kreidberg JA, Hemler ME (2002) Palmitoylation of tetraspanin proteins: modulation of CD151 lateral interactions, subcellular distribution, and integrin-dependent cell morphology. Mol Biol Cell 13:767–781

Yang XH, Richardson AL, Torres-Arzayus MI, Zhou P, Sharma C, Kazarov AR, Andzelm MM, Strominger JL, Brown M, Hemler ME (2008) CD151 accelerates breast cancer by regulating α6 integrin functions, signaling, and molecular organization. Cancer Res 68:3204–3213

Yauch RL, Berditchevski F, Harler MB, Reichner J, Hemler ME (1998) Highly stoichiometric, stable and specific association of integrin α3β1 with CD151 provides a major link to phosphatidylinositol 4-kinase and may regulate cell migration. Mol Biol Cell 9:2751–2765

Ye X, Wang Y, Nathans J (2010) The Norrin/Frizzled4 signaling pathway in retinal vascular development and disease. Trends Mol Med 16:417–425

Yunta M, Lazo PA (2003) Apoptosis protection and survival signal by the CD53 tetraspanin antigen. Oncogene 22:1219–1224

Zeggini E, et al (2008) Meta-analysis of genome-wide association data and large-scale replication identifies additional susceptibility loci for type 2 diabetes. Nat Genet 40:638–645

Zemni R, Bienvenu T, Vinet MC, Sefiani A, Carrie A, Billuart P, McDonell N, Couvert P, Francis F, Chafey P, Fauchereau F, Friocourt G, Portes V, Cardona A, Frints S, Meindl A, Brandau O, Ronce N, Moraine C, Bokhoven H, Ropers HH, Sudbrak R, Kahn A, Fryns JP, Beldjord C (2000) A new gene involved in X-linked mental retardation identified by analysis of an X;2 balanced translocation. Nat Genet 24:167–170

Zhang XA, Kazarov AR, Yang X, Bontrager AL, Stipp CS, Hemler ME (2002) Function of the tetraspanin CD151–α6β1 integrin complex during cellular morphogenesis. Mol Biol Cell 13:1–11

Zoller M (2009) Tetraspanins: push and pull in suppressing and promoting metastasis. Nat Rev Cancer 9:40–55

Chapter 8
Tetraspanins in Lower Eukaryotes

Hiroki Moribe and Eisuke Mekada

Abstract Tetraspanins are widely distributed among metazoan phyla. Recent progress in various genome projects have enabled us to carry out a comparative analysis of tetraspanins and their relatives in lower eukaryotes including unicellular organisms (e.g. Amoeba and Encephalitozoon). The functions of tetraspanins in lower eukaryotes are largely unknown, but recent comprehensive genetic analyses have provided us with some clues for the roles played by these proteins in these organisms. Here we overview tetraspanin proteins in lower eukaryotes and plants and describe what is known about their specific functions in the context of organ development and differentiation. We also review recent data describing physiological regulators of tetraspanins and their involvement in processes related to parasitism.

8.1 Introduction

The simplicity of lower organisms is a great advantage for their use as model systems in genetic and molecular studies of genes with unknown functions. These model organisms have contributed to the identification and clarification of the fundamental principles of regulatory pathways for cell proliferation, developmental processes, and homeostasis. Comparative genomics has also been useful in shedding light on the interface between gene evolution and organism-specific characteristics.

H. Moribe (✉)
Department of Biology, Kurume University School of Medicine,
67 Asahimachi, Kurume, Fukuoka 830-0011, Japan
e-mail: himoribe@med.kurume-u.ac.jp

E. Mekada
Department of Cell Biology, Research Institute for Microbial Diseases,
Osaka University, 3-1 Yamadaoka, Suita, Osaka 565-0871, Japan
e-mail: emekada@biken.osaka-u.ac.jp

F. Berditchevski and E. Rubinstein (eds.), *Tetraspanins*, Proteins and Cell Regulation 9,
DOI 10.1007/978-94-007-6070-7_8, © Springer Science+Business Media Dordrecht 2013

Tetraspanins are highly structurally conserved molecules, but the divergence of their amino acid sequences makes it difficult to trace their origin. Tetraspanins have not been found in unicellular organisms such as yeast and protists. Hence, tetraspanins are thought to be linked to multicellularity. However, results of recent genome projects have enabled comprehensive knowledge of the genome information of both unicellular and multicellular organisms, resulting in the surprising discovery of tetraspanin(s) in two types of protozoa: parasitic amoeba (*Entamoeba histolytica*) and microsporidia (*Encephalitozoon cuniculi*) (Table 8.1). In addition, tetraspanin-like sequences lacking the CCG motif are also present in other protozoa such as *Trypanosoma brucei* and *Leishmania infantum*, but have not been found so far in the protozoa *Plasmodium falciparum*, *Trichomonas vaginalis*, and *Toxoplasma gondii* (Moribe, unpublished observation). Tetraspanins exist in primitive metazoans Porifera (sponge) and in non-bilaterian animals (e.g., Cnidaria) (Huang et al. 2005; Garcia-España et al. 2009). Genome studies also revealed presence of tetraspanin-like molecules in the cellular slime mold (*Dictyostelium discoideum*) (Huang et al. 2005). The functions of tetraspanins in lower eukaryotes are largely unknown.

Here, we summarize recent research on tetraspanins in lower eukaryotes, including animals, fungi, protists, and plants, focusing on their biological functions and intriguing relation to parasitism.

8.2 The Roles of Tetraspanins Revealed by Genetic Studies in *C. elegans*

8.2.1 Tetraspanins in C. elegans

Tetraspanins are conserved in free-living nematodes of the *Caenorhabditis* family (*C. elegans*, *C. briggsae*, *C. remanei*, *C. japonica*, and *C. brenneri*) and also in parasitic nematodes such as *Brugia malayi* and *Wuchereria bancrofti* (see 8.4.1). A large body of results of genetic studies carried out over 40 years using *C. elegans* as the model organism is archived in the online database "WormBase" (http://wormbase.org). A gene with the hallmarks of tetraspanin was first found in 1996 (Tomlinson and Wright 1996). In 1998, whole genome sequencing of *C. elegans*, the first completed genome sequence of a multicellular organism, revealed that *C. elegans* has a 97 Mb[*1] (*1; ultimately 100.274 Mb) genome containing over 19,000 genes The *C. elegans* Sequencing Consortium 1998). Sequence analyses have revealed 21 tetraspanin-like sequences in *C. elegans*, annotated as *tsp-1* to *tsp-21*[*2] (*2; *tsp-6* is supposed to be a pseudogene) (Todres et al. 2000). Unlike the single group of *Drosophila* tetraspanin genes (see 8.4.2.), the *C. elegans* genes are dispersed on six chromosomes without clustering. It is difficult to determine counterparts of the *C. elegans* tetraspanins in other phyla due to poor sequence

Table 8.1 Tetraspanins in lower eukaryotes

Scientific name	Popular name	Cellularity	Life style	No. (ref)	Database
Drosophila melanogaster	Fruit fly	M	FR	37 (Todres et al. 2000)	http://flybase.org/
Bombyx mori	Silkworm	M	FR	17[*1]	http://silkdb.org/
Anopheles gambiae	Mosquito	M	FR	13[*1]	http://agambiae.vectorbase.org/
Ixodes scapularis	Mite	M	FR	19[*1]	http://iscapularis.vectorbase.org/
Daphnia pulex	Water flea	M	FR	14[*1]	http://wfleabase.org/
Caenorhabditis elegans	Nematode	M	FR	21 (Todres et al. 2000)	http://wormbase.org/
Brugia malayi	Nematode	M	P	10[*2]	http://wormbase.org/
Schistosoma mansoni	Trematode	M	P	>25 (Huang et al. 2005)	http://schistodb.net/
Magnaporthe griesea	Fungus	M	P	2 (Lambou et al. 2008a)	http://www.broadinstitute.org/annotation/genome/magnaporthe_grisea/MultiHome.html
Podospora anserina	Fungus	M	FR	2 (Lambou et al. 2008a)	http://podospora.igmors.u-psud.fr/
Saccharomyces cerevisiae	Fungus (yeast)	U	FR	0	
Encephalitozoon cuniculi	Fungi (microsporidia)	U	P	1 (Huang et al. 2005)	http://microsporidiadb.org/
Plasmodium falciparum	Protozoa	U	P	0	http://plasmodb.org/
Trypanosoma brucei	Protozoa	U	P	1	http://tritrypdb.org
Entamoeba histolytica	Protozoa	U	P	6 (Huang et al. 2005)	http://amoebadb.org/
Dictyostelium discoideum	Fungi (cellular slime mold)	U/M	FR	5 (Huang et al. 2005)	http://dictybase.org/
Arabidopsis thaliana	Plant (thale cress)	M	FR	>17 (Cnops et al. 2006)	http://www.arabidopsis.org/
Oryza sativa	Plant (rice)	M	FR	16	http://www.plantgdb.org/OsGDB/

M multicellular organism, *U* unicellular organism, *FR* free-living organism, *P* parasitic organism

[*1]The number of the genes showing homology with *D. melanogaster* tetraspanins searched in orthodb: http://cegg.unige.ch/orthodb4

[*2]The number of the genes showing homology with *C. elegans* tetraspanins searched in WormBase

similarity (Huang et al. 2005; Todres et al. 2000). Among them, however, several nematode tetraspanins showed relatively high homology with mammalian tetraspanins. For instance, *tsp-7* showed 34% identity with human CD63, and *tsp-12* has 45% identity with human TSPAN5, in their amino acid sequences, and these might be among the earliest emerged tetraspanins (Huang et al. 2005). Analysis of transcriptional fusions with GFP reporters revealed that each *tsp* gene has an overlapping but distinct expression pattern in various tissues/organs, suggesting that tetraspanins in *C. elegans* may have unique roles (Moribe, unpublished result). However, gene silencing analysis by RNAi could not establish the physiological roles of tetraspanins, as tetraspanin depletion, with the exception of *tsp-15*, did not affect morphology, motility, viability, or fertility in *C. elegans* (Moribe et al. 2004).

8.2.2 TSP-15 Is Essential for H_2O_2 Generation by the DUOX Pathway in C. elegans

tsp-15 knockdown animals showed abnormal body morphology manifested as Dumpy (short and fat) and Blister (blistering of exoskeletal elements) phenotypes (Moribe et al. 2004). These are typical of deficiencies in the exoskeleton caused by mutations in the collagen genes or components of the collagen biosynthetic pathway (Page and Johnstone 2007; Johnstone 2000). The exoskeleton of *C. elegans*, or cuticle, is a collagenous extracellular matrix (ECM) and contains additional insoluble proteins, as well as lipids (Page and Johnstone 2007). Cuticle has both toughness and flexibility which protects the animal from adverse environments, maintains body morphology and integrity, and facilitates locomotion via attachments to body wall muscle. Collagens are synthesized and processed in underlying epidermal cells, called the hypodermis, and this is followed by apical secretion and cross-linking (Page and Johnstone 2007). TSP-15 is expressed in hypodermis, and *tsp-15* hypomorph mutants show multiple deficiencies in cuticle morphology, but the molecular mechanisms underlying these phenotypes remain unknown (Moribe et al. 2004).

Recently, a screen for mutants with phenotypes similar to *tsp-15* resulted in simultaneous isolation of genes related to the process of collagen cross-linking directed by the Dual oxidase (DUOX) pathway. BLI-3, DOXA-1, and MLT-7 correspond to nematode homologues of DUOX, Duox activator/DUOX maturation factor (DUOXA), and peroxidase, respectively (Moribe et al. 2012). DUOX is a conserved NADPH oxidase (NOX) family enzyme which generates reactive oxygen species (ROS) (Lambeth 2004; Sumimoto 2008). ROS plays multiple physiological roles in protein modification, cell differentiation, apoptosis, and cellular signaling as well as host defense (Bedard and Krause 2007). Mammalian DUOXs (DUOX1 and DUOX2) were originally identified as hydrogen peroxide (H_2O_2) generators that are essential for iodination of thyroid hormone in mammals (De Deken et al. 2000; Dupuy et al. 1999). H_2O_2 produced by DUOX also has a conserved role in innate immunity to eliminate pathogens in many eukaryotes (Allaoui et al. 2009; Ha et al. 2005; Chávez et al. 2009; Flores et al. 2010). In *C. elegans*, RNAi analysis revealed that CeDuox-1 (BLI-3) carried out collagen cross-linking via tyrosyl residues (Edens et al. 2001). DUOXAs (DUOXA1 and DUOXA2) dimerize with DUOXs and play pivotal roles in maturation and targeting of DUOX proteins to the cell surface (Grasberger et al. 2007; Grasberger and Refetoff 2006; Luxen et al. 2009; Morand et al. 2009). Finally, MLT-7 encodes an extracellular peroxidase which is an integral component of the DUOX system in H_2O_2-mediated oxidation of substrates, like thyroperoxidase and myeloperoxidase in mammals (Thein et al. 2009; Donkó et al. 2005). In addition to their similar phenotypes of decreased tyrosine cross-linking, restoration of a *tsp-15* hypomorph mutant by co-expression of *bli-3* and *doxa-1* indicated that TSP-15 acts in the same pathway (Moribe et al. 2012). Furthermore, TSP-15 as well as DOXA-1 were essential for BLI-3-dependent H_2O_2 production when reconstituted in mammalian cells (Moribe et al. 2012). Although the role of TSP-15 in H_2O_2 generation remains uncertain, BLI-3, DOXA-1, and TSP-15 form protein complexes in vitro and in vivo, indicating that TSP-15 may establish or maintain a specialized microdomain at the plasma membrane to facilitate BLI-3 activity (Moribe et al. 2012).

8.2.3 Tetraspanins and Notch Signaling

Recent genetic studies in *C. elegans* have revealed a requirement of tetraspanins in Notch signaling, and this role is also conserved in mammalian cells (Dunn et al. 2010). Notch is a key molecule which exerts multiple functions in cell-cell communication and cell fate decision in metazoan development. Notch signaling followed by ligand stimulation is regulated by sequential proteolytic processing of Notch directed by ADAMs (a disintegrin and metalloproteases) and the γ-secretase complex (Kopan and Ilagan 2009). Tetraspanins are possible modulators of Notch activity since they interact with ADAM10 and γ-secretase, and promote cleavage of β-amyloid precursor protein (β-APP) (Wakabayashi et al. 2009; Xu et al. 2009). The positive role of tetraspanins in Notch signaling was elucidated by phenotypic rescue of a *glp-1/lin-12/Notch* gain-of-function mutant by *tsp-12* or a *tsp-14* knockdown in *C. elegans* (Dunn et al. 2010). This was further substantiated by inhibition of a constitutively active (ca) form of NOTCH1 by depletion of TSPAN33 and TSPAN5, an RDS subgroup of tetraspanins in mammalian cells (Dunn et al. 2010). Silencing of tetraspanins had no effect on activity of a membrane-untethered form of caNOTCH1, implicating a role for tetraspanins in facilitating the γ-secretase cleavage step in Notch activation (Dunn et al. 2010). It is most likely that tetraspanins are involved in the specification of γ-secretase substrate preference, rather than subsequent γ-secretase processing of Notch, since knockdown of the tetraspanins CD81 and CD9 has no effect on Notch activity (Dunn et al. 2010). As elevated Notch signaling is often associated with cancers, including T cell acute lymphoblastic leukemia (T-ALL), targeting of tetraspanins has potential as a novel anti-cancer therapeutic approach to reduce γ-secretase activity.

8.3 Tetraspanins in Fungi

8.3.1 PLS1, a Tetraspanin Essential for Fungal Pathogenicity

Plant fungal diseases are still serious agricultural threat for perennial crop supply. Over 80% of crop damage is thought to be caused by pathogenic fungi. The process of infection of many plant-parasitic fungi involves the attachment of conidia (asexual spores) to the host cell wall. The conidia then differentiate into specialized cells called appressoria. Appressoria then develop a penetration peg, which can perforate the cell wall of the host tissues via a combination of enormous mechanical pressure (8.0 MPa in *Magnaporthe grisea*) and enzymatic breaching (Howard et al. 1991). *M. grisea* is a major plant-pathogenic fungus which causes rice blast disease. A non-pathogenic *M. grisea* mutant named *punchless* was identified by insertional mutagenesis (Clergeot et al. 2001). The mutant appressorium differentiate normally, but are then unable to form a penetration peg. Cloning of the inactivated gene from the mutant resulted in the first identification of a tetraspanin in fungi, *PLS1* (*MgPLS1*) (Clergeot et al. 2001). As a result fungi have become the preferred genetic model to study tetraspanin function in lower eukaryotes.

PLS1 is a fungal specific tetraspanin, and the gene homologous to *MgPLS1* has also been found in other plant-pathogenic fungi such as *Botrytis cinerea* and *Colletotrichum lindemuthianum* (Gourgues et al. 2002). These pathogenic fungi are in different clades of ascomycetes; *M. grisea* and *C. lindemuthianum* are in the *Sordariomycetes*, and *B. cinerea* in the *Leotiomycetes*. However, they infect their hosts similarly via appressorium-mediated penetration, suggesting that appressorial infection has arisen by convergent evolution (see below). Gene disruption of their corresponding *PLS1* gene results in loss of pathogenicity due to penetration failure in spite of appressorial differentiation comparable to wild type (Gourgues et al. 2004; Veneault-Fourrey et al. 2005, 2006). Therefore, *PLS1* has conserved appressorial function widespread among these pathogenic fungi. Since the emerging site of penetration peg was abortive and mislocalized in *Clpls1* mutant appressoria, *PLS1* may have a role in formation and/or positioning of the site where the penetration peg emerges (Veneault-Fourrey et al. 2005).

PLS1 is widely conserved in both of ascomycetes and basidiomycetes. Genomic analysis revealed other tetraspanin genes in higher fungi (Huang et al. 2005; Lambou et al. 2008a). *TSP2* is restricted in basidiomycetes, whereas *TSP3* is an ascomycetes-specific tetraspanin. In addition, *TPL1* (tetraspanin-like family) shows different cysteine patterning from canonical tetraspanin also found in ascomycetes (Lambou et al. 2008a). In *M. grisea*, *TSP3* and *TPL1* gene were involved in pathogenicity, at least in part, since the corresponding deletion mutants showed host-dependent decreased infectivity (Lambou et al. 2008a).

8.3.2 Recruitment of Tetraspanin and ROS Generator in Different Morphogenetic Processes by Convergent Evolution

PLS1 orthologs were also identified in other fungi with saprophytic lifestyles such as *Podospora anserina* and *Neurospora crassa*. These do not differentiate appressoria, suggesting another cellular function of *PLS1* (Lambou et al. 2008a). Using targeted gene disruption in the coprophilous fungus *P. anserina*, the role of *PaPls1* was addressed. *PaPls1* null mutation resulted in loss of ascospore germination due to absence of germination pegs (Lambou et al. 2008b). Despite complementation of germination defect in *PaPls1* mutant by exogenous *MgPLS1*, germination of ascospore in *M. grisea* was unaffected in *MgPLS1* mutant (Clergeot et al. 2001; Lambou et al. 2008b). Although *PLS1* is required for apparently different developmental steps in these fungi, the germination of ascospores in *P. anserina* and emergence of penetration pegs from appressoria in *M. grisea* share several similarities. Indeed, *P. anserina* develops heavily melanized ascospores to withstand their passage through digestive tract of herbivores. Ascospores of *M. grisea* are not melanized, whereas the appressoria are reinforced by melanization to withstand high internal turgor pressure to penetrate into the host (Howard et al. 1991). Other pathogenic fungus *C. lindemuthianum* also produce melanized appressoria. *PLS1* may be recruited in different morphogenetic processes by convergent evolution. This hypothesis is substantiated by the evidence from other molecular machinery, ROS generators. Among several *NOX* (NADPH oxidase) isoforms, the developmental role of *NOX2* gene was independently demonstrated in *P. anserina*, *M. grisea* and *B. cinerea* by gene disruption analysis (Lambou et al. 2008b; Malagnac et al. 2004; Egan et al. 2007; Segmüller et al. 2008). Similar to *PLS1* mutants, *NOX2* knockout resulted in impairment of the germination process in the ascospores of *P. anserina*, and the appressoria of *M. grisea* and *B. cinerea*. In addition, mutants of *PaNoxR*, a homologue of regulatory subunit of NOX complex, also showed the same phenotype as *PaNox2* mutant (Brun et al. 2009). Taken together, in spite of different morphogenetic processes in different clades of fungi, they share similar characteristics, i.e., directional emergence of specialized hyphae from reinforced structures, and require the same molecular machinery, *PLS1* and *NOX2*. Genetic and pharmacologic inhibition of melanin biosynthesis suppressed the germination defects in *PaPls1* and *PaNox2* mutants, indicating that *PLS1* and *NOX2* are in the same pathway for weakening of germ pore. On the other hand, non-pathogenicity in *MgPLS1* mutants was not restored in the absence of melanin (Lambou et al. 2008b; Malagnac et al. 2004). Thus, the functional links between melanin, tetraspanin and ROS generated by Nox enzyme in these morphogenetic processes *M. grisea* remains unclear. Importantly however, co-occurrence and recurrent recruitment of both *NOX2* and *PLS1* during the evolutionary process in higher fungi implies their involvement in the same cellular function by convergent evolution (Lambou et al. 2008b; Malagnac et al. 2008).

8.4 Roles of Tetraspanins in Other Lower Eukaryotes

8.4.1 Parasitic Diseases and Tetraspanins

As found in certain fungi, tetraspanins in lower organisms are often related to parasitism. Generally in higher eukaryotes such as vertebrates, tetraspanins are exploited as host receptors for invasion by viruses and pathogenic protists (see Chaps. 14 and 15). But conversely, in lower eukaryotes having a parasitic life style, endogenous tetraspanins are often pivotal for infectivity and survival in their hosts. This should stimulate development of new control strategies targeting tetraspanins for prevention and eradication of parasitic diseases.

Schistosomiasis and filariasis are prevalent but neglected tropical diseases in tropical and subtropical regions, caused by infections of human parasites. The World Health Organization has estimated that more than 200 million and 120 million people are infected, and 0.7 billion and 1.3 billion people are at risk of infection, by schistosomiasis and filariasis, respectively (World Health Organization 2009). Schistosomiasis is caused by infection by platyhelminths of the genus *Schistosoma*, such as *S. mansoni*, *S. japonicum* and *S. haematobium*, which are blood flukes (trematodes) having two-host life cycles (Hotez et al. 2010). Schistosomiasis is a health threat particularly in endemic areas with poor sanitation. Infection of humans is begun by exposure to contaminated fresh-water with free-swimming larvae (cercariae). Cercariae percutaneously penetrate humans and following a developmental change in the lung, migrate to liver and become sexually mature adults. They eventually reside in blood vessels surrounding the intestine and liver (*S. mansoni* and *S. japonicum*) or bladder (*S. haematobium*), and produce eggs. Eggs are passed out in the feces and urine, and hatched larvae (miracidium) infect fresh-water snails where they transform into cercariae. During chronic infection of humans, parasite eggs elicit continuous host immune responses, often resulting in granulomatous lesions leading to fibrosis in the affected organs (Hotez et al. 2010).

The adult schistosomes are entirely covered by a unique double lipid bilayered syncytium, or tegument. The tegument constitutes the host-parasite interface and contributes to parasite survival by an immunoevasive mechanism through the adsorption of host molecules onto the teguments to mask their non-self-status. Targeting of molecules onto teguments may be a feasible approach to enable sustainable control of schistosomiasis (Loukas et al. 2007). Recent genomic, proteomic and transcriptomic analyses have revealed potential candidates for vaccine antigens including tetraspanins (Berriman et al. 2009; Braschi et al. 2006; Gobert et al. 2010). *Sm23* and *Sm25* in *S. mansoni*, and *Sj23* in *S. japonicum*, are some of the earliest reported tetraspanins in any organism (Wright et al. 1990; Davern et al. 1991). More than 20 tetraspanin-like sequences have been found in the *S. mansoni* genome (SchistoDB: http://schitodb.net). In *S. mansoni*, Sm-TSP-1 and Sm-TSP-2 were originally identified as surface-expressed molecules, and they were actually highly expressed in the outer tegument (Smyth et al. 2003; Tran et al. 2006; Sepulveda et al. 2010). Moreover, the sera from schistosomiasis-resistant individuals showed higher levels of anti-Sm-TSP-2 IgG1 and IgG3 than chronically infected patients (Tran et al. 2006). Vaccination of mice with recombinant Sm-TSP-1 and Sm-TSP-2 showed efficacy in reducing parasites (Tran et al. 2006). Similarly, the Sm-TSP-2 ortholog in *S. japonicum* (Sj-TSP-2) conferred the same protective effect in a mouse vaccination model (Yuan et al. 2010). The role of tetraspanins in schistosomes was recently suggested by RNAi-mediated knockdown analysis. Silencing of *Sm-tsp-1* or *Sm-tsp-2* resulted in malformation of the tegument, and tetraspanin-depleted parasites were defective in survival in the host (Tran et al. 2010). This indicates that these tetraspanins have integral structural roles in tegumental development and maintenance.

Filariasis is caused by parasitic nematodes classified in the Filarioidea superfamily including *Brugia malayi*, *Wuchereria bancrofti*, *Onchocerca volvulus* and *Loa loa* (Taylor et al. 2010). All of these parasites shuttle between two-host species of hematophagous insects (mosquitoes and flies) and humans. Unlike the malaria-causing *Plasmodium*, human is the primary host for these worms. Mature adult organisms in humans produce microfilariae which are taken up by the vector insects and then develop into the infectious stage. Infectious larvae are injected into human skin by bites of other insects. *B. malayi* and *W. bancrofti* lodge in the lymphatic system, and elicit chronic immune responses and dysfunction of lymphatic vessels often resulting in elephantiasis and lymphedema of the limbs and genitals. On the other hand, *O. volvulus* initially infects the subcutaneous layer. Microfilariae then migrate to various tissues including the eye where microfilarial death invokes inflammatory responses which lead to ocular lesions called river blindness (onchocerciasis) (Taylor et al. 2010).

Compared with *Schistosoma*, little is known about tetraspanins in the filaria. In the draft genome of *B. malayi* (Ghedin et al. 2007), there are 10 tetraspanins homologous to the *C. elegans tsp* genes (Moribe, unpublished observation), but their physiological functions are still unclear. Similar to their use in control of schistosomiasis, tetraspanins are potential candidates for vaccine antigens against filariasis (Gnanasekar et al. 2008). However, this has not yet worked in nematodes, presumably because they are covered by an exoskeleton (cuticle) composed of collagens, so tetraspanins are not exposed at the body surface. Nevertheless, if tetraspanin(s) have pivotal roles in the worm life cycle, their targeting may present an alternate method to control filariasis. In particular, similar to *tsp-15* of *C. elegans*, molecules indispensable for cuticle development may be good candidates, as the cuticle is an essential barrier enabling the parasite to survive in the host (Blaxter and Bird 1997).

8.4.2 Drosophila melanogaster

Thirty-seven tetraspanins are found in the genome of the fruit fly *D. melanogaster*, more than in any vertebrates (Flybase: http://flybase.org). This is due to a contiguous array of 18 tetraspanin genes in region 42E of chromosome 2 (Todres et al. 2000). This cluster seems to be specific to the genus *Drosophila*, and is not found in other insects such as *Anopheles gambiae* (mosquito), *Bombyx mori* (silkworm), and

Apis mellifera (honeybee), indicating that rapid and repeated duplication occurred in the *Drosophila* genome (Huang et al. 2005). In spite of the large number of genes, specific or conserved functions of *Drosophila* tetraspanins are basically unknown. Sporadic reports have provided genetic evidence for the functions of tetraspanins and their functional redundancy.

The characterization of the *late bloomer* (*lbm*) mutation was an early example of genetic analysis using lower organisms as models to clarify physiological roles of tetraspanins (Kopczynski et al. 1996). The *lbm* mutant was isolated by classical forward genetics, and exhibited a delay in synaptic formation at neuromuscular junctions. *lbm* is located in the Tsp42E cluster, and two other tetraspanins expressed in motoneurons (Tsp42Ee and Tsp42Ej) were functionally redundant with *lbm* since the neuronal deficiency was enhanced by simultaneous deletion of these three tetraspanins (Fradkin et al. 2002). However, a large deletion encompassing nine tetraspanin genes in this cluster did not affect fertility or viability (Fradkin et al. 2002). Independently, Tsp42Ej was also identified as one of the genes predominantly expressed in the eye by DNA microarray analysis (Xu et al. 2004). This gene was referred to as *sunglasses* (*sun*), and the null mutant of *sun* showed light-induced degeneration of photoreceptor due to impairment of degradation of rhodopsin (Rh1), a G-protein coupled receptor (GPCR). Sun protein associated with Rh1 and was co-localized in lysosomes (Xu et al. 2004). Sun is related to mammalian CD63, which is also known as a late endosome/lysosome-enriched protein. Conservation of the role of CD63 in degradative pathways of GPCRs in mammalian cells should be considered.

Finally, the function of Tsp68C was revealed serendipitously by gain-of-function mutants which suppress excess proliferation and abnormal differentiation of hemocytes caused by mutation in *yantar* (*ytr*), a conserved arginine-rich nuclear protein (Sinenko and Mathey-Prevot 2004). The regulatory mechanisms controlling proliferation of hemocytes by the interaction of Ytr and Tsp68C are not clear, but the Ras/Raf/MAPK pathway may be involved since overexpression of Tsp68C was also able to oppose the proliferative signal of activated Ras and Raf (Sinenko and Mathey-Prevot 2004). It remains to be determined whether the mechanisms are conserved, but interestingly, in mice deficient in TSSC6, CD37, CD81, and CD151 hyperproliferation of T-cells is observed (Miyazaki et al. 1997; van Spriel et al. 2004; Wright et al. 2004; Tarrant et al. 2002).

8.4.3 Plants

There are 17 tetraspanins termed *TETRASPANINs* (*TET1* to *TET17*), and a few additional tetraspanin-like sequences present in the genome of *Arabidopsis thaliana* (Huang et al. 2005; Olmos et al. 2003; Cnops et al. 2006) (Moribe, unpublished observation). Tetraspanins are well conserved in a wide range species of the kingdom Plantae from the Spermatophyta to the Bryophyta including rice (*Oryza sativa*), maize (*Zea mays*), wheat (*Triticum aestivum*), wine grape (*Vitis vinifera*), lotus (*Lotus japonicus*), and moss (*Physcomitrella patens*) (Cnops et al. 2006).

Tetraspanins in plants have a distinctive pattern of cysteines in the second extracellular loop, lacking the CCG sequence (Olmos et al. 2003; Desalle et al. 2010), which distinguishes them from tetraspanins in other species. In spite of this difference, other structural hallmarks and conservation of intron-exon boundaries indicate that the sequences are part of the tetraspanin family and that they have a common origin with the ancestral tetraspanins of animals and fungi (Huang et al. 2005; Garcia-España et al. 2009; Olmos et al. 2003) (see Chap. 2).

Currently, there are only two tetraspanin mutants described for plants: *ekeko* and *tornado2(trn2)*. Both mutants carry mutations in the *TET1* gene of *A. thaliana* (Olmos et al. 2003; Cnops et al. 2006; Chiu et al. 2007). The mutation identified as *ekeko* was isolated by T-DNA insertional gene disruption. *EKEKO* is ubiquitously expressed and loss of function of *EKEKO* results in pleiotropic effects with aberrant pattern formation and differentiation of leaves, roots, and flowers (Olmos et al. 2003). *TRN2* was identified as the gene responsible for *tornado* (*trn*) mutants showing twisted organs (Cnops et al. 2006, 2000). Although the mechanistic insights into this phenotype are uncertain, they may emerge from other mutants. The other *trn* alleles resulted from mutations in *TRN1*, a plant-specific cytoplasmic protein containing a leucine-rich repeat (LRR) motif that is presumably involved in cellular signaling (Cnops et al. 2006). Mutants in *trn1* and *trn2*, exhibit the same phenotype, and a *trn1 trn2* double mutant also showed a similar phenotype to each single mutant, indicating that TRN1 and TRN2 function in the same pathway (Cnops et al. 2006). Also, it was suggested that aberrant leaf venation and asymmetry was caused by altered distribution of auxin in *trn* mutants (Cnops et al. 2006).

TRN genes are conspicuous by their higher expression in shoot apical meristem (SAM), the tissue containing undifferentiated stem cells responsible for outgrowth and primordia development of leaves and floral components (Cnops et al. 2006). A *trn2* mutant showed enhanced expression of the SAM marker gene *SHOOT MERISTEMLESS* (*STM*) and partially restored the phenotype of a *stm* mutation (Chiu et al. 2007). *TRN* genes are proposed to control cell fate decisions in conjunction with *STM* and *ASYMMETRIC LEAVES1*, both of which are key transcriptional factors for maintenance and function of SAM (Cnops et al. 2006; Chiu et al. 2007; Girin et al. 2009).

8.5 Concluding Remarks

Because of their simplicity, lower organisms have proved to be excellent in vivo models providing new insights into tetraspanin functions. A major obstacle to clarification of tetraspanin functions using lower organisms is the divergence of sequences particularly between far-diverged phyla. Hence, a simple comparison of amino acid sequences has been less informative for inferring their functions. Genetic clues from mutant phenotypes have been important in revealing the specific functions of tetraspanins and also in elucidating the conserved genetic pathway in which tetraspanins play crucial roles.

It should be noted that, tetraspanin families consists of many members so that conventional screens for recessive mutants are often defeated by functional redundancy and exhibit subtle phenotypes even in lower organisms (Fradkin et al. 2002). To overcome such problems, we need to design and develop other screening strategies such as genetic modifier screens and utilization of cell/tissue specific markers or chemicals for uncovering the masked phenotypes.

References

Allaoui A, Botteaux A, Dumont JE, Hoste C, De Deken X (2009) Dual oxidases and hydrogen peroxide in a complex dialogue between host mucosae and bacteria. Trends Mol Med 15:571–579

Bedard K, Krause K-H (2007) The NOX family of ROS-generating NADPH oxidases: physiology and pathophysiology. Physiol Rev 87:245–313

Berriman M, Haas BJ, LoVerde PT, Wilson RA, Dillon GP et al (2009) The genome of the blood fluke *Schistosoma mansoni*. Nature 460:352–358

Blaxter M, Bird D (1997) Parasitic nematodes. In: Riddle DL, Blumenthal T, Meyer BJ, Priess JR (eds) *C. elegans* II. Cold Spring Harbor Laboratory Press, Cold Spring Harbor, pp 851–878

Braschi S, Borges WC, Wilson RA (2006) Proteomic analysis of the schistosome tegument and its surface membranes. Mem Inst Oswaldo Cruz 101(Suppl 1):205–212

Brun S, Malagnac F, Bidard F, Lalucque H, Silar P (2009) Functions and regulation of the Nox family in the filamentous fungus *Podospora anserina*: a new role in cellulose degradation. Mol Microbiol 74:480–496

Chávez V, Mohri-Shiomi A, Garsin DA (2009) Ce-Duox1/BLI-3 generates reactive oxygen species as a protective innate immune mechanism in *Caenorhabditis elegans*. Infect Immun 77:4983–4989

Chiu W-H, Chandler J, Cnops G, Van Lijsebettens M, Werr W (2007) Mutations in the *tornado2* gene affect cellular decisions in the peripheral zone of the shoot apical meristem of *Arabidopsis thaliana*. Plant Mol Biol 63:731–744

Clergeot PH, Gourgues M, Cots J, Laurans F, Latorse MP et al (2001) *PLS1*, a gene encoding a tetraspanin-like protein, is required for penetration of rice leaf by the fungal pathogen *Magnaporthe grisea*. Proc Natl Acad Sci USA 98:6963–6968

Cnops G, Wang X, Linstead P, Van Montagu M, Van Lijsebettens M et al (2000) *Tornado1* and *tornado2* are required for the specification of radial and circumferential pattern in the *Arabidopsis* root. Development 127:3385–3394

Cnops G, Neyt P, Raes J, Petrarulo M, Nelissen H et al (2006) The *tornado1* and *tornado2* genes function in several patterning processes during early leaf development in *Arabidopsis thaliana*. Plant Cell 18:852–866

Davern KM, Wright MD, Herrmann VR, Mitchell GF (1991) Further characterisation of the *Schistosoma japonicum* protein Sj23, a target antigen of an immunodiagnostic monoclonal antibody. Mol Biochem Parasitol 48:67–75

De Deken X, Wang D, Many MC, Costagliola S, Libert F et al (2000) Cloning of two human thyroid cDNAs encoding new members of the NADPH oxidase family. J Biol Chem 275:23227–23233

Desalle R, Mares R, Garcia-España A (2010) Evolution of cysteine patterns in the large extracellular loop of tetraspanins from animals, fungi, plants and single-celled eukaryotes. Mol Phylogenet Evol 56:486–491

Donkó A, Péterfi Z, Sum A, Leto T, Geiszt M (2005) Dual oxidases. Philos Trans R Soc Lond B Biol Sci 360:2301–2308

Dunn CD, Sulis ML, Ferrando AA, Greenwald I (2010) A conserved tetraspanin subfamily promotes Notch signaling in *Caenorhabditis elegans* and in human cells. Proc Natl Acad Sci USA 107:5907–5912

Dupuy C, Ohayon R, Valent A, Noël-Hudson MS, Dème D et al (1999) Purification of a novel flavoprotein involved in the thyroid NADPH oxidase. Cloning of the porcine and human cDNAs. J Biol Chem 274:37265–37269

Edens WA, Sharling L, Cheng G, Shapira R, Kinkade JM et al (2001) Tyrosine cross-linking of extracellular matrix is catalyzed by Duox, a multidomain oxidase/peroxidase with homology to the phagocyte oxidase subunit gp91phox. J Cell Biol 154:879–891

Egan MJ, Wang Z-Y, Jones MA, Smirnoff N, Talbot NJ (2007) Generation of reactive oxygen species by fungal NADPH oxidases is required for rice blast disease. Proc Natl Acad Sci USA 104:11772–11777

Flores MV, Crawford KC, Pullin LM, Hall CJ, Crosier KE et al (2010) Dual oxidase in the intestinal epithelium of zebrafish larvae has anti-bacterial properties. Biochem Biophys Res Commun 400:164–168

Fradkin LG, Kamphorst JT, DiAntonio A, Goodman CS, Noordermeer JN (2002) Genomewide analysis of the *Drosophila* tetraspanins reveals a subset with similar function in the formation of the embryonic synapse. Proc Natl Acad Sci USA 99:13663–13668

Garcia-España A, Mares R, Sun T-T, Desalle R (2009) Intron evolution: testing hypotheses of intron evolution using the phylogenomics of tetraspanins. PLoS One 4:e4680

Ghedin E, Wang S, Spiro D, Caler E, Zhao Q et al (2007) Draft genome of the filarial nematode parasite *Brugia malayi*. Science 317:1756–1760

Girin T, Sorefan K, Ostergaard L (2009) Meristematic sculpting in fruit development. J Exp Bot 60:1493–1502

Gnanasekar M, Anand SB, Ramaswamy K (2008) Identification and cloning of a novel tetraspanin (TSP) homologue from *Brugia malayi*. DNA Seq 19:151–156

Gobert GN, Tran MH, Moertel L, Mulvenna J, Jones MK et al (2010) Transcriptional changes in *Schistosoma mansoni* during early schistosomula development and in the presence of erythrocytes. PLoS Negl Trop Dis 4:e600

Gourgues M, Clergeot PH, Veneault C, Cots J, Sibuet S et al (2002) A new class of tetraspanins in fungi. Biochem Biophys Res Commun 297:1197–1204

Gourgues M, Brunet-Simon A, Lebrun MH, Levis C (2004) The tetraspanin BcPls1 is required for appressorium-mediated penetration of *Botrytis cinerea* into host plant leaves. Mol Microbiol 51:619–629

Grasberger H, Refetoff S (2006) Identification of the maturation factor for dual oxidase. Evolution of an eukaryotic operon equivalent. J Biol Chem 281:18269–18272

Grasberger H, De Deken X, Miot F, Pohlenz J, Refetoff S (2007) Missense mutations of dual oxidase 2 (*DUOX2*) implicated in congenital hypothyroidism have impaired trafficking in cells reconstituted with DUOX2 maturation factor. Mol Endocrinol 21:1408–1421

Ha E-M, Oh C-T, Bae YS, Lee W-J (2005) A direct role for dual oxidase in *Drosophila* gut immunity. Science 310:847–850

Hotez PJ, Bethony JM, Diemert DJ, Pearson M, Loukas A (2010) Developing vaccines to combat hookworm infection and intestinal schistosomiasis. Nat Rev Microbiol 8:814–826

Howard RJ, Ferrari MA, Roach DH, Money NP (1991) Penetration of hard substrates by a fungus employing enormous turgor pressures. Proc Natl Acad Sci USA 88:11281–11284

Huang S, Yuan S, Dong M, Su J, Yu C et al (2005) The phylogenetic analysis of tetraspanins projects the evolution of cell-cell interactions from unicellular to multicellular organisms. Genomics 86:674–684

Johnstone IL (2000) Cuticle collagen genes. Expression in *Caenorhabditis elegans*. Trends Genet 16:21–27

Kopan R, Ilagan MXG (2009) The canonical Notch signaling pathway: unfolding the activation mechanism. Cell 137:216–233

Kopczynski CC, Davis GW, Goodman CS (1996) A neural tetraspanin, encoded by late bloomer, that facilitates synapse formation. Science 271:1867–1870

Lambeth JD (2004) NOX enzymes and the biology of reactive oxygen. Nat Rev Immunol 4:181–189

Lambou K, Tharreau D, Kohler A, Sirven C, Marguerettaz M et al (2008a) Fungi have three tetraspanin families with distinct functions. BMC Genomics 9:63

Lambou K, Malagnac F, Barbisan C, Tharreau D, Lebrun M-H et al (2008b) The crucial role of the Pls1 tetraspanin during ascospore germination in *Podospora anserina* provides an example of the convergent evolution of morphogenetic processes in fungal plant pathogens and saprobes. Eukaryot Cell 7:1809–1818

Loukas A, Tran M, Pearson MS (2007) Schistosome membrane proteins as vaccines. Int J Parasitol 37:257–263

Luxen S, Noack D, Frausto M, Davanture S, Torbett BE et al (2009) Heterodimerization controls localization of Duox-DuoxA NADPH oxidases in airway cells. J Cell Sci 122:1238–1247

Malagnac F, Lalucque H, Lepère G, Silar P (2004) Two NADPH oxidase isoforms are required for sexual reproduction and ascospore germination in the filamentous fungus *Podospora anserina*. Fungal Genet Biol 41:982–997

Malagnac F, Bidard F, Lalucque H, Brun S, Lambou K et al (2008) Convergent evolution of morphogenetic processes in fungi: role of tetraspanins and NADPH oxidases 2 in plant pathogens and saprobes. Commun Integr Biol 1:180–181

Miyazaki T, Müller U, Campbell KS (1997) Normal development but differentially altered proliferative responses of lymphocytes in mice lacking CD81. EMBO J 16:4217–4225

Morand S, Ueyama T, Tsujibe S, Saito N, Korzeniowska A et al (2009) Duox maturation factors form cell surface complexes with Duox affecting the specificity of reactive oxygen species generation. FASEB J 23:1205–1218

Moribe H, Yochem J, Yamada H, Tabuse Y, Fujimoto T et al (2004) Tetraspanin protein (TSP-15) is required for epidermal integrity in *Caenorhabditis elegans*. J Cell Sci 117:5209–5220

Moribe H, Konakawa R, Koga D, Ushiki T, Nakamura K et al (2012) Tetraspanin is required for generation of reactive oxygen species by the dual oxidase system in *Caenorhabditis elegans*. PLoS Genet 8:e1002957

Olmos E, Reiss B, Dekker K (2003) The ekeko mutant demonstrates a role for tetraspanin-like protein in plant development. Biochem Biophys Res Commun 310:1054–1061

Page AP, Johnstone IL (2007) The cuticle. In: The *C. elegans* Research Community (ed) Wormbook. WormBook, doi:/10.1895/wormbook.1.7.1, http://www.wormbook.org

Segmüller N, Kokkelink L, Giesbert S, Odinius D, van Kan J et al (2008) NADPH oxidases are involved in differentiation and pathogenicity in *Botrytis cinerea*. Mol Plant Microbe Interact 21:808–819

Sepulveda J, Tremblay JM, DeGnore JP, Skelly PJ, Shoemaker CB (2010) *Schistosoma mansoni* host-exposed surface antigens characterized by sera and recombinant antibodies from schistosomiasis-resistant rats. Int J Parasitol 40:1407–1417

Sinenko SA, Mathey-Prevot B (2004) Increased expression of *Drosophila* tetraspanin, Tsp68C, suppresses the abnormal proliferation of *ytr*-deficient and Ras/Raf-activated hemocytes. Oncogene 23:9120–9128

Smyth D, Mcmanus DP, Smout MJ, Laha T, Zhang W et al (2003) Isolation of cDNAs encoding secreted and transmembrane proteins from *Schistosoma mansoni* by a signal sequence trap method. Infect Immun 71:2548–2554

Sumimoto H (2008) Structure, regulation and evolution of Nox-family NADPH oxidases that produce reactive oxygen species. FEBS J 275:3249–3277

Tarrant JM, Groom J, Metcalf D, Li R, Borobokas B et al (2002) The absence of Tssc6, a member of the tetraspanin superfamily, does not affect lymphoid development but enhances in vitro T-cell proliferative responses. Mol Cell Biol 22:5006–5018

Taylor MJ, Hoerauf A, Bockarie M (2010) Lymphatic filariasis and onchocerciasis. Lancet 376:1175–1185

The *C. elegans* Sequencing Consortium (1998) Genome sequence of the nematode *C. elegans*: a platform for investigating biology. Science 282:2012–2018

Thein M, Winter A, Stepek G, Mccormack G, Stapleton G et al (2009) Combined extracellular matrix cross-linking activity of the peroxidase MLT-7 and the duox BLI-3 are critical for post-embryonic viability in *Caenorhabditis elegans*. J Biol Chem 284:17549–17563

Todres E, Nardi JB, Robertson HM (2000) The tetraspanin superfamily in insects. Insect Mol Biol 9:581–590

Tomlinson MG, Wright MD (1996) A new transmembrane 4 superfamily molecule in the nematode, Caenorhabditis elegans. J Mol Evol 43:312–314

Tran MH, Pearson MS, Bethony JM, Smyth DJ, Jones MK et al (2006) Tetraspanins on the surface of *Schistosoma mansoni* are protective antigens against schistosomiasis. Nat Med 12:835–840

Tran MH, Freitas TC, Cooper L, Gaze S, Gatton ML et al (2010) Suppression of mRNAs encoding tegument tetraspanins from *Schistosoma mansoni* results in impaired tegument turnover. PLoS Pathog 6:e1000840

van Spriel AB, Puls KL, Sofi M, Pouniotis D, Hochrein H et al (2004) A regulatory role for CD37 in T cell proliferation. J Immunol 172:2953–2961

Veneault-Fourrey C, Parisot D, Gourgues M, Lauge R, Lebrun MH et al (2005) The tetraspanin gene *ClPLS1* is essential for appressorium-mediated penetration of the fungal pathogen *Colletotrichum lindemuthianum*. Fungal Genet Biol 42:306–318

Veneault-Fourrey C, Lambou K, Lebrun M-H (2006) Fungal Pls1 tetraspanins as key factors of penetration into host plants: a role in re-establishing polarized growth in the appressorium? FEMS Microbiol Lett 256:179–184

Wakabayashi T, Craessaerts K, Bammens L, Bentahir M, Borgions F et al (2009) Analysis of the gamma-secretase interactome and validation of its association with tetraspanin-enriched microdomains. Nat Cell Biol 11:1340–1346

World Health Organization (2009) Neglected tropical diseases, hidden successes, emerging opportunities. WHO Press, Geneva

Wright MD, Henkle KJ, Mitchell GF (1990) An immunogenic Mr 23,000 integral membrane protein of *Schistosoma mansoni* worms that closely resembles a human tumor-associated antigen. J Immunol 144:3195–3200

Wright MD, Geary SM, Fitter S, Moseley GW, Lau L-M et al (2004) Characterization of mice lacking the tetraspanin superfamily member CD151. Mol Cell Biol 24:5978–5988

Xu H, Lee S-J, Suzuki E, Dugan KD, Stoddard A et al (2004) A lysosomal tetraspanin associated with retinal degeneration identified via a genome-wide screen. EMBO J 23:811–822

Xu D, Sharma C, Hemler ME (2009) Tetraspanin12 regulates ADAM10-dependent cleavage of amyloid precursor protein. FASEB J 23:3674–3681

Yuan C, Y-j F, Li J, Y-f Y, L-l C et al (2010) *Schistosoma japonicum*: efficient and rapid purification of the tetraspanin extracellular loop 2, a potential protective antigen against schistosomiasis in mammalian. Exp Parasitol 126:456–461

Chapter 9
The Role of Tetraspanin Complexes in Egg-Sperm Fusion

Virginie Barraud-Lange and Claude Boucheix

Abstract Cell fusion occurs when cells unit their membranes and share their cytoplasm. Cell fusion may occur between genetically identical or different cells. Gamete fusion is a short event of the organism life cycle but it is essential for all organisms that depend on sexual reproduction for the maintenance of the species. The fusion of gametes having inherited one half of the genetic material of their respective parents ensures the diversity of individuals within a population. This diversity is increased by the exchange of genetic material between homologous chromosomes during meiosis. Therefore gamete fusion is a critical biological step in the life cycle that has recently raised more interest due to the possibility of treating patients with a fertility defect by in vitro methods. Molecular mechanisms underlying the fusion process are far from evident since (1) gene-knock out technologies developed to test in vivo the data from in vitro fertilization (IVF) studies failed to confirm the essential role of most of previously candidate molecules and (2) the surface proteins involved are deprived of fusogen properties. Indeed, among the large number of in vivo tested proteins, only CD9 and CD81 tetraspanins on egg and Izumo on sperm have been shown to be essential in mammalian sperm-egg membrane fusion and none of these molecules contain a fusion peptide.

V. Barraud-Lange (✉)
Inserm U1016, 24 rue du Fbg Saint-Jacques, Paris 75014, France

Université Paris Descartes, UFR médecine,
15 rue de l'école de médecine, Paris 75006, France

Laboratoire de Biologie de la Reproduction, Hôpital Cochin, AP-HP,
27 rue du Fbg Saint-Jacques, Paris 75014, France
e-mail: virginie.barraud-lange@cch.aphp.fr

C. Boucheix
Inserm U1004, 14 Avenue Paul-Vaillant-Couturier, Villejuif 94807, France

Université Paris-Sud 11, Villejuif, France
e-mail: claude.boucheix@inserm.fr

F. Berditchevski and E. Rubinstein (eds.), *Tetraspanins*, Proteins and Cell Regulation 9, 203
DOI 10.1007/978-94-007-6070-7_9, © Springer Science+Business Media Dordrecht 2013

9.1 Fertilization and Gamete Membrane Interaction

Fertilization is defined as the process by which the sperm and the egg fuse to form a new individual, the zygote. Both gametes must undergo cellular and nuclear maturation before making contact and fusing. During meiotic maturation, the genome of diploid germ cells undergoes two rounds of division (first and second meiosis) resulting in haploid gametes which carry half of the genetic content of the original cell. The meiosis is a continuous process during spermiogenesis which begins at the time of puberty, whereas nuclear maturation of the oocyte starts during fetal life and undergoes two arrest points. At birth, the female germ cells are called primary oocytes and get arrested at diplotene stage of meiotic prophase I. This meiotic arrest lasts till puberty that marks the beginning of the menstrual cycles during which and following a Luteinizing Hormone (LH) peak, one primary oocyte (or several, depending on the species) grows and completes the first meiosis with extrusion of the first polar body. This is immediately followed by the initiation of the second meiosis arrested at the metaphase II stage. The mature secondary oocyte is ovulated. The second meiosis division will end at the time of fertilization triggered by the sperm penetration.

In mammals, male and female gametes are two physically separated cells. Their meeting takes place in the female oviduct. Depending on the species, semen is deposited in the vagina or directly in the uterus during coitus and only a tiny fraction of spermatozoa migrates successfully to the fertilization site. Ejaculated spermatozoa are capable of moving actively, yet they do not have the ability for fertilization. They acquire the competence to fuse with eggs through a maturation process, called capacitation, during the ascent of the female genital tract (Austin 1951; Ikawa et al. 2010). The capacitation is a biochemical event resulting in the destabilization of the sperm head plasma and acrosomal membranes rendering them more fusogenic. Mouse spermatozoa collected from epididymis for experimental purpose are capacitated by in vitro incubation in culture medium previous to their use for fertilization. Capacitating sperm that reach the oviductal ampulla readily penetrate the cumulus layer surrounding the ovulated mature oocyte (Fig. 9.1a.1). They bind, then, in a species-specific manner to the zona pellucida (ZP), the last barrier the spermatozoa must pass before fertilizing the egg (Fig. 9.1a.2). The ZP is an extracellular coat protecting the egg and the future embryo from physical damages. In mice, the most widely studied mammalian model, the ZP comprises three sulphate glycoproteins: ZP1 (181–200 kDa), ZP2 (120–140 kDa) and ZP3 (83 kDa) (Wassarman 1988) synthesized and secreted by the oocyte. In the human, a fourth glycoprotein has been identified, ZP4 (or ZPB), which is thought to be dysfunctional in the mouse (Lefievre et al. 2004). Ultrastructural evidence suggests that the mouse zona pellucida is composed of filaments constructed by head-to-tail association of globular proteins. ZP2–ZP3 heterodimers are the basic repeating units of the filament, with cross-linking of filaments mediated by dimeric ZP1 (Wassarman and Mortillo 1991). ZP3 is the primary ligand that binds to the plasma membrane over the acrosomal cap of acrosome intact sperm. ZP3 functions also as an acrosome-reaction inducer. The acrosome reaction consists in the fusion between sperm plasma membrane and outer acrosomal membrane leading to acrosomal content exocytosis (Fig. 9.1a.2). ZP2, in turn, has

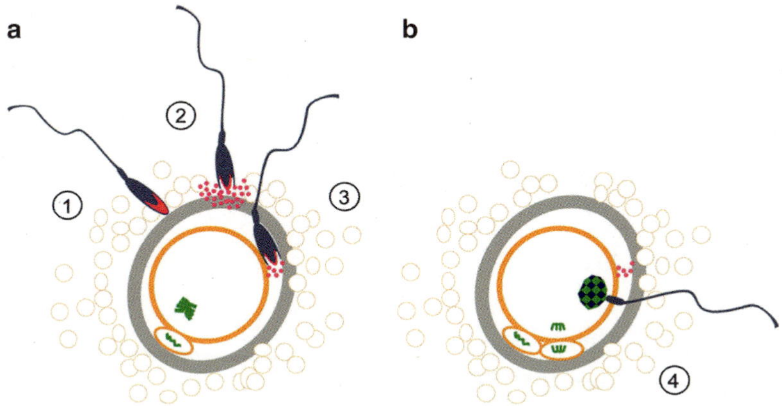

Fig. 9.1 (**a**) Schematic representation of the gamete fertilization. (*1*) Sperm penetration of the cumulus layer and binding to the zona pellucida (ZP). (*2*) Sperm acrosomal reaction during which the acrosome content is released. (*3*) Sperm passing through the ZP and entering the perivitelline space where it adheres to the oocyte and starts the fusion process. (*4*) Oocyte activation induced by sperm fusion and penetration leading to resumption of the second meiosis, expulsion of the second polar body and sperm head decondensation. (**b**) Schematic representation of the acrosome reaction. (*1*) Acrosome intact sperm. (*2*) Sperm head after complete acrosome reaction. *Ac* acrosome, *PM* plasma membrane, *OAM* outer acrosomal membrane, *IAM* inner acrosomal membrane, *N* nucleus, *ER* equatorial region

been postulated to serve as a secondary ligand of the inner acrosomal membrane of sperm, ensuring close contact between the ZP and the penetrating spermatozoon (Wassarman 2002). The fertilizing acrosome-reacted sperm reaches quickly the perivitelline space. The sperm head then binds and fuses with the egg plasma membrane (oolemma) (Fig. 9.1a.3). Early insights into gamete plasma membrane interaction came from scanning and transmission electron microscopy performed by R. Yanagimachi (for review (Yanagimachi 1994)). It appears that sperm interactions with the oolemma occur in a spatially restricted manner. Indeed, the egg membrane is covered with microvilli except in the region overlying the meiotic spindle and sperm egg fusion rarely occurs in this microvilli-free (amicrovillar) region (Runge et al. 2007). Furthermore, the fertilizing sperm makes the first contact with the oolemma via its inner acrosomal membrane, but binds and fuses with the egg microvilli via the plasma membrane of its equatorial region (Fig. 9.1b) (Yanagimachi 1994). If some acrosome-intact sperm are able to bind experimentally to oolemma of ZP-free eggs, they never fuse. The acrosome reaction is required to complete the fusion process. Following gamete fusion, the sperm tail movements decrease and stop within a few seconds. Electron microscopy analysis shows that the inner acrosomal membrane doesn't fuse with the oolemma but is later engulfed by the oocyte (Fig. 9.1a.4). The sperm tail is also eventually incorporated in the egg cytoplasm. Gamete membrane fusion marks the end of the fertilization step and the beginning of the egg activation. The events associated with the egg activation include the initiation of oscillations in intracellular calcium concentration, the completion of the second meiosis with extrusion of a second polar body and the formation of a block to polyspermy via the release of ZP proteolysis enzyme from the egg's cortical granules preventing further ZP crossing by other sperm.

9.2 Sperm-Egg Fusion Candidate Molecules

9.2.1 Mouse Model

9.2.1.1 Essential Molecules

Tetraspanins

The first data suggesting the involvement of CD9 tetraspanin in fusogenic properties of the oolemma were provided by experiments performed in vitro using an anti-CD9 monoclonal antibody, JF9 (Chen et al. 1999). This work revealed (1) the presence of CD9 homogeneously distributed at the egg surface, with exception of the microvillar-free region and (2) the inhibition by JF9 mAb of sperm-egg binding and fusion, in a dose dependent manner. The essential role of CD9 was then firmly confirmed by three different teams who generated mutant mice in which the CD9 gene was disrupted (Le Naour et al. 2000; Miyado et al. 2000; Kaji et al. 2000). Homozygous adults $CD9^{-/-}$ were healthy, however, whereas $CD9^{-/-}$ male mice were normally fertile, $CD9^{-/-}$ female mice presented a severely reduced fertility. Only 50–60% of $CD9^{-/-}$ female mice produced litters after mating with fertile male mice with an extension of the delay to begin a successful pregnancy and a reduction of the litter size leading to 95% global reduction of fertility. CD9-deleted mice produced mature oocytes quantitatively and morphologically comparable to the wild type female, spontaneously as well as after hormones-induced superovulation (Fig. 9.2a–d). In vivo analysis revealed a normal mating behaviour of the CD9-deleted mice and the presence of numerous sperm in their oviducts. However, the mature eggs recovered in $CD9^{-/-}$ genital female tract 0.5 days after mating were not fertilized presenting several sperm into the perivitelline space and showing that the fusion didn't occur resulting in persistence of ZP permeability to sperm (Fig. 9.2e, f). The fertilization tests performed in vitro confirmed the inability of the egg membrane to fuse with sperm (Miyado et al. 2000). For each team, the fertilization rate of zona free oocyte collected from $CD9^{-/-}$ female mice were equivalent, ranging from 0% to 4% versus 96% to 98% for wild type oocyte (Le Naour et al. 2000;

Fig. 9.2 (continued) by the presence of a metaphase spindle (*) and a first polar body containing residual DNA-tubulin material (*solid arrowheads*). (**e** and **f**) An oocyte from a $CD9^{-/-}$ naturally ovulated mated female blocked in metaphase of meiosis II with sperm in the perivitelline space (*open arrowheads*). (**g** and **h**) An oocyte from a CD91/1 naturally ovulated mated female that has progressed to telophase and has extruded the second polar body (*arrow*). Inset shows partially decondensed sperm DNA in a different focal plane. Scale bar, 10 μm (Reproduced from Le Naour et al. 2000)

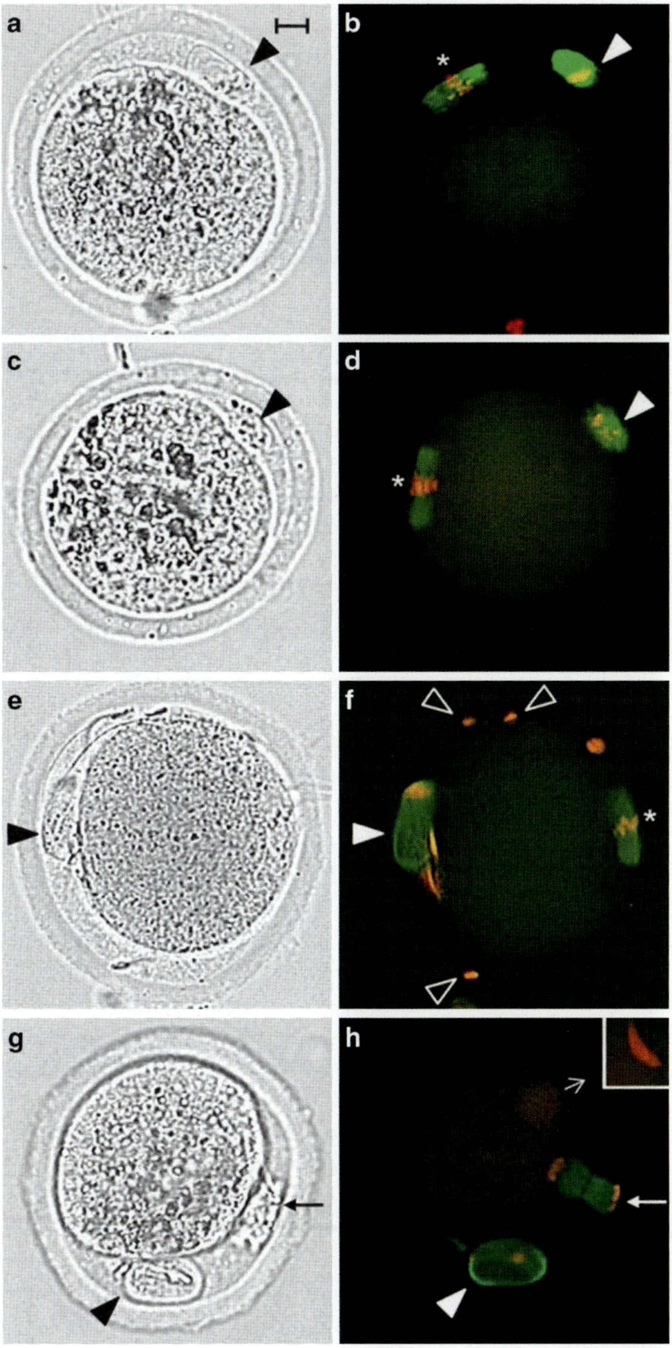

Fig. 9.2 Oocyte DNA-tubulin labeling (Propidium iodide and mAb YL1/2 followed by FITC Goat antibody to mouse Ig on fixed oocytes). Oocytes from nonmated superovulated CD9$^{-/-}$ (**a** and **b**) or CD9$^{+/+}$ (**c** and **d**) females are identically blocked in metaphase of meiosis II as confirmed

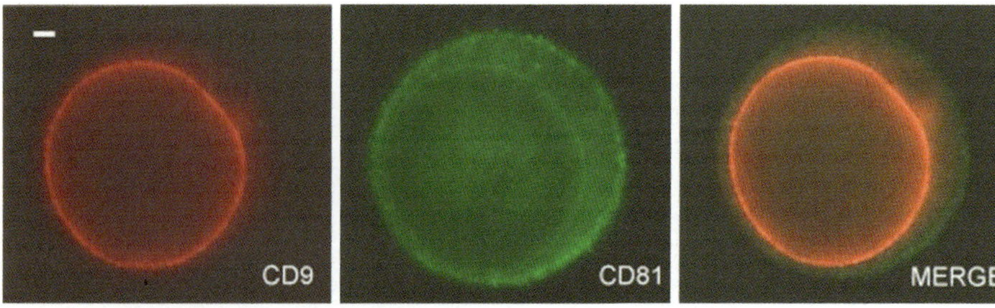

Fig. 9.3 CD9 and CD81 expression on oocytes from wild-type mice. Double labeling was performed with a rat mAb to mouse CD9 4.1 F12 and hamster mAb to CD81 Eat1 followed by Texas Red- and FITC-coupled goat antibodies to rat and hamster respectively. Scale bar, 10 μm

Miyado et al. 2000; Kaji et al. 2000) despite a normal sperm binding to CD9-defective membrane. The development of rare but healthy pups suggested that CD9 would be strictly involved in fusion process. This was confirmed by intra-cytoplasmic injection of wild type sperm in CD9-deleted eggs which showed normal embryo development and uterus implantation with birth of healthy offspring (Miyado et al. 2000). Moreover, the fertilization defect of CD9-deficient eggs was reversed by exogenous mouse CD9 expressed by mRNA intra-cytoplasmic injection (Kaji et al. 2002). Interestingly, overexpression of human CD9 restored the fertilization rate of $CD9^{-/-}$ mouse oocytes to 90% indicating the absence of species specificity of the CD9 molecule, at least between mouse and human. Another tetraspanin, CD81 that shares more than 40% amino acid identities with CD9, is also expressed at the mouse egg surface. The pattern of labelling is different from CD9 since it partially colocalizes at cell surface where its level of expression is much lower and mAbs to CD81 also display a punctuated labelling of zona pellucida that is not observed in oocytes form $CD81^{-/-}$ mice (Fig. 9.3). CD81-deficient female mice were generated (Maecker et al. 1998) and have been shown to present an overall reduction of fertility of 40% (Rubinstein et al. 2006a). Indeed, 60% of CD81-mutant female produced litters after mating with a normal delay and a normal litter size, but an increased rate of postnatal mortality. This high level of dead animals in the first hours (Kelic et al. 2001; Rubinstein et al. 2006a) is still unexplained but could be due to a defect of maternal care. Surprisingly, only 10% of mature eggs recovered 0.5 day after $CD81^{-/-}$ female mouse mating presented a meiosis resumption. and non-fertilized oocytes had several sperm motile into the perivitelline space, indicating a defect of membrane fusion. There is no obvious explanation for the discrepancy between the residual fertility of CD81-deficient mouse (60%) and the low percentage of successful egg fertilization (10%) observed in vivo. One hypothesis could be an increase of the delay necessary for fusion after the crossing of ZP in the absence of CD81 so that the fusion had not occurred at 0.5 day but occurred later. The important role of CD81 in fertilization process appeared when both CD9 and CD81 gene were disrupted (Rubinstein et al. 2006a). Indeed, $CD9^{-/-}$ $CD81^{-/-}$ female mice were totally sterile (0% pregnancy after 2 months of mating). None of the CD9/CD81-deleted mature oocytes observed in vivo or in vitro after sperm contact was fertilized. This shows that CD9 and CD81 may play a complementary role in the fusion process.

Interestingly, the increased defect of gamete fusion in double knockout mice and the ability of CD81 to partially compensate the CD9 function in CD9$^{-/-}$ mouse oocytes (50% fertilization upon overexpression of CD81) (Kaji et al. 2002) suggest a partially redundant role for these two tetraspanins. It has to be noticed that although CD151 may play a role in human gamete fusion (Ziyyat et al. 2006), CD151$^{-/-}$ mutant mice are normally fertile (Wright et al. 2004).

CD9 was not detected on sperm in early studies (Chen et al. 1999), however a recent work reported its presence at the sperm surface following acrosomal reaction (Ito et al. 2010). Nothing is known about its role in gamete membrane interaction; however, sperm CD9 is definitely not essential as CD9 null male mice are fully fertile (Le Naour et al. 2000).

GPI-Anchored Proteins

GPI-anchored proteins possess a covalently linked glycosylated phosphatidylinositol moiety which serves to attach the protein portion of the molecule to the cell surface lipid bilayer (Low and Saltiel 1988). They are thought to be preferentially located in the lipid rafts and known to be involved in a wide variety of cellular functions including T cell activation, hydrolysis of extracellular matrix proteins, transduction of extracellular stimuli, and cell–cell adhesion. GPI-anchored proteins can be released from the outer cell membrane by treatment with the highly specific enzyme phosphatidyinositol specific phospholipase C (PI-PLC) (Low and Finean 1978). Therefore, treatment of intact cells with PI-PLC has become a useful tool to characterize the released proteins and to investigate the role of GPI-anchored proteins in cell function. This strategy has been used to investigate the involvement of GPI-anchored proteins in fertilization process (Coonrod et al. 1999). Co-culture of mouse gametes in presence of PI-PLC led to a dramatic reduction of the fertilization rate (from 59.6% to 2.8%) despite a normal sperm zona binding. Interestingly, unfertilized mature eggs presented numerous motile sperm accumulated in the perivitelline space, suggesting a specific egg membrane fertilization defect. Zona free in vitro fertilization tests associated with alternative gamete pre incubation showed that PI-PLC pretreated sperm kept intact binding and fusion properties whereas pretreated eggs completely lost the ability to bind and fuse with sperm in a dose dependent manner (Coonrod et al. 1999). Electrophoresis analysis revealed that PI-PLC egg treatment induced the release of proteins with 70 kDA and 35–45 kDA apparent MW. Altogether, these observations support the hypothesis that one or more GPI-anchored egg surface proteins would be essential for sperm-egg binding and fusion. This was confirmed by the defect of gamete fusion in Pig-a deleted oocytes (Alfieri et al. 2003). The phosphatidylinositol glycan class-A (Pig-a) gene encodes a subunit of N-acetylglucosaminyltransferase that controls early steps of GPI anchor biosynthesis (Tiede et al. 2000). Since the deletion is embryonic lethal (Kawagoe et al. 1996), a conditional knockout with Cre/loxP recombinaison driven by the oocyte specific ZP3 promoter was used to obtain viable female mice that are infertile. None of the mature eggs recovered in vivo was fertilized despite the presence of perivitelline sperm. Zona-free in vitro fertilization tests confirmed the oolemma defect of fusion with a fertilization rate (percentage of fertilized eggs) eightfold lower and a fertilization index (number of fused sperm per fertilized egg) ninefold lower in *GPI-AP*$^{-/-}$ oocytes compared to the wild type. The 70 kDa component released from the egg surface after PI-PLC treatment could be the decay accelerating factor (DAF), also called CD55, since polyclonal goat anti-CD55 antibodies bind to wild type zona free egg surface. However, knockout of CD55 gene has no effect on female mice fertility (Sun et al. 1999), as well as the targeted deletion of CD59 gene (Holt et al. 2001), another GPI-anchored protein identified at the egg surface (Taylor and Johnson 1996).

Izumo

The protein Izumo 1 is the most recently sperm surface protein identified as an essential actor in sperm-egg binding and fusion. Izumo 1, named after a fertility shrine dedicated to marriage in the Shimane prefecture of Japan, is an Ig superfamily type 1 membrane protein expressed at the equatorial segment of acrosome reacted sperm (Fig. 9.4). Female mice with deleted Izumo 1 gene have normal fertility whereas male are completely sterile despite a normal mating behaviour and a normal sperm production (Inoue et al. 2005). Unfertilized eggs recovered in vivo after mating between wild type female and *Izumo 1*$^{-/-}$ male showed sperm that had crossed the ZP but were accumulated in the perivitelline space unable to fuse with the oolemma. These results have confirmed initial data that have shown the inhibitory effect of OBF13, a monoclonal antibody raised against Izumo 1, on sperm-egg fusion (Okabe et al. 1988). Izumo 1 is exclusively involved in fusion since intracytoplasmic injection of *Izumo 1*$^{-/-}$ sperm in wild type oocytes leads to normal egg activation and full development after the transfer in uterus of pseudo-pregnant female mice. Izumo 1 belongs to a protein family all having a so-called Izumo domain which form large complexes on sperm surface (Ellerman et al. 2009). The role of Izumo 1 in the fertilization process is still unknown even if these latter results raise the possibility that Izumo 1 could organize at the sperm surface protein complexes involved in membrane fusion machinery. Up to now, there are no reports describing a biochemical link between Izumo and egg tetraspanins.

9.2.1.2 Non Essential Molecules

The Controversial ADAM2/α6β1 Couple

Before the discovery of Izumo protein, the best candidate egg ligand on sperm was an ADAM (A Disintegrin and Metalloprotease) family protein. ADAM2 was the first specific sperm ADAM identified in guinea pig, expressed at the equatorial region of the sperm head (Primakoff et al. 1987). ADAM2 is also present in other mammalian species such as mouse, macaque and human. In the mouse, antibodies and peptide of the disintegrin domain of ADAM2 (fertilin β) and ADAM3 (cyritestin), another testis specific ADAM, have been reported to inhibit sperm oocyte binding and fusion

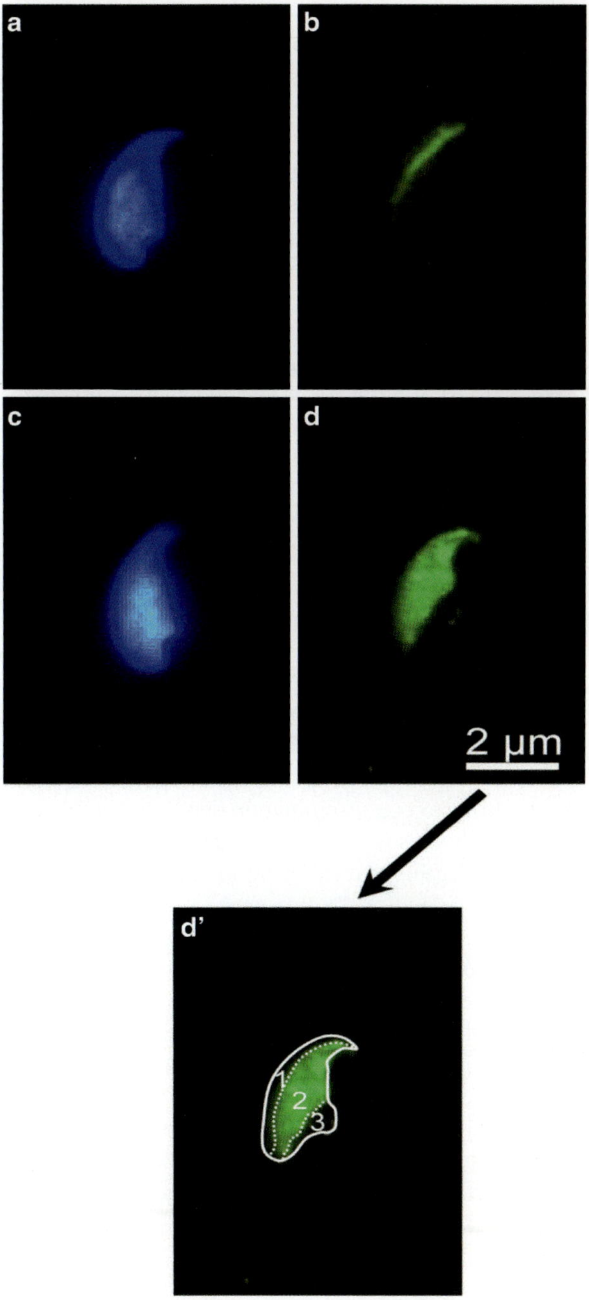

Fig. 9.4 Izumo1 distribution at sperm head surface. After 1 h of in vitro capacitation, live sperm have been exposed to a rabbit polyclonal antibody anti-mouse Izumo revealed by a FITC anti-rabbit antibody and analysed by epifluorescence. Izumo 1 is an acrosomal membrane protein which is exposed during the acrosome reaction and relocates from the acrosomal cap (**b**) to the equatorial region (**d**). (**a** and **c**): Dapi sperm DNA staining. The different regions of the sperm head are represented in **d'** acrosomal cap (*1*); equatorial region (*2*); post-acrosomal cap (*3*). Scale bar, 2 μm

(Primakoff et al. 1987; Evans et al. 1998; Zhu et al. 2000; Yuan et al. 1997). Male mice with deletion of ADAM2 and ADAM3 genes are infertile (Cho et al. 1998; Shamsadin et al. 1999). Even though these experimental data explained the sterility by an impaired migration of sperm into female oviduct and a defect in zona pellucida binding of mutant sperm, they didn't prove that ADAM2 and ADAM3 were also involved in membrane interaction. Since ADAM2 and ADAM3 proteins contain a tripetide sequence that mimics the Arg-Gly-Asp (RGD) and that controls sperm-oocyte membrane interactions (Blobel et al. 1992; Myles et al. 1994; Evans et al. 1995), this suggests that oolemma integrins are involved. Indeed, several integrins have been identified on the oolemma of mammalian eggs and in particular in mouse (Sengoku et al. 2004). The integrin α6β1 has been the primary candidate proposed to bind to sperm ADAM2 since (1) mouse sperm adhere to somatic cells when they express α6β1 and this binding is blocked by anti-mouse ADAM2 antibody (Almeida et al. 1995; Bigler et al. 2000), (2) soluble form of ADAM2 disintegrin domain binds to the egg microvillar surface in a distribution comparable to α6β1 (Evans et al. 1997; Bigler et al. 2000) and this adhesion is inhibited by the function blocking mAb for α6 subunit, GoH3 (Bigler et al. 2000). However, gene deletion experiments failed to confirm that α6β1 play a key role in sperm-egg membrane interaction. Indeed, α6 integrin null oocytes were normally fertilized in vitro by wild type sperm (Miller et al. 2000). Moreover, oocytes lacking all β1 integrins were competent to fuse with sperm in vivo as well as in vitro (He et al. 2003). The discrepancies between gene deletion and peptide/antibodies experiments found a possible explanation in a work which suggests that sperm integrins may also be involved (Barraud-Lange et al. 2007b). Acrosome reacted mouse sperm express α6β1 integrin in equatorial and post-acrosomal regions and treatment of sperm with α6 or β1 function blocking antibodies prior to egg insemination decreased cumulus intact fertilization rate by 50%. When antibodies were present in the insemination medium, thus blocking both sperm and egg integrins, the fertilization rate felt to 10%. Therefore, optimal conditions for fusion may require the presence of α6β1 on the two gamete membranes but the presence of α6β1 on only one of the gametes may be sufficient for fusion to occur. An example of such compensation is provided by β1 defective myoblasts where fusion is rescued by coculture with wild type myoblasts (Schwander et al. 2003). The analysis of fertilization performance of sperm and egg that are both inactivated for α6 or β1 is still lacking to validate this hypothesis.

The association of integrins and tetraspanins within the tetraspanin web (see below) suggesting that it could act as a mediator of the tetraspanins CD9 and CD81 effects (Berditchevski 2001; Boucheix and Rubinstein 2001; Charrin et al. 2009; Hemler 2005) reinforces the possibility of the involvement of α6β1 in sperm-egg membrane interaction.

Other Molecules

The Epididymal Cysteine-RIch secretory Proteins (CRISP) from seminal plasma adhere to sperm surface and one of them CRISP1 has been implicated in sperm-oolemma interaction (Cohen et al. 2007). *Crips1 null* mice have reduced sperm-oolemma fusion in vitro (Da Ros et al. 2008). CRISP2 is a sperm protein that seems to compete with CRISP1 for oolemma binding (Busso et al. 2007). Their ligands on oolemma are unknown.

Most of the other gamete surface proteins involved in gamete interaction act at the level of sperm-ZP interaction and don't seem to participate in the fusion process (for review (Ikawa et al. 2010)).

9.2.2 Available Data in Human

Since human oocyte is an extremely rare material and the laws that authorize the human research programmes are very restrictive, there are very few experimental data available on fertilization process in human. However, based on what has been shown in other species, some molecules have been identified at the surface of human gametes and involved in in vitro sperm-egg membrane interaction.

9.2.2.1 Integrins and Their Ligands

Several integrin subunits have been identified in the human oolemma by immunostaining, particularly, α5β1, αvβ3 and α6β1 (Fusi et al. 1993; Campbell et al. 1995; Ji et al. 1998; Sengoku et al. 2004; Ziyyat et al. 2006). The involvement of the RGD-binding subfamily of integrins has been suggested by the inhibition of interaction of human sperm with zona-free hamster or human oocyte induced by RGD peptide (Bronson and Fusi 1990; Ji et al. 1998; Ziyyat et al. 2005). Furthermore, fibronectin and vitronectin, ligands of α5β1 and αvβ3 integrins respectively, via the RGD sequence, have been shown to be expressed by human sperm. If fibronectin is present at the head surface of capacitated human sperm, vitronectin is contained in the acrosomal vacuole, released in the extracellular medium following acrosome reaction and relocated to the equatorial segment (Fusi and Bronson 1992; Fusi et al. 1994). These data strengthened the hypothesis of a potential involvement of RGD-binding integrins in the recognition of human sperm and egg membranes. Interestingly, α5β1 and αvβ3 are also expressed at the human sperm surface, most of the time depending on sperm status (Fusi et al. 1996b). The presence of vitronectin receptor on both gametes leads to the hypothesis that vitronectin, released by human sperm during acrosome reaction, could participate in sperm/egg adhesion (Fusi et al. 1996a). The pair ADAM/α6β1 integrin was also proposed as a potential actor in human sperm/egg binding and fusion. On the one hand, a FEE containing peptide, mimicking the adhesion site of the disintegrin domain of human ADAM2, has been shown to inhibit or increase (depending on whether a linear or cyclic form was used) the fusion with human sperm to human zona-free eggs (Bronson et al. 1999; Ziyyat et al. 2005). On the other hand, the mAb for α6 subunit, GoH3, was shown to strongly inhibit human sperm-egg fusion in a dose dependent manner up to 96%, suggesting the involvement of α6β1 in the control of human gamete fusion (Ziyyat et al. 2006). Finally, α6β1 has been also described at the human sperm surface and proposed to be a potential clinical marker to evaluate sperm fertilizing ability in men (Reddy et al. 2003). Regardless of these experimental data, limited information is available concerning the involvement of integrins in human physiological fertilization conditions. Indeed, owing to bioethical laws, removal of the ZP is necessary to perform in vitro fertilization assays in human. This leads to the modification of integrin membrane distribution and to the binding and fusion of the oolemma with numerous sperm. This model of in vitro human gamete interaction could mask or bypass molecules necessary for membrane interaction (Ji et al. 1998; Sengoku et al. 2004).

9.2.2.2 Tetraspanins and Izumo

CD9, CD81 and CD151 tetraspanins have been identified on human egg surface (Ziyyat et al. 2006). These molecules are evenly distributed on zona intact oocyte. Surprisingly, on zona-free eggs, CD81 and CD151 form patches and co-localize while CD9 remains unchanged. CD151 and CD9 have been involved in human gamete membrane fusion since sperm and egg co-culture in presence of anti-CD151 and anti-CD9 monoclonal antibodies leads to a reduction of 50% and 78% of fusion, respectively (Ziyyat et al. 2006). No effect on sperm egg fusion was observed with the anti-human CD81 monoclonal antibodies (Ziyyat et al. 2006).

On the human sperm side, Izumo protein has been detected on sperm head surface after acrosome reaction as well as mouse sperm (Inoue et al. 2005). Using the hamster test, an heterologous system of human sperm insemination with zona free hamster egg, no fusion was observed in presence of the anti-human Izumo antibody, suggesting the involvement of Izumo in human sperm fertilization ability (Inoue et al. 2005).

Both molecules, egg CD9 tetraspanin and sperm Izumo protein, are expressed by human gametes and their function is supported by in vitro fertilization assays, suggesting that they take part in human fertilization mechanism. However, despite these experimental observations, no case of women and men infertility could be attributed to a genetic defect of CD9 and Izumo. The complete sequencing of the coding region of the CD9 gene in 87 women with unexplained infertility (Nishiyama et al. 2010) as well as the analysis of 9 exons encoding for the Izumo protein in 36 infertile men (Granados-Gonzalez et al. 2008) did not show any mutation. Nevertheless, the absence of mutations in the coding sequence does not exclude a defect in the regulation of the gene expression and when possible the levels of protein expression on sperm (Hayasaka et al. 2007) and oocytes should be monitored.

9.2.3 Other Species

The expression of CD9 on gametes of other mammalian species is compatible with its involvement in fertilization as confirmed by antibody (Li et al. 2004; Zhou et al. 2009) or soluble CD9 large extracellular loop inhibition of fertilization

(Tang et al. 2008). Despite the fact that the C. elegans genome encodes at least 20 tetraspanins, none of them appear to be involved in gamete fusion (for review (Marcello and Singson 2010)). The explanation may rely on the observation that a phylogenetic comparison of mammalian tetraspanins with the multiple superfamily members found in C. elegans and drosophila showed that they are highly divergent and that no reliable orthologs of mammalian tetraspanins can be found in nematodes or insects with the exception of Tspan5/NET4 (Todres et al. 2000). Other proteins were shown to be involved in gamete fusion of non mammalian species. As shown in Table 9.1, they have variable structures, membrane topology and contain different protein motifs. In C. elegans, four sperm proteins and two oocyte proteins appear to be required for cell fusion. The two egg proteins are related and belong to the LDLR family of proteins (for reviews (Marcello and Singson 2010; Oren-Suissa and Podbilewicz 2010)). Also, an important ancestral gamete fusion apparatus is represented by the GCS1/HAP2 gene that has disappeared from the genome of recently diverged animals. The gene product of GCS1/HAP2, a type I membrane protein expressed on male gametocyte, is necessary for gamete fusion of plants like Arabidopsis thaliana (von Besser et al. 2006) and of two types of protists (Liu et al. 2008; Hirai et al. 2008; Liu et al. 2008).

The diversity of molecules at stake indicates that gamete fusion has evolved entirely different mechanisms during evolution to fulfill the aim of mixing haploid genomes in the process of reproduction.

9.3 Tetraspanins Function in Fertilization: Current Hypotheses

From a mechanistic point of view, cell fusion processes fall in two categories: the one for which an identified fusogenic protein is involved and the others for which proteins required for fusion, may (or may not) have been identified but underlying fusion mechanisms are completely unknown. Gamete fusion falls in the second category. In spite of the discovery of numerous molecules involved in gamete fusion and their structural diversity, the absence of fusogenic peptide renders elusive the way they permit fusion to occur. Despite the involvement of other tetraspanins in fertilization, CD9 prominent effects have attracted most of the studies and various mechanisms have been described and explored as contributing to the fusion process.

9.3.1 CD9 as a Sperm Receptor

In mouse macrophages, CD9 has been shown to be a receptor of PSG17 (Waterhouse et al. 2002). Pregnancy specific glycoproteins (PSGs) form a family of 14 proteins specifically secreted in the serum of pregnant female by the placenta. These secreted proteins belong to the Ig superfamily and to the carcinoembryonic antigen (CEA) subfamily. The high level of PSGs measured in the serum until term is necessary

Table 9.1 Membrane proteins involved in gamete fusion and yeast mating

Protein	Species	Cell type	Membrane topology	A.A.	Shared domains	Phenotype	References
CD9	Mouse	Oocyte	4-pass	228	Tetraspanin	Female unfertile	See the text for details
CD81	Mouse	Oocyte	4-pass	236	Tetraspanin	Female partially unfertile	
Izumo-1	Mouse	Sperm	Single-pass	350	IgSF, Izumo	Male unfertile	
SPE-9	C. elegans	Sperm	Single-pass	660	EGF repeats (similarity Notch)	Hermaphrodites and males unfertile. Hermaphrodites fertile when mated with wild type males	Putiri et al. (2004), Singson et al. (1998), Zannoni et al. (2003)
SPE-38	C. elegans	Sperm	4-pass	179	–	Idem	Chatterjee et al. (2005)
SPE-41 (Trp-3)	C. elegans	Sperm	6 pass	854	TRPC-type (transient receptor potential canonical) calcium-conducting ion channel	Idem	Xu and Sternberg (2003), Kahn-Kirby and Bargmann (2006), Marcello and Singson (2010)
SPE-42	C. elegans	Sperm	7-pass	774	DC-STAMP	Idem	Kroft et al. (2005), Kahn-Kirby and Bargmann (2006)

EGG-1	C. elegans	Oocyte	Single-pass	550	LDLR	Unfertile	Kadandale et al. (2005)
EGG-2	C. elegans	Oocyte	Single-pass	547	LDLR	Unfertile	
PbGCS1	Plasmodium (Protist) Chlamydomonas (Protist)	Male gamete	Single-pass	828–889	HAP2-GCS1	Male gametes are fusion defective	Hirai et al. (2008) Liu et al. (2008)
HAP2	Arabidopsis—Lilium (Plant)	Sperm		658	HAP2-GCS1		von Besser et al. (2006)
Prm1	Yeast	–	5-pass SGD	661	–	Enhance membrane fusion—expressed on both partners	Jin et al. (2004), Aguilar et al. (2007)
Fig1	Yeast	–	4-pass	298	Claudin-related		Aguilar et al. (2007)

for successful pregnancy probably in modulating the maternal immune system (Ha et al. 2005). Considering the key role CD9 has in fertilization, the presence of a PSG17-related ligand on sperm has been investigated (Ellerman et al. 2003). Interestingly, convincing in vitro assays revealed that PSG17 bound specifically to the large extracellular loop (EC2) of CD9 and that the sequence SFQ, required for the functionality of CD9 in fertilization process (see below), was necessary for the recognition between PSG17 and CD9. Furthermore, the recombinant form of PSG17 adhered to the microvillar region of wild type eggs while the soluble protein did not bind to CD9 deleted eggs. Finally, PSG17 fixed at the egg surface inhibited sperm-egg fusion affecting fertilization rate and fertilization index. In light of these experimental data, the most attractive hypothesis is that CD9 may bind in *trans* to a PSG17-related protein on the sperm surface which still remains to be identified. A CEA protein called sperad/AH-20 has been identified on hamster sperm and has been implicated in sperm-egg fusion (Primakoff and Myles 1983; Allen and Green 1995; Ilayperuma 2002) but it remains to be shown if this protein is related to PSG17 (Ellerman et al. 2003).

In spite of the in vitro binding data, the preincubation of sperm with a recombinant protein corresponding to the EC2 domain of CD9 did not inhibit gamete fusion (Zhu et al. 2002). This experiment argues against a receptor function of CD9 but it is possible that a motif that would be critical for interaction of the EC2 domain with sperm is not presented in an adequate conformation.

Interestingly, CD81 EC2 has a phenylalanine residue F186 in a solvent-exposed, low polarity patch that is required for CD81 binding to the hepatitis C virus (Higginbottom et al. 2000). CD9 F174 is present in a corresponding head domain region and its mutation resulted in a fourfold reduction of the fusion rescue ability of CD9 mRNA injected into *Cd9 null* oocytes. The mutation of the tripeptide SFQ to AAA (encompassing the two F174 adjacent a.a.) resulted in an even greater inefficiency of the transcript to rescue the fusion despite a similar level of expression of CD9 at the surface of the oocyte and its recognition by all tested monoclonal antibodies (Zhu et al. 2002). The observation that the mutation of a sequence known to act in trans in a closely related tetraspanin supports a receptor function of CD9 in gamete fusion.

9.3.2 CD9 Positions a Partner Molecule Required for Fusion Within the Tetraspanin Web

9.3.2.1 Tetraspanins Organize a Network of Molecular Interactions at the Cell Surface

The key function attributed to tetraspanins is to organize extensive network of *cis*-partner proteins in the so-called Tetraspanin Web or Tetraspanin Enriched Microdomains (TEMs) within the plasma membrane (Berditchevski 2001; Boucheix and Rubinstein 2001; Charrin et al. 2009; Hemler 2005; Levy and Shoham 2005;

Yanez-Mo et al. 2009). It has been proposed that within these microdomains each tetraspanin interact with their primary partners (like EWI2 for CD9) but heterotypic tetraspanin-tetraspanin interactions allow the assembly of higher molecular complexes that link molecules with diverse functions. For instance α6β1 integrin associates directly to the tetraspanin CD151 but indirectly to CD9 through a CD9-CD151 interaction. The loss of one of the structural components, in the present case CD9, may disorganize these membrane microdomains and deregulate the functions of associated proteins. Based on this view, the defect of fusion ability of *Cd9*-deleted oocyte may be due to the deregulation of one or more of the proteins which belong to the TEMs.

9.3.2.2 First Level Interactions: Primary Partners

Two primary partners of CD9, CD9P-1/EWI-F/CD315 and EWI2/CD316, were described in various cellular types (Charrin et al. 2001, 2003; Clark et al. 2001; Stipp et al. 2001, 2003). They are structurally related and, with EWI-3 and EWI-101, form a novel Ig domain subfamily that includes four members. EWI-2 and CD9P-1 are expressed on oocytes and as in other cellular types both molecules are primary CD9 partners (Rubinstein et al. 2006b; Runge et al. 2007; He et al. 2009; Glazar and Evans 2009). Luminescence assays revealed that the oocyte expression level of EWI2 was decreased by more than 90% in the absence of CD9 (He et al. 2009). The mechanism of this deregulation is unknown. It does not require palmitoylation of CD9 since Cd9 null mice positive for the depalmitoylated CD9 transgene ($CD9^{plm}$ Tg^{+} $Cd9^{-/-}$) had oocytes which expressed normal level of $CD9^{plm}$ and EWI2 and were fully functional in sperm-egg fusion. Three studies have addressed the potential role of EWI-2. In one study, an anti-EWI-2 antibody showed a discrete inhibitory effect on sperm-egg binding but not on fusion (Glazar and Evans 2009). In another study, recombinant EWI-2 ectodomain did not result in an inhibition of sperm-egg binding and fusion despite the fact that it bound to the surface of 82% of acrosome-reacted sperm and to 22% of acrosome intact sperm (He et al. 2009). Finally, a recent work reporting the generation of EWI-2 deleted mice demonstrates surprisingly that EWI-2 is dispensable for sperm-egg fusion (Inoue et al. 2012). A compensatory mechanism during development can't be excluded from these results but appears unlikely.

9.3.2.3 Second Level Interactions

The importance of these interactions in oocytes relied on the belief that sperm/egg adhesion was mediated by the binding of the disintegrin domain of ADAM2 to egg α6β1, a component of TEM (Chen et al. 1999; Zhu and Evans 2002). Moreover, CD9 was shown to control the lateral diffusion of α6β1 within the oolemma (Ziyyat et al. 2006) since (1) anti-CD9 monoclonal antibodies prevent the reorganization in patches of α6β1 induced by zona removal procedure on human oocytes (2) the patches induced by the antibody-mediated aggregation of α6 subunit were larger and more dispersed in absence of CD9 at the mouse egg surface. These observations led to the proposal that CD9 is necessary for the maintenance of a tetraspanin web to which the integrin α6β1 is linked through the tetraspanin CD151 (Serru et al. 1999), thus controlling its lateral mobility (Ziyyat et al. 2006). The cooperation of CD9 with egg membrane components was supported by another team which has generated a soluble form of the large extracellular loop of CD9 (EC2). The preincubation of eggs with the EC2 constructs, before insemination, significantly reduced the fertilization and index rates while sperm pre incubation had no effect (Zhu et al. 2000). This report strengthens the idea that CD9 function in gamete fusion is the consequence of its interaction with partners in *cis* on the egg surface.

9.3.3 CD9 Is Transferred from Egg to Sperm

Recent findings offer a new view on the potential role that CD9 would play in gamete membrane interaction. It originated from an unexpected observation describing a transfer of CD9 from oolemma to the head of sperm present in the perivitelline space of *Cd81* null oocytes (Rubinstein et al. 2006a). This transfer of CD9 has been next confirmed between wild type gametes and proposed to be driven by an egg membrane fragment released during gamete contact prior to fusion (Barraud-Lange et al. 2007a). The authors proposed that this protein transfer might be trogocytosis-related. Trogocytosis is a phenomenon involved in lymphocyte activation, in which lymphocytes actively capture plasma membrane fragments of antigen-presenting cells containing MHC-peptide complexes (Joly and Hudrisier 2003). Another group showed transfer of oocyte material to the sperm using CD9-EGFP Tg+ $CD9^{-/-}$ oocytes (Miyado et al. 2008). The authors monitored IVF assays and showed that the fertilizing sperm which has reached the perivitelline space acquired CD9-EGFP by direct interaction with CD9-EGFP containing-material present in this space. Confocal and electron microscopy analyses allowed observation of the release from the oolemma of vesicles which contained CD9 and presented the morphological characteristics of exosomes. The functional significance of such a phenomenon is still an open question. Does this egg-derived material acquired by the fertilizing sperm contain information necessary for activation to sperm fusion ability? The required acquisition of CD9 by sperm before fusion in order to organize a multimolecular complex of fusion has been proposed (Barraud-Lange et al. 2007a) but is questioned by recent findings which proved that mature sperm express endogenous CD9 (Ito et al. 2010). Miyado's group was able to restore *Cd9 null* oocytes fusion ability by CD9-containing vesicles recovered from wild type eggs. They proposed that sperm that had previously interacted with CD9-containing material released by wild type eggs became competent for *Cd9 null* egg fertilization. Based on these data, they concluded that egg CD9 bestows upon sperm its fusion ability. This model is questioned by others who failed to reproduce the *CD9 null* oocytes rescue experiments (Gupta et al. 2009; Lefevre et al. 2010). If the interaction of sperm with egg membrane material was confirmed by several groups, its requirement for fertilization needs to be proven.

9.3.4 CD9 Structures the Oocyte Membrane

CD9 has been proposed to be involved in the architecture of the oocyte plasma membrane, explaining the fertilization defect of *Cd9 null* eggs by the dimorphism of the oolemma since CD9 expression is restricted to the microvillar area where fusion occurs. By scanning electron microscopy (e.m.), it was shown that the microvilli of mutant *Cd9 null oocytes* appeared shorter, thicker, denser and more uniform than in wild type oocytes (Runge et al. 2007). The authors also showed by immunogold labeling that CD9 was congregated on the microvilli and not on the planar membrane regions. Interestingly, the loss of EWI2 consequent to the loss of CD9 renders this observation particularly relevant. Indeed, in somatic cells, EWI-2 has been shown to bind directly, through its N-terminal domain, to actin-linking ezrin-radixin-moesin (ERM) proteins (Sala-Valdes et al. 2006). ERM proteins bind in turn to the actin filaments in the microvillar core. Thus, a network composed by CD9-EWI-2-ERM-actin has been proposed to regulate the microvillar morphology. Based on these data, one can speculate that in the absence of CD9 and EWI-2, the link between the plasma membrane and the cytoskeleton is lost leading to the disruption of the microvillar structure and, consequently, rendering oolemma fusion incompetent.

There are some examples in which organ specific tetraspanins structure particular areas of the membrane like RDS/ROM for the rim of photoreceptors outer segment disk or uroplakins 1a and 1b for the urothelial plaques of the bladder. Also the tetraspanin CD151 is in epithelial cells a constituent of hemidesmosomes where it is targeted via its interaction with the integrin $\alpha6\beta4$. These cellular structures may be considered as independent or as an extension of the tetraspanin web. No specific structure is seen with CD9 on oocytes apart from a pronounced membrane curvature at the tip of the microvilli. Although morphologically this part of the microvilli resembles the rim of photoreceptors outer segment disk, these two types of structures are not directly comparable since RDS/ROM have a particular mode of organization forming tetramers and intermolecular disulfide bonds and a restricted expression (Goldberg 2006), properties that are not shared by CD9.

9.3.5 Relationship Between Rafts and TEM in Gamete Fusion

Given the various observations showing that tetraspanins CD9 and CD81 are critical players in the gamete fusion, their function should also be considered in view of their involvement in the assembly of TEM. On the other hand the dramatic effect of

GPI anchored protein depletion (see above) on gamete fusion suggests a role for lipid rafts. Until now these two types of membrane domains have been viewed as different based on biochemical (Le Naour et al. 2006) and dynamic membrane studies (single particle tracking or fluorescence recovery after photobleaching) (Barreiro et al. 2008; Espenel et al. 2008). Combining these observations leads to the hypothesis that cooperation between rafts and TEM is required for gamete fusion to occur. The requirement is to bring the gamete membranes in tight apposition (less than 10 nm) to overcome the energetic barrier that prevents spontaneous fusion to occur and allow lipid mixing between sperm and oocyte membranes. This could be achieved by creating patches on the membrane where repulsive forces are lowered and exposing these patches to the partner's membrane, thus building a kind of fusion synapse.

Consequently, depending on associated membrane lipids and/or proteins, tetraspanins may either promote or inhibit cell fusion, explaining the contradictory effects of the CD9/CD81 double deletion on oocytes and monocytes/macrophages (see below).

Virus budding offers another example where lipid rafts and TEM may be cooperating. There is a colocalization of rafts with Gag and Env associated proteins during virus assembly (Holm et al. 2003; Ono and Freed 2005). Gag and TEM were also shown to colocalize (Jolly and Sattentau 2007; Mazurov et al. 2006; Nydegger et al. 2006). Furthermore rafts and TEM components are incorporated into viruses (Ott 2008). A GPI-anchored raft protein (BST2) inhibits virus assembly whereas the precise role of TEM associated proteins in virus assembly remains controversial (Chen et al. 2008; Grigorov et al. 2009; Krementsov et al. 2009; Ruiz-Mateos et al. 2008). A recent analysis of the dynamic relationship between viral components and these microdomains suggests that it is Gag assembly that creates a local microenvironment enriched in raft lipids and proteins and in which CD9 and presumably other tetraspanins are trapped. In these different types of microdomains, tetraspanins have reduced motility whereas raft markers diffuse freely (Krementsov et al. 2010). The ability of Gag to regroup membrane molecules localized in different microdomains raises the question of yet an unknown molecule at the surface of the oocyte that could play a similar role in building appropriate sites of fusion at the tip of microvilli.

9.3.6 Implication of Tetraspanins in Other Cell Fusion Processes

Instead of illuminating the mechanism of sperm-egg fusion, the involvement of tetraspanins in other cell fusion types has added another level of complexity. Indeed, it was reported that anti-CD9 or anti-CD81 antibodies delay the fusion of murine myoblastic cells to form multinucleated myotubes during muscle differentiation and that ectopic expression of CD9 increases the fusion of rhabdomyosarcoma cells (Tachibana and Hemler 1999). This observation is compatible with the properties of

CD9 and CD81 in gamete fusion. The situation differs for osteoclasts and giant cells that are two types of multinucleated cells issued from the fusion of monocytes/macrophages and are involved in bone resumption and in the immune response respectively (Vignery 2005). Silencing of Tspan-5 and CD9 reduces the formation of giant multinuclear osteoclast-like cells by RANKL treated RAW264.7 cells, whereas silencing of Tspan-13/NET-6 has the opposite effect (Ishii et al. 2006; Iwai et al. 2007). However the $CD9^{-/-}$/$CD81^{-/-}$ mice have greater numbers of osteoclasts associated with reduced bone-mineral density. In addition multinucleated giant cells are found spontaneously in the lung of these mice (Takeda et al. 2003). This in vivo observation was confirmed in vitro since the fusion of macrophages lacking these two tetraspanins is enhanced. Thus as in gamete fusion, the tetraspanins CD9 and CD81 play both a role in the fusion of macrophages, but in that case they regulate negatively the fusion process. Interestingly, tetraspanins CD9, CD63, CD81, CD82 and CD231 inhibit HIV-1 induced cell to cell fusion (Gordon-Alonso et al. 2006; Sato et al. 2008). It was further demonstrated that tetraspanins CD9 and CD63 inhibited cell fusion mediated by HIV1 Env protein, whereas CD82 effect was dependent on the coexpression of Gag (Weng et al. 2009). CD82 has also been shown to inhibit cell-cell fusion mediated by the envelope glycoprotein of human T-cell leukemia virus type 1, another retrovirus (Pique et al. 2000).

These findings indicate a complex role for tetraspanins in the fusion process where the same tetraspanin, depending on the cellular type and the molecular context may either promote or inhibit cell fusion, and where different tetraspanins in the same cellular type may have opposite effects on cell fusion. This suggests that the role of tetraspanins in cell fusion is dependent of their molecular environment within the Tetraspanin Web in which they exert a structuring and regulatory role.

9.4 Perspectives

Current knowledge on the mechanisms of cell fusion or more generally of membrane fusion have been of little help to understand the way two gametes unite to form a single cell. None of the gamete fusion proteins reported in the litterature are fusogenic. They don't contain a fusion peptide and when expressed ectopically they don't induce cell fusion as it is observed with FAST fusogens or C. elegans eff-1 and aff-1 (Oren-Suissa and Podbilewicz 2010). In addition, the formal proofs that they may interact with membrane proteins of the other gamete are still lacking. Most of the critical factors were discovered by genetic approach, either by systematic gene screening or by sheer luck with gene-manipulated animals that provided non anticipated phenotypes. Therefore, other surprises may be expected from additional genetic approaches, especially with the global effort of knocking out systematically all mouse genes. If the mouse gene deletion models have also eliminated some genes like $\alpha6\beta1$ from the first line candidates, that doesn't definitely rule out their implication in the process that may be more subtle. For instance, by increasing gametes binding rates, they may give a selection advantage that would appear after several generations of free competition mating (Sutovsky 2009). In addition, human biological mechanisms are not necessarily a copy of what happens in mice.

Ongoing and future studies will continue to focus on:

1. Research of molecular candidates that fit with present models of membrane fusion
2. Understanding of how membrane microdomains are critically involved
3. Elaboration of new models of cell fusion that would modify or go beyond current paradigms and take into account the specific microdomain organization of the gamete membranes
4. Development of new tools for studying gamete interaction with improvement of imaging and of biophysical techniques that would allow an easier appreciation of what occurs directly at the interaction/fusion site.

One of these new tools, called Biomembrane Force Probe (BFP), which allows the measurement of binding nano-forces involved in gamete membrane adhesion, was developed recently by a biophysics team (Jegou et al. 2008). They have also provided relevant information on the mechanical properties of the oolemma. Experiments using wild type cells revealed that some domains of the egg membrane presented an elastic deformation under sperm traction corresponding to strong interaction between gamete membranes (S-adhesion site). While others domains gave rise to a tether via a viscoelastic deformation corresponding to weak interaction (W-adhesion site). Applied to gametes from mutant mice, the BFP is of particular interest to decipher the role played by molecular candidates in gamete membrane fusion. Indeed, a very recent work revealed that the adhesion properties of CD9-deleted egg membrane are deeply modified (Jegou et al. 2011). Force measurement assays recorded sperm-egg adhesion events but showed a loss of membrane S-adhesion site in absence of CD9. Considering the organizer function often attributed to tetraspanins in many membrane cell types, the authors proposed that CD9 induced assembling of part of sperm receptors into multiprotein patches at the egg surface. A receptor involved in a patch is strongly anchored to the cytoskeleton and forms a S-adhesion site at the egg surface. Conversely, in absence of CD9, isolated receptor which is individually connected to the cytoskeleton, provide weak interaction. The authors proposed then that S-adhesion allows the tight sperm-egg contact necessary to induce fusion and finally that CD9 generates fusion competent adhesion sites on eggs. This model describes the fertilization as a direct consequence of CD9 controlled sperm-egg adhesion leading to fusion. To go further, it might be possible to investigate the nature of receptors anchoring to the egg membrane cytoskeleton as well as its involvement in fusion process.

References

Aguilar PS, Engel A, Walter P (2007) The plasma membrane proteins prm1 and fig 1 ascertain fidelity of membrane fusion during yeast mating. Mol Biol Cell 18(2):547–556

Alfieri JA, Martin AD, Takeda J, Kondoh G, Myles DG, Primakoff P (2003) Infertility in female mice with an oocyte-specific knockout of gpi-anchored proteins. J Cell Sci 116(Pt 11):2149–2155

Allen CA, Green DP (1995) Monoclonal antibodies which recognize equatorial segment epitopes presented de novo following the a23187-induced acrosome reaction of guinea pig sperm. J Cell Sci 108(Pt 2):767–777

Almeida EA, Huovila AP, Sutherland AE, Stephens LE, Calarco PG, Shaw LM, Mercurio AM, Sonnenberg A, Primakoff P, Myles DG et al (1995) Mouse egg integrin alpha 6 beta 1 functions as a sperm receptor. Cell 81(7):1095–1104

Austin CR (1951) Observations on the penetration of the sperm in the mammalian egg. Aust J Sci Res B 4(4):581–596

Barraud-Lange V, Naud-Barriant N, Bomsel M, Wolf JP, Ziyyat A (2007a) Transfer of oocyte membrane fragments to fertilizing spermatozoa. FASEB J 21(13):3446–3449

Barraud-Lange V, Naud-Barriant N, Saffar L, Gattegno L, Ducot B, Drillet AS, Bomsel M, Wolf JP, Ziyyat A (2007b) Alpha6beta1 integrin expressed by sperm is determinant in mouse fertilization. BMC Dev Biol 7(1):102

Barreiro O, Zamai M, Yanez-Mo M, Tejera E, Lopez-Romero P, Monk PN, Gratton E, Caiolfa VR, Sanchez-Madrid F (2008) Endothelial adhesion receptors are recruited to adherent leukocytes by inclusion in preformed tetraspanin nanoplatforms. J Cell Biol 183(3):527–542

Berditchevski F (2001) Complexes of tetraspanins with integrins: more than meets the eye. J Cell Sci 114(Pt 23):4143–4151

Bigler D, Takahashi Y, Chen MS, Almeida EA, Osbourne L, White JM (2000) Sequence-specific interaction between the disintegrin domain of mouse adam 2 (fertilin beta) and murine eggs. Role of the alpha(6) integrin subunit. J Biol Chem 275(16):11576–11584

Blobel CP, Wolfsberg TG, Turck CW, Myles DG, Primakoff P, White JM (1992) A potential fusion peptide and an integrin ligand domain in a protein active in sperm-egg fusion. Nature 356(6366):248–252

Boucheix C, Rubinstein E (2001) Tetraspanins. Cell Mol Life Sci 58(9):1189–1205

Bronson RA, Fusi F (1990) Sperm-oolemmal interaction: role of the arg-gly-asp (rgd) adhesion peptide. Fertil Steril 54(3):527–529

Bronson RA, Fusi FM, Calzi F, Doldi N, Ferrari A (1999) Evidence that a functional fertilin-like adam plays a role in human sperm-oolemmal interactions. Mol Hum Reprod 5(5):433–440

Busso D, Cohen DJ, Maldera JA, Dematteis A, Cuasnicu PS (2007) A novel function for crisp1 in rodent fertilization: involvement in sperm-zona pellucida interaction. Biol Reprod 77(5):848–854

Campbell S, Swann HR, Seif MW, Kimber SJ, Aplin JD (1995) Cell adhesion molecules on the oocyte and preimplantation human embryo. Hum Reprod 10(6):1571–1578

Charrin S, Le Naour F, Labas V, Billard M, Le Caer JP, Emile JF, Petit MA, Boucheix C, Rubinstein E (2003) Ewi-2 is a new component of the tetraspanin web in hepatocytes and lymphoid cells. Biochem J 373(Pt 2):409–421

Charrin S, Le Naour F, Oualid M, Billard M, Faure G, Hanash SM, Boucheix C, Rubinstein E (2001) The major cd9 and cd81 molecular partner. Identification and characterization of the complexes. J Biol Chem 276(17):14329–14337

Charrin S, le Naour F, Silvie O, Milhiet PE, Boucheix C, Rubinstein E (2009) Lateral organization of membrane proteins: tetraspanins spin their web. Biochem J 420(2):133–154

Chatterjee I, Richmond A, Putiri E, Shakes DC, Singson A (2005) The caenorhabditis elegans spe-38 gene encodes a novel four-pass integral membrane protein required for sperm function at fertilization. Development 132(12):2795–2808

Chen H, Dziuba N, Friedrich B, von Lindern J, Murray JL, Rojo DR, Hodge TW, O'Brien WA, Ferguson MR (2008) A critical role for cd63 in hiv replication and infection of macrophages and cell lines. Virology 379(2):191–196

Chen MS, Tung KS, Coonrod SA, Takahashi Y, Bigler D, Chang A, Yamashita Y, Kincade PW, Herr JC, White JM (1999) Role of the integrin-associated protein cd9 in binding between sperm adam 2 and the egg integrin alpha6beta1: implications for murine fertilization. Proc Natl Acad Sci USA 96(21):11830–11835

Cho C, Bunch DO, Faure JE, Goulding EH, Eddy EM, Primakoff P, Myles DG (1998) Fertilization defects in sperm from mice lacking fertilin beta. Science 281(5384):1857–1859

Clark KL, Zeng Z, Langford AL, Bowen SM, Todd SC (2001) Pgrl is a major cd81-associated protein on lymphocytes and distinguishes a new family of cell surface proteins. J Immunol 167(9):5115–5121

Cohen DJ, Da Ros VG, Busso D, Ellerman DA, Maldera JA, Goldweic N, Cuasnicu PS (2007) Participation of epididymal cysteine-rich secretory proteins in sperm-egg fusion and their potential use for male fertility regulation. Asian J Androl 9(4):528–532

Coonrod SA, Naaby-Hansen S, Shetty J, Shibahara H, Chen M, White JM, Herr JC (1999) Treatment of mouse oocytes with pi-plc releases 70-kda (pi 5) and 35- to 45-kda (pi 5.5) protein clusters from the egg surface and inhibits sperm-oolemma binding and fusion. Dev Biol 207(2):334–349

Da Ros VG, Maldera JA, Willis WD, Cohen DJ, Goulding EH, Gelman DM, Rubinstein M, Eddy EM, Cuasnicu PS (2008) Impaired sperm fertilizing ability in mice lacking cysteine-rich secretory protein 1 (crisp1). Dev Biol 320(1):12–18

Ellerman DA, Ha C, Primakoff P, Myles DG, Dveksler GS (2003) Direct binding of the ligand psg17 to cd9 requires a cd9 site essential for sperm-egg fusion. Mol Biol Cell 14(12):5098–5103

Ellerman DA, Pei J, Gupta S, Snell WJ, Myles D, Primakoff P (2009) Izumo is part of a multiprotein family whose members form large complexes on mammalian sperm. Mol Reprod Dev 76(12):1188–1199

Espenel C, Margeat E, Dosset P, Arduise C, Le Grimellec C, Royer CA, Boucheix C, Rubinstein E, Milhiet PE (2008) Single-molecule analysis of cd9 dynamics and partitioning reveals multiple modes of interaction in the tetraspanin web. J Cell Biol 182(4):765–776

Evans JP, Kopf GS, Schultz RM (1997) Characterization of the binding of recombinant mouse sperm fertilin beta subunit to mouse eggs: evidence for adhesive activity via an egg beta1 integrin-mediated interaction. Dev Biol 187(1):79–93

Evans JP, Schultz RM, Kopf GS (1995) Mouse sperm-egg plasma membrane interactions: analysis of roles of egg integrins and the mouse sperm homologue of ph-30 (fertilin) beta. J Cell Sci 108(Pt 10):3267–3278

Evans JP, Schultz RM, Kopf GS (1998) Roles of the disintegrin domains of mouse fertilins alpha and beta in fertilization. Biol Reprod 59(1):145–152

Fusi FM, Bernocchi N, Ferrari A, Bronson RA (1996a) Is vitronectin the velcro that binds the gametes together? Mol Hum Reprod 2(11):859–866

Fusi FM, Bronson RA (1992) Sperm surface fibronectin. Expression following capacitation. J Androl 13(1):28–35

Fusi FM, Lorenzetti I, Mangili F, Herr JC, Freemerman AJ, Gailit J, Bronson RA (1994) Vitronectin is an intrinsic protein of human spermatozoa released during the acrosome reaction. Mol Reprod Dev 39(3):337–343

Fusi FM, Tamburini C, Mangili F, Montesano M, Ferrari A, Bronson RA (1996b) The expression of alpha v, alpha 5, beta 1, and beta 3 integrin chains on ejaculated human spermatozoa varies with their functional state. SO -. Mol Hum Reprod 2(3):169–175

Fusi FM, Vignali M, Gailit J, Bronson RA (1993) Mammalian oocytes exhibit specific recognition of the rgd (arg-gly-asp) tripeptide and express oolemmal integrins. Mol Reprod Dev 36(2):212–219

Glazar AI, Evans JP (2009) Immunoglobulin superfamily member igsf8 (ewi-2) and cd9 in fertilisation: evidence of distinct functions for cd9 and a cd9-associated protein in mammalian sperm-egg interaction. Reprod Fertil Dev 21(2):293–303

Goldberg AF (2006) Role of peripherin/rds in vertebrate photoreceptor architecture and inherited retinal degenerations. Int Rev Cytol 253:131–175

Gordon-Alonso M, Yanez-Mo M, Barreiro O, Alvarez S, Munoz-Fernandez MA, Valenzuela-Fernandez A, Sanchez-Madrid F (2006) Tetraspanins cd9 and cd81 modulate hiv-1-induced membrane fusion. J Immunol 177(8):5129–5137

Granados-Gonzalez V, Aknin-Seifer I, Touraine RL, Chouteau J, Wolf JP, Levy R (2008) Preliminary study on the role of the human izumo gene in oocyte-spermatozoa fusion failure. Fertil Steril 90(4):1246–1248

Grigorov B, Attuil-Audenis V, Perugi F, Nedelec M, Watson S, Pique C, Darlix JL, Conjeaud H, Muriaux D (2009) A role for cd81 on the late steps of hiv-1 replication in a chronically infected t cell line. Retrovirology 6:28

Gupta S, Primakoff P, Myles DG (2009) Can the presence of wild-type oocytes during insemination rescue the fusion defect of cd9 null oocytes? Mol Reprod Dev 76(7):602

Ha CT, Waterhouse R, Wessells J, Wu JA, Dveksler GS (2005) Binding of pregnancy-specific glycoprotein 17 to cd9 on macrophages induces secretion of il-10, il-6, pge2, and tgf-beta1. J Leukoc Biol 77(6):948–957

Hayasaka S, Terada Y, Inoue N, Okabe M, Yaegashi N, Okamura K (2007) Positive expression of the immunoglobulin superfamily protein izumo on human sperm of severely infertile male patients. Fertil Steril 88(1):214–216

He ZY, Brakebusch C, Fassler R, Kreidberg JA, Primakoff P, Myles DG (2003) None of the integrins known to be present on the mouse egg or to be adam receptors are essential for sperm-egg binding and fusion. Dev Biol 254(2):226–237

He ZY, Gupta S, Myles D, Primakoff P (2009) Loss of surface ewi-2 on cd9 null oocytes. Mol Reprod Dev 76(7):629–636

Hemler ME (2005) Tetraspanin functions and associated microdomains. Nat Rev Mol Cell Biol 6(10):801–811

Higginbottom A, Quinn ER, Kuo CC, Flint M, Wilson LH, Bianchi E, Nicosia A, Monk PN, McKeating JA, Levy S (2000) Identification of amino acid residues in cd81 critical for interaction with hepatitis c virus envelope glycoprotein e2. J Virol 74(8):3642–3649

Hirai M, Arai M, Mori T, Miyagishima SY, Kawai S, Kita K, Kuroiwa T, Terenius O, Matsuoka H (2008) Male fertility of malaria parasites is determined by gcs1, a plant-type reproduction factor. Curr Biol 18(8):607–613

Holm K, Weclewicz K, Hewson R, Suomalainen M (2003) Human immunodeficiency virus type 1 assembly and lipid rafts: Pr55(gag) associates with membrane domains that are largely resistant to brij98 but sensitive to triton x-100. J Virol 77(8):4805–4817

Holt DS, Botto M, Bygrave AE, Hanna SM, Walport MJ, Morgan BP (2001) Targeted deletion of the cd59 gene causes spontaneous intravascular hemolysis and hemoglobinuria. Blood 98(2):442–449

Ikawa M, Inoue N, Benham AM, Okabe M (2010) Fertilization: a sperm's journey to and interaction with the oocyte. J Clin Invest 120(4):984–994

Ilayperuma I (2002) Identification of the 48-kda g11 protein from guinea pig testes as sperad. J Exp Zool 293(6):617–623

Inoue N, Ikawa M, Isotani A, Okabe M (2005) The immunoglobulin superfamily protein izumo is required for sperm to fuse with eggs. Nature 434(7030):234–238

Inoue N, Nishikawa T, Ikawa M, Okabe M (2012) Tetraspanin-interacting protein igsf8 is dispensable for mouse fertility. Fertil Steril 98(2):465–470

Ishii M, Iwai K, Koike M, Ohshima S, Kudo-Tanaka E, Ishii T, Mima T, Katada Y, Miyatake K, Uchiyama Y, Saeki Y (2006) Rankl-induced expression of tetraspanin cd9 in lipid raft membrane microdomain is essential for cell fusion during osteoclastogenesis. J Bone Miner Res 21(6):965–976

Ito C, Yamatoya K, Yoshida K, Maekawa M, Miyado K, Toshimori K (2010) Tetraspanin family protein cd9 in the mouse sperm: unique localization, appearance, behavior and fate during fertilization. Cell Tissue Res 340(3):583–594

Iwai K, Ishii M, Ohshima S, Miyatake K, Saeki Y (2007) Expression and function of transmembrane-4 superfamily (tetraspanin) proteins in osteoclasts: reciprocal roles of tspan-5 and net-6 during osteoclastogenesis. Allergol Int 56(4):457–463

Jegou A, Pincet F, Perez E, Wolf JP, Ziyyat A, Gourier C (2008) Mapping mouse gamete interaction forces reveal several oocyte membrane regions with different mechanical and adhesive properties. Langmuir 24(4):1451–1458

Jegou A, Ziyyat A, Barraud-Lange V, Perez E, Wolf JP, Pincet F, Gourier C (2011) Cd9 tetraspanin generates fusion competent sites on the egg membrane for mammalian fertilization. Proc Natl Acad Sci USA 108(27):10946–10951

Ji YZ, Wolf JP, Jouannet P, Bomsel M (1998) Human gamete fusion can bypass beta1 integrin requirement. Hum Reprod 13(3):682–689

Jin H, Carlile C, Nolan S, Grote E (2004) Prm1 prevents contact-dependent lysis of yeast mating pairs. Eukaryot Cell 3(6):1664–1673

Jolly C, Sattentau QJ (2007) Human immunodeficiency virus type 1 assembly, budding, and cell-cell spread in t cells take place in tetraspanin-enriched plasma membrane domains. J Virol 81(15):7873–7884

Joly E, Hudrisier D (2003) What is trogocytosis and what is its purpose? Nat Immunol 4(9):815

Kadandale P, Stewart-Michaelis A, Gordon S, Rubin J, Klancer R, Schweinsberg P, Grant BD, Singson A (2005) The egg surface ldl receptor repeat-containing proteins egg-1 and egg-2 are required for fertilization in caenorhabditis elegans. Curr Biol 15(24):2222–2229

Kahn-Kirby AH, Bargmann CI (2006) Trp channels in c. Elegans. Annu Rev Physiol 68:719–736

Kaji K, Oda S, Shikano T, Ohnuki T, Uematsu Y, Sakagami J, Tada N, Miyazaki S, Kudo A (2000) The gamete fusion process is defective in eggs of cd9-deficient mice. Nat Genet 24(3):279–282

Kaji K, Oda S, Miyazaki S, Kudo A (2002) Infertility of cd9-deficient mouse eggs is reversed by mouse cd9, human cd9, or mouse cd81; polyadenylated mrna injection developed for molecular analysis of sperm-egg fusion. Dev Biol 247(2):327–334

Kawagoe K, Kitamura D, Okabe M, Taniuchi I, Ikawa M, Watanabe T, Kinoshita T, Takeda J (1996) Glycosylphosphatidylinositol-anchor-deficient mice: implications for clonal dominance of mutant cells in paroxysmal nocturnal hemoglobinuria. Blood 87(9):3600–3606

Kelic S, Levy S, Suarez C, Weinstein DE (2001) Cd81 regulates neuron-induced astrocyte cell-cycle exit. Mol Cell Neurosci 17(3):551–560

Krementsov DN, Weng J, Lambele M, Roy NH, Thali M (2009) Tetraspanins regulate cell-to-cell transmission of hiv-1. Retrovirology 6:64

Krementsov DN, Rassam P, Margeat E, Roy NH, Schneider-Schaulies J, Milhiet PE, Thali M (2010) Hiv-1 assembly differentially alters dynamics and partitioning of tetraspanins and raft components. Traffic 11(11):1401–1414

Kroft TL, Gleason EJ, L'Hernault SW (2005) The spe-42 gene is required for sperm-egg interactions during c. Elegans fertilization and encodes a sperm-specific transmembrane protein. Dev Biol 286(1):169–181

Le Naour F, Rubinstein E, Jasmin C, Prenant M, Boucheix C (2000) Severely reduced female fertility in cd9-deficient mice. Science 287(5451):319–321

Le Naour F, Andre M, Boucheix C, Rubinstein E (2006) Membrane microdomains and proteomics: lessons from tetraspanin microdomains and comparison with lipid rafts. Proteomics 6(24):6447–6454

Lefevre B, Wolf JP, Ziyyat A (2010) Sperm-egg interaction: is there a link between tetraspanin(s) and gpi-anchored protein(s)? Bioessays 32(2):143–152

Lefievre L, Conner SJ, Salpekar A, Olufowobi O, Ashton P, Pavlovic B, Lenton W, Afnan M, Brewis IA, Monk M, Hughes DC, Barratt CL (2004) Four zona pellucida glycoproteins are expressed in the human. Hum Reprod 19(7):1580–1586

Levy S, Shoham T (2005) The tetraspanin web modulates immune-signalling complexes. Nat Rev Immunol 5(2):136–148

Li YH, Hou Y, Ma W, Yuan JX, Zhang D, Sun QY, Wang WH (2004) Localization of cd9 in pig oocytes and its effects on sperm-egg interaction. Reproduction 127(2):151–157

Liu Y, Tewari R, Ning J, Blagborough AM, Garbom S, Pei J, Grishin NV, Steele RE, Sinden RE, Snell WJ, Billker O (2008) The conserved plant sterility gene hap2 functions after attachment of fusogenic membranes in chlamydomonas and plasmodium gametes. Genes Dev 22(8):1051–1068

Low MG, Finean JB (1978) Specific release of plasma membrane enzymes by a phosphatidylinositol-specific phospholipase c. Biochim Biophys Acta 508(3):565–570

Low MG, Saltiel AR (1988) Structural and functional roles of glycosyl-phosphatidylinositol in membranes. Science 239(4837):268–275

Maecker HT, Do MS, Levy S (1998) Cd81 on b cells promotes interleukin 4 secretion and antibody production during t helper type 2 immune responses. Proc Natl Acad Sci USA 95(5):2458–2462

Marcello MR, Singson A (2010) Fertilization and the oocyte-to-embryo transition in c. Elegans. BMB Rep 43(6):389–399

Mazurov D, Heidecker G, Derse D (2006) Htlv-1 gag protein associates with cd82 tetraspanin microdomains at the plasma membrane. Virology 346(1):194–204

Miller BJ, Georges-Labouesse E, Primakoff P, Myles DG (2000) Normal fertilization occurs with eggs lacking the integrin alpha6beta1 and is cd9-dependent. J Cell Biol 149(6):1289–1296

Miyado K, Yamada G, Yamada S, Hasuwa H, Nakamura Y, Ryu F, Suzuki K, Kosai K, Inoue K, Ogura A, Okabe M, Mekada E (2000) Requirement of cd9 on the egg plasma membrane for fertilization. Science 287(5451):321–324

Miyado K, Yoshida K, Yamagata K, Sakakibara K, Okabe M, Wang X, Miyamoto K, Akutsu H, Kondo T, Takahashi Y, Ban T, Ito C, Toshimori K, Nakamura A, Ito M, Miyado M, Mekada E, Umezawa A (2008) The fusing ability of sperm is bestowed by cd9-containing vesicles released from eggs in mice. Proc Natl Acad Sci USA 105(35):12921–12926

Myles DG, Kimmel LH, Blobel CP, White JM, Primakoff P (1994) Identification of a binding site in the disintegrin domain of fertilin required for sperm-egg fusion. Proc Natl Acad Sci USA 91(10):4195–4198

Nishiyama S, Kishi T, Kato T, Suzuki M, Nishizawa H, Pryor-Koishi K, Sawada T, Nishiyama Y, Iwata N, Udagawa Y, Kurahashi H (2010) Cd9 gene variations are not associated with female infertility in humans. Gynecol Obstet Invest 69(2):116–121

Nydegger S, Khurana S, Krementsov DN, Foti M, Thali M (2006) Mapping of tetraspanin-enriched microdomains that can function as gateways for hiv-1. J Cell Biol 173(5):795–807

Okabe M, Yagasaki M, Oda H, Matzno S, Kohama Y, Mimura T (1988) Effect of a monoclonal anti-mouse sperm antibody (obf13) on the interaction of mouse sperm with zona-free mouse and hamster eggs. J Reprod Immunol 13(3):211–219

Ono A, Freed EO (2005) Role of lipid rafts in virus replication. Adv Virus Res 64:311–358

Oren-Suissa M, Podbilewicz B (2010) Evolution of programmed cell fusion: common mechanisms and distinct functions. Dev Dyn 239(5):1515–1528

Ott DE (2008) Cellular proteins detected in hiv-1. Rev Med Virol 18(3):159–175

Pique C, Lagaudriere-Gesbert C, Delamarre L, Rosenberg AR, Conjeaud H, Dokhelar MC (2000) Interaction of cd82 tetraspanin proteins with htlv-1 envelope glycoproteins inhibits cell-to-cell fusion and virus transmission. Virology 276(2):455–465

Primakoff P, Myles DG (1983) A map of the guinea pig sperm surface constructed with monoclonal antibodies. Dev Biol 98(2):417–428

Primakoff P, Hyatt H, Tredick-Kline J (1987) Identification and purification of a sperm surface protein with a potential role in sperm-egg membrane fusion. J Cell Biol 104(1):141–149

Putiri E, Zannoni S, Kadandale P, Singson A (2004) Functional domains and temperature-sensitive mutations in spe-9, an egf repeat-containing protein required for fertility in caenorhabditis elegans. Dev Biol 272(2):448–459

Reddy VR, Rajeev SK, Gupta V (2003) Alpha 6 beta 1 integrin is a potential clinical marker for evaluating sperm quality in men. SO Fertil Steril 79(Suppl 3):1590–1596

Rubinstein E, Ziyyat A, Prenant M, Wrobel E, Wolf JP, Levy S, Le Naour F, Boucheix C (2006a) Reduced fertility of female mice lacking cd81. Dev Biol 290(2):351–358

Rubinstein E, Ziyyat A, Wolf JP, Le Naour F, Boucheix C (2006b) The molecular players of sperm-egg fusion in mammals. Semin Cell Dev Biol 17(2):254–263

Ruiz-Mateos E, Pelchen-Matthews A, Deneka M, Marsh M (2008) Cd63 is not required for production of infectious human immunodeficiency virus type 1 in human macrophages. J Virol 82(10):4751–4761

Runge KE, Evans JE, He ZY, Gupta S, McDonald KL, Stahlberg H, Primakoff P, Myles DG (2007) Oocyte cd9 is enriched on the microvillar membrane and required for normal microvillar shape and distribution. Dev Biol 304(1):317–325

Sala-Valdes M, Ursa A, Charrin S, Rubinstein E, Hemler ME, Sanchez-Madrid F, Yanez-Mo M (2006) Ewi-2 and ewi-f link the tetraspanin web to the actin cytoskeleton through their direct association with ezrin-radixin-moesin proteins. J Biol Chem 281(28):19665–19675
Sato K, Aoki J, Misawa N, Daikoku E, Sano K, Tanaka Y, Koyanagi Y (2008) Modulation of human immunodeficiency virus type 1 infectivity through incorporation of tetraspanin proteins. J Virol 82(2):1021–1033
Schwander M, Leu M, Stumm M, Dorchies OM, Ruegg UT, Schittny J, Muller U (2003) Beta1 integrins regulate myoblast fusion and sarcomere assembly. Dev Cell 4(5):673–685
Sengoku K, Takuma N, Miyamoto T, Horikawa M, Ishikawa M (2004) Integrins are not involved in the process of human sperm-oolemmal fusion. Hum Reprod 19(3):639–644
Serru V, Le Naour F, Billard M, Azorsa DO, Lanza F, Boucheix C, Rubinstein E (1999) Selective tetraspan-integrin complexes (cd81/alpha4beta1, cd151/alpha3beta1, cd151/alpha6beta1) under conditions disrupting tetraspan interactions. Biochem J 340(Pt 1):103–111
Shamsadin R, Adham IM, Nayernia K, Heinlein UA, Oberwinkler H, Engel W (1999) Male mice deficient for germ-cell cyritestin are infertile. Biol Reprod 61(6):1445–1451
Singson A, Mercer KB, L'Hernault SW (1998) The c. Elegans spe-9 gene encodes a sperm transmembrane protein that contains egf-like repeats and is required for fertilization. Cell 93(1):71–79
Stipp CS, Kolesnikova TV, Hemler ME (2001) Ewi-2 is a major cd9 and cd81 partner and member of a novel ig protein subfamily. J Biol Chem 276(44):40545–40554
Stipp CS, Kolesnikova TV, Hemler ME (2003) Ewi-2 regulates alpha3beta1 integrin-dependent cell functions on laminin-5. J Cell Biol 163(5):1167–1177
Sun X, Funk CD, Deng C, Sahu A, Lambris JD, Song WC (1999) Role of decay-accelerating factor in regulating complement activation on the erythrocyte surface as revealed by gene targeting. Proc Natl Acad Sci USA 96(2):628–633
Sutovsky P (2009) Sperm-egg adhesion and fusion in mammals. Expert Rev Mol Med 11:e11
Tachibana I, Hemler ME (1999) Role of transmembrane 4 superfamily (tm4sf) proteins cd9 and cd81 in muscle cell fusion and myotube maintenance. J Cell Biol 146(4):893–904
Takeda Y, Tachibana I, Miyado K, Kobayashi M, Miyazaki T, Funakoshi T, Kimura H, Yamane H, Saito Y, Goto H, Yoneda T, Yoshida M, Kumagai T, Osaki T, Hayashi S, Kawase I, Mekada E (2003) Tetraspanins cd9 and cd81 function to prevent the fusion of mononuclear phagocytes. J Cell Biol 161(5):945–956
Tang Y, Tan XM, Yue CW, Li CX, Fan ZX, Zhang YZ (2008) Cloning, sequence, and function analyses of giant panda (ailuropoda melanoleuca) cd9 gene. Mol Reprod Dev 75(9):1418–1425
Taylor CT, Johnson PM (1996) Complement-binding proteins are strongly expressed by human preimplantation blastocysts and cumulus cells as well as gametes. Mol Hum Reprod 2(1):52–59
Tiede A, Nischan C, Schubert J, Schmidt RE (2000) Characterisation of the enzymatic complex for the first step in glycosylphosphatidylinositol biosynthesis. Int J Biochem Cell Biol 32(3):339–350
Todres E, Nardi JB, Robertson HM (2000) The tetraspanin superfamily in insects. Insect Mol Biol 9(6):581–590
Vignery A (2005) Macrophage fusion: are somatic and cancer cells possible partners? Trends Cell Biol 15(4):188–193
von Besser K, Frank AC, Johnson MA, Preuss D (2006) Arabidopsis hap2 (gcs1) is a sperm-specific gene required for pollen tube guidance and fertilization. Development 133(23):4761–4769
Wassarman PM (1988) Zona pellucida glycoproteins. Annu Rev Biochem 57:415–442
Wassarman PM (2002) Sperm receptors and fertilization in mammals. Mt Sinai J Med 69(3):148–155
Wassarman PM, Mortillo S (1991) Structure of the mouse egg extracellular coat, the zona pellucida. Int Rev Cytol 130:85–110

Waterhouse R, Ha C, Dveksler GS (2002) Murine cd9 is the receptor for pregnancy-specific glycoprotein 17. J Exp Med 195(2):277–282
Weng J, Krementsov DN, Khurana S, Roy NH, Thali M (2009) Formation of syncytia is repressed by tetraspanins in human immunodeficiency virus type 1-producing cells. J Virol 83(15):7467–7474
Wright MD, Geary SM, Fitter S, Moseley GW, Lau LM, Sheng KC, Apostolopoulos V, Stanley EG, Jackson DE, Ashman LK (2004) Characterization of mice lacking the tetraspanin superfamily member cd151. Mol Cell Biol 24(13):5978–5988
Xu XZ, Sternberg PW (2003) A c. Elegans sperm trp protein required for sperm-egg interactions during fertilization. Cell 114(3):285–297
Yanagimachi R (1994) Mammalian fertilization. In: Knobil E, Neill JD (eds) The physiology of reproduction, Secondth edn. Raven, New York, pp 189–317
Yanez-Mo M, Barreiro O, Gordon-Alonso M, Sala-Valdes M, Sanchez-Madrid F (2009) Tetraspanin-enriched microdomains: a functional unit in cell plasma membranes. Trends Cell Biol 19(9):434–446
Yuan R, Primakoff P, Myles DG (1997) A role for the disintegrin domain of cyritestin, a sperm surface protein belonging to the adam family, in mouse sperm-egg plasma membrane adhesion and fusion. J Cell Biol 137(1):105–112
Zannoni S, L'Hernault SW, Singson AW (2003) Dynamic localization of spe-9 in sperm: a protein required for sperm-oocyte interactions in caenorhabditis elegans. BMC Dev Biol 3:10
Zhou GB, Liu GS, Meng QG, Liu Y, Hou YP, Wang XX, Li N, Zhu SE (2009) Tetraspanin cd9 in bovine oocytes and its role in fertilization. J Reprod Dev 55(3):305–308
Zhu X, Evans JP (2002) Analysis of the roles of rgd-binding integrins, alpha(4)/alpha(9) integrins, alpha(6) integrins, and cd9 in the interaction of the fertilin beta (adam2) disintegrin domain with the mouse egg membrane. Biol Reprod 66(4):1193–1202
Zhu X, Bansal NP, Evans JP (2000) Identification of key functional amino acids of the mouse fertilin beta (adam2) disintegrin loop for cell-cell adhesion during fertilization. J Biol Chem 275(11):7677–7683
Zhu GZ, Miller BJ, Boucheix C, Rubinstein E, Liu CC, Hynes RO, Myles DG, Primakoff P (2002) Residues sfq (173–175) in the large extracellular loop of cd9 are required for gamete fusion. Development 129(8):1995–2002
Ziyyat A, Naud-Barriant N, Barraud-Lange V, Chevalier F, Kulski O, Lemkecher T, Bomsel M, Wolf JP (2005) Cyclic fee peptide increases human gamete fusion and potentiates its rgd-induced inhibition. Hum Reprod 20(12):3452–3458
Ziyyat A, Rubinstein E, Monier-Gavelle F, Barraud V, Kulski O, Prenant M, Boucheix C, Bomsel M, Wolf JP (2006) Cd9 controls the formation of clusters that contain tetraspanins and the integrin {alpha}6{beta}1, which are involved in human and mouse gamete fusion. J Cell Sci 119(Pt 3):416–424

Chapter 10
Tetraspanins and Immunity

Mark D. Wright and Shoshana Levy

Abstract Studies of tetraspanins in cells of the immune system were the first to reveal the interactions of tetraspanins with each other and with their associated molecular partners. The extensive knowledge of immune cell subsets, the functionally distinct molecules expressed by these cells, and the availability of specific antibody reagents has had a major impact on our understanding of how tetraspanins assemble in cell membranes, and how they affect the function of their partners. Here we briefly introduce the various cell types that partake in innate and adaptive immune functions. We then highlight the role of tetraspanins in both arms of the immune system. Tetraspanins influence immune cell migration and antigen presentation. Moreover, they are present on both sides of immune synapses, namely, on antigen presenting cells and on T cells. Indeed, deficiency of specific tetraspanins in both mice and humans results in immune impairments.

Abbreviations

APC	Antigen presenting cells
BCR	B cell receptor
C	Constant region gene
CTL	Cytotoxic T cells
CVID	Variable immunodeficiency disorder
D	Diversity region gene
DCs	Dendritic cells

M.D. Wright
Department of Immunology, Monash University,
AMREP campus, Melbourne, Victoria, Australia

S. Levy (✉)
Department of Medicine, Division of Oncology,
Stanford University Medical Center, Stanford, CA, USA
e-mail: slevy@stanford.edu

F. Berditchevski and E. Rubinstein (eds.), *Tetraspanins*, Proteins and Cell Regulation 9, DOI 10.1007/978-94-007-6070-7_10,

FO	Follicular
FRET	Fluorescence resonance energy transfer
ICAM-1	Intracellular adhesion molecule-1
IS	Immune synapse
J	Joining region gene
LPS	Lipopolysaccharide
MHC	Major histocompatibility molecules
MIICs	MHC class II enriched compartments
MZ	Marginal zone
NK	Natural killer cells
PAMPs	Pathogen associated molecular patterns
PRR	Pattern recognition receptors
TCR	T cell receptor
TEM	Tetraspanin-enriched microdomains
Th	T helper cells
TLR	Toll-like receptor
Treg	Regulatory T cells
V	Variable region genes
VCAM	Vascular cell adhesion molecule-1

10.1 Introduction

The tetraspanin web concept originally emerged by studying tetraspanins in immune cells (Rubinstein et al. 1996). Results of studies analyzing the associations of tetraspanins with histocompatibility molecules and integrins expressed on the surface of B cells, led Rubinstein and colleagues to postulate "the existence of a tetraspanin network which, by connecting several molecules, may organize the positioning of cell surface proteins and play a role in signal transduction, cell adhesion, and motility" (Rubinstein et al. 1996). Similarly, the concept of tetraspanin-enriched microdomains (TEM) that are distinct from lipid rafts (Kropshofer et al. 2002) was deduced following interrogation of immune cells with specific antibody reagents. The extensive knowledge of immune cell subsets, the functionally distinct molecules expressed by these cells, and the availability of specific antibody reagents has had a major impact on our understanding of how tetraspanins assemble in cell membranes, and regulate molecular function.

10.2 Immune Cells

Cells of the innate immune system initiate immune responses in a non-specific way. Innate immune phagocytic cells, such as macrophages and neutrophils, display several germline encoded pattern recognition receptors (PRR), which recognize molecular motifs in pathogens, termed pathogen associated molecular patterns (PAMPs). Upon the recognition of PAMPs by PRR, innate immune cells initiate immune responses by becoming activated. Activated innate immune cells secrete pro-inflammatory cytokines and granules, migrate to the site of infection or to draining lymph nodes, and, ultimately play a critical role in activating the adaptive immune response.

The most important functional distinction between innate and adaptive immune cells is the acquisition of memory by adaptive immune cells. The major players in the adaptive arm of the immune system are B cells and T cells. During their development, these cells have the unique capability of rearranging their antigen receptor genes, which are comprised of a large set of variable (V) region genes and smaller sets of diversity (D), joining (J) and constant (C) region genes, in a process called somatic recombination. This process generates a vast repertoire of antibody-producing B cells and of T cells, the mediators of cellular immunity.

Mature B cells express B cell receptors (BCR), the membrane form of immunoglobulins. The BCR expressed by individual B cells not only differ by their V(D)J-C combinations, their V region genes are also subject to somatic hypermutation after activation, thereby increasing the binding affinity to a given antigen. Further maturation of B cells leads to the production of plasma cells that secrete soluble immunoglobulins. The secreted immunoglobulins (antibodies) bind directly to antigens.

By contrast to B cells, T cells need to be "presented" antigens by third party antigen presenting cells (APC). T cells express T cell receptors (TCR), which interact with peptides presented on major histocompatibility molecules (MHC) by APC thereby enabling cell-mediated recognition of non-self invaders. The mode of antigen presentation defines the two major types of T cells. CD8 T cells, also called cytotoxic T cells (CTL) recognize peptides presented by MHC class I. CD4 T cells recognize peptides presented by MHC class II. The major CD4 T subpopulations are T helper (Th) cells, which produce cytokines that influence immune cell interactions and regulatory T cells (Treg), which suppress immune responses.

The cells that best initiate T cell responses are specialized APC called dendritic cells (DC). DC have the unique ability to stimulate and activate naïve T cells. DC act as sentinels for the immune system, they are present in the skin and in tissues that contact the external environment. DC express high levels of PRR, and upon their activation also express high levels of molecules required to stimulate T cells such as MHC class I and class II molecules onto which processed peptides are "loaded" for presentation to T cells. DC also produce cytokines that influence immune cell interactions, and express cell surface molecules that are required to costimulate naïve T cells.

These major immune cell types can be subdivided into additional subsets based on functional differences and stages of development and differentiation. Immune cell subsets are well defined by exquisitely discriminating monoclonal antibody (mAb) markers. Most important, tools to monitor immune interactions both in vivo and in vitro have been developed and studied extensively. The considerable knowledge of the immune system has contributed tremendously to understanding the functional role of tetraspanins. Conversely, knowledge of tetraspanin-partner functions could shed light on interactions in the immune system. An example to illustrate the latter has been the identification of a genetic mutation in a tetraspanin gene in a patient diagnosed with an common variable immunodeficiency disorder (CVID), as detailed in "In vivo role of Tetraspanins in Adaptive Immunity", below.

A previous review has documented the expression of at least 20 different tetraspanin family members at the mRNA level, whereas mAbs were available at the time for just a few tetraspanins (Tarrant et al. 2003). Unfortunately, this is still the case, particularly, for non-human species. Nevertheless, the available anti-tetraspanin mAbs have been used extensively to analyze their role in antigen presentation and in immune cell activation.

10.3 Tetraspanins in Innate Immunity

10.3.1 Tetraspanins in Pattern Recognition

The notion that innate immune cells, such as antigen presenting cells, do not have intrinsic activity but require activation by PRR, is comparatively recent and was first proposed by Janeway in 1989 (Janeway 1989). Convincing molecular proof that such molecules existed did not come until the functional discovery of what is now known as toll-like receptor (TLR) 4 in 1997 (Medzhitov et al. 1997). It is now appreciated that innate immune cells express a plethora of PRR that are comprised of proteins of various superfamilies. PRR can recognize PAMPS in diverse locations including the cell surface, intracellular vesicles, and alternatively in the cytoplasm. Activation by signal transduction through PRR is a critical first step in the immune response, and also plays an important role in non-infectious inflammation (Iwasaki and Medzhitov 2010).

Given the recent discovery of PRR, it is not surprising that a possible role for tetraspanins in innate immunity is only now emerging (Figdor and van Spriel 2010). There are now compelling studies that suggest that PRR are molecules whose functions can be regulated by tetraspanins, and given the impressive diversity of molecules involved in pattern recognition, it would not surprise us if more reports on tetraspanins regulating PRR will emerge in the future. Macrophages deficient in the tetraspanin CD9 have exaggerated pro-inflammatory responses to the TLR4 agonist LPS (lipopolysaccharide, a key component of gram negative bacterial cell walls) (Suzuki et al. 2009). In vitro CD9-deficient cells secreted greater TNF-α in response to LPS stimulation, and intranasal administration of LPS to CD9-deficient mice showed an increase in lung inflammation. Similarly, macrophages deficient in the tetraspanin CD37 have exaggerated pro-inflammatory responses to agonists of the C-type lectin fungal PRR Dectin-1 (Meyer-Wentrup et al. 2007). In vitro, triggering of dectin-1 leads to an exaggerated production of the pro-inflammatory cytokine IL-6. Moreover *Cd37*$^{-/-}$ mice are resistant to challenge by the fungal pathogen *Candida albicans*, although whether this is causally related to a dysregulation of Dectin-1 activity has not been determined (Figdor and van Spriel 2010).

The molecular mechanisms by which tetraspanins regulate activation in response to PAMPs have not been fully elucidated. Whether their ability to regulate the TLR4 and Dectin-1, respectively, is unique to CD9 and CD37 or a function shared by other tetraspanin has not yet been determined. CD81 has been reported as being in close proximity to TLR4 by Fluorescence Resonance Energy Transfer in LPS-stimulated human monocytes (Triantafilou et al. 2004), and it is also of note that in vitro LPS stimulation of *Cd81*$^{-/-}$ B cells leads to increase activation and proliferation, compared to their wild type counterparts (Sanyal et al. 2009). Moreover, a molecular interaction between Dectin-1 and CD63 has been observed in immature human DC (Mantegazza et al. 2004). CD9 clearly molecularly interacts with CD14, a TLR4 co-receptor, and, given the stability of this interaction after Triton X-100 solubilization, the interaction may be direct (Suzuki et al. 2009). In CD9-deficient macrophages, CD14 expression is upregulated and its association with TLR4 is enhanced. Given that no data was presented to document a molecular interaction between TLR4 and CD9, the simplest model to explain this data might be that CD9 associates with and regulates CD14, negatively controlling the interaction of CD14 with TLR4 by sequestering CD14 away from TLR4. The absence of CD9 from macrophages also affects the membrane compartmentalization of the TLR4/CD14 complex, leading to a greater incorporation of the complex into low-density membrane fractions, which some have argued enhances TLR4/CD14 signaling (Pfeiffer et al. 2001). By contrast, a different mechanism must be invoked to explain the regulation of Dectin-1 by CD37. Whilst there is evidence for a molecular interaction between CD37 and Dectin-1, CD37-deficiency leads to poor expression of Dectin-1 (in contrast to CD9-deficiency which leads to an excess of CD14). This poor Dectin-1 surface expression belies the excess IL-6 produced by Dectin-1 agonists, suggesting that CD37 plays a role in negatively regulating Dectin-1 signaling (Meyer-Wentrup et al. 2007).

10.3.2 Tetraspanins and Innate Immune Cell Migration

Leukocyte migration is of fundamental importance in almost all aspects of the immune system. It is essential to the efficient development of immune responses against microbial pathogens yet also underlies the pathophysiology of inflammation and immune-mediated diseases such as rheumatoid arthritis, multiple sclerosis and atherosclerosis. In innate immunity, leukocytes must migrate out of the circulation towards the site of infection in the periphery (Ley et al. 2007). Moreover, the innate immune system initiates adaptive immune responses as a consequence of DC capturing antigen in the periphery and migrating to lymphoid organs where antigen presentation to T cells occurs (Shortman and Liu 2002).

In non-immune cells, there is strong evidence, from multiple studies of many physiological systems, that tetraspanins regulate cell migration, primarily through the ability of tetraspanins to regulate the function of their partner integrin molecules (see Chap. 6). We would expect that the same is true for immune cells. However, whilst many tetraspanin-integrin interactions have been detected in leucocytes, it is surprising that there is currently a paucity of data suggesting that tetraspanins regulate leukocyte migration. Monoclonal antibodies (mAbs) against CD151 can inhibit in vitro neutrophil chemo-haptotactic migration (Yauch et al. 1998). Conversely, mAbs against several tetraspanins enhanced in vitro chemotactic migration of DC (Mantegazza et al. 2004), and natural killer (NK) cells (Kramer et al. 2009). Whether this modulation of immune cell migration in any of these studies, involved a modulation of integrin function was not examined. Feigelson et al. (2003) used reverse genetics approaches and showed that in both monocyte cell lines and mouse B cells, CD81 played an important role in promoting outside-in signaling and adhesion strengthening through $\alpha_4\beta_1$ integrin, although whether this corresponded to an effect on in vivo cell migration or inflammation was not determined. There is however strong in vitro evidence that tetraspanins may indirectly play a role in leukocyte trafficking via their ability to promote the presentation of high avidity clusters of the integrin ligands vascular cell adhesion molecule-1 VCAM-1 and intercellular adhesion molecule-1 ICAM-1 on endothelial cells (Barreiro et al. 2005, 2008).

10.3.3 Tetraspanins and Antigen Presentation

There have been numerous reports documenting molecular interactions between tetraspanins and the antigenic peptide presenting MHC molecules. Tetraspanin-MHC interactions occur at the cell surface, CD9, CD37, CD53, CD81, and CD82 have all been reported to interact with MHC, and where most of the data has documented interactions between tetraspanins and MHC II (Angelisova et al. 1994; Engering et al. 2003; Kijimoto-Ochiai et al. 2004; Schick and Levy 1993; Szollosi et al. 1996; Rubinstein et al. 1996; Unternaehrer et al. 2007; Zilber et al. 2005; Hoorn et al. 2012), there are also reports of interactions with MHC I (Szollosi et al. 1996; Lagaudriere-Gesbert et al. 1997). Interactions can also occur intracellularly. For example the tetraspanin CD63 is a well-characterized marker for lysosomes and early endosomes and it has been shown to translocate to MHC class II enriched compartments (MIICs) following endocytosis of antigen (Mantegazza et al. 2004; Artavanis-Tsakonas et al. 2006; Pols and Klumperman 2009) where it has a stable direct interaction with MHC II (Hoorn et al. 2012). However, initial analyses of cells isolated from CD63-deficient mice reported no defect in lysosomal function and endocytosis, nor antigen processing and presentation (Schroder et al. 2009), although CD63 may have a role in regulating MHC II trafficking as silencing CD63 in transformed B cell lines lead to an increase in the production of immunostimulatory MHCII-expressing exosomes (Petersen et al. 2011). CD82 is another tetraspanin that shows a vesicular pattern of expression and is also found in abundance in MIICs where it can interact with not only MHC II, but also directly with the peptide editors HLA-DM (Hoorn et al. 2012) and HLA-DO (Hammond et al. 1998). The functional significance of CD82 in MIICs has not been determined, and reverse genetics analysis of CD82 in antigen presenting cells has not yet been reported. However, given the role that tetraspanins have in regulating protein trafficking (Berditchevski and Odintsova 2007), a role in MHC transport (Vyas et al. 2007) or peptide loading (Rocha and Neefjes 2008) is possible.

Cellular immune responses are initiated by the presentation of peptides by DC to naive T lymphocytes. Some molecular immunologists propose that T cell activation requires crosslinking of the T cell receptor (TCR) by peptide-MHC, as soluble MHC/peptide monomers are not capable of full T cell activation (Boniface et al. 1998; Cochran et al. 2000). Precisely how a DC can display limited amounts of a particular antigenic peptide-MHC complex, in a sea of self-peptide/MHC complexes and still induce TCR crosslinking has not been resolved. It is argued that MHC is not randomly displayed at the DC surface but organized into clusters that have been visualized microscopically (Unternaehrer et al. 2007). Biochemical evidence suggests that MHC-II molecules interact with one another to form dimers or higher order multimers (Brown et al. 1993) and even cell surface multimers with MHC-I (Jenei et al. 1997). The mechanism of MHC clustering has also not been resolved. One suggestion is that MHC is clustered via their incorporation into raft microdomains. In particular, at low peptide concentrations, the biochemical disruption of rafts abolishes efficient antigen presentation (Anderson et al. 2000). However, the relevance of rafts to membrane biology is questionable. Their definition, based on insolubility in various types and concentrations of detergent is nebulous. Moreover precisely how these detergent insoluble fractions relate to structures on native membranes is unclear. It has also been argued lipid rafts are too numerous to concentrate MHC peptide complexes and increase MHC avidity (Huby et al. 2001).

Tetraspanins represent an alternative mechanism by which high avidity peptide/MHC structures are formed at the cell surface. Kropshofer et al. identified a supramolecular complex that included the tetraspanins CD82, CD9 and CD81, MHC II, CD86, and HLA-DM (Kropshofer et al. 2002). These MHC II/tetraspanin microdomains carried a restricted peptide repertoire and were argued to be critically important in T cell activation, as their disruption diminished Ag presentation. Unternaehrer et al. also support the model that tetraspanins promote high avidity MHC clusters (Unternaehrer et al. 2007). They demonstrated that CD9 mediated complexes between I-A and I-E, and argued that the differential expression of CD9 in DC compared to B blasts may underlie the superior Ag presenting capacity of DC. However, surprisingly, there was no investigation of the antigen presenting capacity of *Cd9*$^{-/-}$ DC reported in their paper.

The model that tetraspanins are essential for antigen presentation as they promote high avidity MHC clusters is not without controversy. Firstly, Kropshofer et al. originally identified MHC/tetraspanin complexes using the CDw78 monoclonal antibody to identify tetraspanin/MHC microdomains, and the specificity of this reagent has recently been called into question (Poloso et al. 2006). Secondly, a prediction from the clustering model would suggest that dendritic cells obtained from tetraspanin-deficient mice should be poor presenters of antigen as their ability to present high avidity clusters at the cell surface would be impaired. However, to date, there have only been reports on the antigen presenting capacity of DC from two tetraspanin knockouts CD37 and CD151, and surprisingly, deficiency of either of these tetraspanins resulted in hyperstimulatory DC (Sheng et al. 2009). Here, we can put the phenotype induced by CD151 deficiency aside, as there are no reports that CD151 can interact with MHC, and the phenotype seems to be of an elevated costimulatory activity rather than enhanced MHC/antigen presentation. However,

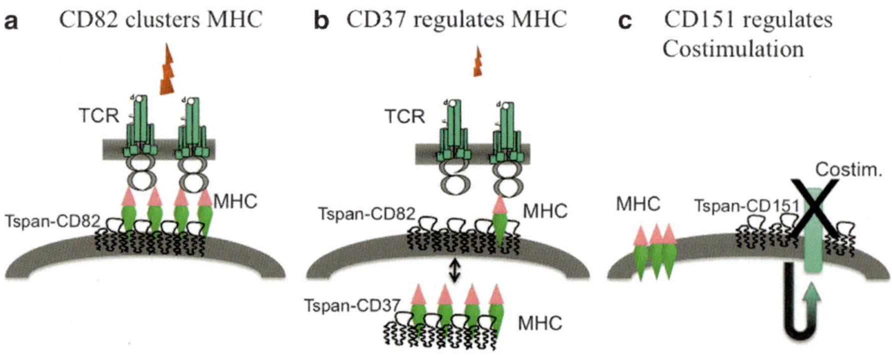

Fig. 10.1 Tetraspanins play distinct roles in antigen presenting cells. This model reconciles biochemical studies with reverse genetics approaches (**a**) CD82 (and also CD9 and CD81) clusters MHC and promotes TCR cross-linking. (**b**) CD37 regulates MHC and inhibits antigen presentation. CD37 may sequester MHC away from the cluster-promoting tetraspanin CD82. (**c**) CD151 negatively regulates co-stimulation. CD151 should not interact with MHC but may interact with co-stimulatory molecules

CD37 is a tetraspanin present in MHC complexes (Angelisova et al. 1994; Escola et al. 1998), and CD37-deficient DC are hyperstimulatory to T cell hybridomas whose activation is generally held to be dependent only on MHC/peptide and independent of costimulatory signals (Sheng et al. 2009). Clearly CD37 has an inhibitory role in antigen presentation rather than a role in promoting MHC clustering and therefore antigen presentation.

What implications then does this result have for the clustering model for tetraspanins in antigen presentation? Firstly it should be considered that the reports that support a role for tetraspanins in promoting MHC clustering studied the tetraspanins CD9, CD82 and CD81, whereas analyses of tetraspanin-deficient cells using a reverse genetics approach focused on CD37. Tetraspanins often associate with one another in the same microdomain, and can often share similar functions. However microdomains with different tetraspanin compositions do exist within the one cell (Nydegger et al. 2006), and the intracellular localization of tetraspanins can differ (Engering et al. 2003). Tetraspanins can also have opposing biological functions: CD151 and CO/029 promote, whereas CD9 and CD82 suppress cancer cell motility (Hemler 2003). Consequently, it is possible that tetraspanin function in APC also varies. For example, CD151 will not molecularly associate with MHC and will regulate costimulatory signals. Some tetraspanins, like CD82, will promote MHC clustering, others, like CD37, will regulate MHC possibly sequestering MHC away from the clustering promoting tetraspanins such as CD82. Several key experiments are required to test this hypothesis. If true, reverse genetic analyses of the tetraspanins biochemically implicated in promoting MHC clustering (e.g., CD9, CD81 and CD82) should reveal that DC deficient in these molecules are poor presenters of antigen. Biochemical and microscopic analysis might also predict different pools of tetraspanins; the MHC interacting with CD37 might be sequestered away from the MHC interacting with the clustering promoting tetraspanins such as CD82 (see model, Fig. 10.1).

Secondly, the hypothesis that T cell activation requires cross-linking of TCR by high avidity MHC is not universally accepted. Whilst soluble monomeric peptide/MHC are incapable of activating T cells, monomers incorporated into lipid bilayers are sufficient to promote T cell activation (Ma et al. 2008a). It has also been argued that the kinetics of TCR interactions with peptide/MHC are too fast to allow for adjacent TCRs to also bind to ligand (Williams and Beyers 1992). Consequently alternative models for T cell activation do exist, and whilst several of these are T lymphocyte centric and do not strongly consider the role of dendritic cells in T cells activation, the potential role of tetraspanins in antigen presentation must also be considered in context of these models. The receptor deformation model argues that TCR signaling is triggered by conformational changes in the TCR induced by mechanical stress provided by "pulling" detaching forces (Ma et al. 2008b). Here it is argued that these detaching forces originate from rearrangement of the cytoskeleton, and whilst the hypothesis most strongly considers T cell cytoskeletal rearrangements it may be possible that rearrangement of the DC cytoskeleton also provides a mechanical force. Tetraspanins are molecules that can regulate cytoskeletal rearrangement as best exemplified by the influence tetraspanins can have on phenomena such as integrin signal strengthening and spreading after cell adhesion (Feigelson et al. 2003; Goschnick et al. 2006; Lammerding et al. 2003; Delaguillaumie et al. 2004). It has also been documented that the dynamic clustering of MHC that occurs after T cell/APC contact is dependent on cytoskeletal rearrangement (de la Fuente et al. 2005). The kinetic segregation model argues that T cell signaling is initiated upon T cell/APC contact by a differential and dynamic segregation of membrane molecules (Davis and van der Merwe 2006). Large transmembrane phosphatases such as CD45 and CD148 are excluded from close contact zones that contain small kinase-associated membrane molecules such as the CD2/CD48 ligand pair and the TCR complex. This results in enhanced phosphorylation at the contact site of T cells and MHC. The model is entirely dependent on the concept of membrane organization and segregation. There are several examples where tetraspanins can regulate the membrane compartmentalization of their partner proteins (Cherukuri et al. 2004; Odintsova et al. 2003). Moreover, several of the molecules that do kinetically segregate in APC upon contact with T cells, are known to interact with tetraspanins including MHC, CD86, and ICAM-1.

10.4 Tetraspanins in Adaptive Immunity

10.4.1 Tetraspanins and T Cell Costimulation

Activation of T cells requires the engagement of the TCR complex, simultaneously with a costimulatory molecule. CD28 is the classical costimulatory molecule expressed on T cells, it binds to molecules (CD80 and CD86) expressed on APCs while MHC molecules present antigenic peptides to the TCR complex. It is possible to simulate in vitro the interaction between a T cell and an APC by antibodies that engage both the TCR and the costimulatory molecule simultaneously, thereby inducing activation and proliferation of T cells in the absence of an APC. Interestingly, several anti-tetraspanins mAbs have been shown to be as potent as anti-CD28 antibodies in T cell costimulation.

In vitro studies have shown that engagement of CD9 or CD81 in mice was as effective as the engagement of CD28 in costimulation of T cells (Tai et al. 1996; Witherden et al. 2000). Here, the availability of *Cd28*$^{-/-}$ mice was crucial in demonstrating that engaging CD9 or CD81 on T cells led to co-stimulation by a mechanism distinct from the CD28 pathway. Subsequent studies have shown that while costimulation via CD28 led to activation of NFκB and IL-2 production, costimulation via CD9 did not activate the NFκB signaling pathway (Zhou et al. 2002).

On the other hand, studies focusing on the activation of human T cells by tetraspanin engagement have benefited from the wider availability of mAbs to family members. Thus, an anti-human CD9 mAb costimulated peripheral blood T cells, albeit, the proliferative effect measured was lower than that induced by the anti-CD28 mAb (Kobayashi et al. 2004). This difference might have been due to selective expression of CD9 on naïve CD4+ T cells, whereas CD28 is expressed on all T cells. Human T cells are also costimulated by anti-CD81 mAbs, as most recently shown (Sagi et al. 2012). Interestingly, CD28 and CD81 costimulated different T cell subsets, where a greater percentage of naïve cells responded to CD81 costimulation. This preferential activation of the naïve subset by CD81 was not due to higher expression level, as CD81 is equally expressed on both naïve and memory T cells. It was due to increased signaling of the most proximal TCR signal transduction molecules, TCRζ, SLP76 and PLCγ (Sagi et al. 2012). An additional study demonstrated that an anti-human CD63 was as effective as the anti-CD28 mAb in delivering a costimulatory signal (Pfistershammer et al. 2004). Studies aimed at understanding the role of CD81 in hepatitis C virus (HCV) infection showed that co-engagement of the T cell receptor complex with CD81 activated T cells (Wack et al. 2001; Tseng et al. 2001; Serra et al. 2008). This coengagement was shown to be mediated by lymphocyte-specific kinase (Lck) (Soldaini et al. 2003) and to induce cytoskeletal rearrangements that were also correlated with increased phosphorylation of the mitogen-activated protein (MAP) kinases Erk1 and Erk2 (Crotta et al. 2006). An anti-CD81 mAb also augmented antigen specific activation—it increased IL-4 production by CD4 T cells that were derived from an allergic individual and were presented with the allergen by the person's B cells (Secrist et al. 1996). Additional studies used super-antigens to study the role of CD81 in T cell–B cell collaboration and showed that engagement of CD81 activates lymphocyte function-associated antigen 1 (LFA-1) on T cells (VanCompernolle et al. 2001) and preferentially induces Th2 cells (Maecker 2003). Coengagement of the costimulatory molecules, CD28 and CD81 on naïve T cells (without activation of the TCR complex) also induced a strong proliferative response, similar in magnitude to CD3/CD28 costimulation. Interestingly, the transition from a naïve to an effector T cell phenotype was more evident in response to CD28/CD81 engagement, which also led to increase in Th2 type cytokines (Serra et al. 2008).

CD82 acted as a costimulatory molecule on peripheral human T cell, this was demonstrated using an anti-CD82 mAb (4F9), which bound mostly CD4+ memory T cells (Nojima et al. 1993; Iwata et al. 2002). A study comparing the costimulatory effect of anti-CD9, CD53, CD81 and CD82 mAbs in a CD4 T cell line (Jurkat) (Lagaudriere-Gesbert et al. 1997) had shown increased IL-2 production, especially by the anti-CD82 mAb. Subsequent studies in Jurkat cells demonstrated a linkage between CD82, Rho GTPases and cytoskeletal actin rearrangements (Delaguillaumie et al. 2002, 2004). Taken together, the engagement of tetraspanins on both mouse and human T cells provides a costimulatory signal by a pathway that is yet to be defined.

10.4.2 Presence of Tetraspanins in Immune Synapses

When an antigen specific T cell (cognate T cell) is presented by an APC with its cognate peptide, the two cells form an immune synapse (IS), where key cell surface and signaling molecules, of both cell types, migrate in a coordinated manner to the point of cellular contact. A tagged CD82 (YFP) was shown to colocalize with filamentous (F) actin in the IS formed between an antigen-specific mouse T cell line presented by its cognate antigen (Delaguillaumie et al. 2004). Experimentally, it is also possible to form conjugates between T cells and APC (and visualize IS formation) by the use of super-antigens, which bind to MHC class II on APC and simultaneously to certain TCR molecules on T cells. One study showed that an anti-CD9 mAb enhanced conjugate formation, as measured by flow cytometry (Zilber et al. 2005). An additional study used microscopy on IS formed between a human T and a B cell line in response to a super-antigen and demonstrated redistribution of CD81 to the interface of the interacting cells (Mittelbrunn et al. 2002). Analysis of IS formation in immune cells lacking tetraspanins has yet to be reported, nevertheless, lack of CD81 affected cognate T cell–B cell interactions (Deng et al. 2002), as detailed in genetic evidence, below.

10.4.3 In Vivo Role of Tetraspanins in Adaptive Immunity

10.4.3.1 Genetic Evidence (Human)

The recent diagnosis of an immunodeficient child, due to a mutation in CD81 (van Zelm et al. 2010) highlights the role of this tetraspanin molecule in B cell function. The offspring of first cousin parents, the patient was diagnosed because of recurrent respiratory tract infection. Further characterization revealed decreased memory-B-cell numbers, impaired specific antibody responses and absence of CD19 expression on B-cells. Unexpectedly, sequence analysis found no mutations in either allele of the CD19 gene.

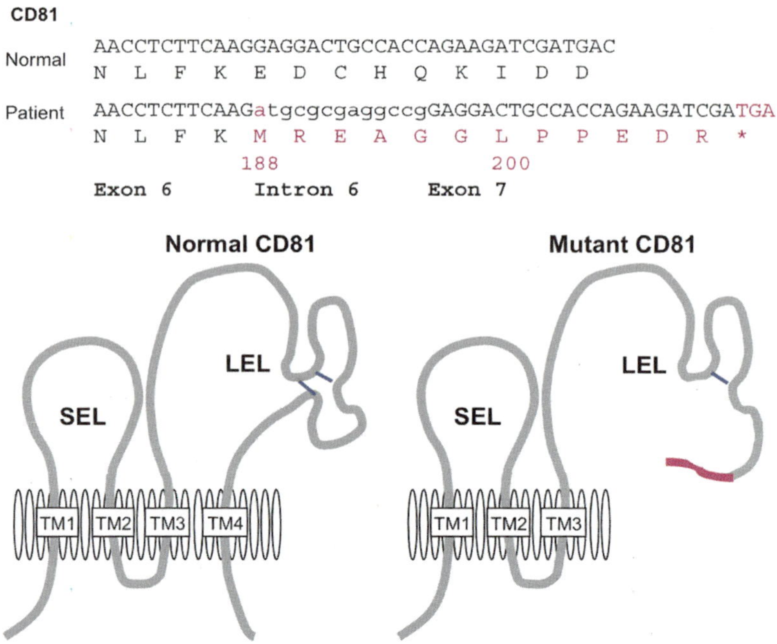

Fig. 10.2 Human CD81 mutant. Normal CD81 contains two disulfide bonds in the LEL (*blue lines*), whereas the mutant protein does not form the second disulfide bond in LEL. It contains a frameshift peptide (*magenta*) and is not anchored in the membrane by TM4

Attention was then focused on associated molecules in the CD19/CD21/CD81 signaling coreceptor complex (Bradbury et al. 1992; Fearon and Carroll 2000). Sequence analysis demonstrated a homozygous splice site mutation in CD81, suggesting that the defective CD81 gene (Fig. 10.2) caused the absence of CD19 surface expression. The human CD81 mutation occurred in a splice site located downstream of exon 6. The use of an alternative cryptic splice site generated a frameshift peptide and a stop codon. The truncated protein lacks the second disulfide bridge in the large extracellular loop (LEL) and the fourth transmembrane domain (TM4). Interestingly, mutations reported for human CD151 also occurred in the LEL (Karamatic Crew et al. 2004, 2008). Further analysis using a B cell lymphoblastoid cell line derived from the patient determined that CD19 protein was produced but sequestered intracellularly in the ER. Moreover, transduction of normal CD81 into this cell line rescued surface expression of both CD81 and CD19 (van Zelm et al. 2010).

This specific case illustrates several aspects of tetraspanin function. First, it emphasizes the function of the partnerships between tetraspanins and their associated molecules. In this particular case, the immunodeficiency is due to the loss of the normal association of CD81 with its B cell partner, CD19. Second, it demonstrates that an association previously characterized using biochemical methods (Bradbury et al. 1992; Shoham et al. 2003, 2006) has functional significance. The third lesson learned from this case is that a mutation in CD81, a widely expressed tetraspanin molecule (Oren et al. 1990), differentially affected B cell, but not T cell function. Thus, the same tetraspanin molecule plays a different role, which is dependent on the cell type. Fourth, the mutation in the human CD81 gene resulted in a more severe phenotype than that seen in three independently derived strains of CD81 knockout mice (Miyazaki et al. 1997; Tsitsikov et al. 1997; Maecker and Levy 1997). The biochemical basis for the difference between the human and the mouse mutants is yet to be determined. It is possible that the affected patient produced a truncated protein that may act as a dominant negative mutant, whereas the mice completely lack CD81. Considering that tetraspanins can molecularly interact both with partner proteins and other tetraspanins, a dominant negative mutant has the potential to disrupt a myriad of other molecular interactions and functions

Recurrent infectious diseases were also reported in a Spanish family with CD53 deficiency (Mollinedo et al. 1997). The immunodeficiency was more severe in the mother, also a product of first cousin parents, than in her two sons, even though all three lacked CD53 (Mollinedo et al. 1997). Unfortunately, genetic information and materials are not available, as the family became "fed up with so many tests and no solutions…and became uncooperative for science" according to Dr. Lazo, the senior author of the report on this family. Thus, whether the immunodeficient phenotype and the failure to express CD53 were causally linked or merely coincident, has not been determined.

10.4.3.2 Genetic Evidence (Mice)

CD9

An attempt to identify markers on splenic marginal zone (MZ) B cells distinguishing them from splenic follicular (FO) B cells found that CD9 is highly expressed in MZ B cells, in antibody producing plasma cells, in the B-1 subset, but not in FO B cell (Won and Kearney 2002). This suggested that CD9 might play a role both in B cell development and in B cell function. However, early development of B cells in the bone marrow of $Cd9^{-/-}$ mice was normal, similarly, MZ and FO splenic B cells and their precursors were present in normal ratios (Cariappa et al. 2005). CD9-deficiency did not affect the peritoneal B-1 B cell population. Immunization by T-dependent and T-independent antigens showed similar antibody production in $Cd9^{-/-}$ and wild type mice. While, non-immunized $Cd9^{-/-}$ mice had a normal distribution of germinal centers, they did show a slight increase in IgM secreting cells (Cariappa et al. 2005). Thus, although CD9 is variably expressed in B cells, its absence does not affect B cell development or the B cell response to immunization.

CD37

The development of both T and B cell lineages is normal, however functional studies indicate immune dysregulation in both lymphocyte lineages.

1. Role in humoral immunity
   T cell dependent IgG antibody responses in *Cd37*$^{-/-}$ mice are poor (Knobeloch et al. 2000). The molecular mechanism for these poor B cell responses are not known as *Cd37*$^{-/-}$ B cells express normal levels of CD19 (unlike *Cd81*$^{-/-}$ B cells) and proliferate normally to B cell mitogens. Conversely, IgA responses are exaggerated, and this phenotype is of pathological relevance as the excess IgA antibodies mediate resistance to the fungal pathogen *Candida albicans* and promote IgA nephropathy in aged mice (Figdor and van Spriel 2010; Rops et al. 2010). Here, the excess production of IgA is a B cell intrinsic phenotype and the likely molecular driver of excess IgA antibody is an increased production of IL-6.

2. Role in cellular immunity
   Cd37$^{-/-}$ T cells are hyperproliferative to stimulation (van Spriel et al. 2004). The phenotype has been observed in T cells stimulated by mitogens, mixed leukocyte reactions, and monoclonal antibodies crosslinking the T cell receptor, particularly in the absence of co-stimulatory signals. The hyperproliferative phenotype, is not unique to *Cd37*$^{-/-}$ T cells and has been observed in T cells deficient in at least three other tetraspanins: CD81 (Miyazaki et al. 1997), Tssc6 (Tarrant et al. 2002) and CD151 (Lau et al. 2004). The molecular mechanisms that underlie this dysregulation in T cell proliferation are not well understood and have been most extensively studied in the *Cd37*$^{-/-}$ mice. Hyperproliferation is not due to a resistance to apoptosis, or a perturbation in TCR internalization and turnover. Cross-linking CD37 with a monoclonal antibody suggests that the molecule may transduce a signal that inhibits proliferation, and biochemical studies suggest that the autophosphorylation of the key tyrosine kinase Lck is exaggerated in the absence of CD37 (van Spriel et al. 2004).

CD63

The development of immune system cells is normal, however, functional studies have not been reported (Schroder et al. 2009).

CD81

1. Role in cell surface expression of CD19
   It has been suggested that tetraspanins' function is redundant. However, CD19 expression in human and in mouse is dependent exclusively on CD81 and not on other tetraspanins, because deficiency in other tetraspanins does not affect CD19 expression. Whereas three independently generated *Cd81*$^{-/-}$ mice display an identical B cell phenotype—reduced cell-surface expression of CD19

(Miyazaki et al. 1997; Tsitsikov et al. 1997; Maecker and Levy 1997). Unlike the homozygous human CD81 mutation, lack of CD81 in mice has a milder effect on CD19 expression.

The introduction of human CD81 into primary *Cd81*$^{-/-}$ B cells (Shoham et al. 2003) and into a B cell line derived from these mice (Shoham et al. 2006) restored CD19 expression, as shown for the human CD81 deficiency (van Zelm et al. 2010). This "add-in" approach was further used to determine whether specific CD81 domain(s) are needed for this function. Because CD81 is the only known tetraspanin required for CD19 expression, chimeric CD81/CD9 molecules were tested for restoration of surface CD19. This analysis identified specific domains of CD81 essential for the intracellular trafficking and processing of CD19 in mouse B cells. Surprisingly, the first transmembrane domain of CD81 (TM1) was sufficient to support the exit of CD19 from the endoplasmic reticulum (ER). The cytoplasmic amino-terminal tail of CD81 was required for the proper maturation of the intracellular CD19 glycoform to a mature, endo-H-resistant glycoform. In addition, CD81 LEL was shown to associate physically with CD19 during biosynthesis in the ER (Shoham et al. 2006).

2. Role in B cell function

   Despite the consistent findings of low CD19 expression in B cells in all three *Cd81*$^{-/-}$ lines, in vitro analyzes of their B cell function has generated inconsistent results (Miyazaki et al. 1997; Tsitsikov et al. 1997; Maecker and Levy 1997). Similarly, conflicting outcomes were reported on the response of these mice to antigenic stimulation (Miyazaki et al. 1997; Tsitsikov et al. 1997; Maecker and Levy 1997). A recent reanalysis of B cell activation, which included measurements of Ca^{2+} influx, phosphorylation of signaling molecules, cell proliferation and antibody secretion demonstrated a hyperactive phenotype of *Cd81*$^{-/-}$ compared to wild-type B cells responding to stimulation both in vitro and in vivo (Sanyal et al. 2009). This differs considerably from the hypo-reactive B cell phenotype observed in the human CD81 mutant (van Zelm et al. 2010). These opposing B cell phenotypes are most likely related to the difference in surface CD19 expression in CD81-deficient human and mice.

3. Role in T cell function

   In vivo studies of the immune response of *Cd81*$^{-/-}$ mice have shown impaired T helper 2 (Th2) immune responses (Maecker 2003). Subsequent studies using an allergen-induced airway hyperactivity model have demonstrated diminished hyper-reactivity (Deng et al. 2000). In vitro studies demonstrated a crucial role for CD81 in cognate T cell–B cell interactions leading to Th2 responses (Deng et al. 2002), as antigen-specific interactions involving *Cd81*$^{-/-}$ transgenic T cells produced less of the Th2-promoting interleukin 4 (IL-4) than wild-type cells, especially when antigen was presented by B cells (Deng et al. 2002). Additional studies comparing *Cd81*$^{-/-}$ and wild type T cells have demonstrated enhanced T cell proliferation in response to stimulation by CD3 and to co-stimulation by CD3 and CD28 in the absence of CD81 (Miyazaki et al. 1997). As detailed above, a similar hyper-proliferative T cell phenotype was observed in *Cd37*$^{-/-}$ mice (Knobeloch et al. 2000).

CD151

Enhanced proliferation was also seen in *Cd151*$^{-/-}$ T cell responding to in vitro stimulation of CD3 and to costimulation of CD3 and CD28 (Lau et al. 2004). The humoral response of these mice to immunization did not differ from that seen in wild type mice (Lau et al. 2004).

TSSC6 (Tspan32)

Humoral responses to immunization are indistinguishable from wild type mice, Tssc6$^{-/-}$ T cells are hyperproliferative to stimulation (Goschnick et al. 2006). Recently the phenotype of mouse lacking both CD37 and TSSC6 has been described, and the data suggests that Tssc6 can functionally cooperate with CD37 in aspects of cellular immunity. CD37$^{-/-}$Tssc6$^{-/-}$ T cells show an exaggerated hyperproliferative phenotype (relative to single knockout T cells) whilst CD37$^{-/-}$Tssc6$^{-/-}$ DC show an exaggeration hyperstimulatory phenotype. Similarly whilst cytotoxic T cell responses to influenza are impaired in single knockout mice, the response is significantly poorer in CD37$^{-/-}$Tssc6$^{-/-}$ mice (Gartlan et al. 2010).

10.4.4 Expression of Tetraspanins in Diseases of the Immune System

A survey of normal and infected human peripheral blood leukocytes (PBL) has shown reduced expression of some tetraspanin molecules in the infected patients (Tohami et al. 2004). A subsequent survey of tetraspanin expression during human B cell development noted different patterns of expression of the individual family members (Barrena et al. 2005a). The same study also noted differences in tetraspanin expression in B cell malignancies. De Bruyne et al. followed up with a study of CD9 expression in a larger number of patients diagnosed with multiple myeloma, a plasma cell malignancy. They found that patients with non-active disease expressed CD9, whereas most cases with active disease were CD9 negative (De Bruyne et al. 2008). Although the mechanisms by which CD9 expression is reduced during the course of the disease in vivo is yet to be determined, studies of human (Drucker et al. 2006) and mouse (De Bruyne et al. 2008) myeloma cell lines have implicated an epigenetic mechanism in the silencing of CD9.

An interesting anecdote reported that the pattern of tetraspanin expression could distinguish two different malignant B cell clones in a single patient where each of the two malignant cell populations showed differential expression of the tetraspanins CD37, CD53 and CD81 (Barrena et al. 2005b).

The finding that CD81 was under-expressed in precursor B cell acute lymphoblastic leukemia (pre-B ALL) (Barrena et al. 2005a) also led to an additional subsequent study that analyzed a larger number of patients. It confirmed the original observation, it also proposed a flow cytometry approach to distinguish pre-B ALL from normal immature pre-B cells termed hematogones (Muzzafar et al. 2009).

10.5 Concluding Remarks

The function of tetraspanins is highly linked to the function of their partner proteins. The study of tetraspanins in the context of the immune system benefitted from the wealth of knowledge of proteins expressed on immune cells, as well as the understanding of their interactions in vivo and in vitro. Studies summarized within highlight the participation of tetraspanins in regulating the response of both innate and adaptive immune cells to pathogens. Tetraspanins partake in the coordination of leukocyte migration. In the immune synapse, the interface of the most important cellular interactions, tetraspanins are located. Importantly, deficiencies in tetraspanins lead to immune impairments. We believe that future studies to unravel the precise mechanism of tetraspanin action may shed light on cellular interactions in the immune system.

Acknowledgements Mark Wright is the recipient of grants from the Australian National Health and Medical Research Council, and the Anti Cancer Council of Victoria. Shoshana Levy is the recipient of grants from the Leukemia Lymphoma Society and the National Institutes of Health. The authors thank Annemiek van Spriel, Po Ki Ho and Frank Alderuccio for their critical reading of the manuscript.

References

Anderson HA, Hiltbold EM, Roche PA (2000) Concentration of MHC class II molecules in lipid rafts facilitates antigen presentation. Nat Immunol 1:156–162

Angelisova P, Hilgert I, Horejsi V (1994) Association of four antigens of the tetraspans family (CD37, CD53, TAPA-1, and R2/C33) with MHC class II glycoproteins. Immunogen 39:249–256

Artavanis-Tsakonas K, Love JC, Ploegh HL, Vyas JM (2006) Recruitment of CD63 to Cryptococcus neoformans phagosomes requires acidification. Proc Natl Acad Sci USA 103:15945–15950, PMCID: 1635107

Barreiro O, Yanez-Mo M, Sala-Valdes M, Gutierrez-Lopez MD, Ovalle S, Higginbottom A, Monk PN, Cabanas C, Sanchez-Madrid F (2005) Endothelial tetraspanin microdomains regulate leukocyte firm adhesion during extravasation. Blood 105:2852–2861

Barreiro O, Zamai M, Yanez-Mo M, Tejera E, Lopez-Romero P, Monk PN, Gratton E, Caiolfa VR, Sanchez-Madrid F (2008) Endothelial adhesion receptors are recruited to adherent leukocytes by inclusion in preformed tetraspanin nanoplatforms. J Cell Biol 183:527–542, PMCID: 2575792

Barrena S, Almeida J, Yunta M, Lopez A, Fernandez-Mosteirin N, Giralt M, Romero M, Perdiguer L, Delgado M, Orfao A, Lazo PA (2005a) Aberrant expression of tetraspanin molecules in B-cell chronic lymphoproliferative disorders and its correlation with normal B-cell maturation. Leukemia 19:1376–1383

Barrena S, Almeida J, Yunta M, Lopez A, Diaz-Mediavilla J, Orfao A, Lazo PA (2005b) Discrimination of biclonal B-cell chronic lymphoproliferative neoplasias by tetraspanin antigen expression. Leukemia 19:1708–1709
Berditchevski F, Odintsova E (2007) Tetraspanins as regulators of protein trafficking. Traffic 8:89–96
Boniface JJ, Rabinowitz JD, Wulfing C, Hampl J, Reich Z, Altman JD, Kantor RM, Beeson C, McConnell HM, Davis MM (1998) Initiation of signal transduction through the T cell receptor requires the multivalent engagement of peptide/MHC ligands [corrected]. Immunity 9:459–466
Bradbury LE, Kansas GS, Levy S, Evans RL, Tedder TF (1992) The CD19/CD21 signal transducing complex of human B lymphocytes includes the target of antiproliferative antibody-1 and Leu-13 molecules. J Immunol 149:2841–2850
Brown JH, Jardetzky TS, Gorga JC, Stern LJ, Urban RG, Strominger JL, Wiley DC (1993) Three-dimensional structure of the human class II histocompatibility antigen HLA-DR1. Nature 364:33–39
Cariappa A, Shoham T, Liu H, Levy S, Boucheix C, Pillai S (2005) The CD9 tetraspanin is not required for the development of peripheral B cells or for humoral immunity. J Immunol 175:2925–2930
Cherukuri A, Shoham T, Sohn HW, Levy S, Brooks S, Carter R, Pierce SK (2004) The tetraspanin CD81 is necessary for partitioning of coligated CD19/CD21-B cell antigen receptor complexes into signaling-active lipid rafts. J Immunol 172:370–380
Cochran JR, Cameron TO, Stern LJ (2000) The relationship of MHC-peptide binding and T cell activation probed using chemically defined MHC class II oligomers. Immunity 12:241–250
Crotta S, Ronconi V, Ulivieri C, Baldari CT, Valiante NM, Abrignani S, Wack A (2006) Cytoskeleton rearrangement induced by tetraspanin engagement modulates the activation of T and NK cells. Eur J Immunol 36:919–929
Davis SJ, van der Merwe PA (2006) The kinetic-segregation model: TCR triggering and beyond. Nat Immunol 7:803–809
De Bruyne E, Bos TJ, Asosingh K, Vande Broek I, Menu E, Van Valckenborgh E, Atadja P, Coiteux V, Leleu X, Thielemans K, Van Camp B, Vanderkerken K, Van Riet I (2008) Epigenetic silencing of the tetraspanin CD9 during disease progression in multiple myeloma cells and correlation with survival. Clin Cancer Res 14:2918–2926
de la Fuente H, Mittelbrunn M, Sanchez-Martin L, Vicente-Manzanares M, Lamana A, Pardi R, Cabanas C, Sanchez-Madrid F (2005) Synaptic clusters of MHC class II molecules induced on DCs by adhesion molecule-mediated initial T cell scanning. Mol Biol Cell 16:3314–3322, PMCID: 1165413
Delaguillaumie A, Lagaudriere-Gesbert C, Popoff MR, Conjeaud H (2002) Rho GTPases link cytoskeletal rearrangements and activation processes induced via the tetraspanin CD82 in T lymphocytes. J Cell Sci 115:433–443
Delaguillaumie A, Harriague J, Kohanna S, Bismuth G, Rubinstein E, Seigneuret M, Conjeaud H (2004) Tetraspanin CD82 controls the association of cholesterol-dependent microdomains with the actin cytoskeleton in T lymphocytes: relevance to co-stimulation. J Cell Sci 117:5269–5282
Deng J, Yeung VP, Tsitoura D, DeKruyff RH, Umetsu DT, Levy S (2000) Allergen-induced airway hyperreactivity is diminished in CD81-deficient mice. J Immunol 165:5054–5061
Deng J, Dekruyff RH, Freeman GJ, Umetsu DT, Levy S (2002) Critical role of CD81 in cognate T-B cell interactions leading to Th2 responses. Int Immunol 14:513–523
Drucker L, Tohami T, Tartakover-Matalon S, Zismanov V, Shapiro H, Radnay J, Lishner M (2006) Promoter hypermethylation of tetraspanin members contributes to their silencing in myeloma cell lines. Carcinogenesis 27:197–204
Engering A, Kuhn L, Fluitsma D, Hoefsmit E, Pieters J (2003) Differential post-translational modification of CD63 molecules during maturation of human dendritic cells. Eur J Biochem 270:2412–2420

Escola JM, Kleijmeer MJ, Stoorvogel W, Griffith JM, Yoshie O, Geuze HJ (1998) Selective enrichment of tetraspan proteins on the internal vesicles of multivesicular endosomes and on exosomes secreted by human B-lymphocytes. J Biol Chem 273:20121–20127

Fearon DT, Carroll MC (2000) Regulation of B lymphocyte responses to foreign and self-antigens by the CD19/CD21 complex. Annu Rev Immunol 18:393–422

Feigelson SW, Grabovsky V, Shamri R, Levy S, Alon R (2003) The CD81 tetraspanin facilitates instantaneous leukocyte VLA-4 adhesion strengthening to vascular cell adhesion molecule 1 (VCAM-1) under shear flow. J Biol Chem 278:51203–51212

Figdor CG, van Spriel AB (2010) Fungal pattern-recognition receptors and tetraspanins: partners on antigen-presenting cells. Trends Immunol 31:91–96

Gartlan KH, Belz GT, Tarrant JM, Minigo G, Katsara M, Sheng KC, Sofi M, van Spriel AB, Apostolopoulos V, Plebanski M, Robb L, Wright MD (2010) A complementary role for the tetraspanins CD37 and Tssc6 in cellular immunity. J Immunol 185:3158–3166

Goschnick MW, Lau LM, Wee JL, Liu YS, Hogarth PM, Robb LM, Hickey MJ, Wright MD, Jackson DE (2006) Impaired "outside-in" integrin alphaIIbbeta3 signaling and thrombus stability in TSSC6-deficient mice. Blood 108:1911–1918

Hammond C, Denzin LK, Pan M, Griffith JM, Geuze HJ, Cresswell P (1998) The tetraspan protein CD82 is a resident of MHC class II compartments where it associates with HLA-DR, -DM and -DO molecules. J Immunol 161:3282–3291

Hemler ME (2003) Tetraspanin proteins mediate cellular penetration, invasion, and fusion events and define a novel type of membrane microdomain. Annu Rev Cell Dev Biol 19:397–422

Hoorn T, Paul P, Janssen L, Janssen H, Neefjes J (2012) Dynamics within tetraspanin pairs affect MHC class II expression. J Cell Sci 125:328–339

Huby R, Chowdhury F, Lombardi G (2001) Rafts for antigen presentation? Nat Immunol 2:3

Iwasaki A, Medzhitov R (2010) Regulation of adaptive immunity by the innate immune system. Science 327:291–295

Iwata S, Kobayashi H, Miyake-Nishijima R, Sasaki T, Souta-Kuribara A, Nori M, Hosono O, Kawasaki H, Tanaka H, Morimoto C (2002) Distinctive signaling pathways through CD82 and beta1 integrins in human T cells. Eur J Immunol 32:1328–1337

Janeway CA Jr (1989) Approaching the asymptote? Evolution and revolution in immunology. Cold Spring Harb Symp Quant Biol 54(Pt 1):1–13

Jenei A, Varga S, Bene L, Matyus L, Bodnar A, Bacso Z, Pieri C, Gaspar R Jr, Farkas T, Damjanovich S (1997) HLA class I and II antigens are partially co-clustered in the plasma membrane of human lymphoblastoid cells. Proc Natl Acad Sci USA 94:7269–7274

Karamatic Crew V, Burton N, Kagan A, Green CA, Levene C, Flinter F, Brady RL, Daniels G, Anstee DJ (2004) CD151, the first member of the tetraspanin (TM4) superfamily detected on erythrocytes, is essential for the correct assembly of human basement membranes in kidney and skin. Blood 104:2217–2223

Karamatic Crew V, Poole J, Long S, Warke N, Colavecchia C, Burton N, Moulds M, Schlanser G, Wilson L, Noumsi G, Moulds JM, Moulds JJ, Daniels G (2008) Two MER2-negative individuals with the same novel CD151 mutation and evidence for clinical significance of anti-MER2. Transfusion 48:1912–1916

Kijimoto-Ochiai S, Noguchi A, Ohnishi T, Araki Y (2004) Complex formation of CD23/surface immunoglobulin and CD23/CD81/MHC class II on an EBV-transformed human B cell line and inferable role of tetraspanin. Microbiol Immunol 48:417–426

Knobeloch KP, Wright MD, Ochsenbein AF, Liesenfeld O, Lohler J, Zinkernagel RM, Horak I, Orinska Z (2000) Targeted inactivation of the tetraspanin CD37 impairs T-cell-dependent B-cell response under suboptimal costimulatory conditions. Mol Cell Biol 20:5363–5369

Kobayashi H, Hosono O, Iwata S, Kawasaki H, Kuwana M, Tanaka H, Dang NH, Morimoto C (2004) The tetraspanin CD9 is preferentially expressed on the human CD4(+)CD45RA+ naive T cell population and is involved in T cell activation. Clin Exp Immunol 137:101–108, PMCID: PMC1809091

Kramer B, Schulte D, Korner C, Zwank C, Hartmann A, Michalk M, Sohne J, Langhans B, Nischalke HD, Coenen M, Mohl C, Vogt A, Hennenberg M, Sauerbruch T, Spengler U, Nattermann J (2009) Regulation of NK cell trafficking by CD81. Eur J Immunol 39:3447–3458

Kropshofer H, Spindeldreher S, Rohn TA, Platania N, Grygar C, Daniel N, Wolpl A, Langen H, Horejsi V, Vogt AB (2002) Tetraspan microdomains distinct from lipid rafts enrich select peptide-MHC class II complexes. Nat Immunol 3:61–68

Lagaudriere-Gesbert C, Le Naour F, Lebel-Binay S, Billard M, Lemichez E, Boquet P, Boucheix C, Conjeaud H, Rubinstein E (1997) Functional analysis of four tetraspans, CD9, CD53, CD81, and CD82, suggests a common role in costimulation, cell adhesion, and migration: only CD9 upregulates HB-EGF activity. Cell Immunol 182:105–112

Lammerding J, Kazarov AR, Huang H, Lee RT, Hemler ME (2003) Tetraspanin CD151 regulates alpha6beta1 integrin adhesion strengthening. Proc Natl Acad Sci USA 100:7616–7621, PMCID: PMC164635

Lau LM, Wee JL, Wright MD, Moseley GW, Hogarth PM, Ashman LK, Jackson DE (2004) The tetraspanin superfamily member CD151 regulates outside-in integrin alphaIIbbeta3 signaling and platelet function. Blood 104:2368–2375

Ley K, Laudanna C, Cybulsky MI, Nourshargh S (2007) Getting to the site of inflammation: the leukocyte adhesion cascade updated. Nat Rev Immunol 7:678–689

Ma Z, Sharp KA, Janmey PA, Finkel TH (2008a) Surface-anchored monomeric agonist pMHCs alone trigger TCR with high sensitivity. PLoS Biol 6:e43, PMCID: 2253636

Ma Z, Janmey PA, Finkel TH (2008b) The receptor deformation model of TCR triggering. FASEB J 22:1002–1008, PMCID: 2679516

Maecker HT (2003) Human CD81 directly enhances Th1 and Th2 cell activation, but preferentially induces proliferation of Th2 cells upon long-term stimulation. BMC Immunol 4:1, PMCID: PMC151668

Maecker HT, Levy S (1997) Normal lymphocyte development but delayed humoral immune response in CD81-null mice. J Exp Med 185:1505–1510

Mantegazza AR, Barrio MM, Moutel S, Bover L, Weck M, Brossart P, Teillaud JL, Mordoh J (2004) CD63 tetraspanin slows down cell migration and translocates to the endosomal-lysosomal-MIICs route after extracellular stimuli in human immature dendritic cells. Blood 104:1183–1190

Medzhitov R, Preston-Hurlburt P, Janeway CA Jr (1997) A human homologue of the Drosophila Toll protein signals activation of adaptive immunity. Nature 388:394–397

Meyer-Wentrup F, Figdor CG, Ansems M, Brossart P, Wright MD, Adema GJ, van Spriel AB (2007) Dectin-1 interaction with tetraspanin CD37 inhibits IL-6 production. J Immunol 178:154–162

Mittelbrunn M, Yanez-Mo M, Sancho D, Ursa A, Sanchez-Madrid F (2002) Cutting edge: dynamic redistribution of tetraspanin CD81 at the central zone of the immune synapse in both T lymphocytes and APC. J Immunol 169:6691–6695

Miyazaki T, Muller U, Campbell KS (1997) Normal development but differentially altered proliferative responses of lymphocytes in mice lacking CD81. EMBO J 16:4217–4225

Mollinedo F, Fontan G, Barasoain I, Lazo PA (1997) Recurrent infectious diseases in human CD53 deficiency. Clin Diagn Lab Immunol 4:229–231, PMCID: PMC170508

Muzzafar T, Medeiros LJ, Wang SA, Brahmandam A, Thomas DA, Jorgensen JL (2009) Aberrant underexpression of CD81 in precursor B-cell acute lymphoblastic leukemia: utility in detection of minimal residual disease by flow cytometry. Am J Clin Pathol 132:692–698

Nojima Y, Hirose T, Tachibana K, Tanaka T, Shi L, Doshen J, Freeman GJ, Schlossman SF, Morimoto C (1993) The 4F9 antigen is a member of the tetra spans transmembrane protein family and functions as an accessory molecule in T cell activation and adhesion. Cell Immunol 152:249–260

Nydegger S, Khurana S, Krementsov DN, Foti M, Thali M (2006) Mapping of tetraspanin-enriched microdomains that can function as gateways for HIV-1. J Cell Biol 173:795–807

Odintsova E, Voortman J, Gilbert E, Berditchevski F (2003) Tetraspanin CD82 regulates compartmentalisation and ligand-induced dimerization of EGFR. J Cell Sci 116:4557–4566

Oren R, Takahashi S, Doss C, Levy R, Levy S (1990) TAPA-1, the target of an antiproliferative antibody, defines a new family of transmembrane proteins. Mol Cell Biol 10:4007–4015

Petersen SH, Odintsova E, Haigh TA, Rickinson AB, Taylor GS, Berditchevski F (2011) The role of tetraspanin CD63 in antigen presentation via MHC class II. Eur J Immunol 41:2556–2561

Pfeiffer A, Bottcher A, Orso E, Kapinsky M, Nagy P, Bodnar A, Spreitzer I, Liebisch G, Drobnik W, Gempel K, Horn M, Holmer S, Hartung T, Multhoff G, Schutz G, Schindler H, Ulmer AJ, Heine H, Stelter F, Schutt C, Rothe G, Szollosi J, Damjanovich S, Schmitz G (2001) Lipopolysaccharide and ceramide docking to CD14 provokes ligand-specific receptor clustering in rafts. Eur J Immunol 31:3153–3164

Pfistershammer K, Majdic O, Stockl J, Zlabinger G, Kirchberger S, Steinberger P, Knapp W (2004) CD63 as an activation-linked T cell costimulatory element. J Immunol 173:6000–6008

Poloso NJ, Denzin LK, Roche PA (2006) CDw78 defines MHC class II-peptide complexes that require Ii chain-dependent lysosomal trafficking, not localization to a specific tetraspanin membrane microdomain. J Immunol 177:5451–5458

Pols MS, Klumperman J (2009) Trafficking and function of the tetraspanin CD63. Exp Cell Res 315:1584–1592

Rocha N, Neefjes J (2008) MHC class II molecules on the move for successful antigen presentation. EMBO J 27:1–5, PMCID: 2206127

Rops AL, Figdor CG, van der Schaaf A, Tamboer WP, Bakker MA, Berden JH, Dijkman HB, Steenbergen EJ, van der Vlag J, van Spriel AB (2010) The tetraspanin CD37 protects against glomerular IgA deposition and renal pathology. Am J Pathol 176:2188–2197, PMCID: 2861084

Rubinstein E, Le Naour F, Lagaudriere-Gesbert C, Billard M, Conjeaud H, Boucheix C (1996) CD9, CD63, CD81, and CD82 are components of a surface tetraspan network connected to HLA-DR and VLA integrins. Eur J Immunol 26:2657–2665

Sagi Y, Landrigan A, Levy R, Levy S (2012) Complementary costimulation of human T-cell subpopulations by cluster of differentiation 28 (CD28) and CD81. Proc Natl Acad Sci USA 109:1613–1618, PMCID: PMC3277132

Sanyal M, Fernandez R, Levy S (2009) Enhanced B cell activation in the absence of CD81. Int Immunol 21:1225–1237

Schick MR, Levy S (1993) The TAPA-1 molecule is associated on the surface of B cells with HLA-DR molecules. J Immunol 151:4090–4097

Schroder J, Lullmann-Rauch R, Himmerkus N, Pleines I, Nieswandt B, Orinska Z, Koch-Nolte F, Schroder B, Bleich M, Saftig P (2009) Deficiency of the tetraspanin CD63 associated with kidney pathology but normal lysosomal function. Mol Cell Biol 29:1083–1094, PMCID: 2643809

Secrist H, Levy S, DeKruyff RH, Umetsu DT (1996) Ligation of TAPA-1 (CD81) or major histocompatibility complex class II in co-cultures of human B and T lymphocytes enhances interleukin-4 synthesis by antigen-specific CD4+ T cells. Eur J Immunol 26:1435–1442

Serra A, Nuti S, Tavarini S, Sammicheli C, Rosa D, Saletti G, Soldaini E, Abrignani S, Wack A (2008) Coligation of the hepatitis C virus receptor CD81 with CD28 primes naive T lymphocytes to acquire type 2 effector function. J Immunol 181:174–185

Sheng KC, van Spriel AB, Gartlan KH, Sofi M, Apostolopoulos V, Ashman L, Wright MD (2009) Tetraspanins CD37 and CD151 differentially regulate Ag presentation and T-cell co-stimulation by DC. Eur J Immunol 39:50–55

Shoham T, Rajapaksa R, Boucheix C, Rubinstein E, Poe JC, Tedder TF, Levy S (2003) The tetraspanin CD81 regulates the expression of CD19 during B cell development in a postendoplasmic reticulum compartment. J Immunol 171:4062–4072

Shoham T, Rajapaksa R, Kuo CC, Haimovich J, Levy S (2006) Building of the tetraspanin web: distinct structural domains of CD81 function in different cellular compartments. Mol Cell Biol 26:1373–1385, PMCID: 1367195

Shortman K, Liu YJ (2002) Mouse and human dendritic cell subtypes. Nat Rev Immunol 2:151–161

Soldaini E, Wack A, D'Oro U, Nuti S, Ulivieri C, Baldari CT, Abrignani S (2003) T cell costimulation by the hepatitis C virus envelope protein E2 binding to CD81 is mediated by Lck. Eur J Immunol 33:455–464

Suzuki M, Tachibana I, Takeda Y, He P, Minami S, Iwasaki T, Kida H, Goya S, Kijima T, Yoshida M, Kumagai T, Osaki T, Kawase I (2009) Tetraspanin CD9 negatively regulates lipopolysaccharide-induced macrophage activation and lung inflammation. J Immunol 182:6485–6493

Szollosi J, Horejsi V, Bene L, Angelisova P, Damjanovich S (1996) Supramolecular complexes of MHC class I, MHC class II, CD20, and teraspan molecules (CD53, CD81 and CD82) at the surface of a B cell line JY. J Immunol 157:2939–2946

Tai XG, Yashiro Y, Abe R, Toyooka K, Wood CR, Morris J, Long A, Ono S, Kobayashi M, Hamaoka T, Neben S, Fujiwara H (1996) A role for CD9 molecules in T cell activation. J Exp Med 184:753–758, PMCID: PMC2192734

Tarrant JM, Groom J, Metcalf D, Li R, Borobokas B, Wright MD, Tarlinton D, Robb L (2002) The absence of Tssc6, a member of the tetraspanin superfamily, does not affect lymphoid development but enhances in vitro T-cell proliferative responses. Mol Cell Biol 22:5006–5018, PMCID: 139789

Tarrant JM, Robb L, van Spriel AB, Wright MD (2003) Tetraspanins: molecular organisers of the leukocyte surface. Trends Immunol 24:610–617

Tohami T, Drucker L, Radnay J, Shapira H, Lishner M (2004) Expression of tetraspanins in peripheral blood leukocytes: a comparison between normal and infectious conditions. Tissue Antigens 64:235–242

Triantafilou M, Brandenburg K, Kusumoto S, Fukase K, Mackie A, Seydel U, Triantafilou K (2004) Combinational clustering of receptors following stimulation by bacterial products determines lipopolysaccharide responses. Biochem J 381:527–536, PMCID: 1133861

Tseng CT, Miskovsky E, Klimpel GR (2001) Crosslinking CD81 results in activation of TCRgammadelta T cells. Cell Immunol 207:19–27

Tsitsikov EN, Gutierrez-Ramos JC, Geha RS (1997) Impaired CD19 expression and signaling, enhanced antibody response to type II T independent antigen and reduction of B-1 cells in CD81-deficient mice. Proc Natl Acad Sci USA 94:10844–10849

Unternaehrer JJ, Chow A, Pypaert M, Inaba K, Mellman I (2007) The tetraspanin CD9 mediates lateral association of MHC class II molecules on the dendritic cell surface. Proc Natl Acad Sci USA 104:234–239

van Spriel AB, Puls KL, Sofi M, Pouniotis D, Hochrein H, Orinska Z, Knobeloch KP, Plebanski M, Wright MD (2004) A regulatory role for CD37 in T cell proliferation. J Immunol 172:2953–2961

van Zelm MC, Smet J, Adams B, Mascart F, Schandene L, Janssen F, Ferster A, Kuo CC, Levy S, van Dongen JJ, van der Burg M (2010) CD81 gene defect in humans disrupts CD19 complex formation and leads to antibody deficiency. J Clin Invest 120:1265–1274

VanCompernolle SE, Levy S, Todd SC (2001) Anti-CD81 activates LFA-1 on T cells and promotes T cell-B cell collaboration. Eur J Immunol 31:823–831

Vyas JM, Kim YM, Artavanis-Tsakonas K, Love JC, Van der Veen AG, Ploegh HL (2007) Tubulation of class II MHC compartments is microtubule dependent and involves multiple endolysosomal membrane proteins in primary dendritic cells. J Immunol 178:7199–7210, PMCID: 2806821

Wack A, Soldaini E, Tseng C, Nuti S, Klimpel G, Abrignani S (2001) Binding of the hepatitis C virus envelope protein E2 to CD81 provides a co-stimulatory signal for human T cells. Eur J Immunol 31:166–175

Williams AF, Beyers AD (1992) T-cell receptors. At grips with interactions. Nature 356:746–747

Witherden DA, Boismenu R, Havran WL (2000) CD81 and CD28 costimulate T cells through distinct pathways. J Immunol 165:1902–1909

Won WJ, Kearney JF (2002) CD9 is a unique marker for marginal zone B cells, B1 cells, and plasma cells in mice. J Immunol 168:5605–5611
Yauch RL, Berditchevski F, Harler MB, Reichner J, Hemler ME (1998) Highly stoichiometric, stable, and specific association of integrin alpha3beta1 with CD151 provides a major link to phosphatidylinositol 4- kinase, and may regulate cell migration. Mol Biol Cell 9:2751–2765
Zhou XY, Yashiro-Ohtani Y, Nakahira M, Park WR, Abe R, Hamaoka T, Naramura M, Gu H, Fujiwara H (2002) Molecular mechanisms underlying differential contribution of CD28 versus non-CD28 costimulatory molecules to IL-2 promoter activation. J Immunol 168:3847–3854
Zilber MT, Setterblad N, Vasselon T, Doliger C, Charron D, Mooney N, Gelin C (2005) MHC class II/CD38/CD9: a lipid-raft-dependent signaling complex in human monocytes. Blood 106:3074–3081

Chapter 11
Tetraspanins in Cancer

Leonie K. Ashman and Margot Zöller

Abstract Tetraspanins play important roles in cancer, especially in metastasis. CD82 and CD9 are frequently down-regulated on progression of epithelial cancers in humans and this has been associated with poor prognosis. In contrast, high levels of CD151 and Tspan 8 are often observed on tumour progression and have also been linked to poor patient outcome. These observations are supported by a large body of evidence from studies in vitro and in animal models. Considerable insights into the mechanisms by which tetraspanins influence tumour behaviour are now emerging. These include effects on cell-matrix and cell-cell interactions which influence migration and invasion of surrounding tissues, as well as angiogenesis. Several tetraspanins influence the function of platelets which can promote metastasis. Tetraspanins are constitutive components of exosomes, which are most important in intercellular communication. This widens the range of tetraspanin activities in physiology and pathology and may well be particularly important during spread and settlement of metastasizing tumor cells. There is hope that the understanding of how tetraspanins contribute to tumour progression indicates novel approaches to therapy.

L.K. Ashman (✉)
School of Biomedical Sciences, University of Newcastle, Room 511,
Medical Sciences Building, Callaghan, NSW 2308, Australia
e-mail: Leonie.Ashman@newcastle.edu.au

M. Zöller (✉)
Department of Tumor Cell Biology, University Hospital of Surgery,
Im Neuenheimer Feld 365, 69120 Heidelberg, Germany
e-mail: margot.zoeller@uni-heidelberg.de

F. Berditchevski and E. Rubinstein (eds.), *Tetraspanins*, Proteins and Cell Regulation 9,
DOI 10.1007/978-94-007-6070-7_11, © Springer Science+Business Media Dordrecht 2013

11.1 Introduction

Several tetraspanins have been defined as markers of human cancer cells. For example, ME491/CD63, a "founder member" of the tetraspanin family (Wright and Tomlinson 1994) was identified as a melanoma-associated antigen (Hotta et al. 1988). CO-029, a monoclonal antibody that recognised a tumour-associated antigen expressed by gastrointestinal tumours, identified the tetraspanin now known as Tspan8 (Szala et al. 1990). MRP-1 (CD9) was identified as the target of an antibody inhibiting cell migration and the cDNA was subsequently cloned from a breast cancer cell line (Miyake et al. 1991). KAI1, which was isolated as a metastasis suppressor gene located on human chromosome 11p11.2 (Dong et al. 1995), was shown to be identical to the leukocyte antigen, CD82. SAS/Tspan31 was identified as a gene that is amplified in human sarcomas (Jankowski et al. 1995). The target of a monoclonal antibody that suppressed metastasis of a human epidermoid tumour cell line in a chick embryo model was shown to be the CD151 protein (Testa et al. 1999).

Multiple studies have demonstrated the prognostic significance of mRNA or protein expression levels of several of these tetraspanins in human cancers. In general, they affect metastasis rather than primary tumour growth. Paradoxically, while some (notably CD82 and CD9) function as tumour suppressors, others (CD151 and Tspan8) appear to promote metastasis. In this article, we review evidence for their involvement in cancer from clinical studies and animal models, how their expression levels are regulated in cancer cells and how they function to modify cancer cell behaviour. Finally, we discuss tetraspanins as targets for therapeutic intervention.

11.2 Evidence for Altered Expression of Tetraspanin Proteins in Human Cancer and Its Prognostic Significance

11.2.1 CD82

CD82 (Tspan 27; also known as KAI1) is the most clear-cut example of a tetraspanin with altered expression in cancer. Following on from its original characterisation as a tumour suppressor in prostate cancer (Dong et al. 1995, 1996), many studies have been conducted linking CD82 down-regulation at the RNA or protein level with invasive and metastatic potential and/or patient outcome in a variety of epithelial cancers. CD82 is widely expressed in human tissues and reduced levels in tumours as well as an inverse relationship between CD82 expression and invasive or metastatic potential have been reported in many solid tumours including prostate, breast, cervix, gastric, colon, lung, pancreatic, liver, skin and thyroid cancers. These data have been reviewed elsewhere (Liu and Zhang 2006; Miranti 2009).

While some studies reported early and progressive down-regulation during tumorigenesis and metastasis, for example in colorectal cancer (Lombardi et al. 1999), other investigators found a biphasic pattern in colorectal and prostate cancers with increased

CD82 expression in low grade tumours, progressively decreasing with tumour stage or grade (Bouras and Frauman 1999; Maurer et al. 1999). In almost all studies, CD82 has been found to be down-regulated or lost in metastases. However, in breast cancer, CD82 expression was dependent on oestrogen receptor (ER) status. In two series of breast cancer patients, down-regulation of CD82 with respect to normal breast epithelium was found in 76–77% of ER-positive tumour specimens. Notably, ER-negative specimens retained CD82 even in metastases (Huang et al. 2005; Christgen et al. 2008, 2009). These results were surprising in view of the association of ER-negative status with metastasis and poor outcome (Weigelt et al. 2005).

Tonoli and Barrett reviewed 64 studies of CD82 expression in cancer conducted prior to 2005 (Tonoli and Barrett 2005). Of these, 52 studies (83%) reported CD82 down regulation including 12/16 gastrointestinal tract, 6/8 prostate, 7/8 non-small cell lung cancer (NSCLC) and 4/5 pancreatic cancer series. Ten studies (16%), scattered across a range of cancer types, showed upregulation of CD82. Prognostic data (statistically significant differences in survival or the development of metastases) were available from 33 of these studies. CD82 expression indicated favourable prognosis in 28 reports including 7/8 gastrointestinal tract series, 5/5 NSCLC, 2/3 oral carcinoma, 3/4 pancreatic carcinoma and 3/3 prostate cancer studies. In bladder and breast cancers CD82 down-regulation was associated with recurrence after treatment (Huang et al. 1998; Su et al. 2004). Thus, there is strong evidence linking loss of CD82 with tumour progression and poor outcome in many types of cancer.

11.2.2 CD9

CD9 (Tspan 29; also known as MRP-1) is also widely expressed in tissues and, like CD82, a large number of studies have examined the changes in CD9 mRNA and/or protein levels in cancer and its relationship to patient prognosis. These have recently been reviewed (Zöller 2009). CD9 levels were reported as being down-regulated relative to the corresponding normal tissue in breast (two series) and lung cancer, but not in ovarian or gastric cancer. Of the studies reviewed, the presence of CD9 was a positive prognostic factor in lung cancer (4/5 studies), breast cancer (2/3 studies), head and neck cancer (2/3 studies), bladder, and uterus (each one study). In general, in these cancers the extent of CD9 down-regulation was related to tumour grade and/or stage. In prostate cancer, CD9 was down-regulated in a proportion of specimens at all stages of progression with a further significant decrease between localised and advanced disease (Wang et al. 2007a). In contrast to the foregoing cancers, CD9 expression appears to be associated with progression and poor prognosis in gastric cancer (Hori et al. 2004; Soyuer et al. 2010). Furthermore, expression of CD9 in small cell lung cancer specimens and cell lines was associated with chemoresistance. Targeting of CD9 by siRNA or a monoclonal antibody induced apoptosis of chemoresistant cell lines (Kohmo et al. 2010).

An early report linked reduced CD9 expression to attributes associated with metastasis in melanoma specimens (Si and Hersey 1993). More recently, CD9 was identified as a gene expressed at lower levels in each of three pairs of metastatic melanomas compared with the corresponding normal melanocytes (Mischiati et al.

2006). This group also examined a series of specimens representing the stages of melanoma development, CD9 was expressed in all (18/18) naevi, and was lost in most (20/28) melanomas (including radial growth phase lesions which have relatively good prognosis) but only 24/52 metastatic lesions. Fan and co-workers studied CD9 expression and function in six human melanoma cell lines. All six lines displayed reduced CD9 mRNA and protein levels relative to normal melanocytes, however transfection of a line derived from a radial growth phase lesion enhanced invasion through Matrigel (Fan et al. 2010). These authors note that blocking CD9 enhances motility of melanocytes (Garcia-Lopez et al. 2005) and suggest that CD9 may play different roles at different stages of melanoma progression. Specifically, down-regulation of CD9 may facilitate early stages of melanoma development, but subsequent re-expression may promote invasion and metastasis. In view of other evidence indicating a role for CD9 in trans-endothelial invasion during metastasis of multiple myeloma (De Bruyne et al. 2006) and cervical cancer (Sauer et al. 2003), this tetraspanin cannot be considered simply as a tumour suppressor, but rather, may have different functions in different tumours and stages of tumour development.

11.2.3 CD151

The initial reports of the metastasis-promoting action of CD151 (Tspan24) in an in vivo model (Testa et al. 1999) were followed by clinical studies in lung, colon and prostate cancers (Tokuhara et al. 2001; Hashida et al. 2003; Ang et al 2004) which showed that high level expression of CD151 in primary tumours was associated with poor prognosis. Tokuhara et al. studied expression of CD9, CD82 and CD151 in specimens from 145 patients with NSCLC by semi-quantitative PCR and immunohistochemistry (IHC). High level CD151 expression was not correlated with tumour size, lymph node status, histological subtype or grade, but in contrast to CD9 and CD82, it was strongly associated with poor survival (Tokuhara et al. 2001). This group also analysed expression of these three tetraspanins in 146 cases of colon cancer with similar results to the NSCLC study (Hashida et al. 2003). In a series of 76 primary prostate cancer and 30 benign prostate hyperplasia (BPH) specimens studied by quantitative IHC, Ang et al. found significantly elevated CD151 expression in prostate cancer relative to BPH (Ang et al. 2004). CD151 levels were related to histologic differentiation status with the highest levels in poorly differentiated tumours. Increased expression of CD151 was strongly associated with overall survival, especially in patients with well- or moderately-differentiated tumours, and was a better predictor of outcome than the Gleason grade.

In the earlier study of primary colon cancer, CD151 levels were compared across tumour specimens of different grade and stage, but were not compared with normal colonic tissue (Hashida et al. 2003). A recent report indicated that CD151 protein levels were reduced in colon cancers relative to the adjacent normal tissue in 137 paired specimens (Chien et al. 2008). Using colon cancer cell lines, this group showed that under hypoxic conditions, CD151 expression was repressed due to binding of hypoxia-inducible factor-1 (HIF-1) to the CD151 promoter. This resulted in detachment of the cells. They propose that hypoxia induced CD151 down-regulation and detachment might play an important role in metastasis of colon cancer.

CD151 has only recently been studied in breast cancer. In normal breast, CD151 expression is largely confined to the myoepithelial-basement membrane interface in both ducts and lobules (Yang et al. 2008; Novitskaya et al. 2010). In a series of 124 unselected breast cancers, CD151 was found by IHC to be elevated relative to normal breast tissue in 31% of patients and high CD151 expression was positively correlated with tumour grade, ER-negativity and basal-like features. No outcome data were available (Yang et al. 2008). CD151 expression levels have been determined, also by IHC, in two further patient series and have been linked to patient survival in one of these. (Sadej et al. 2009) studied 56 specimens of primary invasive ductal carcinoma. Of these, 30% were scored as having elevated CD151 levels, and this was associated with poor overall survival (estimated 5-year survival 45.8% compared with 79.9% for CD151 low/negative cases). In contrast with the study of (Yang et al. 2008), no correlation was found between CD151 expression and tumour grade or ER status. In a second series, this group studied CD151 levels in 87 specimens of ductal carcinoma in situ (DCIS), including 48 with associated invasive disease (Novitskaya et al. 2010). In this study, elevated CD151 expression was associated with high tumour grade. In related experiments in a xenograft model and in Matrigel cultures, CD151 was shown to promote proliferation of the poorly tumorigenic HB2 breast cell line implying that it acts at the level of the primary tumour, not just to promote invasion and metastasis. Clearly, high expression of CD151 is not restricted to basal cancers and it appears that, in some cases, luminal epithelial cells that normally express little or no CD151 strongly upregulate this protein.

A recent study of specimens from 520 patients with hepatocellular carcinoma (HCC) using IHC (Ke et al. 2009) found that over-expression of CD151 relative to normal hepatocytes was a significant, independent predictor of recurrence and overall survival. High level CD151 expression was correlated with vascular invasion, tumour staging, size and differentiation. The prognostic significance was enhanced by also taking into account expression of the receptor tyrosine kinase c-Met, which was previously shown to form complexes with CD151 (Klosek et al. 2005). In an extension of this work, (Shi et al. 2010) showed that high level expression of CD151 and matrix metalloprotease 9 (MMP9) in tumour tissues was associated with increased microvessel density and together these features strongly predicted poor outcome. CD151 on tumour cells may activate MMP9 (Hong et al. 2006) which in turn may trigger an "angiogenic switch" (Bergers et al. 2000). In the experiments of (Sadej et al. 2009), siRNA down-regulation of CD151 in breast cancer cells resulted in reduced angiogenesis when the cells were grown as xenografts in mice. While these studies demonstrate a role for tumour cell CD151 in promoting neoangiogenesis, CD151 on vascular endothelial cells may also be important. CD151 is known to promote angiogenesis in vitro (Sincock et al. 1999), in animal models of ischemia (Zheng and Liu 2006) and in transplanted tumours growing in CD151-knockout mice (Takeda et al. 2007b). Taken together, these studies indicate that CD151 is a potentially important target for inhibiting tumour angiogenesis.

CD151 expression, determined by IHC, has also been linked to malignant transformation and/or prognosis in other tumour types. Increased CD151 protein relative to normal tissue was demonstrated in 30 cases of pancreatic cancer largely independent of grade or stage (Gesierich et al. 2005). More recently, a study of 71 patients with pancreatic ductal carcinoma confirmed overexpression of CD151 relative to normal pancreatic tissue and demonstrated association with elevated c-Met levels, tumour stage and poor survival. CD151 and c-Met were independent prognostic factors (Zhu et al. 2010). High level expression of CD151 was associated with tumour stage and poor survival in a series of 489 cases of clear cell renal carcinoma and was an independent prognostic indicator (Yoo et al. 2011). Similarly elevated CD151 expression was found in intrahepatic cholangiocarcinoma (60 patients) and esophageal squamous cell carcinoma (138 patients). In both of these series high CD151 was associated with tumour stage and predicted poor survival (Huang et al. 2010; Suzuki et al. 2011).

Fewer studies have examined *CD151* mRNA in clinical specimens. A gene expression microarray analysis of 50 brain tumours revealed that, together with other potential mediators of invasion such as integrin α3, CD151 was over-expressed in glioblastomas relative to normal brain (Bredel et al. 2005). In a series of 73 cases of gingival squamous carcinoma, expression of tetraspanins CD9, CD63, CD81, CD82, CD151 and NAG-2 (Tspan4) was studied by Q-PCR. Only CD151 and CD9 were significant prognostic factors, with high CD151 being associated with poor survival, while low CD9 was significantly linked to the presence of lymph node metastases (Hirano et al. 2009).

11.2.4 Tspan8

There are fewer studies of Tspan8 (also known as CO-029, TM4SF3 and D6.1A) than CD82, CD9 or CD151 but most available data indicate that it acts as a promoter of tumour progression. Tspan8 was identified as a marker of gastrointestinal tumours (Szala et al. 1990). Differential display mRNA analysis revealed its overexpression in hepatocellular carcinoma relative to normal liver. Using IHC it was shown that the protein was particularly over-expressed in poorly differentiated tumours, especially those showing intrahepatic spread (Kanetaka et al. 2001). A subsequent study in which a human HCC cell line, transfected to overexpress Tspan8, was orthotopically transplanted into immunocompromised mice revealed no change in primary tumour growth relative to the parent line, but an acquired ability to form intrahepatic metastases (Kanetaka et al. 2003).

Tspan8 was found by IHC to be more highly expressed in 24/30 cases of pancreatic cancer compared with normal pancreas, although ducts in chronic pancreatitis also displayed elevated levels. The intensity of staining was largely independent of tumour grade and stage (Gesierich et al. 2005).

Kuhn and coworkers examined expression of Tspan8 in a series of 104 primary colorectal cancers and 66 liver metastases together with normal colon and liver. IHC staining of normal colon was negative or weak with clear upregulation in the majority of primary lesions and metastases. Staining was not related to tumour stage or grading. Co-expression and complex formation of Tspan8 with claudin, EpCAM and CD44v6 was inversely correlated with disease-free survival and it is proposed that this complex promotes metastasis (Kuhn et al. 2007). Using cell lines derived from primary colorectal cancer and metastases from the same patient, LeNaour et al. (2006a) used proteomic methods to characterise tetraspanin-containing complexes. They found that Tspan8 was strikingly upregulated in the metastatic cell lines. This was followed up by IHC examination of Tspan8 protein levels in matched normal colon, primary tumour and metastases from three patients. In contrast to the findings of Kuhn et al., they reported high expression in normal colon (which was confirmed by western blot), with low expression on both primary tumours and, surprisingly, metastases. The reason for the discrepancy between the two groups in relation to Tspan8 expression in normal colon is not clear although they used different antibodies. In a subsequent study (Greco et al. 2010), Tspan8 expression was examined by IHC using a novel, well validated monoclonal antibody, TS29, in specimens of primary colonic tumours from 52 patients. Intensity of Tspan8 staining was compared between tumorous and adjacent non-tumorous epithelium. Elevated Tspan8 was found to be significantly associated with relapse, especially when combined with cytoplasmic relocalisation of p120 catenin (resulting from E-cadherin down-regulation). This group also demonstrated in vitro that Tspan8 promotes cell migration when E-cadherin is down-regulated, as occurs in aggressive cancers, and propose that it is a potentially important therapeutic target (Greco et al. 2010).

Examination of publicly available gene expression datasets for oesophageal carcinoma indicated upregulation of Tspan8 relative to normal tissue. This was confirmed in 8/14 pairs of normal and cancerous oesophageal specimens by western blotting (Zhou et al. 2008). Transfection of an oesophageal carcinoma cell line with Tspan8 cDNA resulted in acquisition of metastatic ability in a mouse xenograft model.

Overall, these data together with studies of the D6.1A rat tumour model (detailed elsewhere in this Chapter) provide strong evidence for the tumour-promoting, pro-metastatic action of Tspan8.

11.2.5 Other Tetraspanins Implicated in Cancer

11.2.5.1 CD63

Although CD63 (Tspan30) was originally identified as a melanoma antigen (ME491) and has been suggested to be a tumour suppressor, there is a lack of strong evidence from clinical studies to support this. Like the other tetraspanins described above, CD63 is very widely expressed by normal and tumour cells (Pols and Klumperman 2009). In the original reports, monoclonal antibody ME491 was positive on 7/10 melanoma cell lines, 4/4 superficial spreading melanomas and 5/8 melanomas with associated metastases. It was weak or negative on normal melanocytes, but upregulated in culture (Atkinson et al. 1984; Hotta et al. 1988). However, a more recent study did not support the view that CD63 down-regulation is associated with melanoma progression. Specimens from patients (four benign naevi, two primary tumours and 28 metastatic lesions) were analysed by Q-PCR and CD63 was found to be upregulated in melanoma relative to benign lesions (Lewis et al. 2005). Most evidence for a tumour-suppressive function for CD63, especially in melanoma, comes from studies with cell lines (detailed in Sect. 11.4). Some early publications reporting effects of ectopic expression of CD63 in human melanoma (Radford et al. 1997) were compromised by the subsequent demonstration that the cell line used in the study was in fact of rat origin (Moseley et al. 2003).

Two clinical studies in other cancers have provided some support for a tumour-suppressive function of CD63. One series of 90 lung cancer (NSCLC) patients showed down-regulation of CD63 relative to normal tissue in tumours, especially of those of squamous type, and association with tumour stage. In adenocarcinomas, CD63 expression was more variable, but downregulation was associated with poor survival (Kwon et al. 2007). In ovarian cancer, CD63 mRNA levels were shown to be inversely related to tumour grade (Zhijun et al. 2007). However, CD63 mRNA and protein expression were unchanged in series of pancreatic (Sho et al. 1998) and thyroid cancer specimens (Chen et al. 2004) where significant down-regulation of CD82 associated with progression was observed.

11.2.5.2 Tspan1

Although much less studied experimentally than CD151 and Tspan8, there is growing evidence that Tspan1 (also known as NET-1) is also a tumour promoting tetraspanin. At the mRNA level, *NET-1* was over-expressed in cervical neoplasia compared with normal cervical epithelium. It was strongly expressed in all undifferentiated cervical carcinomas examined (Wollscheid et al. 2002). Expression of NET-1 was studied by IHC in a series of 88 patients with colorectal carcinoma (Chen et al. 2009) and 86 cases of gastric carcinoma (Chen et al. 2008). In both series Tspan1 over-expression was correlated with clinical stage and negatively correlated with survival. Consistent with its role as a tumour promoter, knock-down of Tspan1 with siRNA in the squamous cell skin carcinoma cell line, A431, reduced proliferation, migration and infiltration of cells in vitro (Chen et al. 2010).

11.2.5.3 Tspan13

Emerging data indicate that Tspan13 (also known as NET-6) is a tumour suppressor. Gene expression array experiments comparing HER-2 positive and negative breast cancer cells showed that NET-6 levels are related to HER-2 and ER status and are lowest in HER-2-ER-basal-like tumours (Wilson et al. 2002). Transfection of *NET-6* cDNA into MDA-MB-231 breast cancer cells induced apoptosis and reduced growth in vitro and in a mouse xenograft model (Huang et al. 2005, 2007). A recent study of NET-6 mRNA and protein levels in prostate cancer specimens showed that it is over-expressed in prostatic intraepithelial neoplasia and the majority of prostate cancers compared with normal tissue. However, in tumour specimens, NET-6 protein levels showed a significant inverse correlation with Gleason grade consistent with down-regulation in high-grade tumours (Arencibia et al. 2009). This is consistent with the findings of Huang et al. in breast cancer indicating that it acts as a suppressor of tumour progression. Thus NET-6 expression appears to be regulated in a biphasic fashion similar to CD82 in prostate and colon cancer (Bouras and Frauman 1999; Maurer et al. 1999).

11.3 Regulation of Tetraspanin Levels in Cancer Cells

From the previous section it can be seen that levels of several tetraspanin proteins and/or mRNA are correlated with progression and prognosis in many human epithelial cancers. While some of these changes have been identified from gene expression array analyses (for example, Tspan8 (Zhou et al. 2008) and Tspan13 (Wilson et al. 2002; Arencibia et al. 2009), it is perhaps surprising that more examples have not emerged from the large amounts of these data that have been generated in recent years. Protein expression can be regulated at many levels and it seems likely that tetraspanin proteins are regulated in several ways encompassing translation and protein turnover as well as transcription. Although several reports have examined both mRNA and protein expression, newer quantitative studies directly comparing tetraspanin mRNA and protein levels in cancers are required. These will guide development of the most appropriate assays for clinical application. Apart from CD82, little is known about how tetraspanin transcription is regulated and more studies of other tetraspanins are needed.

11.3.1 CD82

Regulation of CD82 transcription and silencing are complex processes (Gao et al. 2003; Tonoli and Barrett 2005; Liu and Zhang 2006). There is no evidence for gene mutation or loss of heterozygosity (Tagawa et al. 1999; Liu et al. 2000) and hypermethylation of CpG islands in the CD82 gene has only been seen in patients with multiple myeloma, where combined de-methylation and de-acetylation induced increased expression of CD82 mRNA (Jackson et al. 2000; Drucker et al. 2006). CD82 down-regulation has also been related to the p53 status. Binding motifs for the transcription factor AP2 in the CD82 promoter function synergistically with p53 and junB such that the absence of wild-type p53 and/or loss of junB and AP2 protein expression correlate with CD82 mRNA down-regulation (Marreiros et al. 2003, 2005). There have been some controversial results on the involvement of NFκB in CD82 transcription, which is likely due to the nature of the recruited cofactors.

In non-metastatic cells, IL-1β supports the recruitment of a Tip60 (HIV-1 TAT-interactive protein 60)/Fe65-Pontin complex, which acts as a co-activator together with NFκB p50 and accounts for the displacement of the co-repressor N-Cor/TAB2 (TAK1-binding adaptor protein)/HDAC3 (histone deacetylase 3) complex from NFκB p50. In metastasizing tumour cells, Tip60 is down-regulated and a β-catenin-reptin complex replaces the Tip60-Pontin complex and represses NFκB activity (Telese et al. 2005). Recently it was shown that HIF1α binds directly to the CD82 promoter leading to increased CD82 protein in hypoxia (Kim et al. 2010).

Alternate splicing has also been proposed as a possible mechanism for regulation of CD82 expression and function. A splice variant lacking exon 7 which codes for part of the second extracellular loop and the fourth transmembrane domain was identified in gastric carcinomas and reported to confer increased metastatic ability in a mouse model of colon cancer (Lee et al. 2003). However, a more recent study in bladder cancer found uniformly low levels of mRNA encoding the splice variant, which was not associated with tumour invasion (Jackson et al. 2007).

A role for protein degradation in control of tetraspanin levels has recently emerged. The E3 ubiquitin ligase, gp78, was shown to functionally interact with CD82 leading to its degradation. In an orthotopic mouse model, knockdown of gp78 in the human HT1080 sarcoma had no effect on growth of primary tumour but blocked lung metastasis. This was accompanied by upregulation of CD82. In a tissue microarray of primary sarcomas, an inverse relationship between CD82 and gp78 staining was observed. (Tsai et al. 2007). Inverse expression of gp78 and CD82 was also observed in human mammary carcinoma cells. Ectopic expression of gp78 in the murine mammary gland resulted in decreased CD82 expression and hyperplasia but was insufficient for tumourigenesis (Joshi et al. 2010). CD82 stability and function may also be regulated by palmitoylation (Zhou et al. 2004).

11.3.2 CD9

There is considerable evidence that CD9 expression is regulated epigenetically. While promoter methylation has been reported as a major mechanism in multiple myeloma (Drucker et al. 2006), other reports have indicated that histone acetylation is more important. In another study of multiple myeloma, CD9 levels were inversely correlated with disease activity, with increased CD9 in patients with inactive disease. High CD9 at diagnosis was associated with increased survival. CD9 expression was regulated primarily by histone acetylation (De Bruyne et al. 2008). CD9 expression was also reported to be regulated by histone acetylation in lung cancer (Zhong et al. 2007), melanoma cell lines (Fan et al. 2010) and B lymphomas (Yoon et al. 2010).

One study of CD9 in prostate cancer (Wang et al. 2007a) found point mutations and/or deletions in cDNA from four adenocarcinomas, one case of prostate intraepithelial neoplasia (PIN), and two prostate cancer cell lines. They suggest that down-regulation of CD9 may result from these mutations. No CD9 mutations were found in cDNA from six normal prostate specimens. However, the generality of these results is uncertain. No mutations were found in CD9 cDNA from six human melanoma lines which had lower levels of CD9 protein than normal melanocytes (Fan et al. 2010).

Recent evidence indicates a role for post-transcriptional regulation of CD9 protein expression. Analysis of the 5'UTR of CD9 cDNA in Merkel cell carcinoma demonstrated two splice variants, the longer of which contained a putative structural pattern that would block translation. There was a shift in favour of this variant in CD9-negative cells suggesting that it may influence CD9 protein expression (Woegerbauer et al. 2010). CUGBP1 is a RNA binding protein that regulates alternate splicing, mRNA stability and translation by binding to the 3'UTR. CUGBP1 binds directly to CD9 mRNA resulting in decreased levels (Le Tonqueze et al. 2010). Another RNA binding protein, HuR, acts by binding to AU-rich sequences in mRNA resulting in stabilisation and enhancement of translation. HuR has been suggested to promote tumour progression, including in breast cancer (Heinonen et al. 2005; Lopez de Silanes et al. 2005). Through co-immunoprecipitation analysis on MCF7 and MDA-MB-231 breast cancer lines, HuR was found to bind CD9 mRNA (Calaluce et al. 2010). Surprisingly, HuR over-expression and knock-down experiments indicated that it decreased CD9 mRNA and protein in MDA-MB-231 cells, but slightly increased their levels in MCF7 cells. Thus, the consequence of HuR binding to CD9 mRNA depends on the cellular context.

CD9 levels may also be regulated post-translationally. CD9 palmitoylation, mediated by the enzyme DHHC2, was shown to protect it from proteasomal and lysosomal degradation (Sharma et al. 2008).

11.3.3 CD151

Recent reports have provided some information about the regulation of *CD151* transcription. The SP1 transcription factor was shown to be required for accessibility and function of the *CD151* promoter (Wang et al. 2010). Elevated SP1 is commonly observed in cancer, particularly in advanced disease, and may regulate expression of a number of genes associated with cancer progression (Safe and Abdelrahim 2005) likely including *CD151*. The *CD151* promoter also binds the hypoxia-inducible factor, HIF-1α, leading to down-regulation of CD151 mRNA and protein levels under hypoxic conditions. (Chien et al. 2008). The authors propose a role for reversible CD151 modulation in metastasis.

It is likely that CD151 protein levels are also regulated post-translationally. The membrane-spanning ubiquitin E3 ligase, GRAIL, binds to CD151 and ubiquitylates its N-terminal cytoplasmic domain promoting its removal from the cell surface and lysosomal degradation (Lineberry et al. 2008). Like CD9, CD151 is palmitoylated by DHHC2 blocking its proteasomal and lysosomal degradation (Sharma et al. 2008).

11.4 Tetraspanins and Metastasis

Metastasis formation is the final result of a cascade of events that primary tumour cells pass through by changing their phenotype and their cross-talk with the tumour environment. In epithelial tumours the metastatic cascade may be initiated through a process called epithelial to mesenchymal transition (EMT) of cancer stem cells/ cancer initiating cells (Brabletz et al. 2005; Yang and Weinberg 2008), followed by migration from the primary tumour, intravasation, extravasation, settlement and growth in distant organs (Geiger and Peeper 2009). Molecules involved in tumour progression are cell-cell and cell-matrix adhesion molecules, matrix degrading enzymes and their inhibitors. In addition, chemotactic factors released from the degraded matrix and chemokine receptors expressed by the metastasizing tumour cell, apoptosis resistance and angiogenesis inducer genes play an important role (Albini et al. 2008). Finally, several tetraspanins can be involved.

Tetraspanins are proposed to contribute to the metastatic cascade by their involvement in cell motility due to their association with integrins. Although there is some evidence that tetraspanins may modulate the ligand binding activity of associated integrins by stabilizing their activated conformation (Nishiuchi et al. 2005), it is proposed that tetraspanins mostly influence cell migration through integrin compartmentalization, their internalization and recycling or by modulating integrin-mediated signalling (Berditchevski 2001; Stipp et al. 2003; Hemler 2005; Levy and Shoham 2005). Besides integrins, the association with EWI proteins influences cell polarity and migration (Sala-Valdés et al. 2006). Several tetraspanins have been shown to regulate invasiveness, possibly due to their association with peptidases (Le Naour et al. 2006; Rana et al. 2011) ADAMs (A disintegrin and metalloproteinase), particularly ADAM10 (Arduise et al. 2008) and matrix metalloproteinases (MMP) (Lafleur et al. 2009; Yanez-Mo et al. 2008). They may also act by modulating MMP transcription and secretion (Hasegawa et al. 2007). By regulating trafficking and biosynthesis of associated molecules, tetraspanins can also influence cell adhesion events (He et al. 2005; Winterwood et al. 2006), which might mediate their actions in inhibiting or promoting metastasis.

11.4.1 Metastasis Suppressing Tetraspanins

As discussed above, CD82/KAI1 is a prototype of a metastasis suppressor gene. Other tetraspanins like CD9, CD81 and CD63 mostly, but not consistently, hamper tumour progression.

11.4.1.1 CD82, CD81 and CD9 Inhibit Tumour Cell Migration

Studies with human and animal cancer cell lines provide strong evidence that metastasis suppression by CD82 may mostly rely on inhibition of tumour cell migration and invasion (Jackson et al. 2005; Tonoli and Barrett 2005; Liu and Zhang 2006). Depending on the associating molecules, several mechanisms have been elaborated through which CD82 could inhibit tumour progression (Liu and Zhang 2006; Miranti 2009).

Firstly, co-internalization of the α6 integrin chain with CD82, which is strengthened by concomitant epidermal growth factor receptor (EGFR) activation, is accompanied by impaired laminin adhesion and migration. Adhesion and migration are abolished by mutating the CD82 sorting motif. The authors suggest that the decrease in α6 integrins in CD82 expressing cells might be responsible for reduced adhesiveness and subsequently attenuated α6 integrins promoted motility (He et al. 2005; Odintsova et al. 2000, 2003). Similarly, the L6 antigen associates with CD82 and CD63 in TEM and may facilitate internalization of these tetraspanins (Lekishvili et al. 2008). It also has been reported that high level CD82 expression correlates with low integrin α6β1 and α6β4 expression (He et al. 2005), which in the case of CD151 was shown to affect integrin-mediated cell migration (Winterwood et al. 2006). CD82 expression can also interfere with integrin αvβ3/vitronectin-mediated tumour cell motility (Ruseva et al. 2009).

Secondly, the functional interplay of CD82 with the Ig superfamily member EWI-2 strengthens the motility inhibitory activity of EWI-2 on laminin and fibronectin (Zhang et al. 2003a). EWI-2 associates with ERM (ezrin, radixin, moesin) proteins and prevents their activation (Sala-Valdés et al. 2006), which is required for the linkage with actin (Louvet-Vallée 2000).

Thirdly, tetraspanins can modify the activity of proteases required for invasion. uPAR (urokinase receptor) co-localizes with integrin α5β1 in focal adhesions only in the presence of CD82. In the presence of the tetraspanin, the stable association between uPAR and α5β1, which prevents binding of uPA to its receptor and pericellular proteolysis, a necessary step in invasion, is strikingly reduced (Bass et al. 2005). In multiple myeloma, CD82 and CD81 over-expression affects motility and invasive potential, which is accompanied by reduced MMP9 secretion (Tohami et al. 2007).

Fourthly, CD82 interferes with c-Met signalling such that hepatocyte growth factor (HGF, also known as scatter factor)-induced cell migration is impaired (Takahashi et al. 2007). In a non-small cell lung cancer line over-expressing CD82, phosphorylation of c-Met by HGF stimulation was not affected, but the presence of CD82 interfered with ligand-induced association of c-Met with Grb2, a key molecule in intracellular signal transduction. Interference of CD82 with Grb2 binding is accompanied by inhibition of downstream signalling via phosphoinositide 3-kinase (PI3K) and the Ras → Raf → MAPK signaling axis, activation of rac and Cdc42 GTPases. As a consequence, lamellipodia formation and cell migration is severely impaired. In contrast, in a prostate cancer cell line (PC3), HGF-induced activation of c-Met and src was impaired in the presence of CD82, which inhibited the formation of the FAK (focal adhesion kinase)–$p130^{CAS}$–Crk complex downstream of Src activation (Sridhar and Miranti 2006). Importantly, the assembly of this complex was linked to increased cell motility. Down-regulation of the $p130^{CAS}$-Crk complex by CD82/KAI1 also has consequences on integrin-mediated cell migration (Zhang et al. 2003b).

Fifthly, some activities of CD82 rely on the contribution of gangliosides in the organization of TEM (Todeschini and Hakomori 2008; Hakomori 2010). For example, the impact of CD82 on EGFR activation varies depending on the presence of ganglioside GD1a, which facilitates the re-localization of the CD82–EGFR complex in TEM (Odintsova et al. 2006). The CD82–integrin α3β1–Met crosstalk is also regulated by gangliosides. Specifically, formation of the complex of GM2/GM3 with CD82 interferes with c-Met activation and c-Met-dependent downstream signaling. This blockade impairs not only cell motility, but also cell proliferation (Todeschini et al. 2008). It has been proposed that the CD82/GM2/GM3 complex inhibits tumour cell proliferation via a pathway similar to the PKCα-mediated inhibition of EGFR-induced proliferation, whereby GM3 together with CD82 controls translocation and phosphorylation of PKCα, and, consequently, induces EGFR phosphorylation and internalization (Wang et al. 2007b).

CD81 shares several features of motility-inhibiting activity with CD82. In hepatocellular carcinoma, the interaction of CD81 with PI4KII may play an important role in suppressing cell motility by promoting the formation of CD81-enriched vesicles that sequester actinin-4. The association of CD81 with PI4KII is accompanied by redistribution to intracellular vesicles, which might negatively affect actin-bundling activity of actinin (Fraley et al. 2003; Janmey and Lindberg 2004; Mazzocca et al. 2008). GPR56 forms a complex with Gαq and CD81 (Little et al. 2004). In melanoma, GPR56 binds tissue transglutaminase 2 (TG2), a major cross-linking enzyme in the ECM. The binding of the GPR56-Gαq-CD81 complex to TG2 could support adhesion and thereby interfere with tumour cell migration (Xu and Hynes 2007).

CD9 can inhibit or promote metastasis (Ikeyama et al. 1993; Ono et al. 1999; Zheng et al. 2005; Kohmo et al. 2010; Sakakura et al. 2002). The opposing activities are likely to depend on the associating molecules in the tetraspanin web. CD9 homoclustering is promoted by integrins α3β1, α6β4 and by palmitoylation of CD9 and the integrin β4 chain. In contrast, EWI-F- and EWI-2-associated or unpalmitoylated CD9 forms heteroclusters, which particularly are seen on malignant epithelial tumours (Yang et al. 2006).

Though CD9 can interfere with tumour progression at several steps of the metastatic cascade, migration inhibiting pathways are so far best described. In ovarian carcinoma cells, expression levels of CD9 and β1, α2, α3, α5 and α6 integrin chains are correlated, and downregulation of CD9 is accompanied by weaker matrix adhesion and dispersed growth in vitro (Ikeyama et al. 1993; Furuya et al. 2005). In addition, CD9 can associate with gangliosides, which can have distinct effects on the cell fate depending on the expression level. A non-invasive bladder cancer line expresses the GM3–CD9 complex at a high level. This correlates with a strong association with α3β1 and low cell motility. The reverse is true for an invasive bladder cancer line. When GM3 is expressed at a low level, it activates Src, whereas a high level GM3 causes Csk (C-terminal Src kinase), an endogenous inhibitor of the Src-family protein tyrosine kinases, translocation into TEM microdomains with subsequent inhibition of Src phosphorylation (Mitsuzuka et al. 2005).

CD9 also can hamper the migration of the isolated metastasizing cells. CD9 associates with the EGFR such that CD9 antibody cross-linking or EGF stimulation promotes EGFR internalization, which results in reduced EGFR autophosphorylation and reduced SHC phosphorylation and recruitment of Grb2. CD9 antibody cross-linking was noted to activate JNK and p38 MAPK and, after 24–48 h, caspase 3. The authors propose that this was due to tyrosine phosphorylation selectively of the p46 Shc isoform and speculate that CD9 might regulate apoptosis in tumor cells through initiating specialized signal transduction pathways (Murayama et al. 2008). In addition, CD9 is associated with the transmembrane form of transforming growth factor (TGF)α and, therefore, may affect the autocrine and juxtacrine activity of the protein (i.e., EGFR-dependent signalling) (Shi et al. 2000). Ectopic overexpression of CD9 in human fibrosarcoma cells correlated with transcriptional down-regulation of WAVE2 (Huang et al. 2006), a member of the WASP (Wiskott-Aldrich syndrome proteins) family of proteins, which act upon actin cytoskeleton and play a critical role in lamellipodium and filipodium formation. CD9 can also affect tumour cell motility through down-regulation of WISP-1 and MMP26, downstream targets of the Wnt signalling pathway associated with aggressive tumour growth (Yamamoto et al. 2004).

Finally, CD9 may affect the transendothelial migration of tumour cells. CD9, CD81 and CD151 co-localize at the tumour cell—endothelial cell contact area, where CD9 promotes strong adhesion via β1 integrins, which hampers transendothelial migration of the tumour cell (Longo et al. 2001). On the other hand, although down-regulated in metastases, high level CD9 expression at tumour cones can support transendothelial migration in cervical carcinoma and recovery of these cone-localized CD9 "hot spots" is a highly significant indicator of lymphangiogenesis (Sauer et al. 2003). Strong CD9 expression is also observed on myeloma cells in close contact to bone marrow endothelial cells (De Bruyne et al. 2006). The reason(s) for these opposing observations likely rely on differences in the CD9–containing "web" of individual tumour cells.

11.4.1.2 Tumour-Related, Migration-Independent Activities of CD82, CD81, CD9 and CD63

For CD82 two additional, migration-independent mechanisms have been described, whereby CD82 interferes with tumour progression. The first of these involves KITENIN, an unrelated four-transmembrane domain protein. Over-expression of KITENIN in a murine colon carcinoma line promotes adhesion to ECM ligands, tumour cell migration and metastasis (Rowe and Jackson 2006). By a not yet fully defined mechanism, binding of KITENIN to the C-terminal tail of CD82 appears to interfere with its metastasis-promoting activity (Lee et al. 2004). Secondly, CD82 interacts in trans with DARC (Duffy antigen receptor for chemokines) on vascular endothelial cells. This induces tumour cell senescence via reduced expression of the senescence related transcription factor TBX2 (T-box 2) gene and up-regulation of the cyclin-dependent kinase inhibitor $p21^{WAF1}$, which is repressed by TBX2 (Prince et al. 2004). Accordingly, the metastasis-suppressor activity of CD82 is significantly reduced in $DARC^{-/-}$ mice (Bandyopadhyay et al. 2006).

While activities of CD82 in tumour cells are largely restricted to the metastatic process, CD9 can have an impact on tumourigenicity. Transformation of chicken or mouse fibroblasts with v-Jun suppresses transcription of GM3 synthase (Miura et al. 2004). Consequently, Jun-induced oncogenic transformation is accompanied by loss of the CD9-GM3 association. This leads to integrin activation, enhanced cell motility, and increased capacity for soft agar colony formation. Transfection with the GM3 synthetase gene, which reverts the oncogenic phenotype, is accompanied by re-establishment of the CD9-GM3 association. Tumour growth inhibition by CD9 may also rely on CD9-dependent regulation of expression of tumour necrosis factor (TNF)α whose production is delayed in $CD9^{-/-}$ mice (Yamane et al. 2005). A similar phenomenon has been described in the hepatic carcinoma cell line H22 (Li et al. 2006). How CD9 influences transmembrane TNFα activity has not been clarified. However, it has been shown that CD9 and ADAM17 can associate and that CD9 negatively regulates ADAM17 sheddase activity on TNFα (Gutiérrez-López et al. 2011; Moss and Bartsch 2004).

CD9 might also interfere with EMT the initiating step of the metastatic cascade. CD9 expression in HT1080 and A549 cells was shown to induce down-regulation of several Wnt family genes, such as *Wnt1*, *Wnt2b1* and *Wnt5a* and their targets including WISP-1, WISP-3, c-Myc, VEGF-A and MMP26. Wnt proteins are a large family of secreted glycoproteins that activate signal transduction pathways to control a wide variety of cellular processes such as determination of cell fate, proliferation, migration, and polarity (Coombs et al. 2008). There is evidence that CD9 is involved in the downregulation of several Wnt family genes, as well as of the rac GTPase regulated WAVE-2, which results in suppression of transformation and EMT (Huang et al. 2004, 2006). More recently it has been shown that by the association of glycoprotein 90 K with CD9 and CD82 the Wnt/β-catenin pathway becomes suppressed via a novel proteasomal-ubiquitination pathway (Lee et al. 2010).

11.4.2 Tumour Progression Promoting Activities of CD151 and Tspan8

In contrast to CD82, two tetraspanins, CD151 and Tspan8, have consistently been reported to promote tumour progression, where the main activity, particularly of CD151 is linked to tumour cell motility and invasiveness.

The first evidence for CD151 as a metastasis promoting molecule derived from a study in which an anti-CD151 antibody inhibited metastasis of a human epidermoid carcinoma line in a chick embryo model. The antibody inhibited cell migration without having any effect on cell adhesion or cell growth (Testa et al. 1999). Subsequently, an association between high CD151 expression and a poor prognosis has been described for many cancers (Sect. 11.2).

The metastasis promoting activity of CD151 mostly relies on its effect on tumour cell migration. Several lines of evidence point towards a link between MMPs and CD151. CD151 contributes to pericellular activation of MMPs by associating with proMMP7. This results in activation of MMP7, a phenomenon which can be prevented by anti-CD151 antibodies (Shiomi et al. 2005). In addition, CD151 has a positive effect on MMP9 expression through the mechanisms involving FAK, Src, p38 and JNK kinases. Signalling is initiated via CD151-associated integrin α3β1 or α6β1 and is stimulated by CD151 homophilic interactions (Hong et al. 2006; Yang et al. 2008). Reduced expression of MMP2, MMP7 and MMP9 in a CD151-knockdown carcinoma line confirmed the involvement of CD151 in MMP expression, complex formation and co-localization at the leading edge of lamellipodia (Shiomi et al 2005; Hasegawa et al. 2007).

Transfection of FAK competent and deficient fibroblasts with CD151 cDNA provided evidence that FAK is needed for CD151 mediated increased migration, Matrigel invasion and metastasis (Kohno et al. 2002). Further studies confirmed that CD151 is important for proper localization of laminin5-binding integrins during tumour cell–stromal cell interactions. Upon EGFR stimulation CD151 and α3β1 become internalized in HSC5 epidermal carcinoma cells. Furthermore, in HSC5-CD151-knockdown cells, α3β1 is partially internalized, α6β4 is redistributed and MMP2, MMP7 and MMMP9 expression is downregulated (Hasegawa et al. 2007). The authors speculate that CD151 might contribute to cell migration by inducing integrin re-localization and MMP production. In line with this is the finding that CD151-knockdown A431 epidermoid carcinoma cells display impaired motility, anomalously persistent adhesive contacts and impaired integrin α3β1 internalization (Winterwood et al. 2006). Notably, too, CD151 regulates glycosylation of α3β1. CD151 knockdown cells with reduced α3β1 glycosylation show strongly impaired migration towards laminin (Baldwin et al. 2008). Confirming the importance of CD151 for integrin traffic, expression of a CD151 molecule with a mutation of the sorting motif in the C-terminal domain markedly attenuates endocytosis of CD151-associated integrins such as α3β1, α5β1 and α6β1 (Liu et al. 2007). Thus, CD151 plays a critical role in integrin recycling as a mechanism to regulate tumour cell migration.

CD151 is also an important regulator of collective tumour cell migration. Monolayers of CD151 knockdown A431 cells display strikingly increased remodelling rates and junctional instability, which is caused by excessive RhoA activation and loss of actin organization at cell-cell junctions. There is evidence that CD151 regulates the stability of tumour cell-cell interaction through its association with integrin α3β1 (Johnson et al. 2009).

Quantitative in vivo assays and intravital imaging using the chicken chorioallantoic membrane model confirmed the impact of CD151 on tumor cell migration. A CD151-specific antibody inhibits matrix-mediated migration, but has no impact on extravasation. Migration inhibition is due to a failure to detach at the rear end. As migration of CD151-knockout cells was not affected, the authors suggested that—when present—CD151 might recruit partner molecules that control de-adhesion, but this process is suppressed in the presence of the CD151 antibody (Zijlstra et al. 2008).

Taken together, CD151 regulates cell migration, mostly through its association with integrins α3β1, α6β4 and MMPs. The TEM location, which facilitates the recruitment of integrins, additional transmembrane and cytosolic proteins in multimolecular complexes, contributes to this dominating theme (Hemler 2005).

Far less is known about the engagement of Tspan8 in tumour cell motility. Tspan8 associates with CD9, CD81, CD151 and several integrins including α3β1 and α6β4, but the integrin associations are probably indirect. Known non-integrin Tspan8-associated molecules are EWI-F, EpCAM, CD13, CD44, PKC and PI4KII (Claas et al. 2005; Zöller 2009).

Tspan8 over-expression in tumours correlates with poor differentiation and metastasis (Sect. 11.2). Tspan8 can support tumour cell proliferation, protection from apoptosis, and induction of angiogenesis and can enhance tumour cell motility. Tspan8-promoted tumour cell motility and liver metastasis may involve its association with integrin α6β4, as it is only seen in tumour cell lines that over-express both Tspan8 and α6β4 (Herlevsen et al. 2003; Gesierich et al. 2005). Tspan8 associates with integrin α6β4 only after PMA stimulation and disassembly of hemidesmosomes, which is accompanied by transient internalization of the Tspan8-α6β4 complex and increased motility (Huerta et al. 2003; Herlevsen et al. 2003). This continuing internalization to the endosomal compartment and rapid recycling back to the cell surface via a short loop recycling machinery under the control of rab4 has been described for several integrins (Caswell and Norman 2008). It may well account for the motility promoting activity of Tspan8.

11.5 CD151 and Tspan8, Tumour Growth and Angiogenesis

11.5.1 The Impact of CD151 and Tspan8 on Tumour Cell Proliferation and Apoptosis Protection

In a cellular model for mammary ductal carcinoma in situ, CD151 was found to support proliferation in a process that does not require direct contact with α3β1 integrin. Depletion of CD151 is accompanied by partial restoration of cell polarity and reduced ERK1/2 and Akt phosphorylation (Novitskaya et al. 2010).

Increased Tspan8 expression in a dedifferentiated rat hepatoma cell line promotes proliferation (Tanaka et al. 2002). Furthermore, interactions with platelets were suggested to provide tumour cells with a shield, which could provide a survival advantage in the hostile environment encountered during metastatic spread (Kanetaka et al. 2003). High Tspan8 expression may be also associated with increased apoptosis resistance (Huerta et al. 2003; Kuhn et al. 2007), which is likely to occur via a Tspan8-associated EpCAM-claudin-7 complex. In human and rat cancer lines, a striking decrease in drug resistance was observed upon knockdown of EpCAM or claudin-7. This was accompanied by reduced PI3K activation and loss of phosphorylation of Akt and downstream anti-apoptotic proteins. Signals are initiated by the recruitment of the EpCAM-claudin-7 complex into TEM, which is accompanied by claudin-7 phosphorylation, possibly via Tspan8-associated PKC (Nübel et al. 2009).

11.5.2 Tetraspanins and Angiogenesis

Angiogenesis defines the process of new capillary formation from a pre-existing vasculature, which is crucial to supply a growing organism with oxygen. Accordingly, the rapid growth of tumour cells essentially requires blood supply (Folkman 2004). Tumour angiogenesis proceeds through several sequential steps. The process is believed to be initiated by angiogenic factors, angiogenin, epidermal growth factor, IL8, TNFα, TGFα, TGFß and VEGF (Hillen and Griffioen 2007), that are produced by tumour cells and bind to endothelial cell (EC) receptors including VEGFR-1, -2, -3 and neuropilins. Stimulated EC grow and secrete matrix degrading enzymes that digest the basement membrane surrounding the vessel. The junctions between EC become altered and EC migrate towards the source of the angiogenic stimulus, e.g., towards the tumour mass. At this stage, sprouting EC are reorganized to form tubes and assemble a new basement membrane. The formation of a lumen is driven by interactions between EC and the extracellular matrix. Molecules involved in this process are, among others, galectin-2, CD31 (PECAM-1) and VE-cadherin (Holderfield and Hughes 2008). An increasing body of recent evidence suggests that tetraspanins may directly regulate the development and functions of the vascular system and the pathogenesis of vascular diseases (Zhang et al 2009).

Several studies have reported that CD151 is important in angiogenesis induction (Dumartin et al. 2010; Takeda et al. 2007b; Zhang et al. 2002, 2009). Though patients with mutations in the CD151 gene and CD151 knockout mice showed no obvious defects in vasculogenesis (Karamatic Crew et al. 2004; Wright et al. 2004; Sachs et al. 2006), defects are seen in angiogenesis. Thus, CD151 expression by the tumour-bearing host facilitates tumour growth due to angiogenesis induction. CD151 supports EC invasiveness, migration, cable formation, matrigel contraction, tube formation and sprouting, activities which are all impaired in CD151 knockout mice (Takeda et al. 2007b). Selective defects in activation on laminin substrates of adhesion-dependent signalling molecules including PKB/c-Akt, e-NOS, Rac and Cdc42 contribute to impaired angiogenesis induction (Takeda et al. 2007b; Zheng and Liu 2007). Also, over-expression of CD151 promotes revascularization and improves blood perfusion in an ischemia model (Lan et al. 2005). Importantly, as in tumour cells, CD151 seems to support functional activity of endothelial cells via the associated integrins, particularly laminin-binding integrins (Liu et al. 2011; Zhang et al. 2009).

In addition to a direct involvement of endothelial CD151, expression of this protein (and other tetraspanins) in tumour cells and tumour-derived exosomes can also play an important role in tumour angiogenesis. In fact, Tspan8 is a strong angiogenesis inducer that contributes to a systemic angiogenic switch by Tspan8 over-expressing tumour cells as well as by exosomes derived thereof (Gesierich et al. 2006). The precise mode of activity of exosomal tetraspanins has not yet been explored. However, we will propose our hypothesis in the following section.

11.5.3 Tetraspanins and Thrombosis

Tumour vessels frequently have thin walls, an incomplete basement membrane and decreased numbers of pericytes, cells that are associated with microvasculature. As a consequence, tumour vessels are leaky, which allows for the extravasation of plasma proteins that form a scaffold for newly migrating EC. The leakiness of the EC layer also facilitates initiation of thrombus formation. Spontaneously occurring focal haemorrhages are a common feature of tumour vessels (Franchini et al. 2007) and a prothrombotic state that can culminate in disseminated intravascular coagulation is frequent in cancer patients, where tumour-initiated angiogenesis and the leakiness of tumour vessels are considered to be important (De Cicco 2004). Knowledge of factors regulating angiogenesis and coagulation has strengthened the expectation that these two systems are closely interconnected. The coagulation cascade is initiated when tissue factor, the principal initiator of coagulation, which is provided by many tumour cells, becomes exposed to plasma components. The cascade ends with platelet bound prothrombin becoming converted to thrombin that initiates clot formation by catalysing fibrinogen cleavage and fibrin polymerization. Tumour angiogenesis facilitates blood clotting through the hyperpermeability of tumour endothelium and the leakage of fibrinogen and other clotting agents. Activated platelets in turn support angiogenesis by releasing pro-angiogenic factors like VEGF and angiopoietin-1. Thrombin also supports angiogenesis by cleaving PAR-1 on EC thereby inducing activation and secretion of proteases including MMPs and uPA (Tsopanoglou and Maragoudakis 2007). Taken together, the particular features of tumour vessels support thrombus formation and the coagulation cascade provides a feedback for angiogenesis induction (Ruf and Mueller 2006), which is supported by platelet-derived tetraspanins.

CD63, CD9 and CD151 are abundantly expressed on platelets (Griffith et al. 1991; Fitter et al. 1995; Schröder et al. 2009). Whereas CD151 is required for efficient platelet activation/aggregation (Lau et al. 2004; Orlowski et al. 2009), $CD9^{-/-}$ mice show alteration in blood coagulation, where CD9 appears to prevent excessive thrombus growth, but does not appear to play a critical role in primary hemostasis (Mangin et al. 2009). From the viewpoint of tetraspanin engagement in tumour cell dissemination, the more interesting aspect relies on the isolated tumour cell within the blood stream taking advantage of CD9 down-regulation. CD9 associates with the platelet aggregation-inducing factor podoplanin. Ectopic expression of CD9 in podoplanin-expressing tumour cells leads to reduced lung metastasis formation accompanied by impaired tumour-induced platelet aggregation (Nakazawa et al. 2008). Platelets bind via CLEC-2 (C-type lectin-like receptor-2) to podoplanin, which induces platelet degranulation (Suzuki-Inoue et al. 2006). Because CLEC-2 is unable to recognize CD9-associated podoplanin (Nakazawa et al. 2008), platelet aggregation will be impaired upon contact with tumour cells expressing both CD9 and podoplanin. Consequently, formation of tumour cell platelet aggregates, which facilitates embolization of the microvasculature and metastasis formation, will also be suppressed. Decrease in the formation of these aggregates will also make tumour cells more susceptible to a host anti-tumour immune attack (Sierko and Wojtukiewicz 2007).

Finally, platelet-derived exosomes constitute about 70–90% of circulating exosomes in the plasma (Berckmans et al. 2001) with a life span of about 30 min (Flaumenhaft 2006). The procoagulant activity of platelet-derived exosomes is well known. Specifically, it has been suggested that exosomes provide negatively charged phospholipids, which are required for factor IXa and Xa activation (Shet et al. 2003). Though still controversial, the therapeutic efficacy of anti-glycoprotein IIb/IIIa could be a consequence of altered platelet exosome formation (Morel et al. 2004; Razmara et al. 2007). The abundance of platelet-derived exosomes and their functional activity in coagulation implies that they may contribute to the prothrombotic state frequently seen in cancer patients. It remains to be explored whether platelet-derived exosomal CD151, CD9, Tspan32 and CD63 contribute to the procoagulant activity. On the other hand, tumour-derived exosomes may also be of utmost importance for platelet activation. Thus, rats transplanted with a Tspan8 over-expressing tumour line develop disseminated intravascular coagulation, which could be prevented by a Tspan8-specific antibody (Claas et al. 1998). Though the underlying mechanism remains to be elaborated, it is tempting to speculate that exosomal Tspan8 contributes to platelet activation.

Taken together, the engagement of tetraspanins in angiogenesis and thrombosis has only recently received attention and work so far covers only few members of the tetraspanin family. Nonetheless, data gathered so far hold promise for a wealth of information in the near future. Since angiogenesis and thrombosis are important parameters in oncology, this knowledge may well lead to new therapeutic options.

11.6 Perspective: Tetraspanins and Exosomes

One feature of tetraspanins, though well known, has received little attention so far. Tetraspanins are enriched in exosomes and we consider it very likely that exosomal tetraspanins play a major role in exosomal message delivery.

11.6.1 Exosomes

Exosomes, small 30–100 nm vesicles, which are believed to derive from fusion of the intraluminal vesicles of multivesicular bodies (MVB) with the plasma membrane (Fevrier and Raposo 2004; de Gassart et al. 2004; Lakkaraju and Rodriguez-Boulan

2008). The molecular composition of exosomes reflects their origin from intraluminal vesicles (Johnstone 2006). Besides a common set of membrane and cytosolic molecules, which includes several tetraspanins, including CD9, CD37, CD53, CD63, CD81, CD82, CD151 and Tspan8, exosomes harbor subsets of proteins, such as adhesion molecules, molecules associated with vesicle transport, cytoskeletal proteins, signal transduction molecules, enzymes and others that are linked to cell type-specific functions (Schorey and Bhatnagar 2008; Mathivanan et al. 2010). Importantly, exosomal proteins maintain their functional activity, including antigen presentation, peptide and protein cleavage (Potolicchio et al. 2005; Stoeck et al. 2006). Another notable feature is the presence of phosphatidylserine at the exosomes' outer membrane leaflet which can trigger exosome uptake by cells expressing phosphatidylserine-binding proteins (scavenger receptors, integrins, complement receptors) (Zakharova et al. 2007). Exosomes contain mRNA and miRNA (so called shuttle RNAs) which are transferred to the target cell, where they can be translated or mediate RNA silencing (Ratajczak et al. 2006; Deregibus et al. 2007; Valadi et al. 2007; Burghoff et al. 2008). Exosome-mediated transfer of DNA to their target cells is specific, so that RNA is transcribed in one, but not another type of cells (Simons and Raposo 2009). In addition, the relative abundance of proteins, mRNA and miRNAs differs between exosomes and the cells from which they are derived. This implies active sorting into MVB (Lakkaraju and Rodriguez-Boulan 2008), which for proteins can be achieved by mono-ubiquitination, localization in cholesterol-rich membrane microdomains, or higher order oligomerization (Gruenberg and Stenmark 2004; Hurley and Emr 2006; Fang et al. 2007; Smalheiser 2007). The mechanisms underlying selective sorting of mRNA and miRNA into exosomes are unknown (Subra et al 2007). Thus, exosomes constitute a most potent mode of intercellular communication that has become appreciated as important in immunity (André et al. 2002), cell-to-cell spread of infectious agents (Johnstone 2006; Schorey and Bhatnagar 2008) and tumour progression (Zöller 2006). Accordingly, therapeutic exploitation of exosomes appears very promising and is already in clinical use as a vaccine strategy (Icro et al. 2008). Exosomes may also be the most potent gene delivery system (Belting and Wittrup 2008; Simpson et al. 2009; Pap et al. 2009; Seow and Wood 2009; Xiao et al. 2009).

11.6.2 Exosomal Tetraspanins

Tetraspanins are abundantly recovered in intracellular vesicles and exosomes (Escola et al. 1998; Sincock et al. 1999; Hemler 2003; Berditchevski and Odintsova 2007; Pols and Klumperman 2009; Zöller 2009). Some tetraspanins possess a tyrosine-based sorting motif, a sequence of Tyr-Xaa-Xaa-ϕ where ϕ stands for an AA with a bulky hydrophobic side chain, in the C-terminal cytoplasmic domain (Marks et al. 1997). By this sorting motif, these tetraspanins are predisposed for delivery to intracellular compartments (Marks et al. 1997; Berditchevski and Odintsova 2007). However, some tetraspanins enriched in exosomes do not possess a sorting motif

(CD9) or have an inappropriately located sorting motif (Tspan8) (Berditchevski and Odintsova 2007). This partial independence of a sorting motif indicates that individual tetraspanins likely follow different routes of internalization. Molecular mechanisms controlling trafficking routes of tetraspanins and associated proteins are reviewed in detail in another chapter of this volume.

Irrespective of the donor cell type, tetraspanins are enriched in exosomes and the tetraspanin web is mostly maintained in them (Abache et al. 2007). Whether tetraspanins are involved in sorting of proteins, mRNA or miRNA to exosomes is currently unknown (Gibbings et al. 2009; Simons and Raposo 2009).

11.6.3 Exosomal Tetraspanins, the Premetastatic Niche and Angiogenesis

Evidence has started to emerge that tetraspanins are important in target cell selection during premetastatic niche formation as well as tumour-angiogenesis. Lodgement of metastasizing tumour cells is facilitated by the establishment of special niches in (pre)metastatic organs (Bissell and Labarge 2005). Niche preparation involves stimulation of local fibroblasts by tumour-derived factors and chemokines that attract tumour cells and hematopoietic progenitors (Kaplan et al. 2006). Nonetheless, information on long-distance communication between a tumour and host organs is still limited and exosomes have been suggested to contribute to premetastatic niche formation as well as tumour-associated angiogenesis and thrombosis (Aharon and Brenner 2009; Al-Nedawi et al. 2009). Notably, under hypoxia tumour cells have been described to secrete exosomes enriched in Tspan15, CD9 and CD81, which have a major impact on the tumour microenvironment such that angiogenesis and metastatic potential becomes increased (Park et al. 2010).

An involvement of exosomes in metastasis was first described for platelet-derived exosomes. These exosomes transferred the αIIb integrin chain to lung cancer cells, stimulated the MAPK pathway and increased expression of MT1-MMP, cyclin D2 and angiogenic factors as well as enhancing adhesion to fibrinogen and human umbilical vein endothelial cells (Janowska-Wieczorek et al. 2005). A direct transfer of metastatic capacity by exosomes was demonstrated for B16 melanoma cells. Exosomes derived from a highly metastatic variant transferred metastatic capacity to low metastatic B16F1 cells. Lung metastasis formation by B16F1 was accompanied by protein uptake from exosomes of the metastasizing subclone (Hao et al. 2006). Tspan8 and/or CD151-containing exosomes also contribute to premetastatic niche formation. After subcutaneous application of CD151- and Tspan8-enriched exosomes together with a soluble tumour matrix, exosomes supported recruitment of hematopoietic progenitors from the bone marrow as well as activation of stroma cells and leukocytes in premetastatic lymph nodes such that a non-metastatic tumour line settled and formed metastases (Jung et al. 2009). Ongoing work aims to define the contribution of exosomal CD151, Tspan8 and associated integrins in target cell selection and binding.

While the question of target cell selection and the contribution of tetraspanins remains to be defined in premetastatic niche preparation, initiation of tumour-angiogenesis was shown to require Tspan8 in a rat adenocarcinoma model. Only Tspan8-expressing exosomes interact with endothelial cells. Furthermore, binding to and uptake by endothelial cells is dependent on the formation of the integrin $\alpha 4\beta 1$-Tspan8 complex. The uptake of Tspan8-bearing exosomes by EC is accompanied by transient recovery of mRNA selectively enriched in the exosomes and initiates transcription of several angiogenesis-related genes, proliferation, migration and sprouting of endothelial cells. Importantly, Tspan8-positive exosomes also bind to endothelial cell progenitors and promote endothelial cell progenitor maturation (Nazarenko et al. 2010).

Exosomes are easy to manipulate and provide a powerful means of protein and gene transfer. Thus, it becomes crucial to further explore the engagement of tetraspanins in target cell selection. This would offer a powerful means to interfere with pathological angiogenesis and metastasis, two major targets in cancer therapy (Pap et al. 2009; Zöller 2009).

11.7 Tetraspanin Based Therapeutic Options

Taking into account the importance of some tetraspanins in tumour progression and angiogenesis, it is important to consider these molecules as therapeutic targets. Due to their mode of activity as molecular facilitators in a wide range of cell types (Maecker et al. 1997), this will not be an easy task. As some tetraspanins function as metastasis suppressors, while others promote metastasis, we will discuss these two aspects of tetraspanin-based therapies.

11.7.1 Rescuing Metastasis Suppressor Genes

CD82 inhibits migration and invasion by associating directly or via bridging integrins with a multitude of different molecules as well as by the recruitment of the partner molecules in TEM. Some of the CD82-based interactions involve transmembrane domains. CD82 inhibits formation of microprotrusions and the release of microvesicles. Mutations of three polar residues in the transmembrane domains of CD82 disrupt these inhibitions (Bari et al. 2009). The authors provide evidence that the transmembrane interactions mediated by these polar residues determine a conformation either in or near the transmembrane regions and that this conformation is needed for the intrinsic activity of CD82. They speculate that a therapeutic perturbation of CD82 transmembrane interactions may open a new avenue to prevent cancer invasion.

Rescuing CD82 gene expression also should prevent tumour progression, which, however, requires an awareness of the regulation of CD82 gene transcription as well as of the mechanisms that down-regulate CD82 expression in tumour cells (Tonoli and Barrett 2005; Liu and Zhang 2006) (Sect. 11.3).

Besides reviving CD82 expression at the transcriptional level, CD82 expression may also be rescued by proteasome inhibitors or by targeting specific components of the ubiquitin system, as ubiquitin ligase gp78 which regulates CD82 expression (Tsai et al. 2007).

The potential therapeutic efficacy of CD82 has already been demonstrated. Thus, nerve growth factor has been shown to rescue CD82 expression, which was accompanied by abrogation of tumourigenicity of prostate cancer cell lines (Sigala et al. 1999). Furthermore, CD82 transfected murine Lewis Lung carcinoma cells lose the capacity to form lymph node metastasis. Even more strikingly, intratracheal administration of adenovirus encoding CD82 or CD9 cDNA in mice orthotopically preimplanted with LLC cells dramatically reduced metastases without affecting growth of the primary tumor (Takeda et al. 2007a).

11.7.2 Interfering with Metastasis and Angiogenesis Promoting Activities of Tetraspanins

Therapeutic approaches aimed at interference with metastasis promoting activities of tetraspanins are mostly based on antibodies, recombinant soluble ECL2 or post-transcriptional gene silencing via siRNA (Hemler 2008; Stipp 2010).

Some tetraspanin-specific antibodies have been shown in several instances to be of potential clinical relevance. Intratumoural application of anti-CD9 inhibited colon carcinoma growth and intravenous application of anti-CD9 inhibited the subcutaneous growth of gastric cancer cell lines (Ovalle et al. 2007; Nakamoto et al. 2009), anti-CD37 improved the survival of B-CLL xenografted mice (Levy et al. 1998) and anti-CD151 interfered with metastasis formation (Testa et al. 1999; Kohno et al. 2002; Zijlstra et al. 2008). Though the underlying mechanisms have not been fully elucidated, it has been suggested that antibodies may interfere with the lateral associations of tetraspanins or promote clustering of tetraspanins and tetraspanin-associated molecules in TEM and thereby interfere with the activity not only of the targeted tetraspanin, but also of associated molecules including cytoplasmic partners. In line with this suggestion, tetraspanin antibodies have in some instances been shown to exert stronger effects than the knockout of an individual tetraspanin, e.g. anti-CD81 has been shown to interfere, besides others, with T and B cell activities, but only the B cell response was impaired in CD81 knockout mice (Oren et al. 1990; Boismenu et al. 1996; Miyazaki et al. 1997; Tsitsikov et al. 1997; Levy et al. 1998). Taking this into account, one has to be aware that the activity of tetraspanin-specific antibodies may vary depending on the recognized epitope (Serru et al. 1999; Yauch et al. 2000; Geary et al. 2001), which may enhance or block the effect of a tetraspanin as demonstrated for anti-CD151 promoting adhesion (Zijlstra et al. 2008) and for anti-CD9 that can amplify the tumour suppressor function (Ovalle et al. 2007).

Besides their blocking or enhancing activity, tetraspanin-specific antibodies repeatedly have been described to induce apoptosis (Murayama et al. 2004), for example in a SCID mouse model, where anti-CD9 interferes with gastric cancer growth by exerting anti-proliferative, pro-apoptotic and anti-angiogenic activity (Nakamoto et al. 2009). Anti-tetraspanins also can support complement and antibody-dependent cellular cytotoxicity (Zhao et al. 2007).

Finally, antibodies can be used as drug transporters as reported for ^{131}I-labelled anti-CD37 (Press et al. 1989) or for transporting nanoparticles with siRNA (Peer et al. 2008), which has not yet been explored for tetraspanins.

Taken together, antibodies have proven in many instances to be a powerful adjuvant cancer therapy (Boyiadzis and Foon 2008). Nonetheless, abundant expression of a molecule, like most tetraspanins, in non-transformed cells can provide a major obstacle (Grünwald et al. 2009). We consider the use of bispecific antibodies that target with both arms the tumour cell as a most promising solution. Such an approach has been used by the group of Hollander for targeting CD44, which is abundantly expressed on many cells. Yet, using anti-CD44/anti-idiotype bispecific antibodies, side effects were avoided and the anti-tumour efficacy was strengthened (Avin et al. 2004). As tetraspanins act as molecular facilitators, this kind of bispecific antibodies can be expected to be highly efficient.

Besides antibodies, the soluble form of the large extracellular domain (ECL2) of tetraspanins as a competitor has mainly been tested with respect to leukocyte endothelial cell interaction via CD9 and CD151 (Barreiro et al. 2005), egg-sperm fusion (Zhu et al. 2002) and virus infectivity, where the ECL2 may be superior to antibodies, as it does not only compete for binding, but additionally exerts functional activity (Molina et al. 2008).

Another therapeutic approach is based on silencing tetraspanins via siRNA. CD9 silencing resulted in pronounced ovarian cancer dissemination (Furuya et al. 2005) and CD151 silencing interfered with integrin-dependent adhesion and migration (Winterwood et al. 2006). Feasibility of this approach has been demonstrated in experiments describing successful lentiviral CD81 shRNA delivery into the nucleus accumbens or the ventral tegmental area of the mesolimbic dopamine system which resulted in a significant decrease in locomotory activity (Bahi et al. 2005).

Therapeutic settings currently being discussed include modulation of amino acids important for transmembrane folding (Tarasova et al. 1999). The authors argue that a therapeutic perturbation of TM interactions may open a new avenue to prevent cancer invasion, which could be far easier approached than a blockade of individual signalling pathways. Modulation of the PDZ domain (Dev 2004; Latysheva et al. 2006), of key interaction sites in the ECL2 (Yauch et al. 2000; Seigneuret 2006), of palmitoylation sites (Berditchevski et al. 2002; Charrin et al. 2002; Yang et al. 2002, 2004; Kovalenko et al. 2005) including targeting of the responsible acyltransferase (Sharma et al. 2008) are additional therapeutic approaches to be discussed. Recently convincing evidence has been provided for different requirements of the CD151-$\alpha3\beta1$ and the CD151-$\alpha6\beta4$ interaction, which would allow to selectively interfere with CD151-$\alpha3\beta1$ adhesion and migration on laminin5 and CD151-$\alpha6\beta4$-mediated stable attachment (Zevian et al. 2011).

Finally, taking into account the increasingly appreciated role of exosomes as intercellular communicators and the strong presence of tetraspanins in exosome membranes, it is tempting to speculate that tetraspanins could be used as an exosome delivery system. This requires further elaboration of the engagement of tetraspanins and the associated molecules that together bind to and become internalized by selective targets (Zöller 2009; Nazarenko et al. 2010). Knowledge of exosome binding and uptake of tetraspanin complexes by selective target cells could enable generation of competitive exosomes carrying desired siRNAs or other drugs that interfere with exosome initiated premetastatic niche preparation, angiogenesis and thrombosis.

In summary, though there are promising concepts, one should be aware that tetraspanin-based therapeutic protocols require sophisticated controls as the composition of TEM may well determine the balance between opposing activities.

11.8 Conclusion

Tetraspanins function as molecular facilitators that assemble a web including many distinct families of transmembrane proteins in specialized membrane microdomains that serve as a scaffold for localised signal transduction and regulation of cytoskeletal dynamics. The reversibility of TEM and their composition, which depends on the cell's activation state, the abundance of associating molecules and their ligands, adds a major constraint in defining tetraspanin functions. Nonetheless, modulation of cell motility, cell fusion and intercellular communication via exosomes may well cover the essential activities of tetraspanins in cancer. The involvement of tetraspanins in these actions basically can follow five routes: (1) Tetraspanins may act as receptors for defined ligands; (2) Tetraspanins are known to directly influence adhesion, signal transduction and/or gene transcription via associated molecules; (3) Tetraspanins indirectly initiate activities via the recruitment of different molecules into TEM, a process that frequently involves gangliosides; (4) Tetraspanins initiate internalization and relocation of associated molecules in distinct membrane regions; (5) Tetraspanins initiate recruitment into MVB and release of TEMs in exosomes, where exosomal tetraspanin and associated molecules may be of major importance in target cell selection and in exosome fusion with the target cell.

Through these different activities, tetraspanins contribute to metastasis inhibition and promotion, to premetastatic niche formation, to angiogenesis and the tumour-associated prothrombotic state. Nonetheless, one of the key questions, why some tetraspanins suppress (CD82) or promote (CD151, Tspan8) tumour progression remains unanswered. In addition, application of tetraspanins as tumour biomarkers and therapeutic targets in human cancer will also require a better understanding of their association with different tumour subsets (e.g., CD82 in ER-positive and negative breast cancer) and the pattern of modulation of their levels at different stages of disease. Answering these questions may provide a solid ground for therapeutic interference with tetraspanin activities in tumour progression.

References

Abache T, Le Naour F, Planchon S, Harper F, Boucheix C, Rubinstein E (2007) The transferrin receptor and the tetraspanin web molecules CD9, CD81, and CD9P-1 are differentially sorted into exosomes after TPA treatment of K562 cells. J Cell Biochem 102:650–664

Aharon A, Brenner B (2009) Microparticles, thrombosis and cancer. Best Pract Res Clin Haematol 22:61–69

Albini A, Mirisola V, Pfeffer U (2008) Metastasis signatures: genes regulating tumor-microenvironment interactions predict metastatic behavior. Cancer Metastasis Rev 27:75–83

Al-Nedawi K, Meehan B, Rak J (2009) Microvesicles: messengers and mediators of tumor progression. Cell Cycle 8:2014–2018

André F, Schartz NE, Chaput N, Flament C, Raposo G, Amigorena S, Angevin E, Zitvogel L (2002) Tumor-derived exosomes: a new source of tumor rejection antigens. Vaccine 20(Suppl 4):A28–A31

Ang J, Lijovic M, Ashman LK, Kan K, Frauman AG (2004) CD151 protein expression predicts the clinical outcome of low-grade primary prostate cancer better than histologic grading: a new prognostic indicator? Cancer Epidemiol Biomarkers Prev 13:1717–1721

Arduise C, Abache T, Li L, Billard M, Chabanon A, Ludwig A, Mauduit P, Boucheix C, Rubinstein E, Le Naour F (2008) Tetraspanins regulate ADAM10-mediated cleavage of TNF-alpha and epidermal growth factor. J Immunol 181:7002–7013

Arencibia JM, Martin S, Perez-Rodriguez FJ, Bonnin A (2009) Gene expression profiling reveals overexpression of TSPAN13 in prostate cancer. Int J Oncol 34:457–463

Atkinson B, Ernst CS, Ghrist BF, Herlyn M, Blaszczyk M, Ross AH, Herlyn D, Steplewski Z, Koprowski H (1984) Identification of melanoma-associated antigens using fixed tissue screening of antibodies. Cancer Res 44:2577–2581

Avin E, Haimovich J, Hollander N (2004) Anti-idiotype x anti-CD44 bispecific antibodies inhibit invasion of lymphoid organs by B cell lymphoma. J Immunol 173:4736–4743

Bahi A, Boyer F, Kolira M, Dreyer JL (2005) In vivo gene silencing of CD81 by lentiviral expression of small interference RNAs suppresses cocaine-induced behaviour. J Neurochem 92:1243–1255

Baldwin G, Novitskaya V, Sadej R, Pochec E, Litynska A, Hartmann C, Williams J, Ashman L, Eble JA, Berditchevski F (2008) Tetraspanin CD151 regulates glycosylation of (alpha)3(beta)1 integrin. J Biol Chem 283:35445–35454

Bandyopadhyay S, Zhan R, Chaudhuri A, Watabe M, Pai SK, Hirota S, Hosobe S, Tsukada T, Miura K, Takano Y, Saito K, Pauza ME, Hayashi S, Wang Y, Mohinta S, Mashimo T, Iiizumi M, Furuta E, Watabe K (2006) Interaction of KAI1 on tumor cells with DARC on vascular endothelium leads to metastasis suppression. Nat Med 12:933–938

Bari R, Zhang YH, Zhang F, Wang NX, Stipp CS, Zheng JJ, Zhang XA (2009) Transmembrane interactions are needed for KAI1/CD82-mediated suppression of cancer invasion and metastasis. Am J Pathol 174:647–660

Barreiro O, Yáñez-Mó M, Sala-Valdés M, Gutiérrez-López MD, Ovalle S, Higginbottom A, Monk PN, Cabañas C, Sánchez-Madrid F (2005) Endothelial tetraspanin microdomains regulate leukocyte firm adhesion during extravasation. Blood 105:2852–2861

Bass R, Werner F, Odintsova E, Sugiura T, Berditchevski F, Ellis V (2005) Regulation of urokinase receptor proteolytic function by the tetraspanin CD82. J Biol Chem 280:14811–14818

Belting M, Wittrup A (2008) J nanotubes, exosomes, and nucleic acid-binding peptides provide novel mechanisms of intercellular communication in eukaryotic cells: implications in health and disease. Cell Biol 183:1187–1191

Berckmans RJ, Neiuwland R, Böing AN, Romijn FP, Hack CE, Sturk A (2001) Cell-derived microparticles circulate in healthy humans and support low grade thrombin generation. Thromb Haemost 85:639–646

Berditchevski F (2001) Complexes of tetraspanins with integrins: more than meets the eye. J Cell Sci 114:4143–4151

Berditchevski F, Odintsova E (2007) Tetraspanins as regulators of protein trafficking. Traffic 8:89–96

Berditchevski F, Odintsova E, Sawada S, Gilbert E (2002) Expression of the palmitoylation-deficient CD151 weakens the association of alpha 3 beta 1 integrin with the tetraspanin-enriched microdomains and affects integrin-dependent signaling. J Biol Chem 277:36991–37000

Bergers G, Brekken R, McMahon G, Vu TH, Itoh T, Tamaki K, Tanzawa K, Thorpe P, Itohara S, Werb Z, Hanahan D (2000) Matrix metalloproteinase-9 triggers the angiogenic switch during carcinogenesis. Nat Cell Biol 2:737–744

Bissell MJ, Labarge MA (2005) Context, tissue plasticity, and cancer: are tumor stem cells also regulated by the microenvironment? Cancer Cell 7:17–23

Boismenu R, Rhein M, Fischer WH, Havran WL (1996) A role for CD81 in early T cell development. Science 271:198–200

Bouras T, Frauman AG (1999) Expression of the prostate cancer metastasis suppressor gene KAI1 in primary prostate cancers: a biphasic relationship with tumour grade. J Pathol 188:382–388

Boyiadzis M, Foon KA (2008) Approved monoclonal antibodies for cancer therapy. Expert Opin Biol Ther 8:1151–1158

Brabletz T, Jung A, Spaderna S, Hlubek F, Kirchner T (2005) Opinion: migrating cancer stem cells—an integrated concept of malignant tumour progression. Nat Rev Cancer 5:744–749

Bredel M, Bredel C, Juric D, Harsh GR, Vogel H, Recht LD, Sikic BI (2005) Functional network analysis reveals extended gliomagenesis pathway maps and three novel MYC-interacting genes in human gliomas. Cancer Res 65:8679–8689

Burghoff S, Ding Z, Gödecke S, Assmann A, Wirrwar A, Buchholz D, Sergeeva O, Leurs C, Hanenberg H, Müller HW, Bloch W, Schrader J (2008) Horizontal gene transfer from human endothelial cells to rat cardiomyocytes after intracoronary transplantation. Cardiovasc Res 77:534–543

Calaluce R, Gubin MM, Davis JW, Magee JD, Chen J, Kuwano Y, Gorospe M, Atasoy U (2010) The RNA binding protein HuR differentially regulates unique subsets of mRNAs in estrogen receptor negative and estrogen receptor positive breast cancer. BMC Cancer 10:126

Caswell P, Norman J (2008) Endocytic transport of integrins during cell migration and invasion. Trends Cell Biol 18:257–263

Charrin S, Manié S, Oualid M, Billard M, Boucheix C, Rubinstein E (2002) Differential stability of tetraspanin/tetraspanin interactions: role of palmitoylation. FEBS Lett 516:139–144

Chen Z, Mustafa T, Trojanowicz B, Brauckhoff M, Gimm O, Schmutzler C, Kohrle J, Holzhausen HJ, Kehlen A, Klonisch T, Finke R, Dralle H, Hoang-Vu C (2004) CD82, and CD63 in thyroid cancer. Int J Mol Med 14:517–527

Chen L, Li X, Wang GL, Wang Y, Zhu YY, Zhu J (2008) Clinicopathological significance of overexpression of TSPAN1, Ki67 and CD34 in gastric carcinoma. Tumori 94:531–538

Chen L, Zhu YY, Zhang XJ, Wang GL, Li XY, He S, Zhang JB, Zhu JW (2009) TSPAN1 Protein expression: a significant prognostic indicator for patients with colorectal adenocarcinoma. World J Gastroenterol 15:2270–2276

Chen L, Zhu Y, Li H, Wang GL, Wu YY, Lu YX, Qin J, Tuo J, Wang JL, Zhu J (2010) Knockdown of TSPAN1 by RNA silencing and antisense technique inhibits proliferation and infiltration of human skin squamous carcinoma cells. Tumori 96:289–295

Chien CW, Lin SC, Lai YY, Lin BW, Lin SC, Lee JC, Tsai SJ (2008) Regulation of CD151 by hypoxia controls cell adhesion and metastasis in colorectal cancer. Clin Cancer Res 14:8043–8051

Christgen M, Bruchhardt H, Ballmaier M, Krech T, Langer F, Kreipe H, Lehmann U (2008) KAI1/CD82 is a novel target of estrogen receptor-mediated gene repression and downregulated in primary human breast cancer. Int J Cancer 123:2239–2246

Christgen M, Christgen H, Heil C, Krech T, Langer F, Kreipe H, Lehmann U (2009) Expression of KAI1/CD82 in distant metastases from estrogen receptor-negative breast cancer. Cancer Sci 100:1767–1771

Claas C, Seiter S, Claas A, Savelyeva L, Schwab M, Zöller M (1998) Association between the rat homologue of CO-029, a metastasis-associated tetraspanin molecule and consumption coagulopathy. J Cell Biol 141:267–280
Claas C, Wahl J, Orlicky DJ, Karaduman H, Schnölzer M, Kempf T, Zöller M (2005) The tetraspanin D6.1A and its molecular partners on rat carcinoma cells. Biochem J 389:99–110
Coombs GS, Covey TM, Virshup DM (2008) Wnt signaling in development, disease and translational medicine. Curr Drug Targets 9:513–531
De Bruyne E, Andersen TL, De Raeve H, Van Valckenborgh E, Caers J, Van Camp B, Delaisse JM, Van Riet I, Vanderkerken K (2006) Endothelial cell-driven regulation of CD9 or motility-related protein-1 expression in multiple myeloma cells within the murine 5T33MM model and myeloma patients. Leukemia 20:1870–1879
De Bruyne E, Bos TJ, Asosingh K, Vande Broek I, Menu E, Van Valckenborgh E, Atadja P, Coiteux V, Leleu X, Thielemans K, Van Camp B, Vanderkerken K, Van Riet I (2008) Epigenetic silencing of the tetraspanin CD9 during disease progression in multiple myeloma cells and correlation with survival. Clin Cancer Res 14:2918–2926
De Cicco M (2004) The prothrombotic state in cancer: pathogenic mechanisms. Crit Rev Oncol Hematol 50:187–196
de Gassart A, Géminard C, Hoekstra D, Vidal M (2004) Exosome secretion: the art of reutilizing nonrecycled proteins? Traffic 5:896–903
Deregibus MC, Cantaluppi V, Calogero R, Lo Iacono M, Tetta C, Biancone L, Bruno S, Bussolati B, Camussi G (2007) Endothelial progenitor cell derived microvesicles activate an angiogenic program in endothelial cells by a horizontal transfer of mRNA. Blood 110:2440–2448
Dev KK (2004) Making protein interactions druggable: targeting PDZ domains. Nat Rev Drug Discov 3:1047–1056
Dong JT, Lamb PW, Rinker-Schaeffer CW, Vukanovic J, Ichikawa T, Isaacs JT, Barrett JC (1995) KAI1, a metastasis suppressor gene for prostate cancer on human chromosome 11p11.2. Science 268:884–886
Dong JT, Suzuki H, Pin SS, Bova GS, Schalken JA, Isaacs WB, Barrett JC, Isaacs JT (1996) Down-regulation of the KAI1 metastasis suppressor gene during the progression of human prostatic cancer infrequently involves gene mutation or allelic loss. Cancer Res 56:4387–4390
Drucker L, Tohami T, Tartakover-Matalon S, Zismanov V, Shapiro H, Radnay J, Lishner M (2006) Promoter hypermethylation of tetraspanin members contributes to their silencing in myeloma cell lines. Carcinogenesis 27:197–204
Dumartin L, Quemener C, Laklai H, Dumartin L, Quemener C, Laklai H (2010) Netrin-1 mediates early events in pancreatic adenocarcinoma progression, acting on tumor and endothelial cells. Gastroenterology 138:1595–1606
Escola JM, Kleijmeer MJ, Stoorvogel W, Griffith JM, Yoshie O, Geuze HJ (1998) Selective enrichment of tetraspan proteins on the internal vesicles of multivesicular endosomes and on exosomes secreted by human B-lymphocytes. J Biol Chem 273:20121–20127
Fan J, Zhu GZ, Niles RM (2010) Expression and function of CD9 in melanoma cells. Mol Carcinog 49:85–93
Fang Y, Wu N, Gan X, Yan W, Morrell JC, Gould SJ (2007) Higher-order oligomerization targets plasma membrane proteins and HIV gag to exosomes. PLoS Biol 5:e158
Fevrier B, Raposo G (2004) Exosomes: endosomal-derived vesicles shipping extracellular messages. Curr Opin Cell Biol 16:415–421
Fitter S, Tetaz T, Berndt MC, Ashman LK (1995) Molecular cloning of cDNA encoding a novel platelet, endothelial tetraspan antigen, PETA-3. Blood 86:1348–1355
Flaumenhaft R (2006) Formation and fate of platelet microparticles. Blood Cells Mol Dis 36:182–187
Folkman J (2004) Endogenous angiogenesis inhibitors. APMIS 112:496–507
Fraley TS, Tran TC, Corgan AM, Nash CA, Hao J, Critchley DR, Greenwood JA (2003) Phosphoinositide binding inhibits alpha-actinin bundling activity. J Biol Chem 278:24039–24045

Franchini M, Montagnana M, Targher G, Lippi G (2007) Reduced von Willebrand factor-cleaving protease levels in secondary thrombotic microangiopathies and other diseases. J Thromb Thrombolysis 24:29–38

Furuya M, Kato H, Nishimura N, Ishiwata I, Ikeda H, Ito R, Yoshiki T, Ishikura H (2005) Down-regulation of CD9 in human ovarian carcinoma cell might contribute to peritoneal dissemination: morphologic alteration and reduced expression of beta1 integrin subsets. Cancer Res 65:2617–2625

Gao AC, Lou W, Dong JT, Barrett JC, Danielpour D, Isaacs JT (2003) Defining regulatory elements in the human KAI1 (CD 82) metastasis suppressor gene. Prostate 57:256–260

Garcia-Lopez MA, Barreiro O, Garcia-Diez A, Sanchez-Madrid F, Penas PF (2005) Role of tetraspanins CD9 and CD151 in primary melanocyte motility. J Invest Dermatol 125:1001–1009

Geary SM, Cambareri AC, Sincock PM, Fitter S, Ashman LK et al (2001) Differential tissue expression of epitopes of the tetraspanin CD151 recognised by monoclonal antibodies. Tissue Antigens 58:141–153

Geiger TR, Peeper DS (2009) Metastasis mechanisms. Biochim Biophys Acta 1796:293–308

Gesierich S, Paret C, Hildebrand D, Weitz J, Zgraggen K, Schmitz-Winnenthal FH, Horejsi V, Yoshie O, Herlyn D, Ashman LK, Zöller M (2005) Colocalization of the tetraspanins, CO-029 and CD151, with integrins in human pancreatic adenocarcinoma: impact on cell motility. Clin Cancer Res 11:2840–2852

Gesierich S, Berezovskiy I, Ryschich E, Zöller M (2006) Systemic induction of the angiogenesis switch by the tetraspanin D6.1A/CO-029. Cancer Res 66:7083–7094

Gibbings DJ, Ciaudo C, Erhardt M, Voinnet O (2009) Multivesicular bodies associate with components of miRNA effector complexes and modulate miRNA activity. Nat Cell Biol 11:1143–1149

Greco C, Bralet MP, Ailane N, Dubart-Kupperschmitt A, Rubinstein E, Le Naour F, Boucheix C (2010) E-cadherin/p120-catenin and tetraspanin Co-029 cooperate for cell motility control in human colon carcinoma. Cancer Res 70:7674–7683

Griffith L, Slupsky J, Seehafer J, Boshkov L, Shaw AR (1991) Platelet activation by immobilized monoclonal antibody: evidence for a CD9 proximal signal. Blood 78:1753–1759

Gruenberg J, Stenmark H (2004) The biogenesis of multivesicular endosomes. Nat Rev Mol Cell Biol 5:317–323

Grünwald V, Soltau J, Ivanyi P, Rentschler J, Reuter C, Drevs J (2009) Molecular targeted therapies for solid tumors: management of side effects. Onkologie 32:129–138

Gutiérrez-López MD, Gilsanz A, Yáñez-Mó M, Ovalle S, Lafuente EM, Domínguez C, Monk PN, González-Alvaro I, Sánchez-Madrid F, Cabañas C (2011) The sheddase activity of ADAM17/TACE is regulated by the tetraspanin CD9. Cell Mol Life Sci 68:3275–3292

Hakomori SI (2010) Glycosynaptic microdomains controlling tumor cell phenotype through alteration of cell growth, adhesion, and motility. FEBS Lett 584:1901–1906

Hao S, Ye Z, Li F, Meng Q, Qureshi M, Yang J, Xiang J (2006) Epigenetic transfer of metastatic activity by uptake of highly metastatic B16 melanoma cell-released exosomes. Exp Oncol 28:126–131

Hasegawa M, Furuya M, Kasuya Y, Nishiyama M, Sugiura T, Nikaido T, Momota Y, Ichinose M, Kimura S (2007) CD151 dynamics in carcinoma-stroma interaction: integrin expression, adhesion strength and proteolytic activity. Lab Invest 87:882–892

Hashida H, Takabayashi A, Tokuhara T, Hattori N, Taki T, Hasegawa H, Satoh S, Kobayashi N, Yamaoka Y, Miyake M (2003) Clinical significance of transmembrane 4 superfamily in colon cancer. Br J Cancer 89:158–167

He B, Liu L, Cook GA, Grgurevich S, Jennings LK, Zhang XA (2005) Tetraspanin CD82 attenuates cellular morphogenesis through down-regulating integrin alpha6-mediated cell adhesion. J Biol Chem 280:3346–3354

Heinonen M, Bono P, Narko K, Chang SH, Lundin J, Joensuu H, Furneaux H, Hla T, Haglund C, Ristimaki A (2005) Cytoplasmic HuR expression is a prognostic factor in invasive ductal breast carcinoma. Cancer Res 65:2157–2161

Hemler ME (2003) Tetraspanin proteins mediate cellular penetration, invasion, and fusion events and define a novel type of membrane microdomain. Annu Rev Cell Dev Biol 19:397–422

Hemler ME (2005) Tetraspanin functions and associated microdomains. Nat Rev Mol Cell Biol 6:801–811

Hemler ME (2008) Targeting of tetraspanin proteins-potential benefits and strategies. Nat Rev Drug Discov 7:747–758

Herlevsen M, Schmidt DS, Miyazaki K, Zöller M (2003) The association of the tetraspanin D6.1A with the alpha6beta4 integrin supports cell motility and liver metastasis formation. J Cell Sci 116:4373–4390

Hillen F, Griffioen AW (2007) Tumour vascularization: sprouting angiogenesis and beyond. Metastasis Rev 26:489–502

Hirano C, Nagata M, Noman AA, Kitamura N, Ohnishi M, Ohyama T, Kobayashi T, Suzuki K, Yoshizawa M, Izumi N, Fujita H, Takagi R (2009) Tetraspanin gene expression levels as potential biomarkers for malignancy of gingival squamous cell carcinoma. Int J Cancer 124:2911–2916

Holderfield MT, Hughes CC (2008) Crosstalk between vascular endothelial growth factor, notch, and transforming growth factor-beta in vascular morphogenesis. Circ Res 102:637–652

Hong IK, Jin YJ, Byun HJ, Jeoung DI, Kim YM, Lee H (2006) Homophilic interactions of tetraspanin CD151 up-regulate motility and matrix metalloproteinase-9 expression of human melanoma cells through adhesion-dependent c-Jun activation signaling pathways. J Biol Chem 281:24279–24292

Hori H, Yano S, Koufuji K, Takeda J, Shirouzu K (2004) CD9 expression in gastric cancer and its significance. J Surg Res 117:208–215

Hotta H, Ross AH, Huebner K, Isobe M, Wendeborn S, Chao MV, Ricciardi RP, Tsujimoto Y, Croce CM, Koprowski H (1988) Molecular cloning and characterization of an antigen associated with early stages of melanoma tumor progression. Cancer Res 48:2955–2962

Huang CI, Kohno N, Ogawa E, Adachi M, Taki T, Miyake M (1998) Correlation of reduction in MRP-1/CD9 and KAI1/CD82 expression with recurrences in breast cancer patients. Am J Pathol 153:973–983

Huang CL, Liu D, Masuya D, Kameyama K, Nakashima T, Yokomise H, Ueno M, Miyake M (2004) MRP-1/CD9 gene transduction downregulates Wnt signal pathways. Oncogene 23:7475–7483

Huang H, Groth J, Sossey-Alaoui K, Hawthorn L, Beall S, Geradts J (2005) Aberrant expression of novel and previously described cell membrane markers in human breast cancer cell lines and tumors. Clin Cancer Res 11:4357–4364

Huang CL, Ueno M, Liu D, Masuya D, Nakano J, Yokomise H, Nakagawa T, Miyake M (2006) MRP-1/CD9 gene transduction regulates the actin cytoskeleton through the downregulation of WAVE2. Oncogene 25:6480–6488

Huang H, Sossey-Alaoui K, Beachy SH, Geradts J (2007) The tetraspanin superfamily member NET-6 is a new tumor suppressor gene. J Cancer Res Clin Oncol 133:761–769

Huang XY, Ke AW, Shi GM, Ding ZB, Devbhandari RP, Gu FM, Li QL, Dai Z, Zhou J, Fan J (2010) Overexpression of CD151 as an adverse marker for intrahepatic cholangiocarcinoma patients. Cancer 116:5440–5451

Huerta S, Harris DM, Jazirehi A, Bonavida B, Elashoff D, Livingston EH, Heber D (2003) Gene expression profile of metastatic colon cancer cells resistant to cisplatin-induced apoptosis. Int J Oncol 22:663–670

Hurley JH, Emr SD (2006) The ESCRT complexes: structure and mechanism of a membrane-trafficking network. Annu Rev Biophys Biomol Struct 35:277–298

Iero M, Valenti R, Huber V, Filipazzi P, Parmiani G, Fais S, Rivoltini L (2008) Tumour-released exosomes and their implications in cancer immunity. Cell Death Differ 15:80–88

Ikeyama S, Koyama M, Yamaoko M, Sasada R, Miyake M (1993) Suppression of cell motility and metastasis by transfection with human motility-related protein (MRP-1/CD9) DNA. J Exp Med 177:1231–1237

Jackson P, Millar D, Kingsley E, Yardley G, Ow K, Clark S, Russell PJ (2000) Methylation of a CpG island within the promoter region of the KAI1 metastasis suppressor gene is not responsible for down-regulation of KAI1 expression in invasive cancers or cancer cell lines. Cancer Lett 157:169–176

Jackson P, Marreiros A, Russell PJ (2005) KAI1 tetraspanin and metastasis suppressor. Int J Biochem Cell Biol 37:530–534

Jackson P, Rowe A, Grimm MO (2007) An alternatively spliced KAI1 mRNA is expressed at low levels in human bladder cancers and bladder cancer cell lines and is not associated with invasive behaviour. Oncol Rep 18:1357–1363

Jankowski SA, De Jong P, Meltzer PS (1995) Genomic structure of SAS, a member of the transmembrane 4 superfamily amplified in human sarcomas. Genomics 25:501–506

Janmey PA, Lindberg U (2004) Cytoskeletal regulation: rich in lipids. Nat Rev Mol Cell Biol 5:658–666

Janowska-Wieczorek A, Wysoczynski M, Kijowski J, Marquez-Curtis L, Machalinski B, Ratajczak J, Ratajczak MZ (2005) Microvesicles derived from activated platelets induce metastasis and angiogenesis in lung cancer. Int J Cancer 113:752–760

Johnson JL, Winterwood N, DeMali KA, Stipp CS (2009) Tetraspanin CD151 regulates RhoA activation and the dynamic stability of carcinoma cell-cell contacts. J Cell Sci 122: 2263–2273

Johnstone RM (2006) Exosomes biological significance: a concise review. Blood Cells Mol Dis 36:315–321

Joshi B, Li L, Nabi IR (2010) A role for KAI1 in promotion of cell proliferation and mammary gland hyperplasia by the gp78 ubiquitin ligase. J Biol Chem 285:8830–8839

Jung T, Castellana D, Klingbeil P, Cuesta Hernández I, Vitacolonna M, Orlicky DJ, Roffler SR, Brodt P, Zöller M (2009) CD44v6 dependence of premetastatic niche preparation by exosomes. Neoplasia 11:1093–1105

Kanetaka K, Sakamoto M, Yamamoto Y, Yamasaki S, Lanza F, Kanematsu T, Hirohashi S (2001) Overexpression of tetraspanin CO-029 in hepatocellular carcinoma. J Hepatol 35:637–642

Kanetaka K, Sakamoto M, Yamamoto Y, Takamura M, Kanematsu T, Hirohashi S et al (2003) Possible involvement of tetraspanin CO-029 in hematogenous intrahepatic metastasis of liver cancer cells. J Gastroenterol Hepatol 18:1309–1314

Kaplan RN, Rafii S, Lyden D (2006) Preparing the "soil": the premetastatic niche. Cancer Res 66:11089–11093

Karamatic Crew V, Burton N, Kagan A, Green CA, Levene C, Flinter F, Brady RL, Daniels G, Anstee DJ (2004) CD151, the first member of the tetraspanin (TM4) superfamily detected on erythrocytes, is essential for the correct assembly of human basement membranes in kidney and skin. Blood 104:2217–2223

Ke AW, Shi GM, Zhou J, Wu FZ, Ding ZB, Hu MY, Xu Y, Song ZJ, Wang ZJ, Wu JC, Bai DS, Li JC, Liu KD, Fan J (2009) Role of overexpression of CD151 and/or c-Met in predicting prognosis of hepatocellular carcinoma. Hepatology 49:491–503

Kim YJ, Yu JM, Joo HJ, Kim HK, Cho HH, Bae YC, Jung JS (2007) Role of CD9 in proliferation and proangiogenic action of human adipose-derived mesenchymal stem cells. Pflugers Arch 455:283–296

Kim B, Boo K, Lee JS, Kim KI, Kim WH, Cho HJ, Park YB, Kim HS, Baek SH (2010) Identification of the KAI1 metastasis suppressor gene as a hypoxia target gene. Biochem Biophys Res Commun 393:179–184

Klosek SK, Nakashiro K, Hara S, Shintani S, Hasegawa H, Hamakawa H (2005) CD151 forms a functional complex with c-Met in human salivary gland cancer cells. Biochem Biophys Res Commun 336:408–416

Kohmo S, Kijima T, Otani Y, Mori M, Minami T, Takahashi R, Nagatomo I, Takeda Y, Kida H, Goya S, Yoshida M, Kumagai T, Tachibana I, Yokota S, Kawase I (2010) Cell surface tetraspanin CD9 mediates chemoresistance in small cell lung cancer. Cancer Res 70:8025–8035

Kohno M, Hasegawa H, Miyake M, Yamamoto T, Fujita S (2002) CD151 enhances cell motility and metastasis of cancer cells in the presence of focal adhesion kinase. Int J Cancer 97:336–343

Kovalenko OV, Metcalf DG, DeGrado WF, Hemler ME (2005) Structural organization and interactions of transmembrane domains in tetraspanin proteins. BMC Struct Biol 5:11

Kuhn S, Koch M, Nübel T, Ladwein M, Antolovic D, Klingbeil P, Hildebrand D, Moldenhauer G, Langbein L, Franke WW, Weitz J, Zöller M (2007) A complex of EpCAM, claudin-7, CD44 variant isoforms, and tetraspanins promotes colorectal cancer progression. Mol Cancer Res 5:553–567

Kwon MS, Shin SH, Yim SH, Lee KY, Kang HM, Kim TM, Chung YJ (2007) CD63 as a biomarker for predicting the clinical outcomes in adenocarcinoma of lung. Lung Cancer 57:46–53

Lafleur MA, Xu D, Hemler ME (2009) Tetraspanin proteins regulate membrane type-1 matrix metalloproteinase-dependent pericellular proteolysis. Mol Biol Cell 20:2030–2040

Lakkaraju A, Rodriguez-Boulan E (2008) Itinerant exosomes: emerging roles in cell and tissue polarity. Trends Cell Biol 18:199–209

Lan RF, Liu ZX, Liu XC, Song YE, Wang DW (2005) CD151 promotes neovascularization and improves blood perfusion in a rat hind-limb ischemia model. J Endovasc Ther 12:469–478

Latysheva N, Muratov G, Rajesh S, Padgett M, Hotchin NA, Overduin M, Berditchevski F (2006) Syntenin-1 is a new component of tetraspanin-enriched microdomains: mechanisms and consequences of the interaction of syntenin-1 with CD63. Mol Cell Biol 26:7707–7718

Lau LM, Wee JL, Wright MD, Moseley GW, Hogarth PM, Ashman LK, Jackson DE (2004) The tetraspanin superfamily member CD151 regulates outside-in integrin alphaIIbbeta3 signaling and platelet function. Blood 104:2368–2375

Le Naour F, André M, Boucheix C, Rubinstein E (2006) Membrane microdomains and proteomics: lessons from tetraspanin microdomains and comparison with lipid rafts. Proteomics 6:6447–6454

Le Tonqueze O, Gschloessl B, Namanda-Vanderbeken A, Legagneux V, Paillard L, Audic Y (2010) Chromosome wide analysis of CUGBP1 binding sites identifies the tetraspanin CD9 mRNA as a target for CUGBP1-mediated down-regulation. Biochem Biophys Res Commun 394:884–889

Lee JH, Seo YW, Park SR, Kim YJ, Kim KK (2003) Expression of a splice variant of KAI1, a tumor metastasis suppressor gene, influences tumor invasion and progression. Cancer Res 63:7247–7255

Lee JH, Park SR, Chay KO, Seo YW, Kook H, Ahn KY, Kim YJ, Kim KK (2004) KAI1 COOH-terminal interacting tetraspanin (KITENIN), a member of the tetraspanin family, interacts with KAI1, a tumor metastasis suppressor, and enhances metastasis of cancer. Cancer Res 64:4235–4243

Lee JH, Bae JA, Lee JH, Seo YW, Kho DH, Sun EG, Lee SE, Cho SH, Joo YE, Ahn KY, Chung IJ, Kim KK (2010) Glycoprotein 90 K, downregulated in advanced colorectal cancer tissues, interacts with CD9/CD82 and suppresses the Wnt/beta-catenin signal via ISGylation of beta-catenin. Gut 59:907–917

Lekishvili T, Fromm E, Mujoomdar M, Berditchevski F (2008) The tumour-associated antigen L6 (L6-Ag) is recruited to the tetraspanin-enriched microdomains: implication for tumour cell motility. J Cell Sci 121:685–694

Levy S, Shoham T (2005) Protein-protein interactions in the tetraspanin web. Physiology (Bethesda) 20:218–224

Levy S, Todd SC, Maecker HT (1998) CD81 (TAPA-1): a molecule involved in signal transduction and cell adhesion in the immune system. Annu Rev Immunol 16:89–109

Lewis TB, Robison JE, Bastien R, Milash B, Boucher K, Samlowski WE, Leachman SA, Dirk Noyes R, Wittwer CT, Perreard L, Bernard PS (2005) Molecular classification of melanoma using real-time quantitative reverse transcriptase-polymerase chain reaction. Cancer 104:1678–1686

Li Q, Li L, Shi W, Jiang X, Xu Y, Gong F, Zhou M, Edwards CK III, Li Z (2006) Mechanism of action differences in the antitumor effects of transmembrane and secretory tumor necrosis factor-alpha in vitro and in vivo. Cancer Immunol Immunother 55:1470–1479

Lineberry N, Su L, Soares L, Fathman CG (2008) The single subunit transmembrane E3 ligase gene related to anergy in lymphocytes (GRAIL) captures and then ubiquitinates transmembrane proteins across the cell membrane. J Biol Chem 283:28497–28505

Little KD, Hemler ME, Stipp CS (2004) Dynamic regulation of a GPCR-tetraspanin-G protein complex on intact cells: central role of CD81 in facilitating GPR56-galpha q/11 association. Mol Biol Cell 15:2375–2387

Liu WM, Zhang XA (2006) KAI1/CD82, a tumor metastasis suppressor. Cancer Lett 240:183–194

Liu FS, Dong JT, Chen JT, Hsieh YT, Ho ES, Hung MJ (2000) Frequent down-regulation and lack of mutation of the KAI1 metastasis suppressor gene in epithelial ovarian carcinoma. Gynecol Oncol 78:10–15

Liu L, He B, Liu WM, Zhou D, Cox JV, Zhang XA (2007) Tetraspanin CD151 promotes cell migration by regulating integrin trafficking. J Biol Chem 282:31631–31642

Liu WF, Zuo HJ, Chai BL, Peng D, Fei YJ, Lin JY, Yu XH, Wang DW, Liu ZX (2011) Role of tetraspanin CD151-α3/α6 integrin complex: implication in angiogenesis CD151-integrin complex in angiogenesis. Int J Biochem Cell Biol 43:642–650

Lombardi DP, Geradts J, Foley JF, Chiao C, Lamb PW, Barrett JC (1999) Loss of KAI1 expression in the progression of colorectal cancer. Cancer Res 59:5724–5731

Longo N, Yáñez-Mó M, Mittelbrunn M, de la Rosa G, Muñoz ML, Sánchez-Madrid F, Sánchez-Mateos P (2001) Regulatory role of tetraspanin CD9 in tumor-endothelial cell interaction during transendothelial invasion of melanoma cells. Blood 98:3717–3726

Lopez de Silanes I, Lal A, Gorospe M (2005) HuR: post-transcriptional paths to malignancy. RNA Biol 2:11–13

Louvet-Vallée S (2000) ERM proteins: from cellular architecture to cell signaling. Biol Cell 92:305–316

Maecker HT, Todd SC, Levy S (1997) The tetraspanin superfamily: molecular facilitators. FASEB J 11:428–442

Mangin PH, Kleitz L, Boucheix C, Gachet C, Lanza F (2009) CD9 negatively regulates integrin alphaIIbbeta3 activation and could thus prevent excessive platelet recruitment at sites of vascular injury. J Thromb Haemost 7:900–902

Marks MS, Ohno H, Kirchnausen T, Bonracino JS (1997) Protein sorting by tyrosine-based signals: adapting to the Ys and wherefores. Trends Cell Biol 7:124–128

Marreiros A, Czolij R, Yardley G, Crossley M, Jackson P (2003) Identification of regulatory regions within the KAI1 promoter: a role for binding of AP1, AP2 and p53. Gene 302:155–164

Marreiros A, Dudgeon K, Dao V, Grimm MO, Czolij R, Crossley M, Jackson P (2005) KAI1 promoter activity is dependent on p53, junB and AP2: evidence for a possible mechanism underlying loss of KAI1 expression in cancer cells. Oncogene 24:637–649

Mathivanan S, Lim JW, Tauro BJ, Ji H, Moritz RL, Simpson RJ (2010) Proteomic analysis of A33-immunoaffinity-purified exosomes released from the human colon tumor cell line LIM1215 reveals a tissue-specific protein signature. Mol Cell Proteomics 9:197–208

Maurer CA, Graber HU, Friess H, Beyermann B, Willi D, Netzer P, Zimmermann A, Buchler MW (1999) Reduced expression of the metastasis suppressor gene KAI1 in advanced colon cancer and its metastases. Surgery 126:869–880

Mazzocca A, Liotta F, Carloni V (2008) Tetraspanin CD81-regulated cell motility plays a critical role in intrahepatic metastasis of hepatocellular carcinoma. Gastroenterology 135:244–256

Miranti CK (2009) Controlling cell surface dynamics and signaling: how CD82/KAI1 suppresses metastasis. Cell Signal 21:196–211

Mischiati C, Natali PG, Sereni A, Sibilio L, Giorda E, Cappellacci S, Nicotra MR, Mariani G, Di Filippo F, Catricala C, Gambari R, Grammatico P, Giacomini P (2006) cDNA-array profiling of melanomas and paired melanocyte cultures. J Cell Physiol 207:697–705

Mitsuzuka K, Handa K, Satoh M, Arai Y, Hakomori S (2005) A specific microdomain ("glycosynapse 3") controls phenotypic conversion and reversion of bladder cancer cells through GM3-mediated interaction of alpha3beta1 integrin with CD9. J Biol Chem 280:35545–35553

Miura Y, Kainuma M, Jiang H, Velasco H, Vogt PK, Hakomori S (2004) Reversion of the Jun-induced oncogenic phenotype by enhanced synthesis of sialosyllactosylceramide (GM3 ganglioside). Proc Natl Acad Sci USA 101:16204–16209

Miyake M, Koyama M, Seno M, Ikeyama S (1991) Identification of the motility-related protein (MRP-1), recognized by monoclonal antibody M31-15, which inhibits cell motility. J Exp Med 174:1347–1354

Miyazaki T, Müller U, Campbell KS (1997) Normal development but differentially altered proliferative responses of lymphocytes in mice lacking CD81. EMBO J 16:4217–4225

Molina S, Castet V, Pichard-Garcia L, Wychowski C, Meurs E, Pascussi JM, Sureau C, Fabre JM, Sacunha A, Larrey D, Dubuisson J, Coste J, McKeating J, Maurel P, Fournier-Wirth C (2008) Serum-derived hepatitis C virus infection of primary human hepatocytes is tetraspanin CD81 dependent. J Virol 82:569–574

Morel O, Hugel B, Jesel L, Mallat Z, Lanza F, Douchet MP, Zupan M, Chauvin M, Cazenave JP, Tedgui A, Freyssinet JM, Toti F (2004) Circulating procoagulant microparticles and soluble GPV in myocardial infarction treated by primary percutaneous transluminal coronary angioplasty. A possible role for GPIIb-IIIa antagonists. J Thromb Haemost 2:1118–1126

Moseley GW, Elliott J, Wright MD, Partridge LJ, Monk PN (2003) Interspecies contamination of the KM3 cell line: implications for CD63 function in melanoma metastasis. Int J Cancer 105:613–616

Moss ML, Bartsch JW (2004) Therapeutic benefits from targeting of ADAM family members. Biochemistry 43:7227–7235

Murayama Y, Miyagawa J, Oritani K, Yoshida H, Yamamoto K, Kishida O, Miyazaki T, Tsutsui S, Kiyohara T, Miyazaki Y, Higashiyama S, Matsuzawa Y, Shinomura Y (2004) CD9-mediated activation of the p46 Shc isoform leads to apoptosis in cancer cells. J Cell Sci 117:3379–3388

Murayama Y, Shinomura Y, Oritani K, Miyagawa J, Yoshida H, Nishida M, Katsube F, Shiraga M, Miyazaki T, Nakamoto T, Tsutsui S, Tamura S, Higashiyama S, Shimomura I, Hayashi N (2008) The tetraspanin CD9 modulates epidermal growth factor receptor signaling in cancer cells. J Cell Physiol 216:135–143

Nakamoto T, Murayama Y, Oritani K, Boucheix C, Rubinstein E, Nishida M, Katsube F, Watabe K, Kiso S, Tsutsui S, Tamura S, Shinomura Y, Hayashi N (2009) A novel therapeutic strategy with anti-CD9 antibody in gastric cancers. J Gastroenterol 44:889–896

Nakazawa Y, Sato S, Naito M, Kato Y, Mishima K, Arai H, Tsuruo T, Fujita N (2008) Tetraspanin family member CD9 inhibits aggrus/podoplanin-induced platelet aggregation and suppresses pulmonary metastasis. Blood 112:1730–1739

Nazarenko I, Rana S, Baumann A, McAlear J, Hellwig A, Trendelenburg M, Lochnit G, Preissner KT, Zöller M (2010) Cell surface tetraspanin Tspan8 contributes to molecular pathways of exosome-induced endothelial cell activation. Cancer Res 70:1668–1678

Nishiuchi R, Sanzen N, Nada S, Sumida Y, Wada Y, Okada M, Takagi J, Hasegawa H, Sekiguchi K (2005) Potentiation of the ligand-binding activity of integrin alpha3beta1 via association with tetraspanin CD151. Proc Natl Acad Sci USA 102:1939–1944

Novitskaya V, Romanska H, Dawoud M, Jones JL, Berditchevski F (2010) Tetraspanin CD151 regulates growth of mammary epithelial cells in three-dimensional extracellular matrix: implication for mammary ductal carcinoma in situ. Cancer Res 70:4698–4708

Nübel T, Preobraschenski J, Tuncay H, Weiss T, Kuhn S, Ladwein M, Langbein L, Zöller M (2009) Claudin-7 regulates EpCAM-mediated functions in tumor progression. Mol Cancer Res 7:285–299

Odintsova E, Sugiura T, Berditchevski F (2000) Attenuation of EGF receptor signaling by a metastasis suppressor, the tetraspanin CD82/KAI-1. Curr Biol 10:1009–1012

Odintsova E, Voortman J, Gilbert E, Berditchevski F (2003) Tetraspanin CD82 regulates compartmentalisation and ligand-induced dimerization of EGFR. J Cell Sci 116:4557–4566

Odintsova E, Butters TD, Monti E, Sprong H, van Meer G, Berditchevski F (2006) Gangliosides play an important role in the organization of CD82-enriched microdomains. Biochem J 400:315–325

Ono M, Handa K, Withers DA, Hakomori S (1999) Motility inhibition and apoptosis are induced by metastasis-suppressing gene product CD82 and its analogue CD9, with concurrent glycosylation. Cancer Res 59:2335–2339

Oren R, Takahashi S, Doss C, Levy R, Levy S et al (1990) TAPA-1, the target of an antiproliferative antibody, defines a new family of transmembrane proteins. Mol Cell Biol 10:4007–4015
Orlowski E, Chand R, Yip J, Wong C, Goschnick MW, Wright MD, Ashman LK, Jackson DE (2009) Platelet tetraspanin superfamily member, CD151 is required for regulation of thrombus stability in vivo. J Thromb Haemost 7:2074–2084
Ovalle S, Gutiérrez-López MD, Olmo N, Turnay J, Lizarbe MA, Majano P, Molina-Jiménez F, López-Cabrera M, Yáñez-Mó M, Sánchez-Madrid F, Cabañas C (2007) The tetraspanin CD9 inhibits the proliferation and tumorigenicity of human colon carcinoma cells. Int J Cancer 121:2140–2152
Pap E, Pállinger E, Pásztói M, Falus A (2009) Highlights of a new type of intercellular communication: microvesicle-based information transfer. Inflamm Res 58:1–8
Park JE, Tan HS, Datta A, Lai RC, Zhang H, Meng W, Lim SK, Sze SK (2010) Hypoxia modulates tumor microenvironment to enhance angiogenic and metastastic potential by secretion of proteins and exosomes. Mol Cell Proteomics 9:1085–1099
Peer D, Park EJ, Morishita Y, Carman CV, Shimaoka M (2008) Systemic leukocyte-directed siRNA delivery revealing cyclin D1 as an anti-inflammatory target. Science 319:627–630
Pols MS, Klumperman J (2009) Trafficking and function of the tetraspanin CD63. Exp Cell Res 315:1584–1592
Potolicchio I, Carven GJ, Xu X, Stipp C, Riese RJ, Stern LJ, Santambrogio L (2005) Proteomic analysis of microglia-derived exosomes: metabolic role of the aminopeptidase CD13 in neuropeptide catabolism. J Immunol 175:2237–2243
Press OW, Eary JF, Badger CC, Martin PJ, Appelbaum FR, Levy R, Miller R, Brown S, Nelp WB, Krohn KA et al (1989) Treatment of refractory non-Hodgkin's lymphoma with radiolabeled MB-1 (anti-CD37) antibody. J Clin Oncol 7:1027–1038
Prince S, Carreira S, Vance KW, Abrahams A, Goding CR (2004) Tbx2 directly represses the expression of the p21(WAF1) cyclin-dependent kinase inhibitor. Cancer Res 64:1669–1674
Radford KJ, Thorne RF, Hersey P (1997) Regulation of tumor cell motility and migration by CD63 in a human melanoma cell line. J Immunol 158:3353–3358
Rana S, Claas C, Kretz CC, Nazarenko I, Zöller M et al (2011) Activation-induced internalization differs for the tetraspanins CD9 and Tspan8: impact on tumor cell motility. Int J Biochem Cell Biol 43:106–119
Ratajczak J, Miekus K, Kucia M, Zhang J, Reca R, Dvorak P, Ratajczak MZ (2006) Embryonic stem cell-derived microvesicles reprogram hematopoietic progenitors: evidence for horizontal transfer of mRNA and protein delivery. Leukemia 20:847–856
Razmara M, Hu H, Masquelier M, Li N (2007) Glycoprotein IIb/IIIa blockade inhibits platelet aminophospholipid exposure by potentiating translocase and attenuating scramblase activity. Cell Mol Life Sci 64:999–1008
Rous BA, Reaves BJ, Ihrke G, Briggs JA, Gray SR, Stephens DJ, Banting G, Luzio JP (2002) Role of adaptor complex AP-3 in targeting wild-type and mutated CD63 to lysosomes. Mol Biol Cell 13:1071–1082
Rowe A, Jackson P (2006) Expression of KITENIN, a KAI1/CD82 binding protein and metastasis enhancer, in bladder cancer cell lines: relationship to KAI1/CD82 levels and invasive behaviour. Oncol Rep 16:1267–1272
Ruf W, Mueller BM (2006) Thrombin generation and the pathogenesis of cancer. Semin Thromb Hemost 32(Suppl 1):61–68
Ruseva Z, Geiger PX, Hutzler P, Kotzsch M, Luber B, Schmitt M, Gross E, Reuning U (2009) Tumor suppressor KAI1 affects integrin alphavbeta3-mediated ovarian cancer cell adhesion, motility, and proliferation. Exp Cell Res 315:1759–1771
Sachs N, Kreft M, van den Bergh Weerman MA, Beynon AJ, Peters TA, Weening JJ, Sonnenberg A (2006) Kidney failure in mice lacking the tetraspanin CD151. J Cell Biol 175:33–39
Sadej R, Romanska H, Baldwin G, Gkirtzimanaki K, Novitskaya V, Filer AD, Krcova Z, Kusinska R, Ehrmann J, Buckley CD, Kordek R, Potemski P, Eliopoulos AG, el Lalani N, Berditchevski

F (2009) CD151 regulates tumorigenesis by modulating the communication between tumor cells and endothelium. Mol Cancer Res 7:787–798
Safe S, Abdelrahim M (2005) Sp transcription factor family and its role in cancer. Eur J Cancer 41:2438–2448
Sakakura C, Hagiwara A, Nakanishi M, Shimomura K, Takagi T, Yasuoka R, Fujita Y, Abe T, Ichikawa Y, Takahashi S, Ishikawa T, Nishizuka I, Morita T, Shimada H, Okazaki Y, Hayashizaki Y, Yamagishi H (2002) Differential gene expression profiles of gastric cancer cells established from primary tumour and malignant ascites. Br J Cancer 87:1153–1161
Sala-Valdés M, Ursa A, Charrin S, Rubinstein E, Hemler ME, Sánchez-Madrid F, Yáñez-Mó M (2006) EWI-2 and EWI-F link the tetraspanin web to the actin cytoskeleton through their direct association with ezrin-radixin-moesin proteins. J Biol Chem 281:19665–19675
Sauer G, Windisch J, Kurzeder C, Heilmann V, Kreienberg R, Deissler H (2003) Progression of cervical carcinomas is associated with down-regulation of CD9 but strong local re-expression at sites of transendothelial invasion. Clin Cancer Res 9:6426–6431
Schorey JS, Bhatnagar S (2008) Exosome function: from tumor immunology to pathogen biology. Traffic 9:871–881
Schröder J, Lüllmann-Rauch R, Himmerkus N, Pleines I, Nieswandt B, Orinska Z, Koch-Nolte F, Schröder B, Bleich M, Saftig P (2009) Deficiency of the tetraspanin CD63 associated with kidney pathology but normal lysosomal function. Mol Cell Biol 29:1083–1094
Seigneuret M (2006) Complete predicted three-dimensional structure of the facilitator transmembrane protein and hepatitis C virus receptor CD81: conserved and variable structural domains in the tetraspanin superfamily. Biophys J 90:212–227
Seow Y, Wood MJ (2009) Biological gene delivery vehicles: beyond viral vectors. Mol Ther 17:767–777
Serru V, Le Naour F, Billard M, Azorsa DO, Lanza F, Boucheix C, Rubinstein E (1999) Selective tetraspan-integrin complexes (CD81/alpha4beta1, CD151/alpha3beta1, CD151/alpha6beta1) under conditions disrupting tetraspan interactions. Biochem J 340:103–111
Sharma C, Yang XH, Hemler ME (2008) DHHC2 affects palmitoylation, stability, and functions of tetraspanins CD9 and CD151. Mol Biol Cell 19:3415–3425
Shet AS, Aras O, Gupta K, Hass MJ, Rausch DJ, Saba N, Koopmeiners L, Key NS, Hebbel RP (2003) Sickle blood contains tissue factor-positive microparticles derived from endothelial cells and monocytes. Blood 102:2678–2683
Shi W, Fan H, Shum L, Derynck R (2000) The tetraspanin CD9 associates with transmembrane TGF-alpha and regulates TGF-alpha-induced EGF receptor activation and cell proliferation. J Cell Biol 148:591–602
Shi GM, Ke AW, Zhou J, Wang XY, Xu Y, Ding ZB, Devbhandari RP, Huang XY, Qiu SJ, Shi YH, Dai Z, Yang XR, Yang GH, Fan J (2010) CD151 modulates expression of matrix metalloproteinase 9 and promotes neoangiogenesis and progression of hepatocellular carcinoma. Hepatology 52:183–196
Shiomi T, Inoki I, Kataoka F, Ohtsuka T, Hashimoto G, Nemori R, Okada Y (2005) Pericellular activation of proMMP-7 (promatrilysin-1) through interaction with CD151. Lab Invest 85:1489–1506
Sho M, Adachi M, Taki T, Hashida H, Konishi T, Huang CL, Ikeda N, Nakajima Y, Kanehiro H, Hisanaga M, Nakano H, Miyake M (1998) Transmembrane 4 superfamily as a prognostic factor in pancreatic cancer. Int J Cancer 79:509–516
Si Z, Hersey P (1993) Expression of the neuroglandular antigen and analogues in melanoma. CD9 expression appears inversely related to metastatic potential of melanoma. Int J Cancer 54:37–43
Sierko E, Wojtukiewicz MZ (2007) Inhibition of platelet function: does it offer a chance of better cancer progression control? Semin Thromb Hemost 33:712–721
Sigala S, Faraoni I, Botticini D, Paez-Pereda M, Missale C, Bonmassar E, Spano P (1999) Suppression of telomerase, reexpression of KAI1, and abrogation of tumorigenicity by nerve growth factor in prostate cancer cell lines. Clin Cancer Res 5:1211–1218
Simons M, Raposo G (2009) Exosomes-vesicular carriers for intercellular communication. Curr Opin Cell Biol 21:575–581

Simpson RJ, Lim JW, Moritz RL, Mathivanan S (2009) Exosomes: proteomic insights and diagnostic potential. Expert Rev Proteomics 6:267–283

Sincock PM, Fitter S, Parton RG, Berndt MC, Gamble JR, Ashman LK (1999) PETA-3/CD151, a member of the transmembrane 4 superfamily, is localised to the plasma membrane and endocytic system of endothelial cells, associates with multiple integrins and modulates cell function. J Cell Sci 112:833–844

Smalheiser NR (2007) Exosomal transfer of proteins and RNAs at synapses in the nervous system. Biol Direct 2:35

Soyuer S, Soyuer I, Unal D, Ucar K, Yildiz OG, Orhan O (2010) Prognostic significance of CD9 expression in loc.lly advanced gastric cancer treated with surgery and adjuvant chemoradiotherapy. Pathol Res Pract 206:607–610

Sridhar SC, Miranti CK (2006) Tetraspanin KAI1/CD82 suppresses invasion by inhibiting integrin-dependent crosstalk with c-Met receptor and Src kinases. Oncogene 25:2367–2378

Stipp CS (2010) Laminin-binding integrins and their tetraspanin partners as potential antimetastatic targets. Expert Rev Mol Med 18(12):e3

Stipp CS, Kolesnikova TV, Hemler ME (2003) Functional domains in tetraspanin proteins. Trends Biochem Sci 28:106–112

Stoeck A, Keller S, Riedle S, Sanderson MP, Runz S, Le Naour F, Gutwein P, Ludwig A, Rubinstein E, Altevogt P (2006) A role for exosomes in the constitutive and stimulus-induced ectodomain cleavage of L1 and CD44. Biochem J 393:609–618

Su JS, Arima K, Hasegawa M, Franco OE, Umeda Y, Yanagawa M, Sugimura Y, Kawamura J (2004) Decreased expression of KAI1 metastasis suppressor gene is a recurrence predictor in primary pTa and pT1 urothelial bladder carcinoma. Int J Urol 11:74–82

Subra C, Laulagnier K, Perret B, Record M (2007) Exosome lipidomics unravels lipid sorting at the level of multivesicular bodies. Biochimie 89:205–212

Suzuki S, Miyazaki T, Tanaka N, Sakai M, Sano A, Inose T, Sohda M, Nakajima M, Kato H, Kuwano H (2011) Prognostic significance of CD151 expression in esophageal squamous cell carcinoma with aggressive cell proliferation and invasiveness. Ann Surg Oncol 18:888–893

Suzuki-Inoue K, Fuller GL, García A, Eble JA, Pöhlmann S, Inoue O, Gartner TK, Hughan SC, Pearce AC, Laing GD, Theakston RD, Schweighoffer E, Zitzmann N, Morita T, Tybulewicz VL, Ozaki Y, Watson SP (2006) A novel Syk-dependent mechanism of platelet activation by the C-type lectin receptor CLEC-2. Blood 107:542–549

Szala S, Kasai Y, Steplewski Z, Rodeck U, Koprowski H, Linnenbach AJ (1990) Molecular cloning of cDNA for the human tumor-associated antigen CO-029 and identification of related transmembrane antigens. Proc Natl Acad Sci USA 87:6833–6837

Tagawa K, Arihiro K, Takeshima Y, Hiyama E, Yamasaki M, Inai K (1999) Down-regulation of KAI1 messenger RNA expression is not associated with loss of heterozygosity of the KAI1 gene region in lung adenocarcinoma. Jpn J Cancer Res 90:970–976

Takahashi M, Sugiura T, Abe M, Ishii K, Shirasuna K (2007) Regulation of c-Met signaling by the tetraspanin KAI-1/CD82 affects cancer cell migration. Int J Cancer 121:1919–1929

Takeda T, Hattori N, Tokuhara T, Nishimura Y, Yokoyama M, Miyake M (2007a) Adenoviral transduction of MRP-1/CD9 and KAI1/CD82 inhibits lymph node metastasis in orthotopic lung cancer model. Cancer Res 67:1744–1749

Takeda Y, Kazarov AR, Butterfield CE, Hopkins BD, Benjamin LE, Kaipainen A, Hemler ME (2007b) Deletion of tetraspanin Cd151 results in decreased pathologic angiogenesis in vivo and in vitro. Blood 109:1524–1532

Tanaka F, Hori N, Sato K (2002) Identification of differentially expressed genes in rat hepatoma cell lines using subtraction and microarray. J Biochem 131:39–44

Tarasova NI, Rice WG, Michejda CJ (1999) Inhibition of G-protein-coupled receptor function by disruption of transmembrane domain interactions. J Biol Chem 274:34911–34915

Telese F, Bruni P, Donizetti A, Gianni D, D'Ambrosio C, Scaloni A, Zambrano N, Rosenfeld MG, Russo T (2005) Transcription regulation by the adaptor protein Fe65 and the nucleosome assembly factor SET. EMBO Rep 6:77–82

Testa JE, Brooks PC, Lin JM, Quigley JP (1999) Eukaryotic expression cloning with an antimetastatic monoclonal antibody identifies a tetraspanin (PETA-3/CD151) as an effector of human tumor cell migration and metastasis. Cancer Res 59:3812–3820
Todeschini RA, Hakomori SI (2008) Functional role of glycosphingolipids and gangliosides in control of cell adhesion, motility, and growth, through glycosynaptic microdomains. Biochim Biophys Acta 1780:421–433
Todeschini AR, Dos Santos JN, Handa K, Hakomori SI (2008) Ganglioside GM2/GM3 complex affixed on silica nanospheres strongly inhibits cell motility through CD82/cMet-mediated pathway. Proc Natl Acad Sci USA 105:1925–1930
Tohami T, Drucker L, Shapiro H, Radnay J, Lishner M (2007) Overexpression of tetraspanins affects multiple myeloma cell survival and invasive potential. FASEB J 21:691–699
Tokuhara T, Hasegawa H, Hattori N, Ishida H, Taki T, Tachibana S, Sasaki S, Miyake M (2001) Clinical significance of CD151 gene expression in non-small cell lung cancer. Clin Cancer Res 7:4109–4114
Tonoli H, Barrett JC (2005) CD82 metastasis suppressor gene: a potential target for new therapeutics? Trends Mol Med 11:563–570
Tsai YC, Mendoza A, Mariano JM, Zhou M, Kostova Z, Chen B, Veenstra T, Hewitt SM, Helman LJ, Khanna C, Weissman AM (2007) The ubiquitin ligase gp78 promotes sarcoma metastasis by targeting KAI1 for degradation. Nat Med 13:1504–1509
Tsitsikov EN, Gutierrez-Ramos JC, Geha RS (1997) Impaired CD19 expression and signaling, enhanced antibody response to type II T independent antigen and reduction of B-1 cells in CD81-deficient mice. Proc Natl Acad Sci USA 94:10844–10849
Tsopanoglou NE, Maragoudakis ME (2007) Inhibition of angiogenesis by small-molecule antagonists of protease-activated receptor-1. Semin Thromb Hemost 33:680–687
Valadi H, Ekström K, Bossios A, Sjöstrand M, Lee JJ, Lötvall JO (2007) Exosome-mediated transfer of mRNAs and microRNAs is a novel mechanism of genetic exchange between cells. Nat Cell Biol 9:654–659
Wang JC, Begin LR, Berube NG, Chevalier S, Aprikian AG, Gourdeau H, Chevrette M (2007a) Down-regulation of CD9 expression during prostate carcinoma progression is associated with CD9 mRNA modifications. Clin Cancer Res 13:2354–2361
Wang XQ, Yan Q, Sun P, Liu JW, Go L, McDaniel SM, Paller AS (2007b) Suppression of epidermal growth factor receptor signaling by protein kinase C-alpha activation requires CD82, caveolin-1, and ganglioside. Cancer Res 67:9986–9995
Wang J, Liu X, Ni P, Gu Z, Fan Q (2010) SP1 is required for basal activation and chromatin accessibility of CD151 promoter in liver cancer cells. Biochem Biophys Res Commun 393:291–296
Weigelt B, Peterse JL, van 't Veer LJ (2005) Breast cancer metastasis: markers and models. Nat Rev Cancer 5:591–602
Wilson KS, Roberts H, Leek R, Harris AL, Geradts J (2002) Differential gene expression patterns in HER2/neu-positive and -negative breast cancer cell lines and tissues. Am J Pathol 161:1171–1185
Winterwood NE, Varzavand A, Meland MN, Ashman LK, Stipp CS (2006) A critical role for tetraspanin CD151 in alpha3beta1 and alpha6beta4 integrin-dependent tumor cell functions on laminin-5. Mol Biol Cell 17:2707–2721
Woegerbauer M, Thurnher D, Houben R, Pammer J, Kloimstein P, Heiduschka G, Petzelbauer P, Erovic BM (2010) Expression of the tetraspanins CD9, CD37, CD63, and CD151 in Merkel cell carcinoma: strong evidence for a posttranscriptional fine-tuning of CD9 gene expression. Mod Pathol 23:751–762
Wollscheid V, Kuhne-Heid R, Stein I, Jansen L, Kollner S, Schneider A, Durst M (2002) Identification of a new proliferation-associated protein NET-1/C4.8 characteristic for a subset of high-grade cervical intraepithelial neoplasia and cervical carcinomas. Int J Cancer 99:771–775
Wright MD, Tomlinson MG (1994) The ins and outs of the transmembrane 4 superfamily. Immunol Today 15:588–594

Wright MD, Geary SM, Fitter S, Moseley GW, Lau LM, Sheng KC, Apostolopoulos V, Stanley EG, Jackson DE, Ashman LK (2004) Characterization of mice lacking the tetraspanin superfamily member CD151. Mol Cell Biol 24:5978–5988

Xiao Z, Blonder J, Zhou M, Veenstra TD (2009) Proteomic analysis of extracellular matrix and vesicles. J Proteomics 72:34–45

Xu L, Hynes RO (2007) GPR56 and TG2: possible roles in suppression of tumor growth by the microenvironment. Cell Cycle 6:160–165

Yamamoto H, Vinitketkumnuen A, Adachi Y, Taniguchi H, Hirata T, Miyamoto N, Nosho K, Imsumran A, Fujita M, Hosokawa M, Hinoda Y, Imai K (2004) Association of matrilysin-2 (MMP-26) expression with tumor progression and activation of MMP-9 in esophageal squamous cell carcinoma. Carcinogenesis 25:2353–2360

Yamane H, Tachibana I, Takeda Y, Saito Y, Tamura Y, He P, Suzuki M, Shima Y, Yoneda T, Hoshino S, Inoue K, Kijima T, Yoshida M, Kumagai T, Osaki T, Eishi Y, Kawase I (2005) Propionibacterium acnes-induced hepatic granuloma formation is impaired in mice lacking tetraspanin CD9. J Pathol 206:486–492

Yanez-Mo M, Barreiro O, Gonzalo P, Batista A, Megías D, Genís L, Sachs N, Sala-Valdés M, Alonso MA, Montoya MC, Sonnenberg A, Arroyo AG, Sánchez-Madrid F (2008) MT1-MMP collagenolytic activity is regulated through association with tetraspanin CD151 in primary endothelial cells. Blood 112:3217–3226

Yang J, Weinberg RA (2008) Epithelial-mesenchymal transition: at the crossroads of development and tumor metastasis. Dev Cell 14:818–829

Yang X, Claas C, Kraeft SK, Chen LB, Wang Z, Kreidberg JA, Hemler ME (2002) Palmitoylation of tetraspanin proteins: modulation of CD151 lateral interactions, subcellular distribution, and integrin-dependent cell morphology. Mol Biol Cell 13:767–781

Yang X, Kovalenko OV, Tang W, Claas C, Stipp CS, Hemler ME (2004) Palmitoylation supports assembly and function of integrin-tetraspanin complexes. J Cell Biol 167:1231–1240

Yang XH, Kovalenko OV, Kolesnikova TV, Andzelm MM, Rubinstein E, Strominger JL, Hemler ME (2006) Contrasting effects of EWI proteins, integrins, and protein palmitoylation on cell surface CD9 organization. J Biol Chem 281:12976–12985

Yang XH, Richardson AL, Torres-Arzayus MI, Zhou P, Sharma C, Kazarov AR, Andzelm MM, Strominger JL, Brown M, Hemler ME (2008) CD151 accelerates breast cancer by regulating alpha 6 integrin function, signaling, and molecular organization. Cancer Res 68:3204–3213

Yauch RL, Kazarov AR, Desai B, Lee RT, Hemler ME (2000) Direct extracellular contact between integrin alpha(3)beta(1) and TM4SF protein CD151. J Biol Chem 275:9230–9238

Yoo SH, Lee K, Chae JY, Moon KC (2011) CD151 expression can predict cancer progression in clear cell renal cell carcinoma. Histopathology 58:191–197

Yoon SO, Zhang X, Freedman AS, Zahrieh D, Lossos IS, Li L, Choi YS (2010) Down-regulation of CD9 expression and its correlation to tumor progression in B lymphomas. Am J Pathol 177:377–386

Zakharova L, Svetlova M, Fomina AF (2007) T cell exosomes induce cholesterol accumulation in human monocytes via phosphatidylserine receptor. J Cell Physiol 212:174–181

Zevian S, Winterwood NE, Stipp CS (2011) Structure-function analysis of tetraspanin CD151 reveals distinct requirements for tumor cell behaviors mediated by alpha3beta1 versus alpha6beta4 integrin. J Biol Chem 286:7496–7506

Zhang XA, Kazarov AR, Yang X, Bontrager AL, Stipp CS, Hemler ME (2002) Function of the tetraspanin CD151-alpha6beta1 integrin complex during cellular morphogenesis. Mol Biol Cell 13:1–11

Zhang XA, He B, Zhou B, Liu L (2003a) Requirement of the p130CAS-Crk coupling for metastasis suppressor KAI1/CD82-mediated inhibition of cell migration. J Biol Chem 278:27319–27328

Zhang XA, Lane WS, Charrin S, Rubinstein E, Liu L (2003b) EWI2/PGRL associates with the metastasis suppressor KAI1/CD82 and inhibits the migration of prostate cancer cells. Cancer Res 63:2665–2674

Zhang F, Kotha J, Jennings LK, Zhang XA (2009) Tetraspanins and vascular functions. Cardiovasc Res 83:7–15

Zhao X, Lapalombella R, Joshi T, Cheney C, Gowda A, Hayden-Ledbetter MS, Baum PR, Lin TS, Jarjoura D, Lehman A, Kussewitt D, Lee RJ, Caligiuri MA, Tridandapani S, Muthusamy N, Byrd JC (2007) Targeting CD37-positive lymphoid malignancies with a novel engineered small modular immunopharmaceutical. Blood 110:2569–2577

Zheng Z, Liu Z (2006) CD151 gene delivery activates PI3K/Akt pathway and promotes neovascularization after myocardial infarction in rats. Mol Med 12:214–220

Zheng ZZ, Liu ZX (2007) Activation of the phosphatidylinositol 3-kinase/protein kinase Akt pathway mediates CD151-induced endothelial cell proliferation and cell migration. Int J Biochem Cell Biol 39:340–348

Zheng R, Yano S, Zhang H, Nakataki E, Tachibana I, Kawase I, Hayashi S, Sone S (2005) CD9 overexpression suppressed the liver metastasis and malignant ascites via inhibition of proliferation and motility of small-cell lung cancer cells in NK cell-depleted SCID mice. Oncol Res 15:365–372

Zhijun X, Shulan Z, Zhuo Z (2007) Expression and significance of the protein and mRNA of metastasis suppressor gene ME491/CD63 and integrin alpha5 in ovarian cancer tissues. Eur J Gynaecol Oncol 28:179–183

Zhong S, Fields CR, Su N, Pan YX, Robertson KD (2007) Pharmacologic inhibition of epigenetic modifications, coupled with gene expression profiling, reveals novel targets of aberrant DNA methylation and histone deacetylation in lung cancer. Oncogene 26:2621–2634

Zhou B, Liu L, Reddivari M, Zhang XA (2004) The palmitoylation of metastasis suppressor KAI1/CD82 is important for its motility- and invasiveness-inhibitory activity. Cancer Res 64:7455–7463

Zhou Z, Ran YL, Hu H, Pan J, Li ZF, Chen LZ, Sun LC, Peng L, Zhao XL, Yu L, Sun LX, Yang ZH (2008) TM4SF3 promotes esophageal carcinoma metastasis via upregulating ADAM12m expression. Clin Exp Metastasis 25:537–548

Zhu GZ, Miller BJ, Boucheix C, Rubinstein E, Liu CC, Hynes RO, Myles DG, Primakoff P (2002) Residues SFQ (173–175) in the large extracellular loop of CD9 are required for gamete fusion. Development 129:1995–2002

Zhu GH, Huang C, Qiu ZJ, Liu J, Zhang ZH, Zhao N, Feng ZZ, Lv XH (2010) Expression and prognostic significance of CD151, c-Met, and integrin alpha3/alpha6 in pancreatic ductal adenocarcinoma. Dig Dis Sci 56(4):1090–1098, Oct 7—Epub ahead of print

Zijlstra A, Lewis J, Degryse B, Stuhlmann H, Quigley JP (2008) The inhibition of tumor cell intravasation and subsequent metastasis via regulation of in vivo tumor cell motility by the tetraspanin CD151. Cancer Cell 13:221–234

Zöller M (2006) Gastrointestinal tumors: metastasis and tetraspanins. Z Gastroenterol 44:573–586

Zöller M (2009) Tetraspanins: push and pull in suppressing and promoting metastasis. Nat Rev Cancer 9:40–55

Chapter 12
Uroplakins as Unique Tetraspanin Networks

Tung-Tien Sun, Gert Kreibich, Angel Pellicer, Xiang-Peng Kong, and Xue-Ru Wu

Abstract A major class of tetraspanins are uroplakins (UP's) Ia and Ib that, together with their associated (non-tetraspanin) uroplakins II and IIIa, form two dimensional crystals of 16-nm particles, known as 'urothelial plaques'. Interconnected by small hinge areas, urothelial plaques cover almost the entire apical surface of mammalian bladder urothelium, and contribute to the remarkable urothelial permeability barrier function. UPIa and Ib bind selectively to UPII and IIIa, respectively, to form a UPIa/II and UPIb/IIIa heterodimer, which constitutes one of the six dumbbell-shaped, heterotetramer subunit of a stellate-shaped 16-nm uroplakin particle. Ultrastructural studies indicate that UPIa/II and UPII/IIIa dimers are associated with the inner and outer subdomains of the 16-nm particles, respectively. In vitro gel overlay assay suggests that the high mannose glycan anchored on the second, large extracellular

T.-T. Sun (✉)
Department of Cell Biology, New York University School of Medicine, New York University Langone Medical Center, 550 First Avenue, New York, NY 10016, USA
e-mail: sunt01@nyumc.org

Department of Dermatology, New York University Langone Medical Center, 550 First Avenue, New York, NY 10016, USA

Department of Pharmacology, New York University Langone Medical Center, 550 First Avenue, New York, NY 10016, USA

Department of Urology, New York University Langone Medical Center, 550 First Avenue, New York, NY 10016, USA

New York University Cancer Institute, New York University Langone Medical Center, 550 First Avenue, New York, NY 10016, USA

G. Kreibich
Department of Cell Biology, New York University School of Medicine, New York University Langone Medical Center, 550 First Avenue, New York, NY 10016, USA

New York University Cancer Institute, New York University Langone Medical Center, 550 First Avenue, New York, NY 10016, USA

F. Berditchevski and E. Rubinstein (eds.), *Tetraspanins*, Proteins and Cell Regulation 9, DOI 10.1007/978-94-007-6070-7_12,

loop of UPIa may serve as the urothelial receptor for the type 1-fimbriated E. coli that causes over 85% of urinary tract infection. Moreover, uroplakin defects may play a role in renal adysplasia and overreactive bladder. Although uroplakins are expressed mainly as major differentiation products of bladder urothelium, small amounts of uroplakins are present in some nonurothelial tissues including oocytes and may play a diverse range of important biological functions.

12.1 Introduction

Uroplakins are integral membrane proteins made by mammalian bladder urothelial cells during terminal differentiation. These proteins are highly conserved during mammalian evolution, and they constitute the 16-nm protein particles that are organized hexagonally into two-dimensional crystals called urothelial plaques that cover almost the entire apical surface of the mammalian bladder urothelium. Since, in cross-sections, the luminal leaflet of urothelial plaques is nearly twice as thick as the cytoplasmic one, these plaques are also called asymmetric unit membrane (AUM). Urothelial plaques purified from bovine bladders consist of four major uroplakins (UP's): UPIa (27-kDa), UPIb (28-kDa), UPII (15-kDa) and UPIIIa (47-kDa). Since UPIa and UPIb are tetraspanins, the crystalline urothelial plaques that consist of UP's represent a unique tetraspanin network (web) that is structurally well-defined and can reach 500–1,000 nm in diameter. In this chapter, we discuss the structure, function and disease implications of the uroplakins.

A. Pellicer
Department of Pathology, New York University Langone Medical Center,
550 First Avenue, New York, NY 10016, USA

New York University Cancer Institute, New York University Langone Medical Center,
550 First Avenue, New York, NY 10016, USA

X.-P. Kong
Department of Biochemistry, New York University Langone Medical Center,
550 First Avenue, New York, NY 10016, USA

New York University Cancer Institute, New York University Langone Medical Center,
550 First Avenue, New York, NY 10016, USA

X.-R. Wu
Department of Urology, New York University Langone Medical Center,
550 First Avenue, New York, NY 10016, USA

Department of Pathology, New York University Langone Medical Center,
550 First Avenue, New York, NY 10016, USA

New York University Cancer Institute, New York University Langone Medical Center,
550 First Avenue, New York, NY 10016, USA

Veterans Affairs New York Harbor Healthcare Systems,
Manhattan Campus, New York, NY 10010, USA

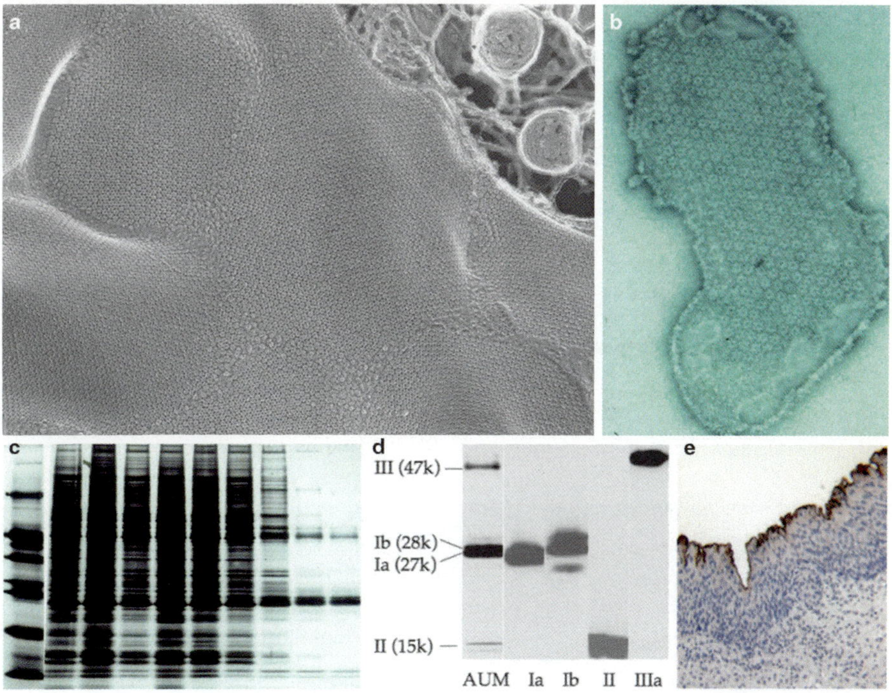

Fig. 12.1 Mono-specific antibodies to uroplakins of bovine urothelial plaques. (**a**) The apical surface of urothelial umbrella cells of the mouse bladder is covered by 2D crystals of 16-nm protein particles (urothelial plaques) interconnected by the (relatively particle-free) hinge areas, as visualized by quick-freeze deep-etch imaging (Kachar et al. 1999). (**b**) A bovine urothelial plaque purified using a discontinuous sucrose-density gradient coupled with detergent-wash (Wu et al. 1990). (**c**) Protein composition of various intermediate steps of bovine urothelial plaque purification showing that highly purified plaques contained four major protein bands that are 47-, 28-, 27- and 15-kDa. (**d**) Generation of rabbit antisera that are monospecific for uroplakin (UP) Ia (27-K-kDa), UPIb (28-kDa), UPII (15-kDa) and UPIIIa (470 kDa). (**e**) Immunohistochemical staining of bovine urothelium using a rabbit antibody to total uroplakins, showing the differentiation-dependent expression of uroplakins in terminally differentiated bovine urothelial umbrella cells (Wu et al. 1990)

12.2 Uroplakins as Major Urothelial Differentiation Products

Although the urothelial plaques cover almost the entire mammalian urothelium (Fig. 12.1) and are therefore major urothelial differentiation products (Porter and Bonneville 1963; Hicks 1965; Hicks and Ketterer 1969; Koss 1969; Vergara et al. 1969; Staehelin et al. 1972; Robertson and Vergara 1980), it was difficult to pinpoint their protein composition due to difficulties in obtaining highly purified plaques or antibodies against any of the putative plaque-associated proteins (Ketterer and Hicks 1971; Ketterer et al. 1973; Vergara et al. 1974). We overcame these technical hurdles by purifying milligram quantities of bovine urothelial

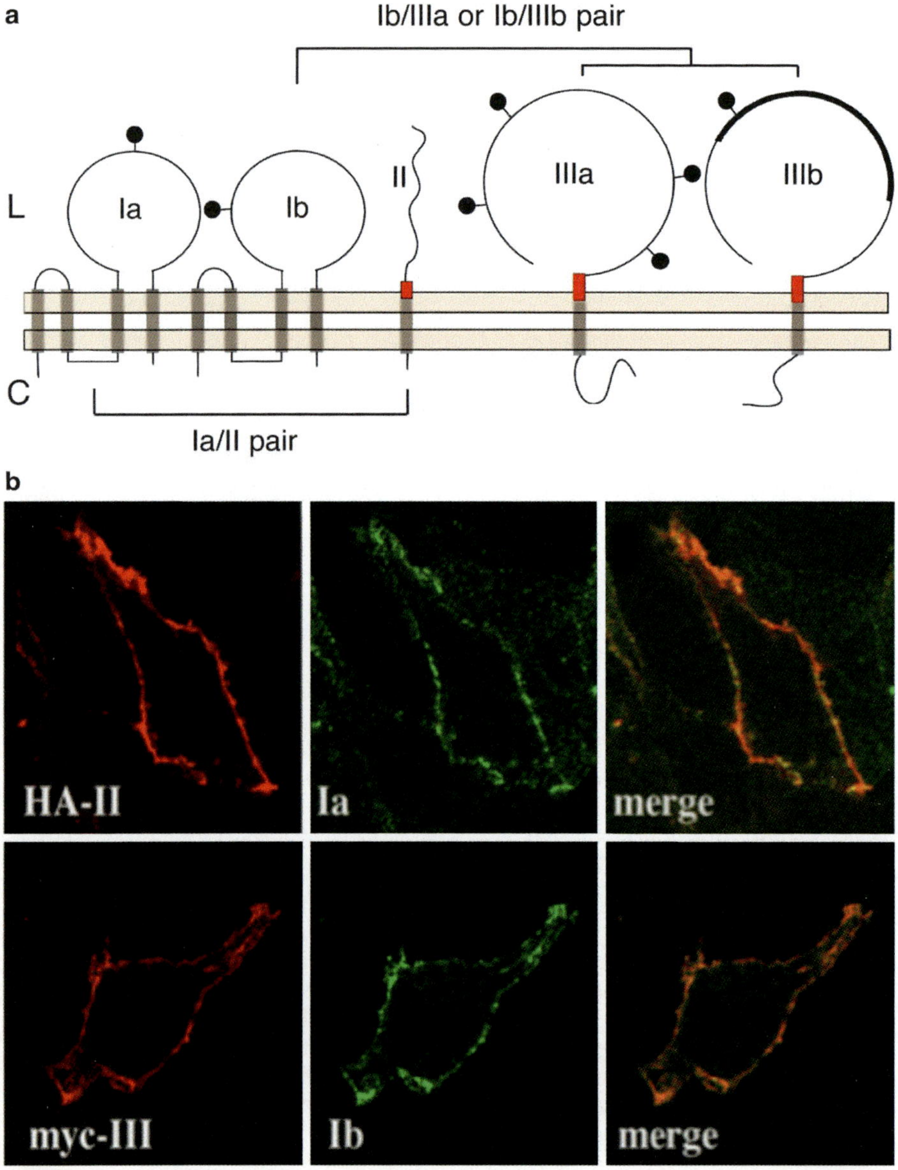

Fig. 12.2 Formation of uroplakin heterodimers. (**a**) A schematic representation of the transmembrane topology of uroplakin (UP) Ia, UPIb (both tetraspanins), UPII, UPIIIa and UPIIIb (an isoform of UPIIIa present in minor amounts in urothelial plaques). Note that UPIa and UPIb form heterodimers with UPII and UPIIIa, respectively, to form heterodimers (based on Wu et al. 1990; Yu et al. 1990; Lin et al. 1994). (**b**) Double transfection of 293T cells with UPII (HA-tagged) and UPIa, or with UPIIIa (myc-tagged) and UPIb, allows the UP heterodimers to exit from the ER to reach the cell surface (Tu et al. 2002; Hu et al. 2005)

plaques (Wu et al. 1990), which showed typical two-dimensional crystals comprised of 16-nm hexagonal particles (Fig. 12.1). These purified plaques contained four major proteins with molecular masses of 15-, 27-, 28- and 47-kDa (Fig. 12.1). We later generated mono-specific antibodies to (Fig. 12.1) and cloned the cDNAs of these proteins (Wu and Sun 1993; Lin et al. 1994; Yu et al. 1994). Since electron microscopy localized each of these proteins to AUM in situ and since they represented major protein components of urothelial plaques, we named these plaque-associated membrane proteins uroplakins Ia (27-kDa), Ib (28-kDa), II (15-kDa) and IIIa (47-kDa) (Fig. 12.1) (Wu and Sun 1993; Lin et al. 1994; Yu et al. 1994).

Uroplakins Ia and Ib are 40% identical in amino acid sequences, and both possess four transmembrane domains (TMD) with a minor (first) and a major (second) hydrophilic domain, which are both extracellular (Fig. 12.2a; Yu et al. 1994). These two uroplakins have very short cytoplasmic domains and are members of the tetraspanin family that consists of a number of leukocyte differentiation antigens (e.g., CD9, CD27, CD63, CD81 and CD82) (Hemler 2005; Levy and Shoham 2005; Berditchevski and Odintsova 2007; Charrin et al. 2009; Rubinstein 2011). Uroplakin II has a signal peptide at the N-terminus followed by a 59 amino acid, glycosylated pro-sequence terminating in RGRR—a cleavage site for a trans-Golgi-associated furin processing enzyme. The mature UPII is un-glycosylated and has one transmembrane domain at its C-terminus serving as a membrane anchor, again with a very little cytoplasmic domain (Lin et al. 1994). UPIIIa also has only one TMD, which separates the protein into an N-terminal luminal domain (189 amino acids) and a C-terminal cytoplasmic domain (52 residues; Fig. 12.2a; Wu and Sun 1993). The N-terminal domain is modified with 20-kDa equivalents of complex-type sugars, while the cytoplasmic domain contains clusters of serines and threonines in an excellent context for phosphorylation. UPIIIa and UPII share a stretch of about 12 amino acids, located on the N-terminal side of their single TMDs, suggesting that the two uroplakins are related (Wu and Sun 1993; Lin et al. 1994). Since UPIIIa is the only major uroplakin that possesses a significant cytoplasmic domain, we speculated that UPIIIa may be involved in anchoring the AUMs to an underlying cytoskeleton (Staehelin et al. 1972) and that its phosphorylation may be involved in regulating this process (Wu and Sun 1993). Finally, there is a minor UPIII isoform, UPIIIb, which is evolutionarily highly conserved but with no known biological function (Deng et al. 2002). Overall, the mass of uroplakins' extracellular domains significantly exceeds that of their cytoplasmic ones. This asymmetric mass distribution of uroplakins across the lipid bilayer may explain why the luminal leaflet of AUM is nearly twice as thick as the cytoplasmic one (Wu and Sun 1993; Lin et al. 1994; Yu et al. 1994).

12.3 The Four Uroplakins Can Be Divided into Two Heterodimer Pairs

Several lines of evidence suggest that the four major uroplakins can be divided into two pairs. *First*, UPIa and UPIb can be specifically cross-linked to UPII and UPIIIa, respectively, in purified bovine urothelial plaques by bifunctional crosslinking reagents (Fig. 12.2), suggesting that UPIa/II and UPIb/IIIa exist as heterodimers. *Second*, UPIa/II and UPIb/IIIa can be isolated as complexes by ion exchange chromatography (Liang et al. 2001). *Third*, transfection studies showed that, even though when 293T cells were transfected with single uroplakin cDNAs, the expressed uroplakins were mostly retained in the ER (except UPIb which can exist by itself), double transfection of UPIa/II or UPIb/IIIa allowed the heterodimers to reach the cell surface (Fig. 12.2) (Tu et al. 2002). These results indicate that the dimerization of UPIa/II or UPIb/IIIa is a prerequisite for uroplakin assembly and is required for ER-exit of the uroplakins (Tu et al. 2002; Hu et al. 2005).

12.4 The UPIa/II and UPIb/IIIa Heterodimer Pairs Are Associated with the Inner and Outer Subdomains of the 16-nm Uroplakin Particle, Respectively

We studied the structure of the purified bovine and mouse urothelial plaques, that consisted of two-dimensional crystals of 16-nm uroplakin particles, using quick-freeze deep-etch (Fig. 12.1; Kachar et al. 1999), negative staining coupled with image processing (Fig. 12.4; Walz et al. 1995; Min et al. 2003, 2006), scanning transmission electron microscopy (Walz et al. 1995), atomic force microscopy and cryo-EM (Min et al. 2002, 2003, 2006). These analyses resolved each 16-nm particle into six inner and six outer subdomains forming two concentric rings. One inner and one outer subdomain interconnect to form a subunit (six of which forming a 16-nm particle). Consistently, scanning transmission electron microscopy of the surface-exposed, extracellular domains of the 16-nm particle suggests that each outer or inner subdomain can accommodate roughly the mass of a uroplakin heterodimer (Walz et al. 1995). This model predicted that each subdomain contains five transmembrane polypeptide chains (Fig. 12.4c). Our recent, high-resolution cryo-EM data at 6 Å resolution showed that this prediction indeed turned out to be correct (Min et al. 2006). We next asked whether the UPIa/II and UPIb/IIIa pairs are associated with the inner and outer subdomains, respectively, or vice versa. To study this, we made an important finding by showing that FimH, a lectin located at the very tip of the type 1-fimbria of the uropathogenic E. coli, can bind selectively to the high mannose moieties associated with the large extracellular loop of uroplakin Ia (Fig. 12.3a; Zhou et al. 2001). This finding not only established UPIa as a putative bacterial receptor, but also enabled us to use FimH as a probe to localize UPIa (Fig. 12.4). We performed negative staining of purified mouse urothelial plaques in the presence and absence of saturating amounts of FimH, and obtained the difference map between the two images (Fig. 12.4a). The result showed that the bound FimH, hence UPIa, was associated exclusively with the inner subdomains of the 16-nm particle (Fig. 12.4b; Min et al. 2002). Taken together, the results indicate that UPIa/II and UPIb/IIIa pairs occupy the inner and outer subdomains of the 16-nm particle, respectively (Fig. 12.4d).

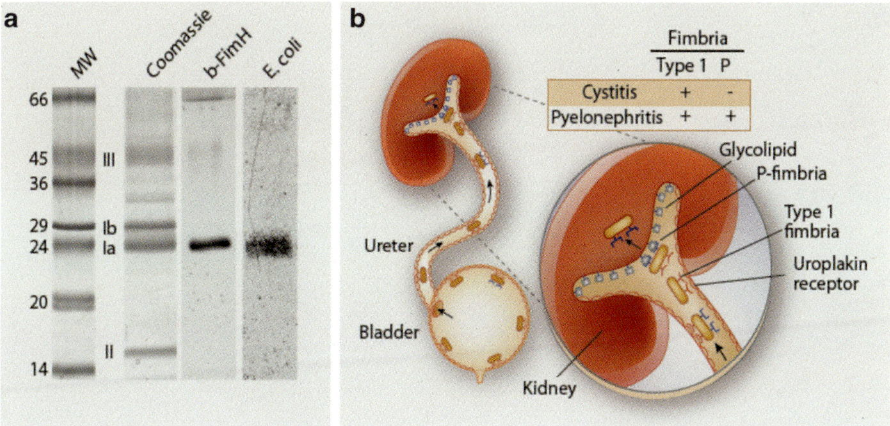

Fig. 12.3 Role of uroplakin Ia in the pathogenesis of urinary tract infections. (**a**) The four major mouse uroplakins (Ia, Ib, II, and IIIa), resolved by SDS-PAGE, were stained with silver nitrate (lane 2), reacted with biotin-tagged, bacterial FimH/C complex (lane 3) or with ^{35}S-methionine-labeled type 1-fimbriated E. coli (lane 4). Note that UPIa was the only uroplakin that reacted with FimH and the uropathogenic E. coli. MW denotes molecular weight standards. (**b**) A diagram depicting the interactions between uropathogenic E. coli and its receptors in the lower and upper urinary tract. Type 1 fimbriae of the uropathogenic E. coli adhere to the UPIa receptor in the urinary tract causing cystitis. Uropathogenic E. coli expressing type 1 as well as P fimbriae, the latter of which is capable of binding to the glycolipids in the renal pelvis, can however ascend to the upper tract and cause pyelonephritis (from Wu et al. 2009)

12.5 Uroplakin Assembly and Trafficking

As mentioned, UPII possesses a heavily glycosylated pro-sequence that we hypothesized to play a key role in regulating UP oligomerization and targeting to the urothelial apical surface. We showed by transfecting cultured cells that the disulfide formation within the prosequence of UPII is a prerequisite for heterodimer formation of UPIa/UPII and their exit from the ER; that aberrant glycosylation of the UPII's prosequence hampers the tetramerization of UPs, hence the formation of the two-dimensional UP crystals; and that post-cleaved UPII prosequence remains associated with the urothelial plaques on the apical surface (Hu et al. 2008). These data underscore the critical importance of the UPII prosequence in the assembly of the UP super-molecular complex and its targeting to the apical surface (Hu et al. 2008).

We also studied the role of several cytoplasmic proteins in the apical targeting and degradation of UPs. We found that Rab27b, known to be essential for the targeting of lysosome-related organelles (LROs) (Barral et al. 2002; Starcevic et al. 2002; Raposo et al. 2007; Tolmachova et al. 2007), is localized to the fusiform vesicles of the urothelial umbrella cells (Chen et al. 2003). Because these organelles have an acidified lumen but are CD63-negative (Guo et al. 2009), they are not typical of LROs. They may be considered, however, as specialized granules that deliver the

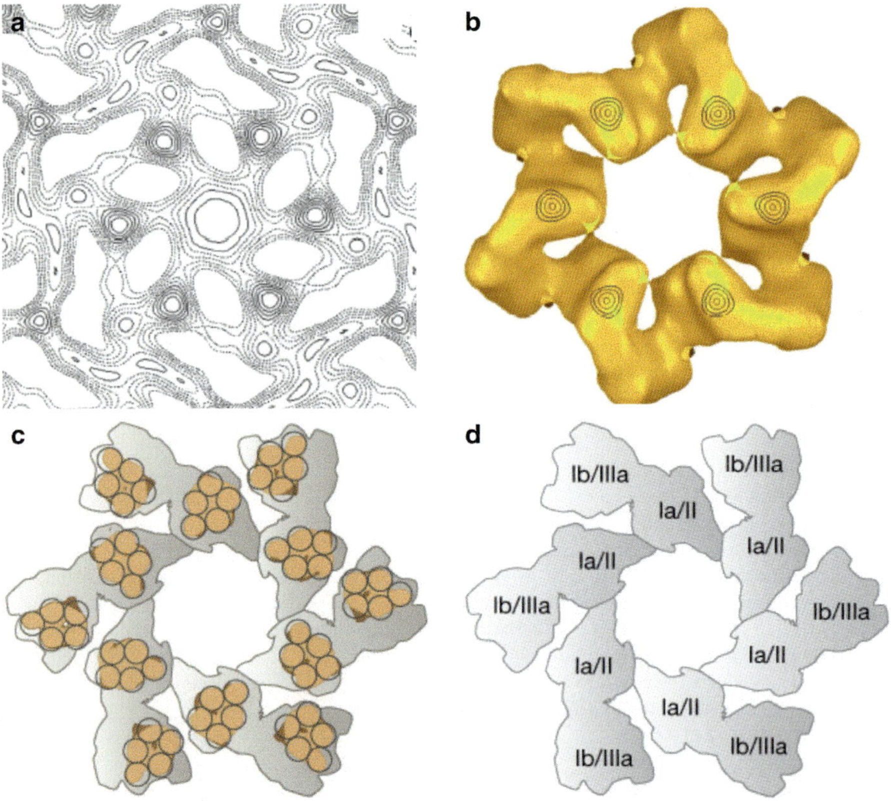

Fig. 12.4 Association of specific uroplakin heterodimers with the inner and outer subdomains of the 16-nm uroplakin particle. (**a**) A difference image after the highly purified mouse urothelial plaques that were negative-stained with or without the presence of FimH adhesin that specifically reacts with UPIa. (**b**) Superimposing **a** onto a 16-nm uroplakin particle demonstrates the association of FimH (specific for UPIa) with the inner subdomains of the uroplakin particle (**c**) Locations of the transmembrane helices of the four major uroplakins (*orange color*). (**d**) A hypothetic model showing that the UPIa/II pair and the UPIb/IIIa pair occupy the inner and outer subdomains of the 16-nm particle, respectively (Adapted from Min et al. 2002, 2003)

crystalline arrays of uroplakins to the apical cell surface. We also found that Vps33a, a Sec1-related protein, plays a key part in mediating the fusion of multi-vesicular bodies with the lysosomal compartment in urothelial umbrella cells. Deficiency in Vps33a, such as loss-of-function mutations, can therefore impair the lysosomal degradation of uroplakins (Guo et al. 2009).

To understand the relationship between UPIa and UPIb genes and other members of the tetraspanin superfamily, we studied, in collaboration with Antonio Garcia-España (Universitat Rovira i Virgili, Tarragona, Spain) and Robert DeSalle (American Museum of Natural History), the evolution of these two genes. Our results suggest that the tetraspanin superfamily can be subclassified into four families, i.e., the CD family, CD63 family, uroplakin family, and RDS family. The uroplakin family can be divided into several ortholog groups, that evolved from the sequence divergence of one or more ancestral gene(s), rather than the convergence from multiple unrelated genes (Garcia-Espana et al. 2006, 2008, 2009).

12.6 A Model of Uroplakin Assembly

Using a large panel of antibodies to peptide-defined uroplakin epitopes coupled with transfection studies, we have demonstrated that uroplakins undergo major conformational changes when they form dimers and tetramers; that differentiation-dependent glycosylation of the UPII prosequence may play a role in allowing the two heterodimers to form tetramers; and that furin-mediated cleavage of the prosequence of UPII may play a key role in triggering the formation of the 16-nm particle (Hu et al. 2005). Figure 12.5 shows the hypothetical steps of how the four major uroplakins (UPIa, Ib, II and IIIa) form the 16-nm particles of the 2D crystals and how this process is regulated (Hu et al. 2005). Briefly, UPIa and UPIb form heterodimers with UPII and UPIIIa, respectively, which can then exit from the ER; the prosequence of UPII undergoes in the Golgi apparatus a differentiation-dependent glycosylation that induces conformational changes in the UPIa/UPII dimer enabling it to bind UPIb/IIIa to form heterotetramer. Finally, the prosequence of UPII is cleaved by a trans-Golgi network-associated furin allowing six heterotetramers to form a 16-nm particle in which UPIa/II and UPIb/IIIa occupy the inner and outer subdomains of a subunit, respectively (Min et al. 2002), and to later form 2D crystals (Hu et al. 2005). In cultured urothelial cells which mimic regenerative urothelium (Sun 2006), the differentiation-dependent glycosylation of UPII pro-sequence is defective thus blocking the assembly at the heterodimer stage (Hu et al. 2005, 2008) (Fig. 12.5).

12.7 Biological Functions of Uroplakins

To assess the biological functions of uroplakins, we chose to genetically ablate mouse genes encoding uroplakin II and IIIa because the knockout of these two uroplakins should abolish the formation of UPIa/II and UPIb/III pair, respectively (Hu et al. 2000; Kong et al. 2004). We have also generated the UPII-UPIIIa double knockout. From these studies we learned that:

(a) *Uroplakins are the integral subunit proteins of the urothelial plaques.* Ablation of the UPIIIa gene resulted in the loss of about 70–80% of the apical urothelial plaques (Hu et al. 2000); there is still some plaque formation due to the remaining minor UPIII isoform, UPIIIb, which we identified from a bovine urothelial subtraction cDNA library (Deng et al. 2002). However, knockout of the UPII gene, which is not known to have an isoform, resulted in the complete loss of

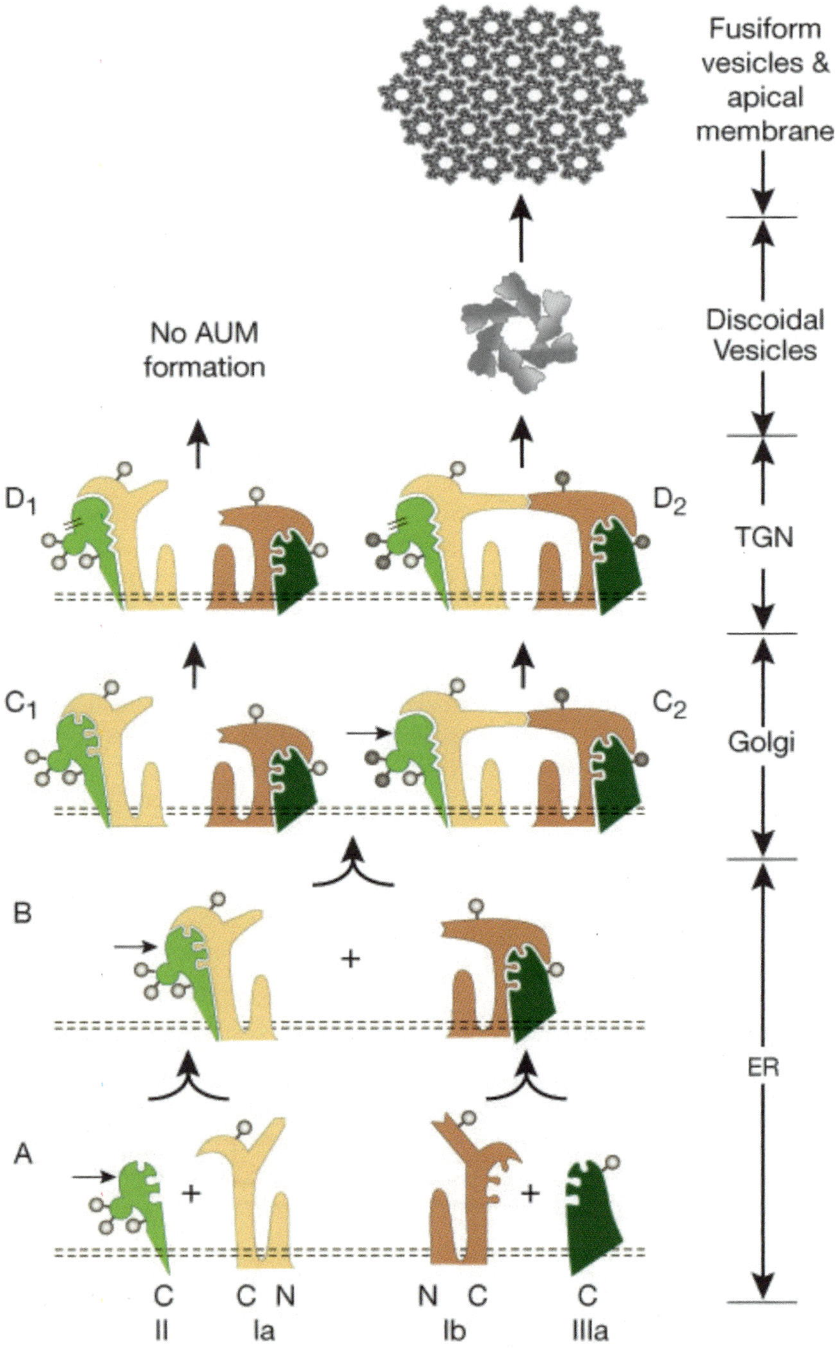
Fusiform vesicles & apical membrane
Discoidal Vesicles
TGN
Golgi
ER
No AUM formation
D_1
D_2
C_1
C_2
B
A
C
C N
N C
C
II
Ia
Ib
IIIa

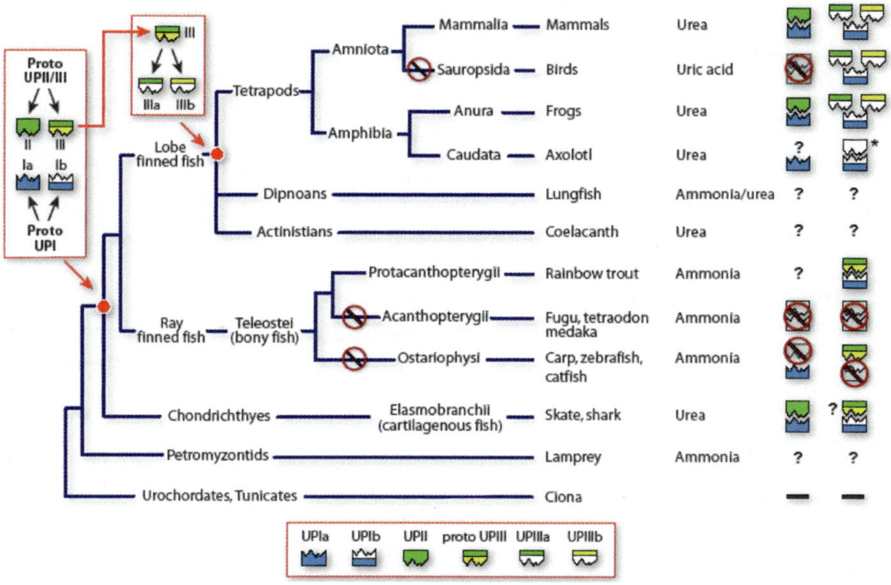

Fig. 12.6 A hypothetic model depicting the evolutionary events of the uroplakin gene family. The divergent evolution of uroplakin genes is illustrated using cartoons on the different lineages. The proto-UPI and proto-UPII/III genes were duplicated in the ancestor of vertebrates. UPII sequences first appeared in the Cartilagenous fish. Symbols: a *red circle* with a slash indicates losses of UP genes; a *dash* indicates the absence of UP-related genes in sea squirt ciona; a *question mark* indicates the ambiguity of whether the UP sequences were present, due in part to small EST databases and/or incomplete genomic sequences; an *asterisk* indicates a truncated Axolotl UPIII protein. Also included in the diagram were the key forms of nitrogen waste for each lineage including ammonia, urea or uric acid (From Garcia-Espana et al. 2008)

Fig. 12.5 Schematic representation of the sequential steps leading to the assembly of the urothelial plaques. Steps A and B: The four major uroplakins (UPIa, UPIb, UPII and UPIIIa) are glycosylated with high-mannose moieties in the ER and interact specifically to form two heterodimers (UPIa/II and UPIb/IIIa). *Arrows* denote the cleavage site of the UPII-prosequence mediated by furin. The *open* and *closed circles* denote high-mannose- and complex-type glycans, respectively. In normal in vivo urothelium (C2), complex-type glycosylation occurs in the Golgi apparatus on two of the three N-glycosylation sites. After the cleavage of the UPII prosequence (D2), oligomerization of the four uroplakins takes place, forming a 16-nm particle. This (default) normal pathway is defective in cultured urothelial cells (*left* pathway: C1 and D1), due to altered glycosylation of the UPII prosequence. This prevents the formation of the uroplakin heterotetramers, the 16-nm particle or the urothelial plaques (Adapted from Hu et al. 2005)

urothelial plaques (Kong et al. 2004). These results establish uroplakins as the integral subunits of the urothelial plaques.

(b) *Uroplakins contribute to the urothelial permeability barrier function*. In collaboration with Mark Zeidel (Harvard University), we showed that the barrier function of the uroplakin-deficient urothelium was compromised, suggesting that the crystalline network of uroplakin proteins may impose structural constraints to the lipid molecules reducing their ability to move laterally, thus enhancing the barrier function (Negrete et al. 1996; Hu et al. 2002; Min et al. 2003).

(c) *Uroplakins may stabilize mechanically the urothelial surface*. A prominent feature of uroplakin knockout urothelium is a greatly reduced umbrella cell size. This raises the interesting possibility that continued insertion of uroplakin plaques into the apical surface may stabilize the surface and contribute to its expansion during the formation of umbrella cells (Kong et al. 2004).

(d) *Small amounts of uroplakins may perform versatile functions in some non-urothelial epithelia*. Although uroplakins are thought to be 'mammalian urothelium-specific', our recent phylogenetic studies of all the available genomic sequences, and a systematic analysis of the expression pattern of the uroplakin-related genes in several animal taxa (Fig. 12.6), suggest that: (i) the primitive UPIa/UPIb and UPII/UPIII genes evolved through gene duplication in the common ancestor of vertebrates; (ii) uroplakins are lost in various combinations in vertebrates such that chicken has only the UPIb/IIIa, but no UPIa/UPII pair; and (iii) UPIa and UPIb likely co-evolved with their partners UPII and UPIIIa/UPIIIb, respectively (Fig. 12.6; (Garcia-Espana et al. 2006, 2008)). Such a co-evolutionary relationship between the tetraspanin UPIa and UPIb and their associated proteins, may be important for optimizing the structure and function of the uroplakins, making them capable of performing species—and tissue-specific functions. The presence of uroplakins in *Xenopus* kidney, oocytes and fat body suggests that these proteins perform much more diverse functions than previously appreciated (Mahbub Hasan et al. 2005, 2007; Sakakibara et al. 2005; Garcia-Espana et al. 2006, 2008; Hasan et al. 2011).

12.8 Clinical Significance of Uroplakins

(a) *Urinary tract infection*: We have shown previously that the high mannose glycans of mouse uroplakin Ia, but not those of its structurally related UPIb, serve as the receptor for the uropathogenic, type 1-fimbriated E. coli (Fig. 12.3) (Wu et al. 1996; Zhou et al. 2001). This finding has implications for how bacteria survive host defenses by invading the host urothelial umbrella cells (Fig. 12.3) forming a biofilm (Anderson et al. 2003; Wright et al. 2007; Justice et al. 2008), and for the design of novel high-affinity inhibitors of the E. coli FimH adhesin (Bouckaert et al. 2005; Wellens et al. 2008). We made several additional findings: (i) In collaboration with Cathy Costello (Boston University Medical

School), we showed that mouse UPIa bears a series of high mannose glycan isoforms, while UPIb bears complex glycans (Xie et al. 2006); and (ii) Recombinant FimH also selectively binds to human UPIa, indicating that the same UPIa receptor may also mediate UTI in human (Xie et al. 2006). These findings can explain the differential binding of the type 1-fimbriated E. coli to UPIa versus UPIb, and provide a basis for assessing the relative affinities of various glycan isoforms to FimH (Xie et al. 2006). (iii) In a collaborative study with David Klumpp's group (Northwestern University Medical School), we demonstrated that UPIIIa, the only major UP with a significant cytoplasmic domain, has a novel signaling role for bacterial invasion and apoptosis. Upon FimH adhesin binding, a specific threonine in the cytoplasmic tail of UPIIIa, undergoes phosphorylation by casein kinase 2 (CK2). This is accompanied by an increase of intracellular calcium and bacterial invasion, events that can be abolished by CK2 inhibitors. These results raised the interesting possibility that UPIIIa plays an essential role in mediating the transmembrane signaling upon adhesion of uropathogenic E. coli to the urothelial cells (Thumbikat et al. 2009). (iv) Data from the above in vitro and in vivo biological analyses are supported by our recent structural studies of FimH/UP interaction. Using cryo-electron microscopy, we found that binding of FimH to UPIa in highly purified urothelial plaques can trigger a marked change in the conformation of the uroplakin complex, including the transmembrane helices (Wang et al. 2009). This may be responsible for the lateral movement of the cytoplasmic tail of UPIIIa, leading to an activation of the downstream signals. These data are important not only for understanding the molecular underpinning of UTI pathogenesis, but also for the possible mechanisms on how pathogens exploit natural receptor complexes to gain entry into host cells (Wang et al. 2009).

(b) *Vesicoureteral reflux and renal adysplasia*: Since genetic ablation of both uroplakins II and IIIa in mice yielded widespread abnormalities in lower urinary tract including vesicoureteral reflux (Hu et al. 2000; Kong et al. 2004), we examined whether uroplakin defects are involved in vesicoureteral reflux (VUR). In collaboration with Anthony Atala (Wake Forest University Medical School), Francis Schneck (University of Pittsburgh), Garth Ehrlich (Drexel University Medical School), Ellen Shapiro (NYU Medical School) and Jun Yu (Beijing Genomic Institute), we assembled a cohort of 76 clinically documented VUR patients and 90 race-matched controls, and determined whether single nucleotide polymorphisms (SNPs) of the four major uroplakins were significantly associated with VUR. We found that most SNPs were not significantly associated with human VUR, although a few of them were marginally associated (Jiang et al. 2004). No truncation or frame shift mutations were found in the VUR patients. This, along with the fact that some UPIII knockout mouse litters exhibited VUR as well as hydronephrosis and neonatal death (Hu et al. 2000; Kong et al. 2004), suggests that major uroplakin mutations in humans may not be tolerated (Jiang et al. 2004). Other investigators studying the SNPs of only the UPIIIa gene have shown similar results (Giltay et al. 2004; Kelly et al. 2005). Jenkins et al., in collaboration with us, demonstrated that de novo heterozygous mutations in the UPIIIa gene were associated with human renal adysplasia, which can lead to renal failure (Jenkins et al. 2005; also see Schonfelder et al. 2006).

(c) *Overactive bladder*: We have shown previously with Mark Zeidel (Harvard Medical School) that UPIIIa knockout led to increased bladder permeability to water and urea (Hu et al. 2002). In a more recent collaboration with M. Zeidel and G. Apodaca (University of Pittsburgh Medical School), we found that UPII knockout and UPII/UPIIIa double knockout led to even higher permeability thus further establishing the importance of uroplakins in bladder barrier function. In another collaboration with George Christ (Wake Forest University), we assessed the bladder physiology of the uroplakin-deficient mice using continuous cystometry in conscious, freely moving mice. The results revealed many gender-specific changes in bladder physiology including nonvoiding contractions, indicating that urothelial defects due to the loss of uroplakins can cause overactive bladder (Hodges et al. 2008; Aboushwareb et al. 2009).

(d) *Urothelial metaplasia and heterogeneity*: Although it is well known that vitamin A deficiency can cause urothelium to keratinize (bladder metaplasia), it has been mysterious why the results were quite variable sometimes with areas showing full keratinization interspersed by normal-looking urothelia. We found recently that mouse and bovine urothelia can be divided into at least three distinct lineages that cover (i) renal pelvis/ureter, (ii) bladder and (iii) proximal urethra/bladder neck, based on their uroplakin content, in vitro growth/differentiation features, and embryological origins. Moreover, we found that proximal urothelium is particularly susceptible to vitamin A-deficiency forming a keratinized epithelium that can invade and replace the surrounding, normal-looking bladder urothelium. These results suggest that 'expansion and invasion' is the cellular basis for urothelial metaplasia (Liang et al. 2005). Although the term 'urothelium' has been used to describe indiscriminatively all the epithelia covering the entire lower urinary tract, our finding that it actually consists of at least three lineages means that we need to carefully re-examine whether urothelia of different lineages may function somewhat differently as a source for bladder reconstruction, and whether neoplasms of different urothelial origins may have different biological and clinical behaviors (Liang et al. 2005).

(e) *Bladder cancer diagnosis and prognosis*: In a multi-group collaborative study, we showed that antibodies to uroplakins can be used for the positive identification of local as well as metastatic transitional cell carcinoma, and that the detection of uroplakin proteins or mRNA in the blood of bladder cancer patients can be used to monitor the circulating bladder cancer cells (Moll et al. 1995; Li et al. 1999; Osman et al. 2004). We have identified several novel blood cell markers that can distinguish bladder cancer from renal and testicular cancer (Osman et al. 2006). We also showed that loss of uroplakin expression correlated with, lymph node metastases, late stages, disease recurrence and cancer-specific mortality of muscle-invasive bladder cancers (Huang et al. 2007).

(f) *The use of uroplakin promoter for studying urothelial tumorigenesis and other urinary tract diseases.* We have isolated the genes of mouse UPIa, UPII and UPIIIa (Lin et al. 1995) and showed that a 3.6-Kb UPII promoter can drive the expression of a bacterial LacZ reporter gene and a human growth hormone gene mainly in suprabasal urothelial cells (Fig. 12.7; Lin et al. 1995; Kerr et al. 1998). We have been able to use this UPII promoter to drive the urothelium-specific expression of SV40 T antigen, Ha-ras, EGF receptor, survivin and to achieve urothelium-specific inactivation of tumor suppressor genes including Rb and p53 (Zhang et al. 1999, 2001; Cheng et al. 2002, 2003; Gao et al. 2004; Salz et al. 2005; Wu 2005, 2009; Mo et al. 2007; He et al. 2009; Zhou et al. 2010). Luo and coworkers have used the same promoter to express ovalbumin in urothelium as a 'self antigen' to generate a model for interstitial cystitis (Liu et al. 2007, 2008). Using the uroplakin II promoter, we have conducted a series of studies in which we drive the mouse urothelial expression of various oncogenes to study their effects on bladder tumorigenesis. In a recent study, we showed that the dosage of Ha-ras activation plays a decisive role in the tumorigenicity of this oncogene. These results have important implications, as targeting ras and its downstream effectors with specific inhibitors, could be important therapeutic strategies to treat low-grade, non-invasive bladder tumors, by far the most prevalent type of urothelial carcinoma (Mo et al. 2007). In another study, we examined the effects of Rb and p53 genes in bladder tumorigenesis. We showed that loss of Rb and/or p53 in urothelial cells is necessary but not sufficient to provoke urothelial tumors; that p107, a member of the Rb family, is important for tumor suppression in the absence of Rb and p53; and that Rb, p53 and p107 together might be more reliable prognostic indicators than Rb and/or p53 alone (He et al. 2009).

To study gene expression or ablation in a urothelium-specific as well as inducible manners, we developed second-generation transgenic systems (Fig. 12.7). We used a 3.6 Kb mouse uroplakin II promoter to drive the urothelial expression of a modified reverse tetracycline transactivator (rtTA-M2), which was much more sensitive than the unmodified version (rtTA) that can mediate doxycycline-mediated reporter gene expression. This UPII/rtTA-M2 transgenic line should be useful for expressing any gene of choice in a urothelium-specific and inducible fashion to study urothelial biology and diseases. Additionally, we have demonstrated the proof-of-principle of inducible gene knockout in the urothelium by generating a double transgenic line harboring the UPII/rtTA-M2 and TRE-Cre (tetracycline-response elements driving Cre recombinase) transgenes (Fig. 12.7). When the double transgenic line was crossed with a reporter line in which a ROSA viral promoter-driven EGFP expression is placed under the control of loxP-flanked sequences, gene knockout occurred specifically in the urothelium upon doxycycline treatment. This system should therefore be widely applicable for ablating any gene of interest in urothelia of adult animals (Zhou et al. 2010).

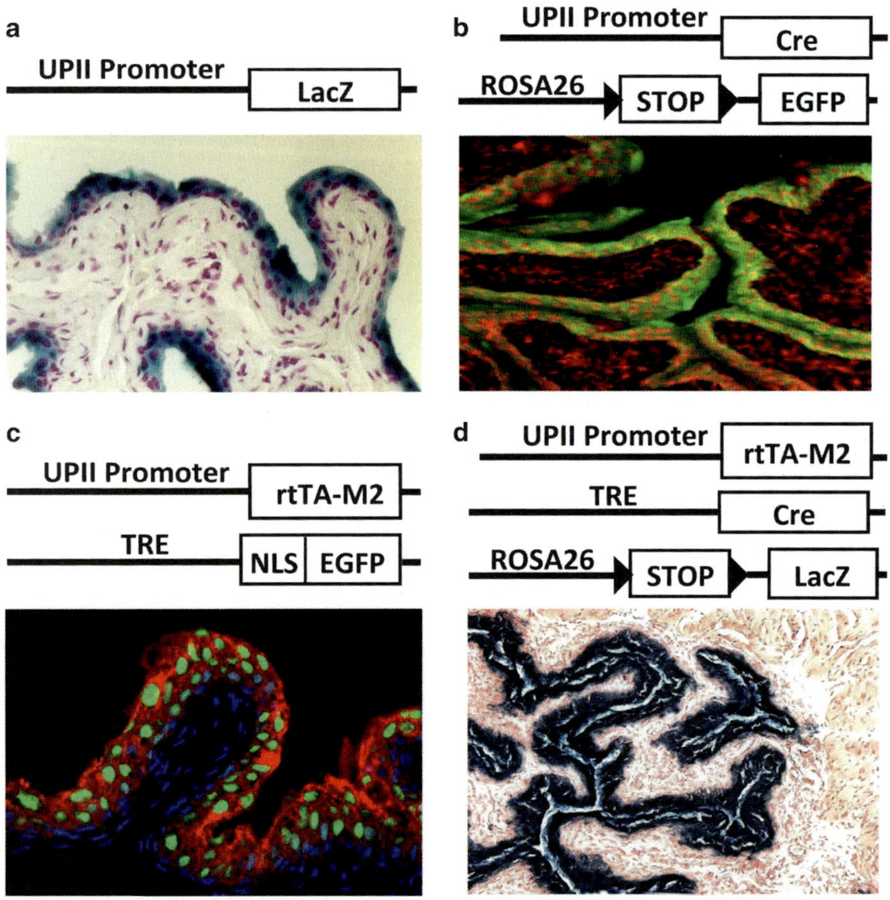

Fig. 12.7 Constitutive and inducible gene expression and knockout mouse systems for urothelium. (**a**) Constitutive urothelial gene expression. A 3.6-Kb mouse UPII promoter was capable of driving constitutive expression of a bacterial reporter gene (LacZ) gene in suprabasal urothelial cells (*blue color*; Lin et al. 1995). (**b**) Consitutive urothelial gene knockout. UPII promoter-driven Cre recombinase could mediate constitutive excision of a loxP-flanked STOP sequence and consequently urothelial expression of an enhanced green fluorescence protein (EGFP) reporter gene (Mo et al. 2005). (**c**) Inducible urothelial gene expression. UPII promoter-driven, reverse modified tetracycline transactivator (rtTA-M2) could induce the urothelial expression of the EGFP reporter under the control of the tetracycline response elements (TRE), upon doxycycline treatment. Note that EGFP was fused to a nuclear localization signal (NLS), thus the nuclear location (Zhou et al. 2010). (**d**) Inducible urothelial gene knockout. When a double transgenic line bearing UPII/rtTA-M2 and TRE/Cre transgenes was crossed with a reporter line in which the LacZ reporter was driven by the ROSA promoter and controlled by a loxP-flanked STOP sequence, urothelium-specific and inducible gene knockout was achieved (as evidenced by LacZ gene expression—*blue color*), upon doxycycline exposure (Zhou et al. 2010)

12.9 Future Directions

Many important questions about uroplakins remain to be answered. For example, what are the specific roles of these membrane proteins in the stabilization, enlargement and repair of the urothelial apical surface? What are the roles of molecular machineries including Rab27b and Vps33a in regulating uroplakin targeting? What is the detailed structure of urothelial plaques and how do they anchor into an underlying cytoskeleton? What are the roles of individual uroplakins and their subdomains in the uroplakin receptor complex in mediating the bacterial binding-induced signals in host umbrella cells? And what are the exact roles of cell cycle regulators, oncoproteins and tumor suppressors, alone or in concert, in driving urothelial tumorigenesis along divergent pathways? Results from such studies can lead to a better understanding of uroplakin structure and function, and have implications on a number of important clinical problems affecting the urinary tract including urinary tract infection, overactive bladder and urothelial carcinomas.

Acknowledgments We thank our collaborators who contributed to this work, including Ueli Abei (University of Basel, Switzerland), Gerard Apodaca (University of Pittsburgh Medical School), George Christ (Wake Forest University), Cathy Costello (Boston University Medical School), Robert DeSalle (American Museum of Natural History), Garth Ehrlich (Drexel University Medical School), Antonio Garcia-Expana (Universitat Rovira i Virgili, Tarragona, Spain), Bechara Kachar (NIDCD, National Institutes of Health), David Klumpp (Northwestern University Medical School), David Sabatini and Ellen Shapiro (NYU Medical School), Francis Schneck (University of Pittsburgh), Jun Yu (Beijing Genomic Institute) and Mark Zeidel (Harvard Medical School). This work was supported by NIH grants DK52206, DK39753, DK69688; a Merit Review Award from the Veterans Affairs Administration; and a grant-in-aid from the Goldstein Fund for Urological Research of the New York University School of Medicine.

References

Aboushwareb T, Zhou G, Deng FM, Turner C, Andersson KE, Tar M, Zhao W, Melman A, D'Agostino R Jr, Sun TT, Christ GJ (2009) Alterations in bladder function associated with urothelial defects in uroplakin II and IIIa knockout mice. Neurourol Urodyn 28(8):1028–1033. doi:10.1002/nau.20688

Anderson GG, Palermo JJ, Schilling JD, Roth R, Heuser J, Hultgren SJ (2003) Intracellular bacterial biofilm-like pods in urinary tract infections. Science 301(5629):105–107

Barral DC, Ramalho JS, Anders R, Hume AN, Knapton HJ, Tolmachova T, Collinson LM, Goulding D, Authi KS, Seabra MC (2002) Functional redundancy of Rab27 proteins and the pathogenesis of Griscelli syndrome. J Clin Invest 110(2):247–257. doi:10.1172/JCI15058

Berditchevski F, Odintsova E (2007) Tetraspanins as regulators of protein trafficking. Traffic 8(2):89–96. doi:10.1111/j.1600-0854.2006.00515.x

Bouckaert J, Berglund J, Schembri M, De Genst E, Cools L, Wuhrer M, Hung CS, Pinkner J, Slattegard R, Zavialov A, Choudhury D, Langermann S, Hultgren SJ, Wyns L, Klemm P, Oscarson S, Knight SD, De Greve H (2005) Receptor binding studies disclose a novel class of high-affinity inhibitors of the Escherichia coli FimH adhesin. Mol Microbiol 55(2):441–455

Charrin S, le Naour F, Silvie O, Milhiet PE, Boucheix C, Rubinstein E (2009) Lateral organization of membrane proteins: tetraspanins spin their web. Biochem J 420(2):133–154. doi:10.1042/BJ20082422

Chen Y, Guo X, Deng FM, Liang FX, Sun W, Ren M, Izumi T, Sabatini DD, Sun TT, Kreibich G (2003) Rab27b is associated with fusiform vesicles and may be involved in targeting uroplakins to urothelial apical membranes. Proc Natl Acad Sci USA 100(24):14012–14017

Cheng J, Huang H, Zhang ZT, Shapiro E, Pellicer A, Sun TT, Wu XR (2002) Overexpression of epidermal growth factor receptor in urothelium elicits urothelial hyperplasia and promotes bladder tumor growth. Cancer Res 62(14):4157–4163

Cheng J, Huang H, Pak J, Shapiro E, Sun TT, Cordon-Cardo C, Waldman FM, Wu XR (2003) Allelic loss of p53 gene is associated with genesis and maintenance, but not invasion, of mouse carcinoma in situ of the bladder. Cancer Res 63(1):179–185

Deng FM, Liang FX, Tu L, Resing KA, Hu P, Supino M, Hu CC, Zhou G, Ding M, Kreibich G, Sun TT (2002) Uroplakin IIIb, a urothelial differentiation marker, dimerizes with uroplakin Ib as an early step of urothelial plaque assembly. J Cell Biol 159(4):685–694

Gao J, Huang HY, Pak J, Cheng J, Zhang ZT, Shapiro E, Pellicer A, Sun TT, Wu XR (2004) p53 deficiency provokes urothelial proliferation and synergizes with activated Ha-ras in promoting urothelial tumorigenesis. Oncogene 23(3):687–696. doi:10.1038/sj.onc.1207169

Garcia-Espana A, Chung PJ, Zhao X, Lee A, Pellicer A, Yu J, Sun TT, Desalle R (2006) Origin of the tetraspanin uroplakins and their co-evolution with associated proteins: implications for uroplakin structure and function. Mol Phylogenet Evol 41(2):355–367

Garcia-Espana A, Chung PJ, Sarkar IN, Stiner E, Sun TT, Desalle R (2008) Appearance of new tetraspanin genes during vertebrate evolution. Genomics 91(4):326–334

Garcia-Espana A, Mares R, Sun TT, Desalle R (2009) Intron evolution: testing hypotheses of intron evolution using the phylogenomics of tetraspanins. PLoS One 4(3):e4680. doi:10.1371/journal.pone.0004680

Giltay JC, van de Meerakker J, van Amstel HK, de Jong TP (2004) No pathogenic mutations in the uroplakin III gene of 25 patients with primary vesicoureteral reflux. J Urol 171(2 Pt 1):931–932

Guo X, Tu L, Gumper I, Plesken H, Novak EK, Chintala S, Swank RT, Pastores G, Torres P, Izumi T, Sun TT, Sabatini DD, Kreibich G (2009) Involvement of vps33a in the fusion of uroplakin-degrading multivesicular bodies with lysosomes. Traffic 10(9):1350–1361. doi:10.1111/j.1600-0854.2009.00950.x

Hasan AK, Fukami Y, Sato K (2011) Gamete membrane microdomains and their associated molecules in fertilization signaling. Mol Reprod Dev 78(10–11):814–830. doi:10.1002/mrd.21336

He F, Mo L, Zheng XY, Hu C, Lepor H, Lee EY, Sun TT, Wu XR (2009) Deficiency of pRb family proteins and p53 in invasive urothelial tumorigenesis. Cancer Res 69(24):9413–9421. doi:10.1158/0008-5472.CAN-09-2158

Hemler ME (2005) Tetraspanin functions and associated microdomains. Nat Rev Mol Cell Biol 6(10):801–811

Hicks RM (1965) The fine structure of the transitional epithelium of rat ureter. J Cell Biol 26:25–48

Hicks RM, Ketterer B (1969) Hexagonal lattice of subunits in the thick luminal membrane of the rat urinary bladder. Nature (Lond) 224:1304–1305

Hodges SJ, Zhou G, Deng FM, Aboushwareb T, Turner C, Andersson KE, Santago P, Case D, Sun TT, Christ GJ (2008) Voiding pattern analysis as a surrogate for cystometric evaluation in uroplakin II knockout mice. J Urol 179(5):2046–2051

Hu P, Deng FM, Liang FX, Hu CM, Auerbach AB, Shapiro E, Wu XR, Kachar B, Sun TT (2000) Ablation of uroplakin III gene results in small urothelial plaques, urothelial leakage, and vesicoureteral reflux. J Cell Biol 151(5):961–972

Hu P, Meyers S, Liang FX, Deng FM, Kachar B, Zeidel ML, Sun TT (2002) Role of membrane proteins in permeability barrier function: uroplakin ablation elevates urothelial permeability. Am J Physiol Renal Physiol 283(6):F1200–F1207

Hu CC, Liang FX, Zhou G, Tu L, Tang CH, Zhou J, Kreibich G, Sun TT (2005) Assembly of urothelial plaques: tetraspanin function in membrane protein trafficking. Mol Biol Cell 16(9):3937–3950

Hu CC, Bachmann T, Zhou G, Liang FX, Ghiso J, Kreibich G, Sun TT (2008) Assembly of a membrane receptor complex: roles of the uroplakin II prosequence in regulating uroplakin bacterial receptor oligomerization. Biochem J 414(2):195–203

Huang HY, Shariat SF, Sun TT, Lepor H, Shapiro E, Hsieh JT, Ashfaq R, Lotan Y, Wu XR (2007) Persistent uroplakin expression in advanced urothelial carcinomas: implications in urothelial tumor progression and clinical outcome. Hum Pathol 38(11):1703–1713

Jenkins D, Bitner-Glindzicz M, Malcolm S, Hu CC, Allison J, Winyard PJ, Gullett AM, Thomas DF, Belk RA, Feather SA, Sun TT, Woolf AS (2005) De novo Uroplakin IIIa heterozygous mutations cause human renal adysplasia leading to severe kidney failure. J Am Soc Nephrol 16(7):2141–2149

Jiang S, Gitlin J, Deng FM, Liang FX, Lee A, Atala A, Bauer SB, Ehrlich GD, Feather SA, Goldberg JD, Goodship JA, Goodship TH, Hermanns M, Hu FZ, Jones KE, Malcolm S, Mendelsohn C, Preston RA, Retik AB, Schneck FX, Wright V, Ye XY, Woolf AS, Wu XR, Ostrer H, Shapiro E, Yu J, Sun TT (2004) Lack of major involvement of human uroplakin genes in vesicoureteral reflux: implications for disease heterogeneity. Kidney Int 66(1):10–19

Justice SS, Hunstad DA, Cegelski L, Hultgren SJ (2008) Morphological plasticity as a bacterial survival strategy. Nat Rev Microbiol 6(2):162–168

Kachar B, Liang F, Lins U, Ding M, Wu XR, Stoffler D, Aebi U, Sun TT (1999) Three-dimensional analysis of the 16nm urothelial plaque particle: luminal surface exposure, preferential head-to-head int.raction, and hinge formation. J Mol Biol 285(2):595–608

Kelly H, Ennis S, Yoneda A, Bermingham C, Shields DC, Molony C, Green AJ, Puri P, Barton DE (2005) Uroplakin III is not a major candidate gene for primary vesicoureteral reflux. Eur J Hum Genet 13(4):500–502

Kerr DE, Liang F, Bondioli KR, Zhao H, Kreibich G, Wall RJ, Sun TT (1998) The bladder as a bioreactor: urothelium production and secretion of growth hormone into urine. Nat Biotechnol 16(1):75–79

Ketterer B, Hicks RM (1971) Proteins of the plasma membrane lining the lumen of the rat bladder. Biochem J 122(5):66

Ketterer B, Hicks RM, Christodoulides L, Beale D (1973) Studies of the chemistry of the luminal plasma membrane of rat bladder epithelial cells. Biochim Biophys Acta 311(2):180–190

Kong XT, Deng FM, Hu P, Liang FX, Zhou G, Auerbach AB, Genieser N, Nelson PK, Robbins ES, Shapiro E, Kachar B, Sun TT (2004) Roles of uroplakins in plaque formation, umbrella cell enlargement, and urinary tract diseases. J Cell Biol 167(6):1195–1204

Koss LG (1969) The asymmetric unit membranes of the epithelium of the urinary bladder of the rat. An electron microscopic study of a mechanism of epithelial maturation and function. Lab Invest 21:154–168

Levy S, Shoham T (2005) The tetraspanin web modulates immune-signalling complexes. Nat Rev Immunol 5(2):136–148

Li SM, Zhang ZT, Chan S, McLenan O, Dixon C, Taneja S, Lepor H, Sun TT, Wu XR (1999) Detection of circulating uroplakin-positive cells in patients with transitional cell carcinoma of the bladder. J Urol 162(3 Pt 1):931–935

Liang FX, Riedel I, Deng FM, Zhou G, Xu C, Wu XR, Kong XP, Moll R, Sun TT (2001) Organization of uroplakin subunits: transmembrane topology, pair formation and plaque composition. Biochem J 355(Pt 1):13–18

Liang FX, Bosland MC, Huang H, Romih R, Baptiste S, Deng FM, Wu XR, Shapiro E, Sun TT (2005) Cellular basis of urothelial squamous metaplasia: roles of lineage heterogeneity and cell replacement. J Cell Biol 171(5):835–844

Lin JH, Wu XR, Kreibich G, Sun TT (1994) Precursor sequence, processing, and urothelium-specific expression of a major 15-kDa protein subunit of asymmetric unit membrane. J Biol Chem 269(3):1775–1784

Lin JH, Zhao H, Sun TT (1995) A tissue-specific promoter that can drive a foreign gene to express in the suprabasal urothelial cells of transgenic mice. Proc Natl Acad Sci USA 92(3): 679–683

Liu W, Evanoff DP, Chen X, Luo Y (2007) Urinary bladder epithelium antigen induces CD8+ T cell tolerance, activation, and autoimmune response. J Immunol 178(1):539–546

Liu W, Deyoung BR, Chen X, Evanoff DP, Luo Y (2008) RDP58 inhibits T cell-mediated bladder inflammation in an autoimmune cystitis model. J Autoimmun 30(4):257–265

Mahbub Hasan AK, Sato K, Sakakibara K, Ou Z, Iwasaki T, Ueda Y, Fukami Y (2005) Uroplakin III, a novel Src substrate in Xenopus egg rafts, is a target for sperm protease essential for fertilization. Dev Biol 286(2):483–492. doi:10.1016/j.ydbio.2005.08.020

Mahbub Hasan AK, Ou Z, Sakakibara K, Hirahara S, Iwasaki T, Sato K, Fukami Y (2007) Characterization of Xenopus egg membrane microdomains containing uroplakin Ib/III complex: roles of their molecular interactions for subcellular localization and signal transduction. Genes Cells 12(2):251–267. doi:10.1111/j.1365-2443.2007.01048.x

Min G, Stolz M, Zhou G, Liang F, Sebbel P, Stoffler D, Glockshuber R, Sun TT, Aebi U, Kong XP (2002) Localization of uroplakin Ia, the urothelial receptor for bacterial adhesin FimH, on the six inner domains of the 16 nm urothelial plaque particle. J Mol Biol 317(5):697–706

Min G, Zhou G, Schapira M, Sun TT, Kong XP (2003) Structural basis of urothelial permeability barrier function as revealed by Cryo-EM studies of the 16 nm uroplakin particle. J Cell Sci 116(Pt 20):4087–4094

Min G, Wang H, Sun TT, Kong XP (2006) Structural basis for tetraspanin functions as revealed by the cryo-EM structure of uroplakin complexes at 6-A resolution. J Cell Biol 173(6):975–983

Mo L, Cheng J, Lee EY, Sun TT, Wu XR (2005) Gene deletion in urothelium by specific expression of Cre recombinase. Am J Physiol Renal Physiol 289(3):F562–F568

Mo L, Zheng X, Huang HY, Shapiro E, Lepor H, Cordon-Cardo C, Sun TT, Wu XR (2007) Hyperactivation of Ha-ras oncogene, but not Ink4a/Arf deficiency, triggers bladder tumorigenesis. J Clin Invest 117(2):314–325. doi:10.1172/JCI30062

Moll R, Wu XR, Lin JH, Sun TT (1995) Uroplakins, specific membrane proteins of urothelial umbrella cells, as histological markers of metastatic transitional cell carcinomas. Am J Pathol 147(5):1383–1397

Negrete HO, Lavelle JP, Berg J, Lewis SA, Zeidel ML (1996) Permeability properties of the intact mammalian bladder epithelium. Am J Physiol 271(4 Pt 2):F886–F894

Osman I, Kang M, Lee A, Deng FM, Polsky D, Mikhail M, Chang C, David DA, Mitra N, Wu XR, Sun TT, Bajorin DF (2004) Detection of circulating cancer cells expressing uroplakins and epidermal growth factor receptor in bladder cancer patients. Int J Cancer 111(6):934–939

Osman I, Bajorin DF, Sun TT, Zhong H, Douglas D, Scattergood J, Zheng R, Han M, Marshall KW, Liew CC (2006) Novel blood biomarkers of human urinary bladder cancer. Clin Cancer Res 12(11 Pt 1):3374–3380

Porter KR, Bonneville MA (1963) An introduction to the fine structure of cells and tissues. Lea & Febiger, Philadelphia

Raposo G, Marks MS, Cutler DF (2007) Lysosome-related organelles: driving post-Golgi compartments into specialisation. Curr Opin Cell Biol 19(4):394–401. doi:10.1016/j.ceb.2007.05.001

Robertson JD, Vergara J (1980) Analysis of the structure of intramembrane particles of the mammalian urinary bladder. J Cell Biol 86(2):514–528

Rubinstein E (2011) The complexity of tetraspanins. Biochem Soc Trans 39(2):501–505. doi:10.1042/BST0390501

Sakakibara K, Sato K, Yoshino K, Oshiro N, Hirahara S, Mahbub Hasan AK, Iwasaki T, Ueda Y, Iwao Y, Yonezawa K, Fukami Y (2005) Molecular identification and characterization of Xenopus egg uroplakin III, an egg raft-associated transmembrane protein that is tyrosine-phosphorylated upon fertilization. J Biol Chem 280(15):15029–15037. doi:10.1074/jbc.M410538200

Salz W, Eisenberg D, Plescia J, Garlick DS, Weiss RM, Wu XR, Sun TT, Altieri DC (2005) A survivin gene signature predicts aggressive tumor behavior. Cancer Res 65(9):3531–3534. doi:10.1158/0008-5472.CAN-04-4284

Schonfelder EM, Knuppel T, Tasic V, Miljkovic P, Konrad M, Wuhl E, Antignac C, Bakkaloglu A, Schaefer F, Weber S (2006) Mutations in Uroplakin IIIA are a rare cause of renal hypodysplasia in humans. Am J Kidney Dis 47(6):1004–1012

Staehelin LA, Chlapowski FJ, Bonneville MA (1972) Lumenal plasma membrane of the urinary bladder. I. Three-dimensional reconstruction from freeze-etch images. J Cell Biol 53(1):73–91

Starcevic M, Nazarian R, Dell'Angelica EC (2002) The molecular machinery for the biogenesis of lysosome-related organelles: lessons from Hermansky-Pudlak syndrome. Semin Cell Dev Biol 13(4):271–278

Sun TT (2006) Altered phenotype of cultured urothelial and other stratified epithelial cells: implications for wound healing. Am J Physiol Renal Physiol 291(1):F9–F21

Thumbikat P, Berry RE, Zhou G, Billips BK, Yaggie RE, Zaichuk T, Sun TT, Schaeffer AJ, Klumpp DJ (2009) Bacteria-induced uroplakin signaling mediates bladder response to infection. PLoS Pathog 5(5):e1000415. doi:10.1371/journal.ppat.1000415

Tolmachova T, Abrink M, Futter CE, Authi KS, Seabra MC (2007) Rab27b regulates number and secretion of platelet dense granules. Proc Natl Acad Sci USA 104(14):5872–5877. doi:10.1073/pnas.0609879104

Tu L, Sun TT, Kreibich G (2002) Specific heterodimer formation is a prerequisite for uroplakins to exit from the endoplasmic reticulum. Mol Biol Cell 13(12):4221–4230

Vergara JA, Longley W, Robertson JD (1969) A hexagonal arrangement of subunits in membrane of mouse urinary bladder. J Mol Biol 46:593–596

Vergara J, Zambrano F, Robertson JD, Elrod H (1974) Isolation and characterization of luminal membranes from urinary bladder. J Cell Biol 61:83–94

Walz T, Haner M, Wu XR, Henn C, Engel A, Sun T-T, Aebi U (1995) Towards the molecular architecture of the asymmetric unit membrane of the mammalian urinary bladder epithelium: a closed "twisted ribbon" structure. J Mol Biol 248(5):887–900

Wang H, Min G, Glockshuber R, Sun TT, Kong XP (2009) Uropathogenic E. coli adhesin-induced host cell receptor conformational changes: implications in transmembrane signaling transduction. J Mol Biol 392(2):352–361. doi:10.1016/j.jmb.2009.06.077, S0022-2836(09)00808-0 [pii]

Wellens A, Garofalo C, Nguyen H, Van Gerven N, Slattegard R, Hernalsteens JP, Wyns L, Oscarson S, De Greve H, Hultgren S, Bouckaert J (2008) Intervening with urinary tract infections using anti-adhesives based on the crystal structure of the FimH-oligomannose-3 complex. PLoS One 3(4):e2040

Wright KJ, Seed PC, Hultgren SJ (2007) Development of intracellular bacterial communities of uropathogenic Escherichia coli depends on type 1 pili. Cell Microbiol 9(9):2230–2241

Wu XR (2005) Urothelial tumorigenesis: a tale of divergent pathways. Nat Rev Cancer 5(9): 713–725. doi:10.1038/nrc1697

Wu XR (2009) Biology of urothelial tumorigenesis: insights from genetically engineered mice. Cancer Metastasis Rev 28(3–4):281–290. doi:10.1007/s10555-009-9189-4

Wu XR, Sun TT (1993) Molecular cloning of a 47 kDa tissue-specific and differentiation-dependent urothelial cell surface glycoprotein. J Cell Sci 106:31–43

Wu XR, Manabe M, Yu J, Sun TT (1990) Large scale purification and immunolocalization of bovine uroplakins I, II, and III. Molecular markers of urothelial differentiation. J Biol Chem 265(31):19170–19179

Wu XR, Sun T-T, Medina JJ (1996) In vitro binding of type 1-fimbriated Escherichia coli to uroplakins Ia and Ib: relation to urinary tract infections. Proc Natl Acad Sci USA 93(18): 9630–9635

Wu XR, Kong XP, Pellicer A, Kreibich G, Sun TT (2009) Uroplakins in urothelial biology, function, and disease. Kidney Int 75(11):1153–1165. doi:10.1038/ki.2009.73

Xie B, Zhou G, Chan S-Y, Shapiro E, Kong X-P, Wu X-R, Sun T-T, Costello CE (2006) Distinct glycan structures of uroplakins Ia and Ib: structural basis for the selective binding of FimH adhesin to uroplakin Ia. J Biol Chem 281:14644–14653

Yu J, Manabe M, Wu XR, Xu C, Surya B, Sun TT (1990) Uroplakin I: a 27-kD protein associated with the asymmetric unit membrane of mammalian urothelium. J Cell Biol 111(3):1207–1216

Yu J, Lin JH, Wu XR, Sun TT (1994) Uroplakins Ia and Ib, two major differentiation products of bladder epithelium, belong to a family of four transmembrane domain (4TM) proteins. J Cell Biol 125(1):171–182

Zhang ZT, Pak J, Shapiro E, Sun TT, Wu XR (1999) Urothelium-specific expression of an oncogene in transgenic mice induced the formation of carcinoma in situ and invasive transitional cell carcinoma. Cancer Res 59(14):3512–3517

Zhang ZT, Pak J, Huang HY, Shapiro E, Sun TT, Pellicer A, Wu XR (2001) Role of Ha-ras activation in superficial papillary pathway of urothelial tumor formation. Oncogene 20(16): 1973–1980. doi:10.1038/sj.onc.1204315

Zhou G, Mo WJ, Sebbel P, Min G, Neubert TA, Glockshuber R, Wu XR, Sun TT, Kong XP (2001) Uroplakin Ia is the urothelial receptor for uropathogenic Escherichia coli: evidence from in vitro FimH binding. J Cell Sci 114(Pt 22):4095–4103

Zhou H, Liu Y, He F, Mo L, Sun TT, Wu XR (2010) Temporally and spatially controllable gene expression and knockout in mouse urothelium. Am J Physiol Renal Physiol 299(2): F387–F395. doi:10.1152/ajprenal.00185.2010

Chapter 13
Essential Tetraspanin Functions in the Vertebrate Retina

Andrew F.X. Goldberg

Abstract This chapter summarizes current knowledge and reviews recent findings regarding the structure, function and importance of tetraspanins in the vertebrate neural retina. Since inherited defects in several human tetraspanins, including tetraspanin12 and peripherin-2/rds, are well documented to cause sight-robbing ocular diseases, these molecules are of significant interest from both basic science and clinical perspectives. In the retina, as is the case more generally, tetraspanin superfamily members serve diverse biological functions, but have in common the capacity to organize lateral interactions within cellular membranes. Proteins characterized to date can also display several distinctive properties, including: a high degree of cell-type specificity, a reduced tendency to participate in interaction webs, and an inability to compensate for one another functionally. Moreover, these features have facilitated the production of mouse models with robust and predictable phenotypes, and the biophysical/biochemical characterization of individual proteins. The resultant advances demonstrate essential roles for these molecules in building specialized membrane features and facilitating Wnt/β-catenin signaling for angiogenesis. Since tetraspanins are increasingly viewed as potential therapeutic targets, lessons learned from superfamily members with well-documented and essential functions in the retina may prove useful for understanding the roles tetraspanins play in disease more generally.

13.1 Introduction

The vertebrate retina is a well-studied tissue that belongs to the central nervous system (Dowling 1987; Oyster 1999; Rodieck 1998). It develops from embryonic evaginations of the anterior neural tube that invaginate to form optic cups. The

A.F.X. Goldberg (✉)
Eye Research Institute, Oakland University, Rochester, MI 48309, USA
e-mail: goldberg@oakland.edu

F. Berditchevski and E. Rubinstein (eds.), *Tetraspanins*, Proteins and Cell Regulation 9,
DOI 10.1007/978-94-007-6070-7_13,

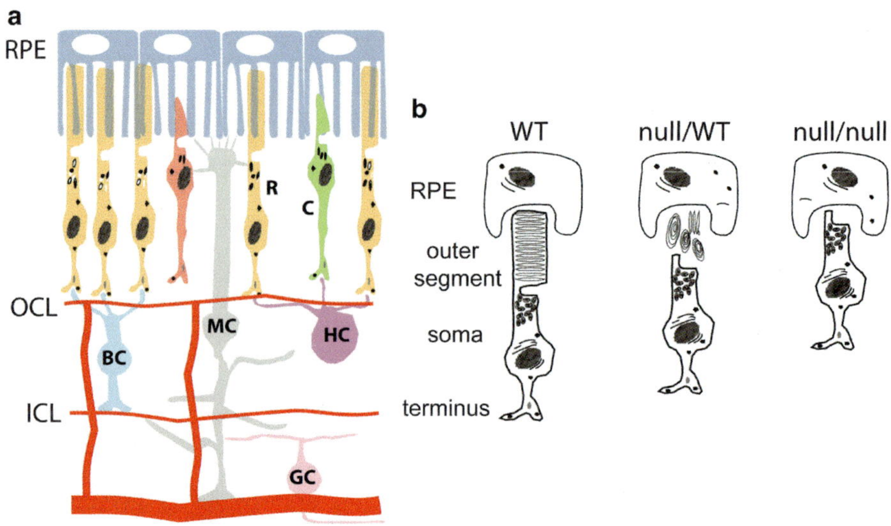

Fig. 13.1 (**a**) Simplified organization of the adult vertebrate retina, including: *RPE* retinal pigment epithelium, *R* rod photoreceptor, *C* cone photoreceptor, *OCL* outer capillary layer, *MC* Müeller glial cell, *HC* horizontal cell, *BC* bipolar cell, *ICL* inner capillary layer, *GC* ganglion cell. (**b**) Loss of OS integrity phenotype in the *rds* mouse. WT rod photoreceptor cell and relationship with an apposing RPE cell (*left*) as compared with *rds* (+/–) and *rds* (–/–) genotypes (*center* and *right*) (Adapted from (Goldberg 2006))

neuroepithelum that lines these cups proliferates and differentiates to become the mature neural retina. This CNS tissue is organized as a laminar structure, which has facilitated studies of its synaptic organization and electrophysiological responses (Fig. 13.1a). Its importance as the biological basis for human vision has brought significant resources to bear on understanding its normal function and breakdown under duress. Advances over the past two decades have made the retina a prime target for clinical applications of gene therapy, and recent early successes bode well for the future (Simonelli et al. 2010). Relatively few tetraspanins have been documented in the vertebrate retina to date; however two proteins, peripherin-2/rds (P/rds) and tetraspanin12 (tspan12), have been shown to play irreplaceable roles in animal models, and figure prominently in human retinal health and disease. This chapter will focus on current knowledge of these molecules, and briefly mention other tetraspanins for which essential function has yet to be demonstrated in the vertebrate retina.

13.1.1 P/rds: A Keystone for Photoreceptor Structure

Vertebrate photoreceptors rely upon a membranous outer segment (OS)—a specialized non-motile cilium—for their physiological function. They include rods, which function for monochrome vision in dim light, and cones, which function for color vision in bright light. Photoreceptor OSs are visible light detectors of exquisite design, sensitivity, and dynamic range; visual information is captured by hundreds of photopigment-containing membranes stacked along the axis of incoming light. Both rod and cone OSs undergo daily renewal, maintaining a roughly constant length by a balanced (basal) addition of newly synthesized disks with a (distal) shedding of disk packets (Young 1976). The shed disks are phagocytosed by an apposed retinal pigment epithelium (RPE), upon which the photoreceptors are also dependent for retinoid recycling and metabolite exchange. Importantly, OS structure does not merely enclose phototransduction components, but appears to actively organize the signaling proteins that support cell function for vision. Since photoreceptor machinery for protein biosynthesis is housed within the inner segment, and nearly 10% of the OS organelle is renewed each day, the trafficking of components through the connecting cilium and their organization into OS membrane structures represents a formidable daily challenge.

P/rds (also known as: peripherin-2, rds, rds/peripherin, and Prph2) is the gene product of the *RDS* gene (also known as *PRPH2* and *TSPAN22*) and a member of the tetraspanin superfamily, and is critical for vertebrate photoreceptor structure and function (Conley and Naash 2009; Farjo and Naash 2006; Boesze-Battaglia and Goldberg 2002; Goldberg 2006). In several ways dissimilar to other tetraspanins, P/rds has nonetheless provided important paradigms for superfamily protein structure. The protein was originally identified independently by convergent technical approaches. Using a discovery process common to many tetraspanins (antigen identification), P/rds was characterized as part of a long-term effort to analyze photoreceptor protein structure/function (Molday et al. 1987). In parallel, an effort to clone the locus responsible for the *retinal degeneration slow* (*rds*) mouse phenotype identified a gene product critical for photoreceptor structure/function (Travis et al. 1989). The investigations ultimately converged, and it became clear that P/rds, the *rds* gene product, was an integral membrane glycoprotein essential for photoreceptor OS structure (Connell et al. 1991).

Since those landmark studies, more than 100 mutations in human RDS have been associated with a wide range of inherited retinal degenerations, including instances of *retinitis pigmentosa* and macular dystrophies (Boon et al. 2008; Kohl et al. 1998). This broad heterogeneity spans more than ten clinically distinct disease phenotypes. Importantly, the mechanisms by which particular mutations generate specific phenotypes remain largely undefined. Although rationales for inheritance patterns are available in some instances, in no case is the pathophysiology of disease well understood in molecular terms. Clinical treatment options are limited, although new therapeutic approaches show promise in animal models (Georgiadis et al. 2010; Ali et al. 2000; Cai et al. 2010a).

Current knowledge and recent findings will be relatively briefly summarized here; the reader is referred to several excellent previous reviews for more extensive and detailed discussion of topics related to P/rds structure, function, and role in disease (Conley and Naash 2009; Farjo and Naash 2006; Boesze-Battaglia and Goldberg 2002; Goldberg 2006; Cai et al. 2010b; Molday 1994).

13.1.2 P/rds Importance at the Cellular Level Is Well-Described

The well-studied heterozygous and homozygous *rds* mouse models, null at one or both *rds* alleles, show a disruption and complete absence of OSs respectively (depicted in Fig. 13.1b). Dysmorphic and disorganized OSs, observed as whorls of membrane derived from the photoreceptor connecting cilia, are produced in heterozygotes (Hawkins et al. 1985). An obvious absence of OSs is observed in homozygotes (Sanyal and Jansen 1981); RPE microvilli freed from their normal relationship with the retina extend into the subretinal space. This striking primary defect appears to be highly specific, since other photoreceptor structures (and the retina more generally) are seemingly unaffected in young adult animals. Furthermore, the robust phenotypic dose-dependence suggests that the gene product has an essential stoichiometric (vs. catalytic) function for OS integrity. Surprisingly, severe disruption or complete loss of the OS organelle has relatively mild consequences for photoreceptor viability. Early studies observed that although hetero- and homozygous *rds* mice animals lost photoreceptors (Hawkins et al. 1985; Sanyal et al. 1980), the degenerative process was slow relative to other genetic defects impacting this cell type.

P/rds protein expression has not been documented in cells other than photoreceptors, and the protein is normally tightly restricted to rod and cone cell OS disk rims (Molday et al. 1987). Disk rims are formed at the periphery of new OS disks as they are internalized from folded sheets of OS plasma membrane (Steinberg et al. 1980; Corless et al. 1987). This discrete cell-type specificity and striking ultrastructural distribution (illustrated in Fig. 13.2), combined with the *rds* loss-of-OS phenotype, led to the suggestion that P/rds plays a role for establishing and maintaining the unique membranous structures that constitute the OS organelle (Connell et al. 1991; Travis et al. 1992).

Although the protein is present both in rods and cones (Arikawa et al. 1992), detailed study of the *rds* and several engineered mouse models suggest that it may function somewhat differently for each. Haploinsufficiency affects rods more than cones (Cheng et al. 1997), and complete loss of P/rds prevents rods, but not cones from generating tubular membranous structures (Farjo et al. 2006). Furthermore, a pathogenic mutation that appears to impact human cones preferentially likewise does so in mice (Ding et al. 2004; Conley et al. 2007). Altogether, these findings suggest that P/rds is utilized by rods and cones in somewhat different ways—potentially contributing to the unique OS structures associated with each cell type.

13.1.3 P/rds Is Essential for Retinal Health and Human Vision

P/rds plays an important role for human photoreceptors and vision more widely. Mutations in the RDS gene have been associated with a broad range of progressive human retinal diseases (Boon et al. 2008; Kohl et al. 1998). The Human Gene Mutation Database (www.hgmd.cf.ac.uk), lists more than 100 unique defects,

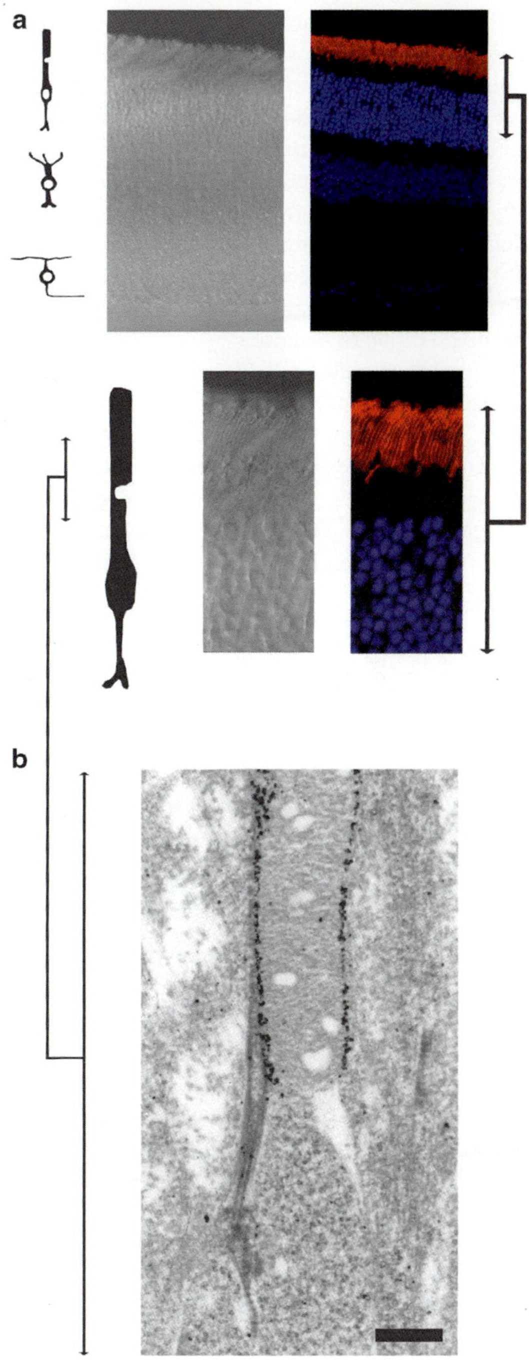

Fig. 13.2 Cell-type specific expression and localization of P/rds in vertebrate retina. (**a**) Indirect immunofluorescence localization (*red*) in a murine ocular cryosection shows P/rds presence in a single retinal strata; DAPI-stained nuclear layers appear *blue*. Laminar positions of retinal cell types and DIC images are provided for reference. (**b**) TEM immunogold localization of P/rds in a murine rod photoreceptor. The image shows shows the IS-OS junction and connecting cilium, and P/rds distributed at the periphery of the OS disks. Scale bar = 1 μm

although this tally may include a number of polymorphisms. Missense changes represent the majority of the pathogenic mutations; however, nonsense and frameshift changes also represent a sizable fraction (roughly 40%). Diseases reported include: autosomal dominant retinitis pigmentosa (RP), digenic RP, retinitis punctata albescens, butterfly-shaped macular dystrophy, adult vitelliform macular dystrophy, foveomacular dystrophy, cone-rod dystrophy, central aerolar choroidal dystrophy, and pattern dystrophy (Boon et al. 2008; Kohl et al. 1998). No simple scheme for correlating phenotype with genotype is available. P/rds multifunctionality has been proposed to generate phenotypic diversity (Goldberg 2006), and it is likely that genetic background and environmental factors contribute to the clinical manifestations of disease.

Numerous studies of the *rds* and engineered mouse models confirm the importance of P/rds for OS structure and normal vision (Ding and Naash 2006). In concert with protein-level investigations, they indicate that both haploinsufficiency and dominant negative mechanisms are important for generating human disease, and provide intriguing clues to P/rds molecular function (discussed below). They also suggest that the pathogenic process in humans may include: shortening of OSs, whorls of OS membranes in the subretinal space, alterations in the normal OS-RPE interface and phagocytosis, and compromised photoreceptor functionality and viability. These primary insults can provoke more general retinal abnormalities and immune responses, disrupt RPE and choroidal integrity, and eventually progress to highly destructive neovascularization (Bramall et al. 2010).

13.1.4 P/rds Structure/Function at the Molecular Level: Commonalities and Differences Relative to Other Tetraspanins

13.1.4.1 A Family Member with Distinctive Features

A recent analysis of the evolutionary relationships between tetraspanins to be found in completely sequenced genomes suggests that RDS-related genes represent one of four major monophyletic families (Garcia-Espana et al. 2008). RDS family genes appear to have evolved primarily as a result of duplication events in ancestor organisms of vertebrates and mammals. RDS orthologs are present in all vertebrates examined to date, and physical clones have been isolated from roughly a dozen species, ranging from skate to human (Conley and Naash 2009; Farjo and Naash 2006; Goldberg 2006). Greater than 70% amino acid identity is seen when comparing vertebrate P/rds orthologs, while upwards of 90% identity is observed when mammalian versions are aligned. The bovine protein, frequently used for biochemical-level studies, possesses 346 amino acids with an approximate molecular weight of 39 kDa. This is significantly larger than most tetraspanins; CD81 for example, weighs in at roughly 26 kDa. The mass difference between P/rds and CD81 lies in the relatively larger EC2 (150 vs. 89 amino acids) and C-terminal (60 vs. 10 amino acids) domains. In addition to its larger mass, P/rds also appears to be far more abundant than other tetraspanins, with the exception of uroplakins (Wu et al. 2009). Its surface density (in a tetrameric form) at the rod OS disk rim has been estimated at 4,100/μm^2 (Goldberg and Molday 1996a).

P/rds topological organization is identical to that suggested for other tetraspanins; primary sequence analysis predicts four hydrophobic α-helical transmembrane domains. The protein adopts a polytopic topology that places N- and C-termini within the cytoplasm, exposing the bulk of the protein's hydrophilic domains to the extracellular (or disk luminal) space (Connell and Molday 1990). Like many other tetraspanins, P/rds has a consensus site for N-linked glycosylation in the EC2 domain, and the protein is known to be glycosylated (Connell and Molday 1990; Travis et al. 1991). A Pfam tetraspanin signature (Sammut et al. 2008) and hallmark EC2 cysteine motifs are present in P/rds. An odd number of cysteine residues within the P/rds EC2 domain is a unique feature amongst superfamily members, and is shared only by a retinal homolog, rom-1. In contrast to other tetraspanins, the protein does not appear to be palmitoylated (Conley et al. 2012).

13.1.5 An Aloof Tetraspanin: Where's the Web?

A central tenant amongst tetraspanin researchers is that these proteins act to organize lateral interactions within membranes. The common finding that a given tetraspanin maintains (both direct and indirect) associations with a large number of other proteins (including other tetraspanins) has led to the prevalent concept of a tetraspanin interaction web or tetraspanin-enriched microdomains (TEMs) (Maecker et al. 1997; Kovalenko et al. 2005; Charrin et al. 2009; Yanez-Mo et al. 2009). Within the context of these microdomains, most tetraspanins appear to function as "molecular facilitators"—their presence lends a regulatory aspect to the biological process of interest (Maecker et al. 1997). This modulatory role, combined with functional redundancy (or presumed compensating mechanisms) can lead to relatively subtle phenotypic effects when tetraspanins are genetically ablated (Levy et al. 1998; Fradkin et al. 2002).

P/rds stands apart in this respect, as there is no evidence that it participates in a dynamic web mediated by weak lateral interactions with a variety of other proteins. Likewise, it has not been proposed to "fine-tune" transient interactions of associated integral membrane proteins participating in a signaling pathway. Instead, P/rds is concentrated within photoreceptor disk rims, where it acts as an essential structural component. In this regard, it bears comparison to uroplakins Ia and Ib—tetraspanins that also may lack some "classic" superfamily behaviours.

Uroplakins resemble P/rds in several ways. Firstly, they display a highly restricted distribution—in this case, localized within the asymmetric unit membrane

(AUM) of the urothelium (Wu et al. 2009). Secondly, they assemble into relatively stable complexes that exclude most other proteins and give rise to a distinct membranous structure with a specialized function. In the case of uroplakins, complexes (16-nm particles) associate to form plaques that line the apical urothelium surface—providing an essential permeability barrier function (Min et al. 2006). Third, knockout models for the uroplakin tetraspanins generate robust, predictable, and discrete phenotypes (Hu et al. 2000; Kong et al. 2004). Finally, consistent with a structural (vs. facilitative) role, uroplakins Ia and Ib function at very high densities (Kachar et al. 1999). A typical plaque of 0.6 μm diameter (roughly 1,400 particles) is estimated to possess a density of $\sim 2 \times 10^4$ tetraspanins/μm^2—a number quite close to the $\sim 1.6 \times 10^4$ tetraspanins/μm^2 to be found in the photoreceptor disk rim. Taken together, these observations suggest the utility of considering uroplakins in urothelium AUM, like P/rds in retinal photoreceptor disk rims, as constituting tetraspanin-dense microdomains (TDMs)—rather than tetraspanin-*enriched* microdomains (TEMs). Although uroplakins are of ancient origin, their structural function as a permeability barrier appears to have been a more recent event (Garcia-Espana et al. 2006). It is interesting to speculate that P/rds may also have been recruited away from the "facilitative" role common for other superfamily members.

13.1.5.1 Non-Covalent Self-Assembly: An Essential but Unresolved Role for Function

P/rds exhibits a behaviour described for other tetraspanins that may be ubiquitous—non-covalent self-assembly. An early hydrodynamic study demonstrated that the protein is solubilized from native photoreceptor membranes in a tetrameric form; this investigation provided the first rigorous evidence that tetraspanins can self-assemble in a defined stoichiometry (Goldberg and Molday 1996a). Subsequent investigations showed that P/rds tetramers form via dimerization of dimeric intermediates (Goldberg and Molday 1996b; Loewen et al. 2001), and that determinants within EC2 domains govern this process (Goldberg et al. 2001; Ding et al. 2005). Three lines of evidence suggest that tetramerization may occur within ISs, prior to trafficking to the OS. First, tetramers can form at the biosynthetic level in cultured cells (Goldberg et al. 1995). Second, tetramers can form in murine retina in the absence of OSs (Chakraborty et al. 2008). Third, non-tetrameric forms of P/rds are not incorporated in *X. laevis* photoreceptor OSs (Loewen et al. 2003). P/rds tetramers are clearly essential to support OS structure; defective subunit assembly appears as the primary cause for instances of inherited retinal degeneration (Goldberg and Molday 1996b; Loewen et al. 2001, 2003; Kedzierski et al. 2001; Molday et al. 2004). In contrast, the structural features governing monomer-to-dimer assembly have not yet been described for this protein, and it is not known whether particular retinal degenerations are associated with defects in this process. It is possible that P/rds monomer-to-dimer interfaces reside in transmembrane regions—akin to those detailed for other tetraspanins (Kovalenko et al. 2005).

13.1.5.2 Disulfide Bonds Stabilize a Relatively Rigid EC2 Domain and Mediate Higher-Order Oligomerization

Like other superfamily members (Wright and Tomlinson 1994; Levy and Shoham 2005), P/rds contains a hallmark CCG motif within its EC2 domain. It also possesses five other cysteine residues within this domain, resulting in a total of seven EC2 cysteines. This uneven quantity is unprecedented elsewhere in the superfamily, and is noteworthy for its implications as regards disulfide bonding potential. An early mutagenesis study proposed that six of the cysteines participate in *intra*molecular disulfide binds, while the seventh (C150) forms an *inter*molecular bridge to another P/rds polypeptide (Goldberg et al. 1998). A subsequent investigation refined this model and demonstrated that intermolecular disulfide bonds function to polymerize P/rds tetramers into higher-order oligomers (Loewen and Molday 2000). These structures appear to be without parallel within the superfamily; other tetraspanins probably do not form *inter*molecular disulfide bonds, since all possess even numbers of EC2 cysteine residues (4, 6, or 8).

Experimental evidence for P/rds EC2 disulfide bonding topology is not available; however, CD81 structural data (Kitadokoro et al. 2001) combined with a modeling study (Seigneuret et al. 2001) provide a scheme that may apply (Goldberg 2006). Regardless of the specific residue linkages, it seems likely that three disulfide bonds function to tightly constrain the P/rds EC2 to create a rigid domain of defined structure—leaving a single sulfhydryl group free to engage other polypeptides. Consistent with this model, inherited defects in P/rds EC2 cysteines (predicted to disrupt normal protein folding) have been reported to cause retinal degeneration in humans (Boon et al. 2008), protein loss and retinal degeneration in a transgenic mouse model (Stricker et al. 2005), and defective protein targeting in transgenic *X. laevis* (Loewen et al. 2003).

The functional role of disulfide-mediated tetramer oligomerization is not as well defined. Integrity of C150 was found important for a microsome flattening activity described for protein translated in vitro (Wrigley et al. 2000). More significantly, *X. laevis* and murine transgenic animal models expressing C150S mutant protein suggest that P/rds unable to form *inter*molecular bonds loses some, but not all functionality (Loewen et al. 2003; Chakraborty et al. 2009, 2010). Additional studies will be required to advance understanding of the role this unique feature plays for protein function in support of OS architecture.

13.1.5.3 An Intrinsically Disordered C-terminus Contains an Amphipathic Helix and May Mediate Interactions with Other OS Constituents

A structural characterization of native and recombinant versions of the P/rds C-terminus revealed that the majority of this ~7 kDa domain is present in a highly extended non-globular form that largely lacks fixed secondary and tertiary structure (Ritter et al. 2005). Thus, beyond a single α-helix (discussed below), the C-terminal domain is intrinsically disordered. Intrinsically disordered (also referred to as intrinsically or natively unstructured/unfolded) domains are regions within proteins that lack fixed three-dimensional structure (Wright and Dyson 1999). A growing body of literature emphasizes the importance of disorder for protein function, and challenges the traditional view that protein function in vivo requires a stable three-dimensional solution structure (Dyson and Wright 2005; Dunker et al. 2005; Tompa 2005). Intrinsic disorder plays varied roles for biological systems; however, function for molecular recognition appears the most common. Conformational flexibility can allow a small region of sequence the ability to recognize multiple targets, or alternatively, a single target in multiple ways (Dunker et al. 2005). Although the significance of intrinsic disorder for the P/rds C-terminus remains to be fully appreciated, it seems reasonable to speculate that it may mediate weak and possibly multivalent interactions with other OS constituents. Importantly, C-terminal flexibility is likely a general feature of the superfamily, since most tetraspanins possess C-termini of lengths insufficient to maintain stable secondary structures.

Perhaps because they are weak or transient, interactions of P/rds with other proteins have been challenging to demonstrate rigorously. In this regard, reciprocal immunoprecipitations have identified only glutamic acid rich proteins (GARPs) as P/rds binding partners to date (Poetsch et al. 2001). Three GARP proteins are encoded by alternatively spliced transcripts from the CNGB1 gene (Korschen et al. 1995; Ardell et al. 2000; Colville and Molday 1996)—the rod photoreceptor cyclic nucleotide-gated cation channel and two truncated soluble splice variants. Each protein includes a glutamic acid and proline rich domain, and like P/rds, is characterized by substantial intrinsic disorder (Batra-Safferling et al. 2005). The regions of P/rds that mediate interaction with GARPs have not yet been described; however, given that the C-terminus represents the bulk of P/rds within the cytoplasm, it seems likely that this domain contributes to these associations. Genetic ablation of GARP proteins in knockout mice produces profound disorganization of OSs (Zhang et al. 2009), a result supporting a role for P/rds-GARP interactions in OS architecture. Recent reports also propose that binding of melanoregulin (Boesze-Battaglia et al. 2007a) and calmodulin (Edrington et al. 2007) to the P/rds C-terminus regulates its interaction with OS membranes and fusogenic activity.

In fact, the first detailed investigation of the P/rds C-terminus discovered a highly conserved and periodic arrangement of charged and hydrophobic residues, and demonstrated that this region forms an amphipathic α-helix (Boesze-Battaglia et al. 1998). That study also showed that a synthetic peptide version of this helix (as well as full-length P/rds derived from OS membranes), could drive membrane fusion in vitro. This membrane-active (fusogenic) property was suggested to catalyze membrane fusion events that underlie the OS renewal process. A variety of studies have shown that fusogenic activity of the P/rds C-terminus can be modulated by: phosphorylation, calmodulin, melanoregulin, and rom-1 (Boesze-Battaglia et al. 2007a, b, 1997; Edrington et al. 2007).

Alternative functions for the amphipathic helix are also plausible, since this motif plays numerous biological roles (Segrest et al. 1994). In this regard, Tam et al. (2004) demonstrated that the P/rds C-terminal helical region (CHR) participates in protein targeting in vivo. A relatively small region within the C-terminus is necessary and sufficient to direct a membrane-associated GFP fusion protein to OS rim regions of transgenic *X. laevis* photoreceptors. The targeting/localization signal encompasses the fusogenic helical region, and a subsequent mutagenesis study showed that the two activities could be uncoupled (Ritter et al. 2004). Finally, since amphipathic helices can remodel membranes (Drin and Antonny 2010), and the P/rds CHR is ideally situated to contribute to the energetically unfavorable small diameter of OS disk rims, we propose that the CHR contributes to disk rim structure. Our recent findings demonstrate that this domain can indeed function in a direct fashion to generate membrane curvature (N. Khattree, L. Ritter and A.F.X. Goldberg; manuscript submitted).

13.1.5.4 A Conserved Acidic Transmembrane Residue Essential for Function

A majority of tetraspanins possess an acidic residue within their third or fourth transmembrane domains (Maecker et al. 1997). P/rds is similar in this regard; a glutamic acid residue is conserved in its fourth transmembrane domain in all currently available ortholog sequences (Goldberg 2006). Modeling studies suggest that the polar/charged residues within tetraspanin membrane spanning regions may be important for intramolecular interactions between the transmembrane helices (Kovalenko et al. 2005; Seigneuret 2006). A recent investigation, using a transgenic mouse model, found that substitution of a polar glutamine for the conserved and ionizable glutamate at position 276 of P/rds indeed maintained normal protein structure and self-assembly. Interestingly however, the E276Q mutant protein was unable to support in vivo function for OS architecture (Goldberg et al. 2007). Thus, the presence of an ionizable (vs. merely polar) residue is required for protein function and suggests that pH sensitivity may be an important aspect of P/rds regulation in vivo. Precisely what aspect of protein action is regulated, how it occurs, and whether it applies to other tetraspanins as well, remains to be determined.

13.1.6 Rom-1 Modulates P/rds Function

Many, but not all vertebrate species possess a P/rds homolog protein called rom-1 (Bascom et al. 1992; Moritz and Molday 1996). A similar organization of *ROM1* (also called TSPAN23) and *RDS* nucleotide sequences at the genomic level suggests an evolutionary duplication event (Bascom et al. 1993). With respect to P/rds, Rom-1 is 35% identical at the amino acid level and is similar as regards major protein features, including topology and EC2 cysteine content. Rom-1, like P/rds, is present in both rod and cone photoreceptors (Moritz and Molday 1996), and may possess a localization signal that directs the protein to the OS membranes of these cells (Lee et al. 2006). In contrast to P/rds, rom-1 is not glycosylated (Moritz and Molday 1996). Although ubiquitous amongst mammals, rom-1 is also apparently not present in several lower vertebrate species (Weng et al. 1998; Kedzierski et al. 1996; Naash et al. 2003).

In mammals, rom-1 appears to work together with P/rds to establish and maintain OS structure. Rom-1 is suggested to occupy membrane rafts (Boesze-Battaglia et al. 2002) and "potentiate" P/rds function for membrane fusion (Boesze-Battaglia et al. 2007b). These two proteins co-assemble, likely at the biosynthetic level (Goldberg et al. 1995), and likely via interactions mediated by their EC2 domains (Loewen et al. 2001; Loewen and Molday 2000). Co-assembly provides a molecular rationale for an unusual digenic pattern of human retinal disease inheritance (Goldberg and Molday 1996b; Kedzierski et al. 2001; Kajiwara et al. 1994). Mouse models show a clear asymmetry of function between the two homologs; loss of P/rds in whole or part severely disrupts OS architecture, but loss of rom-1 (again in whole or part) generates far subtler phenotypes (Clarke et al. 2000). Photoreceptor viability and OS structure are both essentially normal in heterozygous rom-1 knockout mice. In contrast, homozygous rom-1 knockouts display age-dependent loss of photoreceptors (~65% remain after 12 months), and relatively mild changes in OS structure. In particular, OSs become somewhat shorter (~80% of WT), and fatter (~40% greater average diameter than WT). In sum, these findings form the basis for considering rom-1 a modulator of P/rds function (Clarke et al. 2000).

Human retinal diseases have not been unambiguously tied to monogenic defects in *ROM1*. Several pathogenic mutations in *ROM1* have been convincingly documented, including: Gly80fs, Leu114fs, and Gly113Glu. Each however, requires co-inheritance of a mutation in *RDS* (L185P), to generate a disease phenotype (Kajiwara et al. 1994; Dryja et al. 1997). A series of studies suggest that reduced levels of P/rds-containing tetrameric complexes account for the digenic inheritance pattern (Goldberg and Molday 1996b; Loewen et al. 2001; Kedzierski et al. 2001). Interestingly, a recent study (Wistow et al. 2002) suggests that an ENU-induced W182R rom1 mutation can act in a monogenic fashion to reduce levels of rom-1 and P/rds proteins) to impede disk morphogenesis and cause retinal degeneration. These findings reiterate the importance of subunit interactions for P/rds support of OS structure and suggest a scheme by which retinal degeneration could be associated with monogenic defects in *ROM1*.

13.1.7 Tetraspanin12 Functions for Retinal Blood Vessel Development

TSPAN12 was recently identified as a critical player in retinal vascular development. Screening of a large knockout mouse collection (475 lines) by fluorescein angiography identified retinal vasculature defects in TSPAN12 (-/-) animals (Junge et al. 2009). Retinal vascularization in mice is a postnatal event that occurs in several stages; initial development of a primary superficial vessel plexus on the vitreal (nerve fiber layer) surface of the retina is followed by the establishment of two intraretinal capillary beds (Fig. 13.1a). Absence of Tetraspanin12 (Tspan12) causes delayed outgrowth of the primary vessel plexus, and a failure of inner and outer plexiform layer capillaries to form—resulting in pronounced retinal hypovascularity in adult animals (Junge et al. 2009). The otherwise initially normal retinal histology in young animals suggests that more general retinal defects observed in adults may be secondary to avascularity.

The discovery of Tspan12 function for the retina is part of an intriguing and ongoing story that has potentially profound implications for the treatment of retinal diseases. Since vision loss associated with many disorders (including inherited retinal dystrophies, age-related macular degeneration, diabetic retinopathy, retinopathy of prematurity, and others) involves vascular dysfunction, understanding how this process is developmentally regulated may aid efforts to control pathogenic neovascular processes in mature eyes. In addition to the recently elucidated need for Tspan12, defects in three other genes, NDP (Berger et al. 1996; Rehm et al. 2002; Xu et al. 2004), FZD4 (Xu et al. 2004), and LRP5 (Xia et al. 2008), have been documented to cause retinal hypovascularization in mouse models. The underlying mechanisms have recently been described in several elegantly detailed studies (Junge et al. 2009; Xu et al. 2004; Ye et al. 2009). These investigations [reviewed in (Ye et al. 2010)] demonstrate that the Wnt-family receptor Frizzled4 (the FZD4 gene product), a co-receptor Lrp5 (the LRP5 gene product), and Tspan12 appear on the surface of endothelial cells within nascent retinal vessels. In the presence of the Müeller cell–secreted ligand Norrin (the NDP gene product), these molecules together participate in a specialized Wnt/β-catenin signaling pathway that controls blood vessel development (Fig. 13.3a).

The molecular function of Tspan12 remains to be determined precisely, but appears similar to that proposed for other superfamily members—organization of lateral interactions between integral membrane proteins. In this instance, Tspan12 is suggested to bind Frizzled4 to promote its oligomerization and thereby enhance β-catenin signaling to activate transcriptional programs for blood vessel development (Junge et al. 2009). Importantly and in contrast to the relatively broad expression of FZ4, LRP5, and Norrin, Tspan12 expression within the retina appears to be restricted to endothelial cells, and is therefore proposed to provide spatiotemporal specificity for Wnt/β-catenin signaling in retinal angiogenesis.

TSPAN12 appears to be a relatively ancient tetraspanin (Garcia-Espana et al. 2008), and its expression across differing cell types may be quite broad outside the retina (Serru et al. 2000). A recent report suggests that Tspan12 (Xu et al. 2009), like tetraspanins more generally (Dunn et al. 2010; Arduise et al. 2008; Wakabayashi et al. 2009; Lafleur et al. 2009), can participate in membrane-anchored proteolysis and ectodomain shedding processes. Whether tetraspanin-regulated proteolysis functions for signaling pathways important for retinal angiogenesis awaits further investigation.

13.1.8 TSPAN12 Defects Cause Retinal Disease

Mutations in FZD4, LRP5, and NDP are documented to cause aberrant retinal blood vessel development in humans, and defects have most commonly (but not exclusively) been associated with Familial Exudative Vitreoretinopathy (FEVR; reviewed

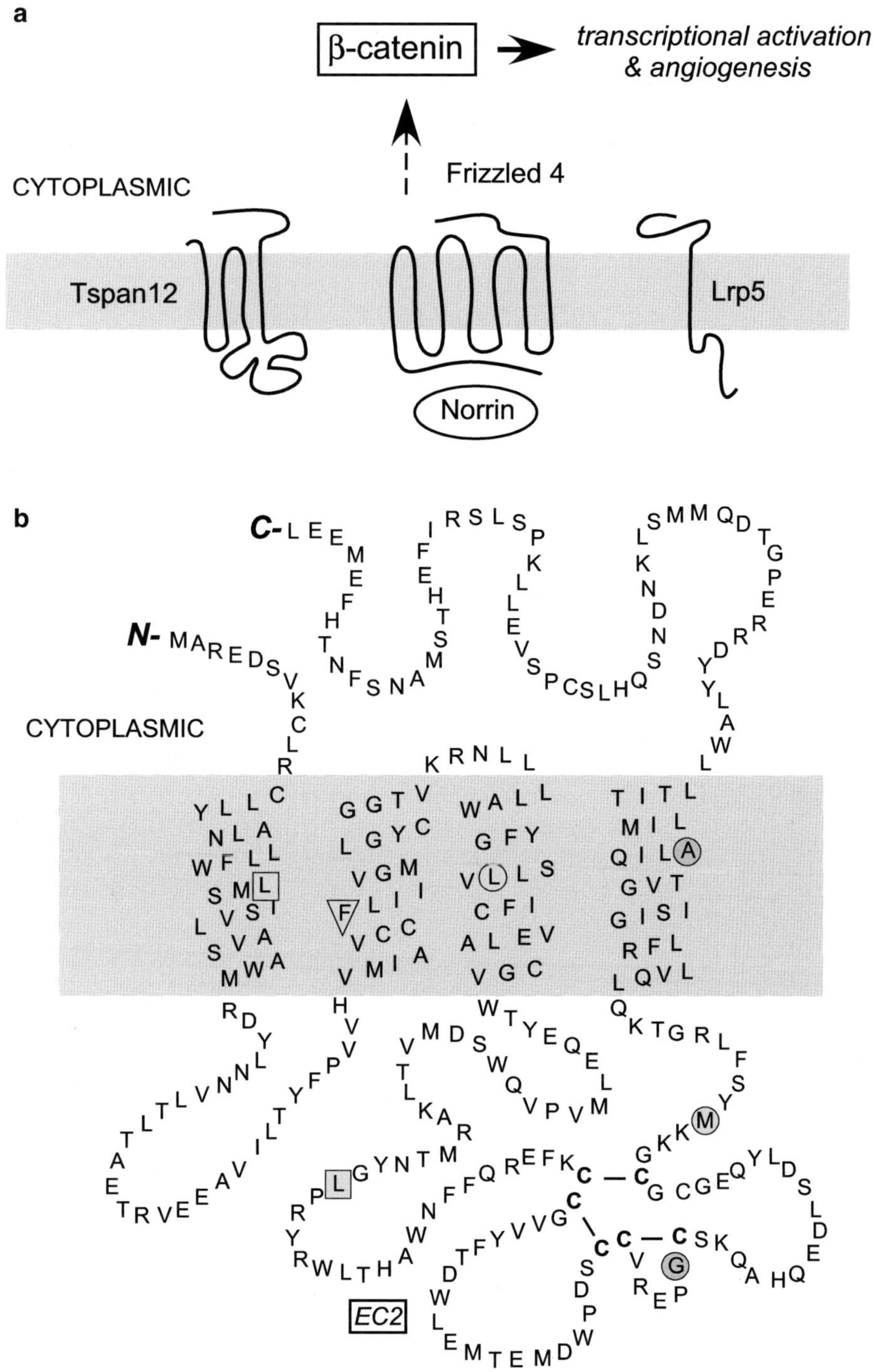
a
β-catenin
transcriptional activation
& angiogenesis
Frizzled 4
CYTOPLASMIC
Tspan12
Lrp5
Norrin
b
C-
N-
CYTOPLASMIC
EC2

in Ye et al. 2010). This inherited blinding disorder causes peripheral retinal avascularity and ischemia from birth; a variety of sight-robbing consequences follow (Pendergast and Trese 1998). Since currently documented defects in FZD4, LRP5, and NDP account for less than half of all FEVR patients, it is likely that other genes remain to be identified. The recent discovery that Tspan12 functions together with Frizzled4, Lrp5 and Norrin for Wnt/β-catenin signaling suggests the possibility that defects in the TSPAN12 gene may also be causative for hypovascular disease in humans.

Two separate studies now confirm that mutations in TSPAN12 can also generate FEVR. In the first investigation (Nikopoulos et al. 2010), a genome-wide linkage analysis of a total of 23 individuals (from two families) with autosomal dominant FEVR (and no defects in FZD4, LRP5, and NDP) identified a 40.5 Mb region containing more than 3,000 exons. Genes within this interval were captured via array hybridization, and high-throughput sequencing provided 14 variants, and three likely candidate genes—including TSPAN12. Conventional Sanger sequencing revealed two missense mutations (G188R and A237P) were present in disease-presenting individuals, but not ethnically matched controls (Fig. 13.3b, highlighted in pink). The recently described role of Tspan12 in mice prompted authors of the second study (Poulter et al. 2010) to screen TSPAN12 in a panel of 70 FEVR patients, for whom potential contributions from known FEVR genes were already excluded. Putatively pathogenic defects were found in up to 10% of these FEVR cases. A total of seven different mutations were detected (Fig. 13.3b, highlighted in blue), including missense (L101H, M210R), nonsense (L12X, L140X), frameshift (F73fs), and splice-site defects (codon149, codon361). Two-hundred ethnically matched control individuals were used for comparison to exclude the possibility that these changes are common polymorphisms. Although further studies are clearly needed to detail how inherited defects affect Tspan12 function, it seems plausible that mutations associated with autosomal dominant patterns of disease (i.e.—G188R, A237P) are mediated via the general tendency of tetraspanins to self-associate. The availability of cell culture systems that report on Wnt/β-catenin signaling, combined with the discrete phenotype produced by TSPAN12 knockout mice should facilitate a better understanding of Tspan12 function for retinal angiogenesis.

Fig. 13.3 (**a**) Essential elements of Wnt/β-catenin signaling pathway regulating retinal angiogenesis during development appear on the surface of vascular endothelial cells. Ligand (Norrin) binding to receptor (Frizzled 4), in the presence of a co-receptor (Lrp5) and tetraspanin facilitator (Tspan12), activates transcriptional programs for vascularization. Inherited defects in any component can result in FEVR or other progressive retinal disease associated with hypovascularity. (**b**) FEVR-associated mutations in TSPAN12 reported by Poulter et al. (2010) and Nikopoulos et al. (2010) mapped onto the predicted protein sequence (*blue* and *pink* respectively). Nonsense (*colored squares*), missense (*colored circles*), and frameshift (*inverted triangle*) are given; potential splice-site defects not shown. Transmembrane boundaries are based on models by Seigneuret et al. (2006) and Kovalenko et al. (2005)

13.2 Other Tetraspanins and 4TM Proteins Reported in Retina

A single recent report suggests that CD9 may function for retinal angiogenesis as well (Deissler et al. 2007); anti-CD9 (but not other anti-tetraspanin) antibodies inhibited migration of retinal endothelial cells in a wound-healing assay.

CD81 (formerly TAPA1) was initially documented in the rat retina as a ~27 kDa antigen upregulated by injury, and most closely associated with glial cells (Clarke and Geisert 1998). The temporal association of CD81 expression with that of retinal glial cell maturation during development is suggested to reflect a functional connection (Pan et al. 2007). This tetraspanin is also robustly expressed in the RPE (Geisert et al. 2002), where it is proposed to function for regulation of cell cycle progression (Song et al. 2004), and phagocytosis of shed photoreceptors OSs (Chang and Finnemann 2007).

Sequencing of a cDNA library from human RPE and choroidal tissues identified a highly enriched transcript predicted to encode a novel gene, determined to reside on chromosome 17 (Wistow et al. 2002). The mRNA endodes an open reading frame of 355 amino acids, from which a protein of 36.4 kDa and pI 5.7 is predicted. Homologies with the tetraspanin superfamily led to the naming of the predicted gene product, oculospanin. Western blotting and immunohistochemistry (using a peptide antisera) both suggest protein presence in RPE, as well as retina, iris, and ciliary body. Presence of oculospanin (also known as TSPAN10) in ocular tissues implies functionality; however, no information is yet available on this point.

Several other proteins reported as tetraspanins have substantial presence in the retina, including Als2cr4 (Zuniga and Craft 2010), and Clarin-1 (Zallocchi et al. 2009; Aarnisalo et al. 2007). Although each has a predicted membrane topology in common with tetraspanins, canonical conserved cysteine motifs within EC2 are not evident, and thus, a clear evolutionary relationship within the superfamily has not been established.

13.3 Summary

To date, inherited defects in two tetraspanin genes have been shown to cause retinal disease leading to significant vision loss and blindness in humans. Mouse models have reiterated the importance of the gene products, P/rds and Tspan12, for normal retina structure and function in vertebrates. Within this tissue, each protein displays a highly restricted expression pattern, showing presence only in photoreceptor and vascular endothelial cells respectively. P/rds appears to function at the molecular level using a non-classical (but precedented) tetraspanin behaviour—building stable membranous features critical for cell structure. Several stages of (stable) homotypic oligomerization within the membrane plane are crucial for this scaffolding behaviour. Recent studies indicate that weak and transient interactions with cytoplasmic elements are also important, and new findings (N. Khattree, L. Ritter and A.F.X. Goldberg; manuscript submitted) suggest that P/rds can remodel membranes to create the high curvature associated with photoreceptor disk rims. Tspan12, only recently documented as an important player in retinal health, displays more classical tetraspanin behaviour—appearing to act as a molecular facilitator in a Wnt/β-catenin signaling signal transduction pathway essential for retinal angiogenesis during development. This exciting development suggests the potential benefits of investigating whether this signaling pathway functions for angiogenesis more generally, since vascular defects figure prominently in a wide variety of retinal diseases. Finally, since tetraspanins are documented to act in both neuronal and non-neuronal cell types, and functional redundancy can mask effects of their genetic ablation, it is possible that significant roles for other superfamily members remain to be discovered in the vertebrate retina.

Acknowledgements The author thanks the members of his laboratory for their conscientious research efforts, and Drs. B. Anand-Apte (Cleveland Clinic) and B.S. Winkler (Oakland University) for their comments on the manuscript. This research was supported by the National Eye Institute and the E. Matilda Zeigler Foundation for the Blind.

References

Aarnisalo AA, Pietola L, Joensuu J, Isosomppi J, Aarnisalo P, Dinculescu A, Lewin AS, Flannery J, Hauswirth WW, Sankila EM, Jero J (2007) Anti-clarin-1 AAV-delivered ribozyme induced apoptosis in the mouse cochlea. Hear Res 230:9–16

Ali RR, Sarra GM, Stephens C, Alwis MD, Bainbridge JW, Munro PM, Fauser S, Reichel MB, Kinnon C, Hunt DM, Bhattacharya SS, Thrasher AJ (2000) Restoration of photoreceptor ultrastructure and function in retinal degeneration slow mice by gene therapy. Nat Genet 25:306–310

Ardell MD, Bedsole DL, Schoborg RV, Pittler SJ (2000) Genomic organization of the human rod photoreceptor cGMP-gated cation channel beta-subunit gene. Gene 245:311–318

Arduise C, Abache T, Li L, Billard M, Chabanon A, Ludwig A, Mauduit P, Boucheix C, Rubinstein E, Le NF (2008) Tetraspanins regulate ADAM10-mediated cleavage of TNF-alpha and epidermal growth factor. J Immunol 181:7002–7013

Arikawa K, Molday LL, Molday RS, Williams DS (1992) Localization of peripherin/rds in the disk membranes of cone and rod photoreceptors: relationship to disk membrane morphogenesis and retinal degeneration. J Cell Biol 116:659–667

Bascom RA, Manara S, Collins L, Molday RS, Kalnins VI, McInnes RR (1992) Cloning of the cDNA for a novel photoreceptor membrane protein (rom-1) identifies a disk rim protein family implicated in human retinopathies. Neuron 8:1171–1184

Bascom RA, Schappert K, McInnes RR (1993) Cloning of the human and murine ROM1 genes: genomic organization and sequence conservation. Hum Mol Genet 2:385–391

Batra-Safferling R, Abarca HK, Korschen HG, Tziatzios C, Stoldt M, Budyak I, Willbold D, Schwalbe H, Klein-Seetharaman J, Kaupp UB (2005) Glutamic acid-rich proteins of rod photoreceptors are natively unfolded. J Biol Chem 281:1449–1460

Berger W, van de Pol D, Bachner D, Oerlemans F, Winkens H, Hameister H, Wieringa B, Hendriks W, Ropers HH (1996) An animal model for Norrie disease (ND): gene targeting of the mouse ND gene. Hum Mol Genet 5:51–59

Boesze-Battaglia K, Goldberg AF (2002) Photoreceptor renewal: a role for peripherin/rds. Int Rev Cytol 217:183–225

Boesze-Battaglia K, Kong F, Lamba OP, Stefano FP, Williams DS (1997) Purification and light-dependent phosphorylation of a candidate fusion protein, the photoreceptor cell peripherin/rds. Biochemistry 36:6835–6846

Boesze-Battaglia K, Lamba OP, Napoli AA Jr, Sinha S, Guo Y (1998) Fusion between retinal rod outer segment membranes and model membranes: a role for photoreceptor peripherin/rds. Biochemistry 37:9477–9487

Boesze-Battaglia K, Dispoto J, Kahoe MA (2002) Association of a photoreceptor-specific tetraspanin protein, ROM-1, with triton X-100-resistant membrane rafts from rod outer segment disk membranes. J Biol Chem 277:41843–41849

Boesze-Battaglia K, Song H, Sokolov M, Lillo C, Pankoski-Walker L, Gretzula C, Gallagher B, Rachel RA, Jenkins NA, Copeland NG, Morris F, Jacob J, Yeagle P, Williams DS, Damek-Poprawa M (2007a) The tetraspanin protein peripherin-2 forms a complex with melanoregulin, a putative membrane fusion regulator. Biochemistry 46:1256–1272

Boesze-Battaglia K, Stefano FP, Fitzgerald C, Muller-Weeks S (2007b) ROM-1 potentiates photoreceptor specific membrane fusion processes. Exp Eye Res 84:22–31

Boon CJ, Den Hollander AI, Hoyng CB, Cremers FP, Klevering BJ, Keunen JE (2008) The spectrum of retinal dystrophies caused by mutations in the peripherin/RDS gene. Prog Retin Eye Res 27:213–235

Bramall AN, Wright AF, Jacobson SG, McInnes RR (2010) The genomic, biochemical, and cellular responses of the retina in inherited photoreceptor degenerations and prospects for the treatment of these disorders. Annu Rev Neurosci 33:441–472

Cai X, Conley SM, Naash MI (2010a) Gene therapy in the retinal degeneration slow model of retinitis pigmentosa. Adv Exp Med Biol 664:611–619

Cai X, Conley SM, Nash Z, Fliesler SJ, Cooper MJ, Naash MI (2010b) Gene delivery to mitotic and postmitotic photoreceptors via compacted DNA nanoparticles results in improved phenotype in a mouse model of retinitis pigmentosa. FASEB J 24:1178–1191

Chakraborty D, Ding XQ, Fliesler SJ, Naash MI (2008) Outer segment oligomerization of Rds: evidence from mouse models and subcellular fractionation. Biochemistry 47:1144–1156

Chakraborty D, Ding XQ, Conley SM, Fliesler SJ, Naash MI (2009) Differential requirements for retinal degeneration slow intermolecular disulfide-linked oligomerization in rods versus cones. Hum Mol Genet 18:797–808

Chakraborty D, Conley SM, Fliesler SJ, Naash MI (2010) The function of oligomerization-incompetent RDS in rods. Retinal degenerative diseases: laboratory and therapeutic investigations. Adv Exp Med Biol 664:39–46

Chang Y, Finnemann SC (2007) Tetraspanin CD81 is required for the alpha v beta5-integrin-dependent particle-binding step of RPE phagocytosis. J Cell Sci 120:3053–3063

Charrin S, Le NF, Silvie O, Milhiet PE, Boucheix C, Rubinstein E (2009) Lateral organization of membrane proteins: tetraspanins spin their web. Biochem J 420:133–154

Cheng T, Peachey NS, Li S, Goto Y, Cao Y, Naash MI (1997) The effect of peripherin/rds haploinsufficiency on rod and cone photoreceptors. J Neurosci 17:8118–8128

Clarke K, Geisert EE Jr (1998) The target of the antiproliferative antibody (TAPA) in the normal and injured rat retina. Mol Vis 4:3

Clarke G, Goldberg AF, Vidgen D, Collins L, Ploder L, Schwarz L, Molday LL, Rossant J, Szel A, Molday RS, Birch DG, McInnes RR (2000) Rom-1 is required for rod photoreceptor viability and the regulation of disk morphogenesis. Nat Genet 25:67–73

Colville CA, Molday RS (1996) Primary structure and expression of the human beta-subunit and related proteins of the rod photoreceptor cGMP-gated channel. J Biol Chem 271:32968–32974

Conley SM, Naash MI (2009) Focus on molecules: RDS. Exp Eye Res 89:278–279

Conley S, Nour M, Fliesler SJ, Naash MI (2007) Late-onset cone photoreceptor degeneration induced by R172W mutation in Rds and partial rescue by gene supplementation. Invest Ophthalmol Vis Sci 48:5397–5407

Conley SM, Stuck MW, Naash MI (2012) Structural and functional relationships between photoreceptor tetraspanins and other superfamily members. Cell Mol Life Sci 69:1035–1047

Connell GJ, Molday RS (1990) Molecular cloning, primary structure, and orientation of the vertebrate photoreceptor cell protein peripherin in the rod outer segment disk membrane. Biochemistry 29:4691–4698

Connell G, Bascom R, Molday L, Reid D, McInnes RR, Molday RS (1991) Photoreceptor peripherin is the normal product of the gene responsible for retinal degeneration in the rds mouse. Proc Natl Acad Sci USA 88:723–726
Corless JM, Fetter RD, Zampighi OB, Costello MJ, Wall-Buford DL (1987) Structural features of the terminal loop region of frog retinal rod outer segment disk membranes: II. Organization of the terminal loop complex. J Comp Neurol 257:9–23
Deissler H, Kuhn EM, Lang GE, Deissler H (2007) Tetraspanin CD9 is involved in the migration of retinal microvascular endothelial cells. Int J Mol Med 20:643–652
Ding XQ, Naash MI (2006) Transgenic animal studies of human retinal disease caused by mutations in peripherin/rds. Adv Exp Med Biol 572:141–146
Ding XQ, Nour M, Ritter LM, Goldberg AF, Fliesler SJ, Naash MI (2004) The R172W mutation in peripherin/rds causes a cone-rod dystrophy in transgenic mice. Hum Mol Genet 13:2075–2087
Ding XQ, Stricker HM, Naash MI (2005) Role of the second intradiscal loop of peripherin/rds in homo and hetero associations. Biochemistry 44:4897–4904
Dowling JE (1987) The retina: an approachable part of the brain. Harvard University Press, Cambridge
Drin G, Antonny B (2010) Amphipathic helices and membrane curvature. FEBS Lett 584:1840–1847
Dryja TP, Hahn LB, Kajiwara K, Berson EL (1997) Dominant and digenic mutations in the peripherin/RDS and ROM1 genes in retinitis pigmentosa. Invest Ophthalmol Vis Sci 38:1972–1982
Dunker AK, Cortese MS, Romero P, Iakoucheva LM, Uversky VN (2005) Flexible nets. The roles of intrinsic disorder in protein interaction networks. FEBS J 272:5129–5148
Dunn CD, Sulis ML, Ferrando AA, Greenwald I (2010) A conserved tetraspanin subfamily promotes Notch signaling in Caenorhabditis elegans and in human cells. Proc Natl Acad Sci USA 107:5907–5912
Dyson HJ, Wright PE (2005) Intrinsically unstructured proteins and their functions. Nat Rev Mol Cell Biol 6:197–208
Edrington TC, Yeagle PL, Gretzula CL, Boesze-Battaglia K (2007) Calcium-dependent association of calmodulin with the C-terminal domain of the tetraspanin protein peripherin/rds. Biochemistry 46:3862–3871
Farjo R, Naash MI (2006) The role of Rds in outer segment morphogenesis and human retinal disease. Ophthalmic Genet 27:117–122
Farjo R, Skaggs JS, Nagel BA, Quiambao AB, Nash ZA, Fliesler SJ, Naash MI (2006) Retention of function without normal disc morphogenesis occurs in cone but not rod photoreceptors. J Cell Biol 173:59–68
Fradkin LG, Kamphorst JT, DiAntonio A, Goodman CS, Noordermeer JN (2002) Genomewide analysis of the Drosophila tetraspanins reveals a subset with similar function in the formation of the embryonic synapse. Proc Natl Acad Sci USA 99:13663–13668
Garcia-Espana A, Chung PJ, Zhao X, Lee A, Pellicer A, Yu J, Sun TT, Desalle R (2006) Origin of the tetraspanin uroplakins and their co-evolution with associated proteins: implications for uroplakin structure and function. Mol Phylogenet Evol 41:355–367
Garcia-Espana A, Chung PJ, Sarkar IN, Stiner E, Sun TT, Desalle R (2008) Appearance of new tetraspanin genes during vertebrate evolution. Genomics 91:326–334
Geisert EE Jr, Abel HJ, Fan L, Geisert GR (2002) Retinal pigment epithelium of the rat express CD81, the target of the anti-proliferative antibody (TAPA). Invest Ophthalmol Vis Sci 43:274–280
Georgiadis A, Tschernutter M, Bainbridge JW, Robbie SJ, McIntosh J, Nathwani AC, Smith AJ, Ali RR (2010) AAV-mediated knockdown of peripherin-2 in vivo using miRNA-based hairpins. Gene Ther 17:486–493
Goldberg AF (2006) Role of peripherin/rds in vertebrate photoreceptor architecture and inherited retinal degenerations. Int Rev Cytol 253:131–175
Goldberg AF, Molday RS (1996a) Subunit composition of the peripherin/rds-rom-1 disk rim complex from rod photoreceptors: hydrodynamic evidence for a tetrameric quaternary structure. Biochemistry 35:6144–6149

Goldberg AF, Molday RS (1996b) Defective subunit assembly underlies a digenic form of retinitis pigmentosa linked to mutations in peripherin/rds and rom-1. Proc Natl Acad Sci USA 93:13726–13730

Goldberg AF, Moritz OL, Molday RS (1995) Heterologous expression of photoreceptor peripherin/rds and Rom-1 in COS-1 cells: assembly, interactions, and localization of multisubunit complexes. Biochemistry 34:14213–14219

Goldberg AF, Loewen CJ, Molday RS (1998) Cysteine residues of photoreceptor peripherin/rds: role in subunit assembly and autosomal dominant retinitis pigmentosa. Biochemistry 37:680–685

Goldberg AF, Fales LM, Hurley JB, Khattree N (2001) Folding and subunit assembly of photoreceptor peripherin/rds is mediated by determinants within the extracellular/intradiskal EC2 domain: implications for heterogeneous molecular pathologies. J Biol Chem 276:42700–42706

Goldberg AF, Ritter LM, Khattree N, Peachey NS, Fariss RN, Dang L, Yu M, Bottrell AR (2007) An intramembrane glutamic acid governs peripherin/rds function for photoreceptor disk morphogenesis. Invest Ophthalmol Vis Sci 48:2975–2986

Hawkins RK, Jansen HG, Sanyal S (1985) Development and degeneration of retina in rds mutant mice: photoreceptor abnormalities in the heterozygotes. Exp Eye Res 41:701–720

Hu P, Deng FM, Liang FX, Hu CM, Auerbach AB, Shapiro E, Wu XR, Kachar B, Sun TT (2000) Ablation of uroplakin III gene results in small urothelial plaques, urothelial leakage, and vesicoureteral reflux. J Cell Biol 151:961–972

Junge HJ, Yang S, Burton JB, Paes K, Shu X, French DM, Costa M, Rice DS, Ye W (2009) TSPAN12 regulates retinal vascular development by promoting Norrin—but not Wnt-induced FZD4/beta-catenin signaling. Cell 139:299–311

Kachar B, Liang F, Lins U, Ding M, Wu XR, Stoffler D, Aebi U, Sun TT (1999) Three-dimensional analysis of the 16 nm urothelial plaque particle: luminal surface exposure, preferential head-to-head int.raction, and hinge formation. J Mol Biol 285:595–608

Kajiwara K, Berson EL, Dryja TP (1994) Digenic retinitis pigmentosa due to mutations at the unlinked peripherin/RDS and ROM1 loci. Science 264:1604–1608

Kedzierski W, Moghrabi WN, Allen AC, Jablonski-Stiemke MM, Azarian SM, Bok D, Travis GH (1996) Three homologs of rds/peripherin in Xenopus laevis photoreceptors that exhibit covalent and non-covalent interactions. J Cell Sci 109(Pt 10):2551–2560

Kedzierski W, Nusinowitz S, Birch D, Clarke G, McInnes RR, Bok D, Travis GH (2001) Deficiency of rds/peripherin causes photoreceptor death in mouse models of digenic and dominant retinitis pigmentosa. Proc Natl Acad Sci USA 98:7718–7723

Kitadokoro K, Bordo D, Galli G, Petracca R, Falugi F, Abrignani S, Grandi G, Bolognesi M (2001) CD81 extracellular domain 3D structure: insight into the tetraspanin superfamily structural motifs. EMBO J 20:12–18

Kohl S, Giddings I, Besch D, Apfelstedt-Sylla E, Zrenner E, Wissinger B (1998) The role of the peripherin/RDS gene in retinal dystrophies. Acta Anat (Basel) 162:75–84

Kong XT, Deng FM, Hu P, Liang FX, Zhou G, Auerbach AB, Genieser N, Nelson PK, Robbins ES, Shapiro E, Kachar B, Sun TT (2004) Roles of uroplakins in plaque formation, umbrella cell enlargement, and urinary tract diseases. J Cell Biol 167:1195–1204

Korschen HG, Illing M, Seifert R, Sesti F, Williams A, Gotzes S, Colville C, Muller F, Dose A, Godde M (1995) A 240 kDa protein represents the complete beta subunit of the cyclic nucleotide-gated channel from rod photoreceptor. Neuron 15:627–636

Kovalenko OV, Metcalf DG, DeGrado WF, Hemler ME (2005) Structural organization and interactions of transmembrane domains in tetraspanin proteins. BMC Struct Biol 5:11

Lafleur MA, Xu D, Hemler ME (2009) Tetraspanin proteins regulate membrane type-1 matrix metalloproteinase-dependent pericellular proteolysis. Mol Biol Cell 20:2030–2040

Lee ES, Burnside B, Flannery JG (2006) Characterization of peripherin/rds and rom-1 transport in rod photoreceptors of transgenic and knockout animals. Invest Ophthalmol Vis Sci 47:2150–2160

Levy S, Shoham T (2005) Protein-protein interactions in the tetraspanin web. Physiology (Bethesda) 20:218–224

Levy S, Todd SC, Maecker HT (1998) CD81 (TAPA-1): a molecule involved in signal transduction and cell adhesion in the immune system. Annu Rev Immunol 16:89–109

Loewen CJ, Molday RS (2000) Disulfide-mediated oligomerization of peripherin/Rds and Rom-1 in photoreceptor disk membranes. Implications for photoreceptor outer segment morphogenesis and degeneration. J Biol Chem 275:5370–5378

Loewen CJ, Moritz OL, Molday RS (2001) Molecular characterization of peripherin-2 and rom-1 mutants responsible for digenic retinitis pigmentosa. J Biol Chem 276:22388–22396

Loewen CJ, Moritz OL, Tam BM, Papermaster DS, Molday RS (2003) The role of subunit assembly in peripherin-2 targeting to rod photoreceptor disk membranes and retinitis pigmentosa. Mol Biol Cell 14:3400–3413

Maecker HT, Todd SC, Levy S (1997) The tetraspanin superfamily: molecular facilitators. FASEB J 11:428–442

Min G, Wang H, Sun TT, Kong XP (2006) Structural basis for tetraspanin functions as revealed by the cryo-EM structure of uroplakin complexes at 6-A resolution. J Cell Biol 173:975–983

Molday RS (1994) Peripherin/Rds and Rom-1—molecular-properties and role in photoreceptor cell degeneration. Prog Retin Eye Res 13:271–299

Molday RS, Hicks D, Molday L (1987) Peripherin. A rim-specific membrane protein of rod outer segment discs. Invest Ophthalmol Vis Sci 28:50–61

Molday RS, Molday LL, Loewen CJ (2004) Role of subunit assembly in autosomal dominant retinitis pigmentosa linked to mutations in peripherin 2. Novartis Found Symp 255:95–112

Moritz OL, Molday RS (1996) Molecular cloning, membrane topology, and localization of bovine rom-1 in rod and cone photoreceptor cells. Invest Ophthalmol Vis Sci 37:352–362

Naash MI, Ding XQ, Li C, O'Brien J, Al Ubaidi MR (2003) Peripherin/rds in skate retina. Adv Exp Med Biol 533:377–383

Nikopoulos K, Gilissen C, Hoischen A, van Nouhuys CE, Boonstra FN, Blokland EA, Arts P, Wieskamp N, Strom TM, Ayuso C, Tilanus MA, Bouwhuis S, Mukhopadhyay A, Scheffer H, Hoefsloot LH, Veltman JA, Cremers FP, Collin RW (2010) Next-generation sequencing of a 40 Mb linkage interval reveals TSPAN12 mutations in patients with familial exudative vitreoretinopathy. Am J Hum Genet 86:240–247

Oyster CW (1999) The human eye: structure and function. Sinauer Associates, Sunderland

Pan Y, Brown C, Wang X, Geisert EE (2007) The developmental regulation of CD81 in the rat retina. Mol Vis 13:181–189

Pendergast SD, Trese MT (1998) Familial exudative vitreoretinopathy. Results of surgical management. Ophthalmology 105:1015–1023

Poetsch A, Molday LL, Molday RS (2001) The cGMP-gated channel and related glutamic acid-rich proteins interact with peripherin-2 at the rim region of rod photoreceptor disc membranes. J Biol Chem 276:48009–48016

Poulter JA, Ali M, Gilmour DF, Rice A, Kondo H, Hayashi K, Mackey DA, Kearns LS, Ruddle JB, Craig JE, Pierce EA, Downey LM, Mohamed MD, Markham AF, Inglehearn CF, Toomes C (2010) Mutations in TSPAN12 cause autosomal-dominant familial exudative vitreoretinopathy. Am J Hum Genet 86:248–253

Rehm HL, Zhang DS, Brown MC, Burgess B, Halpin C, Berger W, Morton CC, Corey DP, Chen ZY (2002) Vascular defects and sensorineural deafness in a mouse model of Norrie disease. J Neurosci 22:4286–4292

Ritter LM, Boesze-Battaglia K, Tam BM, Moritz OL, Khattree N, Chen SC, Goldberg AF (2004) Uncoupling of photoreceptor peripherin/rds fusogenic activity from biosynthesis, subunit assembly, and targeting: a potential mechanism for pathogenic effects. J Biol Chem 279:39958–39967

Ritter LM, Arakawa T, Goldberg AFX (2005) Predicted and measured disorder in peripherin/rds, a retinal tetraspanin. Protein Pept Lett 12:677–686

Rodieck RW (1998) The first steps in seeing. Sinauer Associates, Sunderland

Sammut SJ, Finn RD, Bateman A (2008) Pfam 10 years on: 10,000 families and still growing. Brief Bioinform 9:210–219

Sanyal S, Jansen HG (1981) Absence of receptor outer segments in the retina of rds mutant mice. Neurosci Lett 21:23–26

Sanyal S, De Ruiter A, Hawkins RK (1980) Development and degeneration of retina in rds mutant mice: light microscopy. J Comp Neurol 194:193–207
Segrest JP, Garber DW, Brouillette CG, Harvey SC, Anantharamaiah GM (1994) The amphipathic alpha helix: a multifunctional structural motif in plasma apolipoproteins. Adv Protein Chem 45:303–369
Seigneuret M (2006) Complete predicted three-dimensional structure of the facilitator transmembrane protein and hepatitis C virus receptor CD81: conserved and variable structural domains in the tetraspanin superfamily. Biophys J 90:212–227
Seigneuret M, Delaguillaumie A, Lagaudriere-Gesbert C, Conjeaud H (2001) Structure of the tetraspanin main extracellular domain—a partially conserved fold with a structurally variable domain insertion. J Biol Chem 276:40055–40064
Serru V, Dessen P, Boucheix C, Rubinstein E (2000) Sequence and expression of seven new tetraspans. Biochim Biophys Acta 1478:159–163
Simonelli F, Maguire AM, Testa F, Pierce EA, Mingozzi F, Bennicelli JL, Rossi S, Marshall K, Banfi S, Surace EM, Sun J, Redmond TM, Zhu X, Shindler KS, Ying GS, Ziviello C, Acerra C, Wright JF, McDonnell JW, High KA, Bennett J, Auricchio A (2010) Gene therapy for Leber's congenital amaurosis is safe and effective through 1.5 years after vector administration. Mol Ther 18:643–650
Song BK, Levy S, Geisert EE Jr (2004) Increased density of retinal pigment epithelium in cd81-/- mice. J Cell Biochem 92:1160–1170
Steinberg RH, Fisher SK, Anderson DH (1980) Disc morphogenesis in vertebrate photoreceptors. J Comp Neurol 190:501–508
Stricker HM, Ding XQ, Quiambao A, Fliesler SJ, Naash MI (2005) The Cys214-->Ser mutation in peripherin/rds causes a loss-of-function phenotype in transgenic mice. Biochem J 388:605–613
Tam BM, Moritz OL, Papermaster DS (2004) The C terminus of peripherin/rds participates in rod outer segment targeting and alignment of disk incisures. Mol Biol Cell 15:2027–2037
Tompa P (2005) The interplay between structure and function in intrinsically unstructured proteins. FEBS Lett 579:3346–3354
Travis GH, Brennan MB, Danielson PE, Kozak CA, Sutcliffe JG (1989) Identification of a photoreceptor-specific mRNA encoded by the gene responsible for retinal degeneration slow (rds). Nature 338:70–73
Travis GH, Sutcliffe JG, Bok D (1991) The retinal degeneration slow (rds) gene product is a photoreceptor disc membrane-associated glycoprotein. Neuron 6:61–70
Travis GH, Groshan KR, Lloyd M, Bok D (1992) Complete rescue of photoreceptor dysplasia and degeneration in transgenic retinal degeneration slow (rds) mice. Neuron 9:113–119
Wakabayashi T, Craessaerts K, Bammens L, Bentahir M, Borgions F, Herdewijn P, Staes A, Timmerman E, Vandekerckhove J, Rubinstein E, Boucheix C, Gevaert K, De SB (2009) Analysis of the gamma-secretase interactome and validation of its association with tetraspanin-enriched microdomains. Nat Cell Biol 11:1340–1346
Weng J, Belecky-Adams T, Adler R, Travis GH (1998) Identification of two rds/peripherin homologs in the chick retina. Invest Ophthalmol Vis Sci 39:440–443
Wistow G, Bernstein SL, Wyatt MK, Fariss RN, Behal A, Touchman JW, Bouffard G, Smith D, Peterson K (2002) Expressed sequence tag analysis of human RPE/choroid for the NEIBank project: over 6000 non-redundant transcripts, novel genes and splice variants. Mol Vis 8:205–220
Wright PE, Dyson HJ (1999) Intrinsically unstructured proteins: re-assessing the protein structure-function paradigm. J Mol Biol 293:321–331
Wright MD, Tomlinson MG (1994) The ins and outs of the transmembrane 4 superfamily. Immunol Today 15:588–594
Wrigley JD, Ahmed T, Nevett CL, Findlay JB (2000) Peripherin/rds influences membrane vesicle morphology. Implications for retinopathies. J Biol Chem 275:13191–13194
Wu XR, Kong XP, Pellicer A, Kreibich G, Sun TT (2009) Uroplakins in urothelial biology, function, and disease. Kidney Int 75:1153–1165

Xia CH, Liu H, Cheung D, Wang M, Cheng C, Du X, Chang B, Beutler B, Gong X (2008) A model for familial exudative vitreoretinopathy caused by LPR5 mutations. Hum Mol Genet 17:1605–1612

Xu Q, Wang Y, Dabdoub A, Smallwood PM, Williams J, Woods C, Kelley MW, Jiang L, Tasman W, Zhang K, Nathans J (2004) Vascular development in the retina and inner ear: control by Norrin and Frizzled-4, a high-affinity ligand-receptor pair. Cell 116:883–895

Xu DS, Sharma C, Hemler ME (2009) Tetraspanin12 regulates ADAM10-dependent cleavage of amyloid precursor protein. FASEB J 23:3674–3681

Yanez-Mo M, Barreiro O, Gordon-Alonso M, Sala-Valdes M, Sanchez-Madrid F (2009) Tetraspanin-enriched microdomains: a functional unit in cell plasma membranes. Trends Cell Biol 19:434–446

Ye X, Wang Y, Cahill H, Yu M, Badea TC, Smallwood PM, Peachey NS, Nathans J (2009) Norrin, frizzled-4, and Lrp5 signaling in endothelial cells controls a genetic program for retinal vascularization. Cell 139:285–298

Ye X, Wang Y, Nathans J (2010) The Norrin/Frizzled4 signaling pathway in retinal vascular development and disease. Trends Mol Med 16:417–425

Young RW (1976) Visual cells and the concept of renewal. Invest Ophthalmol Vis Sci 15:700–725

Zallocchi M, Meehan DT, Delimont D, Askew C, Garige S, Gratton MA, Rothermund-Franklin CA, Cosgrove D (2009) Localization and expression of clarin-1, the Clrn1 gene product, in auditory hair cells and photoreceptors. Hear Res 255:109–120

Zhang Y, Molday LL, Molday RS, Sarfare SS, Woodruff ML, Fain GL, Kraft TW, Pittler SJ (2009) Knockout of GARPs and the beta-subunit of the rod cGMP-gated channel disrupts disk morphogenesis and rod outer segment structural integrity. J Cell Sci 122:1192–1200

Zuniga FI, Craft CM (2010) Deciphering the structure and function of Als2cr4 in the mouse retina. Invest Ophthalmol Vis Sci 51:4407–4415

Chapter 14
The Role of CD81 in HCV and *Plasmodium* Infection

Laurence Cocquerel and Olivier Silvie

Abstract Hepatitis C and malaria, two of the most prevalent infectious diseases in the world, are caused by Hepatitis C virus (HCV) and *Plasmodium* parasites, respectively. Both HCV particles and *Plasmodium* sporozoites, the mosquito-transmitted stage of the malaria parasite, infect and replicate in the liver. Whereas HCV enters cells by clathrin-mediated endocytosis, *Plasmodium* sporozoite invasion is a specific active process that relies on the parasite motility machinery. Remarkably, both pathogens critically depend on the host tetraspanin CD81 to enter hepatocytes. In this chapter, we summarize the current knowledge on the role of CD81, tetraspanin-enriched microdomains and CD81-associated partners during HCV and *Plasmodium* liver infection.

Abbreviations

AMA-1	Apical membrane antigen 1
ApoB	Apolipoprotein B
ApoE	Apolipoprotein E
CLDN-1	Claudin-1

L. Cocquerel (✉)
Center for Infection and Immunity of Lille, Inserm U1019 – CNRS UMR8204,
Pasteur Institute of Lille, University Lille North of France,
1 rue Calmette, B.P.447, 59021 Lille Cedex, France
e-mail: laurence.cocquerel@ibl.fr

O. Silvie
INSERM, UMR S 945, Paris 75013, France

Université Pierre et Marie Curie-Paris 6, UMR S 945, Paris 75013, France
e-mail: olivier.silvie@inserm.fr

F. Berditchevski and E. Rubinstein (eds.), *Tetraspanins*, Proteins and Cell Regulation 9,
DOI 10.1007/978-94-007-6070-7_14, © Springer Science+Business Media Dordrecht 2013

CSP	Circumsporozoite protein
EC1	First extracellular loop
EEFs	Exo-erythrocytic forms
ER	Endoplasmic reticulum
hCD81	Human CD81
HCV	Hepatitis C virus
HCVcc	HCV produced in cell culture
HCVpc	HCV from primary culture
HCVpp	HCV pseudoparticles
HCVs	HCV from infectious serum
HDL	High-density lipoproteins
HSPGs	Heparan sulfate proteoglycans
HTLV-1	Human T-lymphotropic virus 1
HVR1	Hypervariable region 1
LDL	Low-density lipoproteins
LDL-R	Low-density lipoproteins receptor
LEL	Large extracellular loop
LS	Liver stages
LVPs	Lipo-viro-particles
mCD81	Mouse CD81
MβCD	Methyl-beta-cyclo-dextrin
NS	Non structural
PCSK9	Proprotein convertase subtilisin/kexin type 9
PHH	Primary human hepatocytes
sE2	Soluble E2
SR-BI	Scavenger receptor class B type I
TEM	Tetraspanin-enriched microdomains
TRAP	Thrombospondin-related anonymous protein
VLDL	Very-low-density lipoproteins

14.1 Introduction

Several pathogens including bacteria, viruses and parasites, must enter mammalian cells to survive, replicate and evade the immune-system. To infect a target cell, the pathogen proceeds through a multistep entry process, during which each step is tightly regulated in time and space. In this chapter, we focused on two major human pathogens for which importance of the tetraspanin CD81 in their life cycle has been extensively demonstrated: the malaria parasite *Plasmodium* and Hepatitis C Virus (HCV).

CD81 was originally identified as a target of an antiproliferative antibody on human B cells (Oren et al. 1990) and has since been found in many different cell types and involved in an astonishing range of physiological processes. For example, it plays a role in B cell activation, immune response, cell adhesion and migration, astrocyte and microglial cell proliferation, and sperm-egg fusion (reviewed in Charrin et al. 2009a; Levy and Shoham 2005b). Importantly, CD81 plays also a role in several infectious diseases. Indeed, in addition to its role in HCV and *Plasmodium* infection (Pileri et al. 1998; Silvie et al. 2003), CD81 was implicated in the infection with *Listeria monocytogenes* (Tham et al. 2009) and is involved in the formation of syncitia produced by HTLV-1 (human T-lymphotropic virus 1) (Imai and Yoshie 1993).

The current knowledge accumulated on the involvement of CD81 and associated factors in the *Plasmodium* and HCV life cycles is summarized in this chapter.

14.2 HCV Infection

HCV infection is a global public health problem affecting over 130 million individuals worldwide. Although the annual incidence of HCV is difficult to determine, it has been estimated that approximately two million individuals are annually infected by HCV worldwide as a result of contaminated injections (Hauri et al. 2004). Since most of the infected individuals (70–80%) fail to spontaneously clear the virus from the liver, HCV infection generally progresses to chronic hepatitis, which can evolve to hepatic steatosis, liver cirrhosis and hepatocellular carcinoma (Lemon et al. 2007). Unfortunately, no vaccine is currently available to prevent new HCV infections. In addition, current treatments based on the use of pegylated interferon-α in combination with ribavirin are not fully efficient and are accompanied by serious side effects (Manns et al. 2006). Clearly, new therapeutic strategies are urgently required.

14.2.1 HCV Particles

HCV is a small, enveloped virus that is a member of the *Hepacivirus* genus in the *Flaviviridae* family (reviewed in Lindenbach et al. 2007). Its genome is composed of a single-stranded, positive sense RNA molecule, approximately 9.6 kb in length. This RNA encodes a unique polyprotein composed of about 3,000 amino acids, which is cleaved by cellular and viral proteases into three structural (Core, E1 and E2) and seven nonstructural (p7, NS2, NS3, NS4A, NS4B, NS5A and NS5B) proteins. Structural proteins are the components of the viral particle whereas non-structural proteins are involved in viral replication, morphogenesis and release of particles. In addition to structural and non structural proteins, another protein, called F, has been proposed to be produced as a result of a ribosomal frameshift in the N-terminal core encoding region (Xu et al. 2001). The function of F and its role in the pathophysiology of HCV remain to be elucidated.

Although the detailed structure of the HCV virion remains unclear, it is known that viral particles have a nucleocapsid containing the viral RNA surrounded by a host cell-derived lipid envelope in which the viral E1 and E2 glycoproteins are embedded (Fig. 14.1a). These virion-associated envelope glycoproteins are organized as large covalent complexes stabilized by disulfide bridges (Vieyres et al. 2010).

It is worth noting that HCV particles isolated from patient sera are heterogeneous for their density and composition (Andre et al. 2005; Popescu and Dubuisson 2009). The majority of HCV circulating in the serum is of low density. These particles are supposed to be the most infectious and they are called Lipo-Viro-Particles (LVPs), which are lipoprotein-like structures composed of triglyceride-rich lipoproteins bearing apolipoprotein B (ApoB), apolipoprotein E (ApoE), viral nucleocapsids, and envelope glycoproteins (Andre et al. 2002; Nielsen et al. 2006). HCV particles produced in cell culture also exhibit heterogeneous densities (Lindenbach et al. 2005). Indeed, infectious HCV particles are associated with low-density lipoproteins (Fig. 14.1a) and are likely assembled and co-secreted with very-low-density lipoproteins (VLDL). Their production in hepatoma cells depends on apoB, apoE, and microsomal triglyceride transfer protein, host factors that are essential for secretion of VLDL (Chang et al. 2007; Gastaminza et al. 2008; Huang et al. 2007). In addition, co-precipitation of cell culture derived HCV with ApoE or ApoC1-specific antibodies as well as neutralization of infection by these antibodies suggest a direct interaction between viral particle and ApoE and ApoC1-comprising lipoprotein complexes (Benga et al. 2010; Chang et al. 2007; Dreux et al. 2007; Jiang and Luo 2009; Meunier et al. 2008; Owen et al. 2009).

14.2.2 Model Systems to Study Early Steps of HCV Life Cycle

Although the cloning of the HCV genome has allowed a rapid analysis of the genomic organization and a biochemical characterization of its proteins, the heterogeneity and low levels of particles in sera as well as the lack of a cell culture system allowing efficient *in vitro* amplification have long been obstacles for the study of HCV entry. Consequently, surrogate models based on the expression of envelope glycoproteins, recombinant virus-like particles produced in insect cells or vesicular stomatitis virus pseudoparticles, have been developed. One model that has been

Fig. 14.1 (continued) with the scavenger receptor class B type I (*SR-BI*), the tetraspanin CD81, and the tight junction proteins Claudin-1 (*CLDN-1*) and Occludin (*OCLN*). Viral particles are then internalized by clathrin-mediated endocytosis. After fusion between the viral membrane and the membrane of an early endosome, viral RNA is released into the cytosol. The viral RNA is then translated into a polyprotein that is processed to generate structural and non structural proteins. NS3 to NS5B proteins form the replication complex in association with ER-derived membranes called membranous web, and they replicate the genome. After accumulation of neosynthesized genomic RNA and viral proteins, new HCV particles are assembled in an ER-related compartment in close connection with the VLDL biogenesis pathway and in the proximity of lipid droplets (*LD*). Finally, lipoprotein-associated virions are released from the cell, likely through the secretory pathway

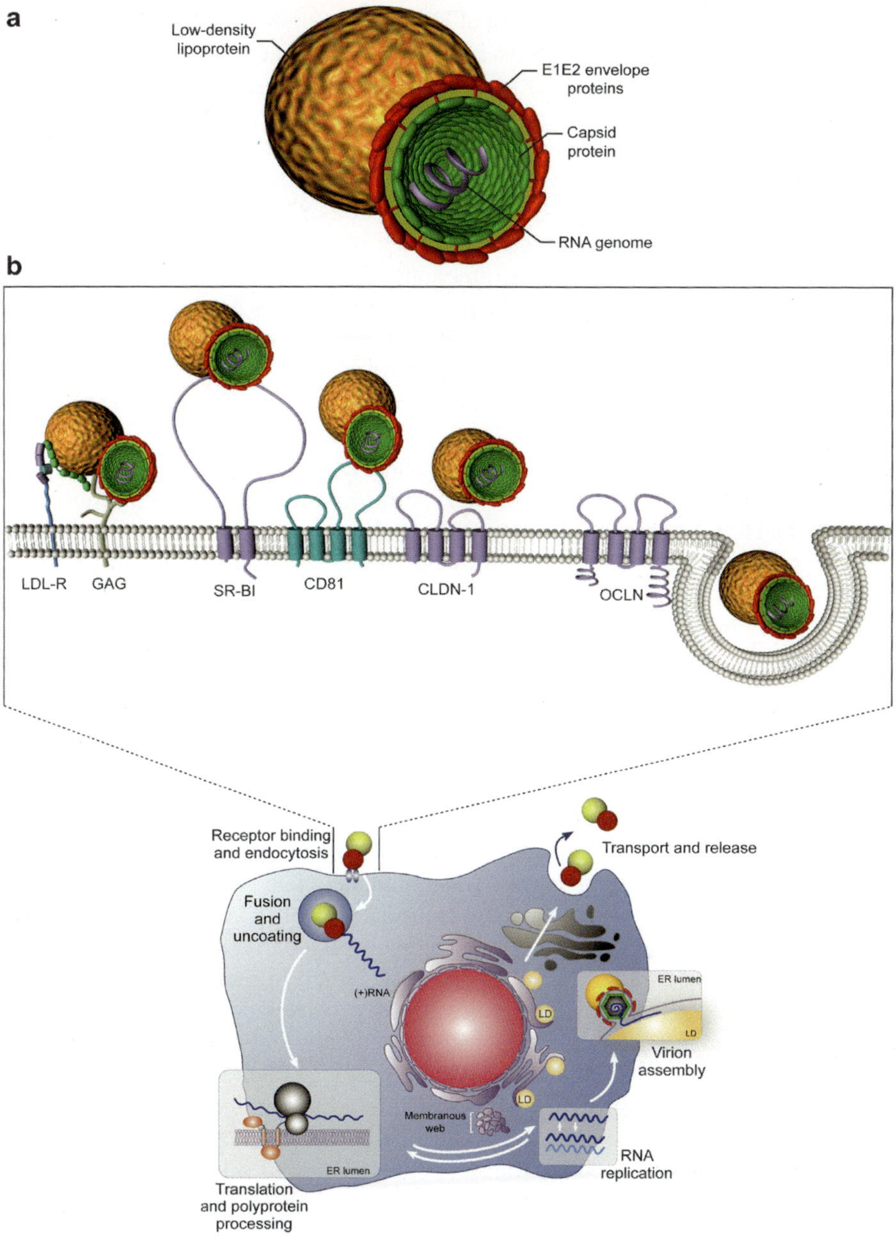

Fig. 14.1 (**a**) Schematic representation of HCV. Viral particles are composed of a nucleocapsid (*green*) bearing the viral RNA (*purple*) surrounded by a host cell-derived lipid envelope (*yellow*) in which the E1 and E2 glycoproteins (*red*) are embedded. Low-density lipopropteins that are associated with particles are represented by *gold spheres* (**b**) Main steps of the HCV life cycle. The initial attachment of particles to the cell involves glycosaminoglycans (*GAG*), which concentrate viruses on the cell surface. Particles next interact with the low-density lipoprotein receptor (*LDL-R*) likely through their associated lipoproteins. After this initial docking, particles interact

very useful is a truncated soluble form of E2 envelope protein (sE2), which aided the identification of cell surface proteins involved in HCV entry. However, neither this tool nor recombinant virus-like particles enabled functional studies of E1E2-mediated cell entry up to the point of membrane fusion (reviewed in Barth et al. 2006; Bartosch and Cosset 2006; Cocquerel et al. 2006). The first breakthrough in investigating HCV entry was made possible by the development of HCV pseudoparticles (HCVpp). This system is based on the incorporation of unmodified envelope glycoproteins onto retroviral core particles (Bartosch et al. 2003b; Drummer et al. 2003; Hsu et al. 2003). These particles are infectious, show a preferential tropism for liver cells, and are neutralized by anti-E2 monoclonal antibodies as well as by sera from infected patients (Bartosch et al. 2003a; Hsu et al. 2003; Op De Beeck et al. 2004). A cryo-transmission electron microscopy analysis has recently shown that HCVpp are regular 100-nm spherical structures containing the dense nucleocapsid surrounded by the lipid bilayer. E1E2 glycoproteins were not readily visible on the membrane surface (Bonnafous et al. 2010). Studying viral entry using HCVpp has shed some light on the role of some cell surface molecules involved in the early steps of the HCV life cycle. However, since HCVpp are not associated with lipoproteins, results obtained with HCVpp may not necessarily reflect the actual entry of native HCV particles.

The most important milestone in HCV field was the development of a cell culture system that enables efficient *in vitro* amplification of HCV (Lindenbach et al. 2005; Wakita et al. 2005; Zhong et al. 2005). These particles named HCVcc, for cell culture derived HCV, allow the study of its entire life cycle in certain hepatoma cells (Lindenbach et al. 2005; Wakita et al. 2005; Zhong et al. 2005) and primary hepatocytes (Molina et al. 2008). HCVcc are produced after the human hepatoma Huh-7 cell line has been transfected with a HCV genome isolated from a patient with a fulminant hepatitis C. In addition to their ability to infect naive Huh-7 cells, HCVcc are also infectious in chimpanzees, in immunodeficient mice with human liver grafts and in immunocompetent mice transduced with adenoviral recombinants expressing the four human cell surface receptors (Dorner et al. 2011; Lindenbach et al. 2006; Wakita et al. 2005; Washburn et al. 2011). In contrast to HCV derived from the plasma of HCV-infected patients, virus recovered from HCVcc-infected animals are capable of replicating robustly in cell culture (Lindenbach et al. 2005). It should be noted that animal-derived particles are characterized by a lower density but higher infectivity than viruses produced in cell culture, indicating that associations with lipoproteins are likely different (Lindenbach et al. 2006).

Several groups have reported the infection of cultured primary human hepatocytes (PHHs) with serum-derived HCV (HCVs) and HCVcc (Buck 2008; Carloni et al. 1993; Fournier et al. 1998; Iacovacci et al. 1993; Lazaro et al. 2007; Molina et al. 2008; Ploss et al. 2010; Rumin et al. 1999). However, replication levels appeared to be low in these systems or difficult to reproduce. Recently, Podevin et al. established a system that allows to maintain PHHs in culture for up to 2 weeks (Podevin et al. 2010). These PHHs could be productively infected with HCVcc and led to the efficient production of infectious progeny virus particles, which were called primary-culture-derived HCV (HCVpc). Compared with HCVcc produced in

Huh-7 cells, HCVpc had lower average buoyant density and higher specific infectivity as particles produced *in vivo*. Thus, assembly of infectious, low-density particles is more efficient in cultured PHHs than in Huh-7 cells. In addition, HCVpc were neutralized by antibodies directed against either E1 envelope glycoprotein or CD81 (Podevin et al. 2010), suggesting that these particles could become a model of choice for the study of HCV interactions with its natural host cell.

14.2.3 Main Steps of HCV Life Cycle

HCV infection of the target cells, the hepatocytes, begins with the particles binding to attachment factors, which helps to concentrate viruses on the cell surface (Fig. 14.1b). As with many other viruses, glycosaminoglycans seem to be an initial docking site for HCV attachment (Barth et al. 2003; Basu et al. 2007; Germi et al. 2002; Koutsoudakis et al. 2006; Morikawa et al. 2007). After this initial attachment to the host cell, a complex process occurs in which the virion interacts with a series of cellular entry factors, including CD81 (see details below). The viral particle is then internalized by clathrin-mediated endocytosis (Blanchard et al. 2006; Coller et al. 2009; Meertens et al. 2006). After the fusion of the viral membrane with the membrane of an early endosome (Coller et al. 2009; Meertens et al. 2006), viral RNA is released into the cytosol. The genome is then translated and processed to generate structural and non structural proteins. The proteins NS3 through NS5B assemble as a macromolecular replicase complex on cytoplasmic membranes forming a membranous web-like structure (reviewed in Lemon et al. 2007). After the accumulation of neosynthesized genomic RNA and viral proteins, new HCV particles are assembled in an ER-related compartment in close connection with the VLDL biogenesis pathway and in the proximity of lipid droplets. Finally, lipoprotein-associated virions are released from the cell, likely through the secretory pathway.

Although a CD81-independent mechanism for cell-to-cell transmission (Jones et al. 2010; Timpe et al. 2008; Witteveldt et al. 2009), and the involvement of CD81 in HCV genome replication (Zhang et al. 2010) have been recently reported, this tetraspanin has been extensively described as an essential entry factor in the early steps of HCV life cycle. The current knowledge of the role of CD81 in the entry of cell-free HCV is detailed below.

14.3 *Plasmodium* Infection

Malaria is one of the most prevalent infectious diseases in the world, affecting more than 200 million people and killing an estimated 1 million every year, most of which are children in sub-saharan Africa (WHO Malaria Report 2009). Malaria is caused by *Plasmodium*, an obligate intracellular protist, which belongs to the phylum *Apicomplexa*. Among the five species that infect humans, *Plasmodium falciparum* accounts for the majority of malaria morbidity and mortality. *Plasmodium* has a complex life cycle that alternates between a mosquito vector and a vertebrate host. Transmission of the parasite occurs during the bite of an infected female *Anopheles* mosquito, which injects *Plasmodium* sporozoites into the skin of the host (Amino et al. 2006; Yamauchi et al. 2007). The motile sporozoites actively migrate in the skin and, after finding a blood vessel, enter the peripheral blood circulation and rapidly reach the liver (Fig. 14.2a). Once there, the sporozoites cross the sinusoidal cell layer and migrate through several hepatocytes before finding one to infect. Infection begins with the formation of a parasitophorous vacuole (PV) (Frevert et al. 2005; Mota et al. 2001) inside which sporozoites transform into replicative liver stages (LS), also called exo-erythrocytic forms (EEFs) (Fig. 14.2a). Over a period of 2–6 days (depending on the *Plasmodium* species), LS undergo a dramatic parasite multiplication, after which thousands of merozoites are released. These merozoites invade erythrocytes and initiate the pathogenic blood stage cycle, causing the symptoms and complications of malaria. Infection of the liver is a necessary and clinically silent phase of the *Plasmodium* life cycle. Therefore, pre-erythrocytic stages (sporozoites and liver stages) are ideal targets for prophylactic antimalarial drug and vaccine interventions. The only malaria vaccine currently in phase III of development, RTS, S, targets the major sporozoite surface protein (CSP), but confers only partial protection (Snounou et al. 2005). Development of novel strategies specifically aimed at blocking infection of hepatocytes clearly requires a better characterization of the molecular mechanisms of sporozoite invasion.

14.3.1 Plasmodium Sporozoite Invasion

Plasmodium sporozoites are elongated polarized cells, 10–15 μm large and 1 μm wide (Fig. 14.2b). They move by gliding motility, a process powered by a subpellicular actomyosin motor (Kappe et al. 1999; Santos et al. 2009). Like all invasive stages of *Apicomplexa,* sporozoites have specialized secretory organelles at their apical end, called micronemes and rhoptries. These vesicles secrete their contents sequentially during parasite movement and invasion of target cells (Baum et al. 2008). Some of the secreted proteins contain adhesive domains that may interact with host cells. Mechanisms of sporozoite entry into hepatocytes remain poorly characterized. According to the current model, which is based primarily on observations made with other *Plasmodium* stages (merozoites) or other apicomplexan parasites such as *Toxoplasma*, host cell invasion involves the formation of a tight junction, through which the sporozoite actively propels itself into the cell, using its own actomyosin motor machinery (Santos et al. 2009) (Fig. 14.2c). It is speculated that the junction is formed because of an interaction between sporozoite adhesive surface molecules and their cognate receptors on the hepatocyte surface; however, as of yet the nature of these molecules has remained elusive. Invasion through a junction results in the formation of the parasitophorous vacuole, the membrane of

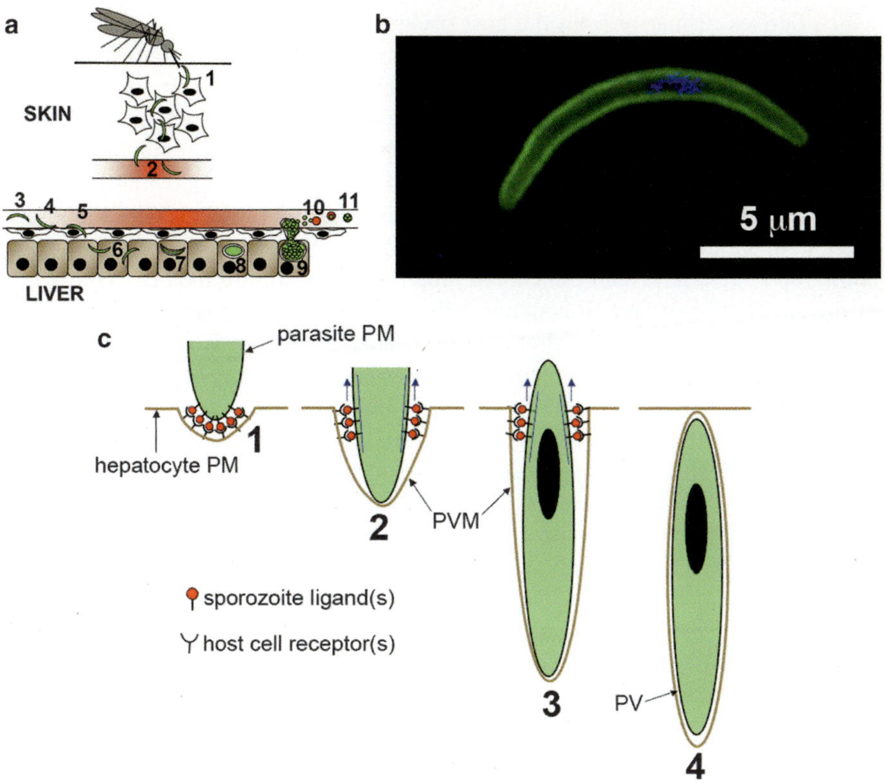

Fig. 14.2 (**a**) Pre-erythrocytic stages of the *Plasmodium* life cycle. *Plasmodium* sporozoites are injected into the skin of the host during the blood meal of an infected anopheline mosquito (*1*). Sporozoites actively migrate through the dermis until they enter a blood vessel (*2*). Once in the blood circulation, sporozoites are transported to the liver (*3*), where they are sequestered through interaction with hepatic heparan sulfate proteoglycans (*4*). Sporozoites then traverse the sinusoidal barrier (*5*) and migrate through several hepatocytes (*6*) before infecting a final one by forming a parasitophorous vacuole (*7*). Inside this compartment, the parasite differentiates into a replicative liver stage (*8*), and after several rounds of nuclear divisions (*9*) release thousands of merozoites in the blood circulation (*10*). Merozoites invade erythrocytes and initiate an exponential phase of parasite multiplication in the blood (*11*), which is responsible for the symptoms and complications of malaria. (**b**) Immunofluorescence microscopy of a *P. berghei* sporozoite. The sporozoite surface was labeled with antibodies against the circumsporozoite protein (*green*) and the nucleus was stained with DRAQ5 (*blue*). (**c**) Model of *Plasmodium* sporozoite entry into host cells. Invasion starts with the formation of a tight junction, resulting from interactions between still unidentified parasite ligands exposed at the apical tip of the sporozoite and their cognate receptors on the hepatocyte surface (*1*). A parasite actomyosin motor located beneath the sporozoite plasma membrane powers the posterior redistribution of the junction (*2*), which propels the sporozoite inside an invagination of the hepatocyte plasma membrane that forms the parasitophorous vacuole membrane (PVM) (*3*). Scission of the PVM leads to complete sporozoite internalization and individualization of the parasitophorous vacuole (PV) (*4*), where the parasite further replicates. *PM* plasma membrane, *PV* parasitophorous vacuole, *PVM* parasitophorous vacuole membrane

which derives primarily from the host plasma membrane (Bano et al. 2007), and which is essential for the further differentiation of the parasite into liver stages (Mota et al. 2001; Mueller et al. 2005; Silvie et al. 2006b).

The mechanisms of sporozoite entry have not been elucidated on a molecular level. The *Plasmodium* genome encodes approximately 5,500 proteins, 60% of which have no similarity to known proteins. Parasite proteins potentially involved during sporozoite invasion include the circumsporozoite protein (CSP), the thrombospondin anonymous protein (TRAP), apical membrane antigen 1 (AMA-1), P36 and P36p/P52. CSP is a GPI-anchored protein that constitutes the main component of the sporozoite surface. It contains a thrombospondin-type domain which interacts with liver HSPGs (Frevert et al. 1993). This interaction is responsible for the selective sequestration of sporozoites in the liver and the initial attachment to host cells (Coppi et al. 2007; Frevert et al. 1993; Pinzon-Ortiz et al. 2001). TRAP and AMA-1 are two transmembrane proteins contained in the sporozoite micronemes, the secretion of which is up-regulated after contact with hepatocytes (Gantt et al. 2000; Silvie et al. 2004). Both proteins possess adhesive domains in their extracellular region, but the exact role of these domains in the interaction with host cells is currently unknown. TRAP is important for sporozoite gliding motility, by connecting the subpellicular actomyosin motor to the parasite surface (Kappe et al. 1999). As of yet, P36 and P36p/P52 are the only sporozoite molecules which have been specifically associated with productive invasion, although it is not yet clear whether they are required for invasion or for maintenance of the intracellular parasites immediately after invasion (Ishino et al. 2005; Labaied et al. 2007; van Dijk et al. 2005; van Schaijk et al. 2008). On the other hand, very few host molecules have been identified as being involved during sporozoite invasion. As mentioned above, HSPGs constitute docking sites, and contribute to activation of sporozoites before invasion (Coppi et al. 2005, 2007), but probably do not participate directly in the process of invasion itself (Frevert et al. 1996). As detailed in this chapter, the tetraspanin CD81 is required for infection of hepatocytes by *Plasmodium* sporozoites, and remains to date the only hepatocyte factor that is clearly essential for sporozoite invasion.

14.3.2 Model Systems to Study Sporozoite Infection

Investigations of *Plasmodium falciparum* sporozoite infection remain technically restrictive. Sporozoites can be obtained from mosquitoes that have been experimentally infected by being fed on blood containing parasite sexual stages, which in the case of *P. falciparum* can be produced *in vitro* from continuous blood stage cultures (Ponnudurai et al. 1982; Trager and Jensen 1976). Only primary human hepatocytes fully support *P. falciparum* sporozoite productive invasion and liver stage development (Mazier et al. 1985). In contrast, *P. vivax*, another highly prevalent human malaria parasite, readily infects cell lines such as HepG2 cells (Hollingdale et al. 1984; Mazier et al. 1984), but the absence of a blood stage culture system constitutes a major obstacle for transmission to mosquitoes and production of sporozoites. Rodent malaria parasites, *P. yoelii* and *P. berghei*, provide excellent models for the study of sporozoite biology. Indeed, the entire life cycle can be reproduced in the laboratory, by cycling between infected mice and mosquitoes. Furthermore, in contrast to *P. falciparum*, rodent parasites infect cell lines *in vitro* (Calvo-Calle et al. 1994; Hollingdale et al. 1983; Mota and Rodriguez 2000). The development of highly effective gene targeting technologies in rodent parasites combined with the possibility of studying the whole cycle of these parasites in experimental animals has enabled the identification of genes with essential function in *Plasmodium* sporozoites or liver stages (Aly et al. 2009; Janse et al. 2006; Menard and Janse 1997).

14.4 CD81 and HCV Entry

14.4.1 Identification of CD81 as a Molecule Involved in HCV Entry

In 1998, using sE2 as a probe to identify cell-surface molecules involved in HCV entry, Pileri et al. discovered CD81 as the first putative receptor for HCV (Pileri et al. 1998). This was done by isolating a subclone of Molt-4, a human T-cell line with very high binding capacity for sE2, which was subsequently used as a source for cloning of the cellular receptor. Multiple rounds of cDNA expression cloning identified the E2-binding entity expressed by this subclone as CD81. A soluble form of human CD81 containing only its LEL was sufficient for E2 binding, whereas a similar construct containing the mouse CD81 LEL did not bind. Soluble human CD81 protein, but not the mouse homologue, enabled the binding of HCV particles from infectious plasma (HCVs). Moreover, binding could be blocked by antisera from chimpanzees that had been vaccinated with the E1 and E2 proteins and were protected from challenge. Thus, anti-E2 antibodies, which were capable of neutralizing HCV infection *in vivo*, could inhibit the binding of HCV to CD81 *in vitro,* supporting the hypothesis that CD81 acts as a receptor for HCV (Pileri et al. 1998).

14.4.2 Affinity of Binding

Truncated soluble forms of E2 proteins from genotype 1 have been reported to bind CD81 with an affinity of 1.8 nM (Petracca et al. 2000), whereas E2 proteins cloned from other genotypes showed minimal interaction with CD81 (Roccasecca et al. 2003; Shaw et al. 2003). However, while full length and truncated forms of E2 were originally reported to fold without E1 (Dubuisson and Rice 1996; Michalak et al. 1997), it has been subsequently demonstrated that the presence of E1 affects the properties of E2. Indeed, folding and membrane insertion of E2 are assisted by E1

(Brazzoli et al. 2005; Cocquerel et al. 2001) and E2 expressed alone is not recognized by a human conformational antibody and shows a weaker binding to CD81 (Cocquerel et al. 2003a, b). Compared to E1E2 complexes, E2 has an increased rate of dissociation from CD81 LEL as measured by surface plasmon resonance (Nakajima et al. 2005). Importantly, co-expression of both E1 and E2 is essential for the production of infectious HCV particles (Bartosch et al. 2003b; Hsu et al. 2003; Wakita et al. 2005). Using a series of CD81 sequences from different species, Flint and colleagues also showed that sE2 interaction with cell surface-expressed CD81 fails to predict virus-receptor interactions leading to HCV entry (Flint et al. 2006). Together, these data suggest that E2 cannot reach a native/fully functional conformation in absence of E1.

The formation of disulfide bonds among the four cysteine residues in the LEL is necessary for sE2 binding to the CD81 LEL (Petracca et al. 2000) and dimeric forms of CD81 were found more effective than the monomeric ones (Drummer et al. 2002, 2005; Nakajima et al. 2005). Like for E2, recombinant CD81 LEL proteins were found not to be predictive of the ability of full-length CD81 to support HCVpp and HCVcc infection (Flint et al. 2006). For example, the mouse CD81 LEL failed to interact with HCV glycoproteins (Flint et al. 1999a, 2006; Higginbottom et al. 2000) or to inhibit HCVpp infection (Flint et al. 2006), whereas the full-length mouse CD81 was able to support infection by HCVpp and HCVcc (Flint et al. 2006; Rocha-Perugini et al. 2009).

Altogether, these data suggest that biochemical studies with recombinant E2 and CD81 do not necessarily reflect the reality of the E2-CD81 interaction as it occurs during virus entry, and recombinant sE2 and CD81 LEL fail to fully mimic their full-length counterparts. Concurring with this, it has been demonstrated that HCVpp and HCVcc bearing diverse glycoproteins infect cells in a CD81-dependent manner, despite their minimal binding affinities of their sE2 for CD81 (Gottwein et al. 2009; Lavillette et al. 2005; McKeating et al. 2004). In addition, it has been very recently shown that functional virion-associated HCV glycoproteins, which interact with CD81, correspond to large covalent complexes containing both E1 and E2 glycoproteins (Vieyres et al. 2010).

14.4.3 CD81, a Key Entry Factor for HCV

Since its identification as a molecule that interacts with sE2, the involvement of CD81 in HCV entry has been confirmed in different models. Indeed, antibodies directed against CD81 as well as CD81 LEL are able to inhibit entry of HCVpp, HCVcc and HCVs (Bartosch et al. 2003b; Cormier et al. 2004; Hsu et al. 2003; Kapadia et al. 2007; Koutsoudakis et al. 2006; Lavillette et al. 2005; Molina et al. 2008; Wakita et al. 2005; Zhang et al. 2004; Zhong et al. 2005) and HCV infection *in vivo* (Meuleman et al. 2008). Although the affinity of E2 glycoprotein for CD81 may differ depending on viral genotype (McKeating et al. 2004; Roccasecca et al. 2003; Shaw et al. 2003; Yagnik et al. 2000), anti-CD81 antibodies are able to block infection of HCV from different genotypes (Gottwein et al. 2009; Lavillette et al. 2005;

McKeating et al. 2004). Moreover, CD81 downregulation using siRNA in hepatoma cells abolishes HCV infection (Akazawa et al. 2007; Koutsoudakis et al. 2007; Molina et al. 2008; Zhang et al. 2004). Although CD81 is normally expressed at the surface of primary hepatocytes and most hepatoma cell lines, it has been observed that HepG2 and HH29 cells do not express this tetraspanin. Interestingly, ectopic expression of CD81 in these non-permissive cell lines confers susceptibility to HCVpp and HCVcc infection (Bartosch et al. 2003c; Cormier et al. 2004; Hsu et al. 2003; Lavillette et al. 2005; Lindenbach et al. 2005; Zhang et al. 2004), providing additional evidence for the importance of CD81 in HCV entry. Other studies have also shown that susceptibility of cells to HCV infection is closely related to the CD81 expression level (Akazawa et al. 2007; Koutsoudakis et al. 2007; Rocha-Perugini et al. 2009) and the ratio between the levels of CD81 and SR-BI at the cell surface also modulates HCV entry (Kapadia et al. 2007).

14.4.4 CD81 and Kinetic of Viral Entry

CD81 is likely to act at a stage after virus binding. Indeed, the kinetics of infection have been studied by incubating the virus with cells at 4°C. Under these conditions, the virus attaches to the cells but does not efficiently enter, thus permitting a synchronous infection when the inoculum is removed and cells are shifted to 37°C. Anti-CD81 antibodies inhibited HCVpp and HCVcc infection when added prior to, during, or after viral attachment to target cells (Cormier et al. 2004; Evans et al. 2007; Koutsoudakis et al. 2006; Morikawa et al. 2007; Zeisel et al. 2007). Antibodies against CD81 were able to potently inhibit HCVcc infection, by 60%, even when following an extended binding phase at 37°C (Koutsoudakis et al. 2006). More precisely, the half-life of the interaction of CD81 with HCVpp and HCVcc was determined to be approximately 17–18 min, respectively (Bertaux and Dragic 2006; Evans et al. 2007).

14.4.5 Determinants in CD81

The CD81 binding site for E2 has been localized within the LEL (Pileri et al. 1998) and characterization of chimeric CD9/CD81 molecules confirmed that the LEL sequence is a determinant of HCV entry (Zhang et al. 2004), with the binding site for E2 having been mapped to the variable double-helix subdomain (Kitadokoro et al. 2001). Specific LEL amino acid residues essential for the CD81-E2 interaction have been identified (Drummer et al. 2002; Higginbottom et al. 2000). However, mutation of these residues, which abrogates binding to sE2, allows entry of HCVpp and HCVcc, suggesting again that the affinity of E1E2 binding to LEL may be higher on virion than in sE2 (Flint et al. 2006). Further investigations with the HCVcc system are needed to determine which residues in CD81 are specifically involved in viral interaction.

14.4.6 CD81 and Species Restriction

It has been shown that sE2 binds to human but not to mouse, rat, or African green monkey CD81 proteins, suggesting that CD81 may be a factor in the species-specificity of HCV infection (Flint et al. 1999a; Higginbottom et al. 2000). However, CD81 cannot be the sole determinant, as transgenic mice expressing human CD81 fail to support HCV infection (Masciopinto et al. 2002) and CD81 from tamarin, which is not susceptible to HCV infection, has been reported to bind sE2 (Allander et al. 2000; Meola et al. 2000). In addition, ectopic expression of mouse CD81 (mCD81) in CD81-deficient hepatoma cells restored to some extent permissivity to HCVpp and HCVcc (Flint et al. 2006; Rocha-Perugini et al. 2009). Thus, CD81 contributes to, but does not alone define the species restriction so additional cellular factors are involved.

14.4.7 Interplay Between CD81 and the Other HCV Entry Factors

HCV shows a restricted tropism for human hepatic cell lines expressing CD81 (Bartosch et al. 2003b; Cormier et al. 2004; Hsu et al. 2003; Zhang et al. 2004). However, CD81 expression in non-hepatic cell lines does not render them permissive to infection (Bartosch et al. 2003c; Hsu et al. 2003), indicating that additional molecules are necessary for HCV infection. Indeed, the human scavenger receptor class B type I (SR-BI, also called CLA-1) is the second cell surface molecule that was identified as another potential receptor for HCV (Scarselli et al. 2002). SR-BI was identified as a HCV receptor because it binds sE2 through interactions with E2 hypervariable region 1 (HVR1) (Scarselli et al. 2002). HCV entry is strongly reduced by antibodies directed against SR-BI or in SR-BI knock-down hepatoma cells (Bartosch et al. 2003c; Catanese et al. 2007, 2010; Dreux et al. 2009; Grove et al. 2007; Kapadia et al. 2007; Zeisel et al. 2007). In contrast, SR-BI overexpression enhances HCV internalization (Grove et al. 2007; Schwarz et al. 2009). The kinetics of inhibition of HCV infection by anti-CD81 antibodies were almost identical to those observed for anti-SR-BI antibodies (Zeisel et al. 2007), indicating that SR-BI might act concomitantly with CD81. However, it is likely that HCV particle encounters SR-BI before CD81. Indeed, HCVcc can bind to CHO cells expressing SR-BI but not to CHO cells expressing CD81, suggesting that a first contact with SR-BI might be necessary for the particle to interact with CD81 (Evans et al. 2007). Species-specific SR-BI protein residues required for sE2 binding have recently been identified (Catanese et al. 2010). Interestingly, SR-BI mutants with reduced binding to sE2 were also impaired in their ability to restore the infectivity of a SR-BI-knocked-down Huh-7.5 cell line. In addition, these SR-BI mutants were still able to form oligomeric structures and bind the physiological ligand HDL and mediate cholesterol efflux, suggesting that distinct protein determinants are responsible for the interaction with HDL and the HCV particle (Catanese et al. 2010).

The observation that expression of CD81 and SR-BI was not sufficient to support HCV entry led to the hypothesis that at least one additional factor might be required (Bartosch et al. 2003c; Hsu et al. 2003). Indeed, the tight junction proteins Claudin-1 (also Claudin-6 and -9) and Occludin have been recently identified as additional HCV entry factors (Evans et al. 2007; Meertens et al. 2008; Ploss et al. 2009; Zheng et al. 2007). No direct interaction between Claudin-1 (CLDN-1) and HCV envelope glycoproteins associated with HCV particles has yet been reported (Evans et al. 2007; Krieger et al. 2010; Zheng et al. 2007), which may reflect the multi-step and sequential nature of the entry process. Recently, it has been shown that infection of human cell lines and primary hepatocytes can be blocked by specific antibodies directed against CLDN-1 (Fofana et al. 2010; Krieger et al. 2010). In addition, the kinetics of inhibition with such antibodies showed that CLDN-1 acts cooperatively with CD81 and SR-BI in HCV entry (Krieger et al. 2010). Similar to CLDN-1, the initial binding steps of HCV to the cell membrane are not affected by occludin gene silencing, suggesting that occludin plays a role in a late entry stage (Benedicto et al. 2009). In addition, occludin depletion that impairs HCV entry does not perturb CLDN-1 expression or localization, suggesting that both entry factors function separately during HCV infection (Benedicto et al. 2009; Liu et al. 2009). Thus, every cell expressing human CD81, SR-BI, Claudin-1 and Occludin is susceptible to HCV infection (Ploss et al. 2009). Although the mouse versions of SR-BI and CLDN-1 function at least as well as the human proteins in promoting HCV entry, both Occludin and CD81 must be of human origin to allow efficient infection (Ploss et al. 2009).

The Claudin and Occludin family of transmembrane proteins form the core of tight junctions that are structures forming firm seals between adjacent cells. These structures prevent the mixing of membrane proteins between the apical and basolateral membranes of a polarized cell and control the paracellular passage of ions and solutes between cells (Hartsock and Nelson 2008). Tight junctions form a fence in the cellular membrane, allowing cells to maintain functionally distinct surface-membrane domains. In a very simplified view of the hepatocytes, tight junctions separate an apical canalicular domain engaged in bile secretion from a basolateral luminal domain that is in constant contact with the blood. Shortly after the identification of CLDN-1 as a HCV entry factor, it was suggested that the virion interacts with tight junction proteins after a lateral migration of the virus-SR-BI/CD81 complex from the basolateral surface to the tight junction facing the bile canaliculi. This model was proposed because a similar mechanism has been described for Group B coxsackie viruses (Coyne and Bergelson 2006). In addition, recent data indicate that distribution of CLDN-1 in tight junctions correlates with permissiveness to HCV infection (Liu et al. 2009; Yang et al. 2008), suggesting that junctional CLDN-1 may play an important role in HCV entry. However, increasing evidence supports a model in which HCV utilizes CLDN-1 molecules that are not associated with tight junctions. Evans and colleagues demonstrated that a CLDN-1 variant lacking the C-terminal region supported HCV entry into hepatoma cells (Evans et al. 2007). Since this region of CLDN-1 mediates interactions with the cytoplasmic and signalling components of the tight junction complex (Heiskala et al. 2001), these data were the first to support a model in which HCV uses nonjunctional CLDN-1 molecules.

In addition, disruption of tight junctions by calcium depletion enhanced HCVpp infection of polarized Caco-2 and HepG2 cells (Mee et al. 2008, 2009). Several studies have highlighted a critical role for the first extracellular loop (EC1) of CLDN-1 in HCV entry (Evans et al. 2007; Meertens et al. 2008; Yang et al. 2008; Zheng et al. 2007), and, using Förster resonance energy transfer and biochemical techniques, several reports have demonstrated that CLDN-1 associates with CD81 (Harris et al. 2008; Mee et al. 2009). Mutation of residues 32 and 48 in CLDN-1 EC1 ablates its association with both CD81 and occludin as well as viral receptor activity. Importantly, mutation of the same residues in the receptor inactive CLDN-7 molecule allows association with CD81 and facilitates viral entry into 293T cells in the absence of any detectable occludin interaction. These data demonstrate an essential role for CLDN-CD81 complexes in HCV infection (Harris et al. 2010). Indeed, neutralizing anti-CLDN-1 antibodies specifically disrupt CD81-CLDN-1 interaction suggesting that CD81-CLDN-1 coreceptor complexes are critical for HCV entry. CLDN-1 may potentiate CD81 association with HCV particles by way of E2 interactions. Supporting this idea, it was found that anti-CLDN-1 antibodies inhibit envelope glycoprotein E2 and virion binding to permissive cells. This suggests that CLDN-1 association with CD81 enhances viral glycoprotein associations to the HCV coreceptor complex that are required for virus internalization (Krieger et al. 2010). Importantly, CLDN-1 has been found to associate with CD81 at the basolateral membrane of polarized HepG2 cells, whereas tight junction associated pools of CLDN-1 demonstrated a minimal association with CD81. These results, together with data from a single-particle tracking analysis of infectious fluorescent HCV (Coller et al. 2009), support the hypothesis that HCV entry does not require tight junctions.

Due to the association of lipoproteins with HCV particles, the LDL-R (LDL receptor) has been proposed to be involved in HCV entry (Agnello et al. 1999; Monazahian et al. 1999). The most important ligand for this receptor is LDL, which is responsible for the transport of most of the plasma cholesterol. It has been shown that cell surface adsorption of HCV particles isolated from patients and accumulation of viral RNA in cells can be inhibited by purified LDLs and VLDLs as well as antibodies directed against the LDL-R. Furthermore, a correlation has been shown between the accumulation of HCV RNA in primary hepatocytes and the expression of LDL-R mRNA and LDL entry (Molina et al. 2007). Finally, the inhibition of HCVcc entry by anti-ApoE or -ApoB antibodies is another argument in favour of a role for the LDL-R in HCV entry (Andreo et al. 2007; Chang et al. 2007). Very recently, it has been suggested that LDL-R acts as a cooperative HCV co-receptor, supporting viral entry and infectivity through interaction with the ApoE ligand present in an infectious HCV/lipoprotein complex comprising the virion (Owen et al. 2009). However, recent investigations with the HCVcc system showed that LDL-R, at least in some conditions, leads to a non-productive internalization of HCV (Albecka et al. 2012). It should be noted that proprotein convertase subtilisin/kexin type 9 (PCSK9), a member of proprotein convertase family, has been shown to regulate cell surface expression of LDL-R, mainly by enhancing degradation of LDL-R protein in the liver (reviewed in Lopez 2008). PCSK9 is found as a soluble protein, which is secreted in the bloodstream, mostly by hepatocytes. This endogenous molecule acts in an autocrine and a paracrine fashion in liver. Inactivation of PCSK9 in mice reduces plasma cholesterol levels by increasing hepatic expression of LDL-R protein and thereby accelerating clearance of circulating LDL cholesterol (Lopez 2008). A recent study analysed the effect of PCSK9 expression on HCV infection of Huh-7 cells (Labonte et al. 2009). Expression of soluble PCSK9 or membrane-bound nonsecreted chimera of PCSK9 not only reduced the LDL-R protein level but also that of CD81. The effect of PCSK9 on CD81 was not dependent on the presence of LDL-R and led to a significant decrease of HCV entry into Huh-7 cells. In conclusion, this study demonstrated that hepatic PCSK9 can down-regulate expression of CD81 and LDL-R *in vitro* and consequently, PCSK9 may modulate liver susceptibility to HCV infection *in vivo*.

14.4.8 Regions of HCV Envelope Glycoproteins Involved in HCV Entry

HCV envelope proteins are type I membrane proteins with a highly glycosylated N-terminal ectodomain and a C-terminal transmembrane domain anchored in the lipid envelope. E1 and E2 envelope proteins form a complex that is the viral component present at the surface of HCV particles (Wakita et al. 2005) and is therefore the obvious candidate ligand for cellular receptors. In particular, the E2 glycoprotein is the subunit involved in interactions with receptors.

During their synthesis, E1 and E2 ectodomains are directed to the lumen of the ER and their transmembrane domains are inserted in the membrane of this compartment. Transmembrane domains of E1 and E2 are involved in multiple functions. Indeed, in addition to serving as a membrane anchor (Michalak et al. 1997), they possess a signal for the reinitiation of translocation in their C-terminal half (Cocquerel et al. 2002; Lavie et al. 2007), they contain heterodimerization sequences (Cocquerel et al. 1998, 2000; Michalak et al. 1997; Op De Beeck et al. 2000; Patel et al. 2001; Selby et al. 1994) and they are responsible for the retention of heterodimers in the ER (Cocquerel et al. 1998, 1999, 2000; Duvet et al. 1998; Flint and McKeating 1999). In addition, specific residues in these domains are involved in the fusion function of E1 and E2 (Ciczora et al. 2005). Interestingly, the mutation of charged residues present in the middle of E1 and E2 transmembrane domains alters all functions and the topology of these domains (Cocquerel et al. 2000), indicating that the multifunctionality of E1 and E2 transmembrane domains is associated with their structural dynamic. Indeed, a study of the topology of these domains has been performed by determining the accessibility of their N- and C-termini in selectively permeabilized cells (Cocquerel et al. 2002). Before the maturation of the viral polyprotein, E1 and E2 transmembrane domains form a hairpin structure that is essential to their function of reinitiation of translocation. Then, after cleavage between E1 and E2 or between E2 and p7, the second C-terminal hydrophobic

Table 14.1 HCV E2 amino acids potentially involved in CD81 interaction

E2 residues/regions[a]	Tools[b]	Assay	Authors
480–493/544–551	E2661	Blocking antibodies[c]	Flint et al. (1999a)
517–535	Cell surface expressed E2715	Blocking antibodies	Forns et al. (2000a)
474–494/522–551	E2683	Modelling	Yagnik et al. (2000)
407–524	E2660	Interstrain chimeras	Patel et al. (2000)
412–423	E2660	Blocking antibodies	Owsianka et al. (2001)
396–407/412–423/528–535	E1E2	Blocking antibodies	Owsianka et al. (2001)
396–407/412–423/432–447/528–535	VLPs[d]	Blocking antibodies	Owsianka et al. (2001)
HVRs/613–618	E2661	Deletions/Mutagenesis	Roccasecca et al. (2003)
HVR1[e]/432–447	HCVpp	Blocking antibodies	Hsu et al. (2003)
HVR1	HCVpp	Deletion	Callens et al. (2005)
G^{436}WLAGLFY^{443}	HCVpp	Mutagenesis	Drummer et al. (2006)
420, 527, 529, 530 and 535	HCVpp	Blocking antibodies/Mutagenesis	Owsianka et al. (2006), (2008)
527, 529, 535, 613, 614, 616 and 617	HCVpp	Mutagenesis	Rothwangl et al. (2008)
412–423	HCVcc	Blocking antibodies/Mutagenesis	Dhillon et al. (2010)
425–443/529–535	HCVpp HCVcc	Blocking antibodies/Mutagenesis	Keck et al. (2011)
H^{421}	HCVpp	Mutagenesis	Boo et al. (2012)

[a]Positions of amino acids in the polyprotein of reference strain H (GenBank accession no. AF009606)
[b]E2661, E2715, E2683, E2660 are for E2s ending at indicated positions
[c]Indicated E2 regions/amino acids correspond to the epitopes of blocking antibodies
[d]Virus-like particles produced in insect cells
[e]HVR1 includes residues 384–411

stretch is reoriented towards the cytosol, leading to the formation of a single membrane-spanning domain (Cocquerel et al. 2002).

HCV envelope glycoproteins are very important at different stages of the HCV life cycle. They are involved in the production of infectious viral particle and they are essential for the entry of virions into target cells. Indeed, HCVpp expressing E1 or E2 separately (Bartosch et al. 2003b; Drummer et al. 2003) or HCVcc lacking E1E2 (Wakita et al. 2005) are non-infectious. The E2 glycoprotein likely plays a major role in the interaction between the virus and its major cellular receptors. The CD81 binding region of E2 requires correctly folded E2 (Flint et al. 1999b) and is comprised of discontinuous sequences that form the binding surface. Indeed, several regions and residues of E2 that are potentially involved in CD81 interaction have been identified (see Table 14.1).

The N-terminus of E2 containing a hypervariable region called HVR1 whose high variability may contribute to HCV escape from the immune system. This region also seems to be involved in HCV entry. Indeed, it contains several basic amino acids which modulate HCVpp infectivity (Callens et al. 2005), and the use of antibodies targeting HVR1 inhibit both cell entry of HCVpp and cellular binding of HCV-like particles (Barth et al. 2005; Bartosch et al. 2003c). Moreover, HVR1 influences the use of SR-BI by HCV since sE2 devoid of this domain fails to interact with SR-BI (Scarselli et al. 2002). In addition, HCVpp or HCVcc lacking HVR1 are no longer susceptible to neutralization by antibodies directed against SR-BI (Bankwitz et al. 2010; Bartosch et al. 2003c). It has been shown that a mutant virus lacking HVR1 remained infectious in hepatoma cells and chimpanzees (Bankwitz et al. 2010; Forns et al. 2000b) but viral infection was clearly attenuated. More precisely, deletion of HVR1 does not seem to affect HCV replication or virus release but impairs virus entry (Bankwitz et al. 2010). In addition, deletion of HVR1 reduced the abundance and infectivity of low density HCV particles and impaired HCV fusion. Interestingly, a compensatory mutation of an amino acid in the vicinity of the transmembrane domain of E1 partially restored infectivity defects. Strikingly, although deletion of HVR1 ablated viral dependence on SR-BI, it also increased exposure of the viral CD81 binding site and conserved epitopes, thus facilitating neutralization (Bankwitz et al. 2010). It should be noted that two other hypervariable regions in the E2 glycoprotein might also play a role in viral entry (Roccasecca et al. 2003; Troesch et al. 2006; Weiner et al. 1991).

E1 and E2 contain 4 and 11 conserved glycosylation sites, respectively, all of which have been shown to be modified by N-glycosylation (Goffard et al. 2005; Meunier et al. 1999; Nakano et al. 1999; Slater-Handshy et al. 2004). Despite variability in sequences, glycosylation sites are highly conserved, suggesting that the glycans associated with these proteins play an essential role in the HCV life cycle. Studies of the function of these glycans have shown that they play a major role in envelope protein folding, in HCV assembly and/or infectivity, in protection against neutralization and in HCV entry (Falkowska et al. 2007; Goffard et al. 2005; Helle et al. 2007, 2010). For instance, at least five glycans on E2 strongly reduce the sensitivity of HCVcc to antibody neutralization with four of them surrounding CD81 binding site, suggesting that this "glycan shield" contributes to the evasion of HCV from the humoral immune response by masking the CD81 binding site.

For many enveloped viruses, an acidic pH induces an irreversible conformational change, which is necessary for fusion between viral and endosomal membranes. Such viruses are generally inactivated by acid pH treatment. Surprisingly, exposure of cell-surface-bound virions to acid pH followed by a return to neutral pH did not affect HCV infectivity (Meertens et al. 2006; Tscherne et al. 2006), suggesting that HCV may need an additional trigger, such as receptor interaction, to become sensitive to low pH. Indeed, it has been very recently shown that CD81 likely promotes conformational changes in E1E2, which then confer the sensitivity to low pH (Sharma et al. 2011). However, a fusion system between HCVpp and liposomes showed that fusion is pH-dependent, temperature-dependent, does not require the presence of any protein at the surface of liposomes and is facilitated by the presence of cholesterol in the target membrane (Haid et al. 2009; Lavillette et al. 2006). Although previous studies identified potential fusion peptides in the sequences of E1 and E2 (Ciccaglione et al. 2001; Drummer et al. 2007; Flint et al. 1999b; Garry and Dash 2003; Li et al. 2009; Pacheco et al. 2006; Perez-Berna et al. 2006, 2008a, b, 2009; Yagnik et al. 2000), three regions (one region in E1 (residues 270–284) and two in E2 (residues 416–430 and 600–620)) have been suggested to be involved in HCV fusion (Lavillette et al. 2007). Interestingly, three point mutations in the region of E1 reduced HCVpp infectivity and severely perturbed HCVcc production in hepatoma cells (Russell et al. 2009). In addition, one residue in E2, identified as being essential for infectivity and fusion of HCVpp, was also shown to have a similar key role in the HCVcc context (Haid et al. 2009). Whether the involvement of these regions in fusion is direct or indirect remains to be determined. A high-resolution structure of HCV envelope proteins will be useful for answering this question. Recently, disulfide connectivity along with functional data and secondary structure predictions allowed Krey et al. to thread the E2 polypeptide chain onto a class II fusion protein template (Krey et al. 2010). This model revealed the distribution of E2 amino acids among three different domains (DI, DII, DIII), mapped the CD81 binding site to the DI/DIII interface, and highlighted a strictly conserved segment (residues 502–520) of the polypeptide chain as a strong candidate for the fusion loop in E2 that is likely a contact region with E1 (Krey et al. 2010). This model will provide an invaluable structural framework for better understanding the function of HCV envelope glycoproteins in the early steps of HCV lifecycle.

14.5 CD81 and *Plasmodium* Sporozoite Entry

14.5.1 Identification of CD81 as a Sporozoite Entry Factor

CD81 was the first host molecule shown to play an essential role in *Plasmodium* sporozoite entry into hepatocytes (Silvie et al. 2003). This was first demonstrated with the rodent malaria parasite *P. yoelii* in an *in vivo* model of infection using CD81 knockout mice (Maecker and Levy 1997). Whereas cd81+/+ mice developed malaria after inoculation of *P. yoelii* sporozoites, cd81–/– mice were totally refractory to infection. In contrast, cd9–/– mice were susceptible to *P. yoelii*, showing the specificity of CD81 function during *Plasmodium* infection. Experiments using primary hepatocytes isolated from wild type and cd81–/– mice demonstrated that the refractoriness of cd81–/– mice to *P. yoelii* sporozoite infection is due to a defect in sporozoite productive invasion, whereas sporozoite cell traversal is not affected by the absence of CD81 (Silvie et al. 2003). Anti-CD81 (but not anti-CD9) monoclonal antibodies (mAbs) completely block parasite invasion when added before or at the same time as sporozoites, but have no effect when they are added after completion of invasion. CD81-deficient mice remained susceptible to *P. yoelii* malaria when infection was initiated by inoculation of infected erythrocytes, thus bypassing the liver. This observation is consistent with the absence of CD81 expression on the surface of red blood cells, and shows that CD81 is specifically required for infection of hepatocytes. CD81 is also required for *P. yoelii* infection in the mouse hepatoma cell line Hepa1-6, where anti-CD81 antibodies and siRNA-induced knockdown of CD81 both inhibit sporozoite entry (Silvie et al. 2006a, b). *P. yoelii* can infect human hepatocytes, at least *in vitro*, and hCD81 was shown to be fully functional for *P. yoelii* sporozoite invasion in primary hepatocytes from hCD81 transgenic mice and in transfected Hepa1-6 cells (Silvie et al. 2006a, b). Importantly, ectopic expression of CD81 in CD81-negative HepG2 cells is sufficient to render them fully susceptible to *P. yoelii* infection (Silvie et al. 2006b). Infection correlates well with the level of expression of CD81 in HepG2/CD81 cells.

CD81 is also required for *P. falciparum* sporozoite infection, as shown in a model of infection of primary human hepatocytes (Mazier et al. 1985; Silvie et al. 2003). Anti-hCD81 mAbs almost completely blocked *P. falciparum* infection in a dose dependent manner, and siRNA-mediated silencing of CD81 expression also inhibited infection. *P. falciparum* sporozoites do not infect mouse hepatocytes, and expression of hCD81 in transgenic mice was not sufficient to confer susceptibility to *P. falciparum*, indicating that other factors account for species restriction (Silvie et al. 2006b). While ectopic expression of CD81 was sufficient to render HepG2 cells susceptible to *P. yoelii* infection, HepG2/CD81 remained completely refractory to *P. falciparum* infection, showing that other molecules are specifically required for infection by the human parasite (Silvie et al. 2006b).

Although closely related to *P. yoelii*, the rodent parasite *P. berghei* infects HepG2 as efficiently as HepG2/CD81, showing that this parasite can use CD81-independent pathways to enter cells (Hollingdale et al. 1983; Silvie et al. 2006a). Consistent with this notion, cd81–/– mouse hepatocytes were susceptible to *P. berghei* sporozoite infection both *in vivo* and *in vitro* (Silvie et al. 2003). Interestingly, CD81 does not seem to be involved at all during infection of human cells by *P. berghei*, as shown by the absence of inhibition induced by anti-CD81 antibodies or siRNA-mediated CD81 silencing in the human hepatocarcinoma cell lines HepG2 and Huh-7, or in non-hepatocytic HeLa cells (Silvie et al. 2006a, 2007). In sharp contrast, infection of mouse hepatoma Hepa1-6 cells by *P. berghei* appears to be strictly CD81-dependent, as shown by complete blockage by anti-CD81 antibodies (Silvie et al. 2007). Interestingly, *P. berghei* can use both CD81-dependent and CD81-independent pathways to infect primary mouse hepatocytes *in vitro*, as shown by partial inhibition of infection by anti-CD81 antibodies (Silvie et al. 2007; Yalaoui et al. 2008b). The relative contribution of the two pathways varies depending on the mouse genetic background.

In contrast with HCV, there is no evidence that CD81 directly interacts with *Plasmodium* sporozoites. In particular, soluble recombinant forms of mouse or human CD81 LEL do not inhibit *P. yoelii* or *P. falciparum* infection, respectively, and immunofluorescence and far western experiments failed to provide evidence for interaction with any sporozoite protein (Silvie et al. 2003). Combined with the data of the structure-function analysis detailed below (Yalaoui et al. 2008b), these observations strongly suggest that CD81 is not a receptor for *Plasmodium* sporozoites, but rather acts indirectly.

Although CD81 is required for sporozoite productive invasion, the precise step in which it is involved in the entry process remains unknown. This is notably due to the fact that sporozoite entry is not synchronous, as parasites first migrate through a variable number of cells before entering a host cell through a junction and forming a parasitophorous vacuole (Mota et al. 2001). This has precluded detailed analysis of the kinetics of sporozoite entry. In particular, it remains unknown whether CD81 acts during initial attachment of sporozoites, formation of the junction, internalization of the parasite or individualization of the parasitophorous vacuole at the end of sporozoite entry (Fig. 14.2c).

14.5.2 CD81 and Species Restriction

Human CD81 can restore susceptibility to *P. yoelii* in mouse cells where CD81 is either absent or neutralized by antibodies (Silvie et al. 2006a, b; Yalaoui et al. 2008b). In addition, *P. yoelii*, like *P. berghei*, readily infects human cells, including human hepatocytes, demonstrating the absence of species restriction for liver infection by these parasites (Calvo-Calle et al. 1994; Hollingdale et al. 1983; Silvie et al. 2006b, 2007). In contrast, *P. falciparum* infects human but not mouse hepatocytes, indicative of species restriction (Silvie et al. 2006b). *P. falciparum* sporozoites do not infect hepatocytes from hCD81 transgenic mice, showing that CD81 alone is not responsible for species restriction, and that other factors are involved (Silvie et al. 2006b).

14.5.3 Determinants in CD81

Based on the observation that human CD81 but not CD9 supports *P. yoelii* sporozoite entry, Yalaoui et al. performed an analysis of the molecular determinants of CD81 activity during infection using hCD9/hCD81 chimeric molecules (Yalaoui et al. 2008b). This study demonstrated that, similarly to HCV, the CD81 LEL is critical for CD81 function during sporozoite infection. However, chimeric constructs with a swap in the middle of the LEL showed that the first half of CD81 LEL, comprising the A and B helices, was essential. The C helix of CD81LEL, although not essential, also contributes to optimal CD81 function. In contrast, the D helix, which is critical for HCV-E2 binding, does not play any role during sporozoite infection. Interestingly, a chimera where the end of the A helix and the entire B helix of CD9 were substituted by the 21 corresponding residues of CD81 could support infection. Site-directed mutagenesis on selected residues showed that mutation of D137 to Alanine reduces CD81 function, whereas mutations of VVD (135–137) and DDD (137–139) abrogate activity. This is the first demonstration that this region of CD81 is functionally important in a biological process. Interestingly, mutations at the A-B junction do not alter binding of anti-CD81 mAbs, except 5A6, the binding of which is reduced threefold in the non-functional VVD mutant. This indicates that the mutants have a correct conformation and that the epitope recognized by 5A6 critically depends on residues located at the A-B junction. Interestingly, 5A6 is the only anti-CD81 mAb that does not inhibit *P. yoelii* infection, possibly because its epitope is masked on the CD81 molecules that participate in sporozoite invasion. This strongly suggests that the A-B region of CD81 does not bind a sporozoite ligand but instead interacts with another host cell factor, which may compete with the antibody for binding to CD81.

14.6 The Role of Tetraspanin-Enriched Microdomains in HCV and *Plasmodium* Entry

14.6.1 Lipids, Tetraspanin-Enriched Microdomains and HCV

SR-BI was initially identified as the major physiological receptor for high-density lipoproteins (HDL) in the liver and is involved in selective lipid uptake (Connelly and Williams 2004). SR-BI is a multiligand receptor and several of its ligands have been found to affect HCV infectivity. Oxidized LDL, serum amyloid A and VLDL have been shown to inhibit HCVpp and HCVs cell entry (Cai et al. 2007; Lavie et al. 2006; Maillard et al. 2006; von Hahn et al. 2006). In contrast, HDL enhances HCV entry, a process which depends on the lipid transfer function of SR-BI and the presence of apolipoprotein C1, a minor component of HDL (Bartosch et al. 2005; Dreux et al. 2007; Meunier et al. 2005; Voisset et al. 2005), suggesting that HCV exploits the physiological functions of SR-BI during the entry process. It should be noted that SR-BI directly interacts with cholesterol and likely localizes in cholesterol-enriched domains of the membrane (Silver et al. 2001). Furthermore, HDL enhances HCVcc infectivity through SR-BI only when CD81 is expressed (Zeisel et al. 2007), suggesting that an interplay between CD81 and SR-BI might occur during HCV entry.

Besides the effects of some lipoproteins and apolipoproteins, the lipid composition of the plasma membrane can also modulate HCV entry into host cells. Indeed, cholesterol depletion by treatment with methyl-beta-cyclo-dextrin (MβCD), a cyclic oligosaccharide that selectively removes cholesterol from the plasma membrane, strongly reduced HCV entry into target cells, and conversely cholesterol replenishment by MβCD-cholesterol complexes restored the infection levels (Kapadia et al. 2007; Rocha-Perugini et al. 2009). Depletion of cholesterol from the plasma membrane affects HCV entry by reducing the cell surface expression of CD81 that is physically associated with cholesterol (Charrin et al. 2003b). Beyond cholesterol, sphingolipids are also known to be important for the organization of the plasma membrane. Among them, sphingomyelin can be converted into ceramide by sphingomyelinase (Smase), and increasing ceramide concentration can lead to lipid microdomain reorganization (Bollinger et al. 2005). Ceramide enrichment of the plasma membrane of Huh-7 cells following sphingomyelin hydrolysis by sphingomyelinase strongly inhibits HCVpp and HCVcc entry (Rocha-Perugini et al. 2009; Voisset et al. 2008). The analysis of the effect of ceramide enrichment of the plasma membrane on CD81, SR-BI and CLDN-1 expression showed a 50% decrease of cell-surface expression of CD81, which was due to its ATP-independent internalization (Voisset et al. 2008). Altogether, these studies provide evidence that HCV entry could be regulated by the lipid content of target cell membranes.

CD81 belongs to the tetraspanin family whose members have the distinctive feature of forming dynamic clusters with numerous partner proteins and with one another, forming extended cholesterol-associated complexes on the cell surface, called tetraspanin-enriched microdomains (TEMs) or tetraspanin webs (Berditchevski et al. 2002; Rubinstein et al. 1996). Within this web, the association of a tetraspanin with its nontetraspanin partner molecule has been referred to as primary complex, and tetraspanins can interact with each other through their associated partner (Boucheix and Rubinstein 2001; Charrin et al. 2009a). In contrast to primary complexes, tetraspanin-tetraspanin interactions are not stoichiometric. Tetraspanin molecules can form different partnerships in different cell types that enable lateral dynamic organization in the membrane and the cross-talk with intracellular signalling and cytoskeletal structures (Charrin et al. 2009a; Levy and Shoham 2005a). Precise mechanisms of interaction within webs still need to be elucidated. However, in addition to specific domains, cellular lipids such as cholesterol, glycosphingolipids and the modification of tetraspanins with palmitate residues seem to play an important role in the interaction of tetraspanins with one another and therefore in the building of the tetraspanin network.

Recently, generation of a human cell line expressing mCD81 permissive to HCV infection has allowed the analysis of the role of TEM-associated CD81 in HCV infection (Rocha-Perugini et al. 2009). Indeed, MT81w antibody, which only recognizes TEM-associated mCD81 (Silvie et al. 2006a), had a modest effect on HCV infection. Furthermore, cholesterol depletion, which inhibits HCV infection and reduces total cell surface expression of CD81, did not affect TEM-associated CD81 levels. In addition, sphingomyelinase treatment, which also reduces HCV infection and cell surface expression of total CD81, raised TEM-associated CD81 levels. Together, these results indicate that TEM-associated CD81 is not preferentially used by HCV for its entry (Rocha-Perugini et al. 2009). HCV entry is a multi-sequential process that is likely highly dynamic. CD81 association with TEM might be affiliated with a kind of sequestration of this tetraspanin impeding the kinetics of interaction with other coreceptors that are necessary for efficient HCV entry. Alternatively, tetraspanins may compete with coreceptors for their interaction with CD81. In addition, lipid composition of TEMs might not be suitable for HCV entry.

14.6.2 Tetraspanin-Enriched Microdomains and Plasmodium Sporozoites

In contrast with HCV, there is evidence showing that localization of CD81 in TEMs plays an important role during *Plasmodium* sporozoite infection. First, the anti-mCD81 mAb MT81w, which selectively binds to CD81 associated with

TEMs, inhibits up to 70% invasion of mouse hepatoma Hepa1-6 cells by *P. yoelii* sporozoites, demonstrating that the pool of CD81 engaged in TEMs is functional (Silvie et al. 2006a). Furthermore, cholesterol depletion by MβCD causes not only a reduction of MT81*w* binding to Hepa1-6 cells, demonstrating the role of cholesterol in CD81 organization at the cell surface in this model, but also a decrease in infection (Silvie et al. 2006a). Importantly, inhibition of infection was observed without a reduction in the level of total CD81 at the surface of MβCD-treated Hepa1-6 cells, which differs from the results obtained in mCD81-transfected Huh-7 cells (Rocha-Perugini et al. 2009), and was completely reversed by cholesterol replenishment with preformed MβCD/cholesterol complexes. Cholesterol depletion also inhibits infection of primary mouse and human hepatocytes by *P. yoelii* and *P. falciparum*, respectively (Silvie et al. 2006a). Importantly, in the case of *P. berghei* sporozoites, which invade cells through either CD81-dependent or CD81-independent pathways depending on the cell type, cholesterol depletion only affects CD81-dependent invasion, demonstrating the functional link between cholesterol and CD81 during *Plasmodium* infection (Silvie et al. 2006a, 2007). Finally, the role of membrane cholesterol in CD81 function during sporozoite invasion was recently confirmed by a study showing that SR-BI is important for infection (Rodrigues et al. 2008; Yalaoui et al. 2008a). Although SR-BI knockout mice remained susceptible to infection *in vivo*, *P. yoelii* infection of SR-BI- knockout primary hepatocytes was reduced *in vitro*, as compared to wild type hepatocytes (Yalaoui et al. 2008a). This lower susceptibility of SR-BI-deficient cells correlates with a reduction of the level of cell membrane cholesterol, and a decrease in CD81 surface expression and localization to TEMs. Anti-SR-BI antibodies block invasion of *P. yoelii* and *P. falciparum* in mouse and human hepatocytes, respectively, and cause a reduction of the membrane cholesterol contents (Rodrigues et al. 2008; Yalaoui et al. 2008a). Interestingly, unlike anti-CD81 antibodies, anti-SRBI antibodies only block infection when added before, but not at the same time as sporozoites (Yalaoui et al. 2008a). Together, these data support an indirect role of SR-BI during *P. yoelii* infection, via the regulation of cholesterol levels, expression and organization of CD81 at the cell surface. The function of SR-BI during *Plasmodium* infection therefore differs from its role as an HCV entry factor. Indeed, as detailed above, SR-BI and CD81 act cooperatively during HCV entry, and there is evidence for direct interaction between SR-BI and the virus (Scarselli et al. 2002). Interestingly, Rodrigues et al. found that siRNA-mediated silencing of SR-BI leads to a reduction of *P. berghei* invasion of Huh-7 cells (Rodrigues et al. 2008), an experimental model where invasion does not depend on CD81 (Silvie et al. 2007). This observation suggests that SR-BI might play additional roles during sporozoite entry, at least in the case of *P. berghei*. It should be noted that SR-BI also plays a role after invasion, during *Plasmodium* liver stage growth (Rodrigues et al. 2008; Yalaoui et al. 2008a). Like other apicomplexan parasites, *Plasmodium* cannot synthesize sterols *de novo* (Bano et al. 2007). By mediating cholesterol transfer to hepatocytes, SR-BI, although not essential, probably participates indirectly in parasite uptake of lipids from the host cell.

14.7 CD81 and Its Associated Partners

Within the network of TEMs, tetraspanins form primary complexes with a limited number of proteins called tetraspanin partners. These tetraspanin-partner interactions are direct and highly specific. In most cell lines, CD81 is associated with a high stoichiometry with EWI-F (also called CD9P-1, FPRP or CD315) and EWI-2 (also called PGRL, IgSF8 or CD316) (Charrin et al. 2001, 2003a; Clark et al. 2001; Stipp et al. 2001a, b). Both are members of the EWI family, a newly discovered family of immunoglobulins sharing a conserved EWI (Glu-Trp-Ile) sequence and containing an ectodomain composed of V-type Ig domains, a transmembrane domain and a short, highly charged cytoplasmic tail. EWI-2 and EWI-F/CD9P-1 contain four and six Ig domains, respectively (Charrin et al. 2003a; Stipp et al. 2001a). Both proteins also associate with the tetraspanin CD9 (Charrin et al. 2001; Stipp et al. 2001b). Co-immunoprecipitation experiments using digitonin lysates from biotin-labeled cells have shown that EWI-2 and EWI-F/CD9P-1 are the two major partners of CD81 in hepatocytes (Charrin et al. 2003a, 2009b). As detailed below, EWI-2 and EWI-F/CD9P-1 are not required for host cell infection, but on the contrary play inhibitory roles during entry of HCV and *Plasmodium* sporozoites, respectively.

14.7.1 CD81 Partners and HCV

Hepatitis C virus has a very narrow host range due to its specificity for the human homologues for some of its receptors, but it also has a very narrow cell tropism in humans, infecting almost exclusively liver cells. Recently, we hypothesized that, in addition to the presence of specific entry factors in the hepatocytes, the lack of a specific inhibitor could contribute to the hepatotropism of HCV (Rocha-Perugini et al. 2008). Indeed, several host restriction factors that protect cells from viral infection have been identified (Strebel et al. 2009). These host factors can target many distinct steps in the viral life cycle including entry, replication, the intracellular movement of viral nucleic acids, and viral gene expression. The existence of these factors offers the potential to design antiviral therapies by developing small molecules that mimic their activities.

In our study (Rocha-Perugini et al. 2008), we first investigated the interaction between the viral envelope glycoproteins and CD81 in several cell lines. Using detergent conditions in which tetraspanin complexes were maintained, we observed that the E1E2-CD81 interaction was inhibited by a cellular protein present in some cell lines such as Daudi and Ramos cells. To identify the CD81 partner blocking the interaction between CD81 and E1E2, we next performed co-immunoprecipitation experiments using lysates from surface biotinylated cells. Using this approach, a 55 kDa surface protein was identified as a specific partner of CD81 in Daudi and

Ramos cells. Following mass spectrometric analyses and N-terminal sequencing, the interacting partner was identified as a cleavage product of EWI-2, that we called EWI-2wint for EWI-2 without its N-terminus. To address the impact of EWI-2wint on HCV tissue tropism, we analyzed EWI-2wint expression in a large panel of cell lines and primary human hepatocytes. EWI-2wint expression was present in many non-HCV-permissive cell lines, but was absent in HCV permissive cells such as primary human hepatocytes and Huh-7 cells. Furthermore, the presence of EWI-2wint expression correlated with the inhibition of CD81-E1E2 binding in co-immunoprecipitation studies. These results suggested that EWI-2wint inhibited CD81-E1E2 interaction in non-HCV-permissive cell lines but not in hepatic cells. To confirm that EWI-2wint plays a functional role in inhibiting HCV infection, we next ectopically expressed EWI-2wint protein in Huh-7 cells. Importantly, these EWI-2wint-expressing Huh-7 cells displayed a significant decrease in HCVpp and HCVcc entry, whereas replication and assembly/release of particles were not affected by EWI-2wint expression. (Rocha-Perugini et al. 2008). These results demonstrate that a pathogen is able to gain entry into a host cell by virtue of the lack of a specific inhibitory factor.

Although determinants in EWI-2/EWI-2wint and CD81 for their interaction have been identified (Montpellier et al. 2011), the mechanism by which EWI-2wint inhibits HCV entry remains to be defined. EWI-2wint may inhibit HCV entry by reducing E1E2-CD81 interactions for a number of possible reasons. EWI-2wint may reduce CD81 accessibility to envelope glycoproteins by steric hindrance. Alternatively, the association of EWI-2wint with CD81 may induce conformational modifications in CD81, blocking the binding of E1E2 heterodimers. CD81 and CLDN-1 associate to form receptor complexes, which preexist within cells and the formation or stability is not dependent upon or promoted by interaction with the viral glycoproteins (Harris et al. 2008). The interaction of EWI-2wint with CD81 may induce a change in the stoichiometry or distance between coreceptor molecules that is necessary for efficient HCV entry. As described in previous sections, the exact role of CD81 in the course of virus infection is not well defined; however, it likely functions as a post-attachment entry co-factor. CD81 may potentially be required for a post-binding step such as escorting the particle into the endocytic pathway or priming it for the pH triggered fusion mechanism. The association of EWI-2wint with CD81 could block such entry stages. HCV is associated with actin following internalization (Coller et al. 2009) and CD81 engagement by anti-CD81 mAbs or recombinant E2 glycoprotein leads to actin rearrangement (Brazzoli et al. 2008; Crotta et al. 2002, 2006; Wack et al. 2001). Since it has been shown that CD81 and EWI-2 interact with Ezrin, an actin-linking ERM protein (Sala-Valdes et al. 2006), we can also hypothesize that EWI-2wint may interfere with actin polymerization during viral entry or block signalling pathways necessary for viral entry. However, recent results indicate that EWI-2wint induces a reorganization of CD81 molecules leading to a reduced diffusion in the plasma membrane (Potel et al. 2013).

14.7.2 CD81 Partners and Plasmodium

Although the role of EWI-2wint during *Plasmodium* infection has not yet been reported, EWI-2 does not seem to be involved during sporozoite entry, based on the absence of effect of monoclonal antibodies or siRNA targeting EWI-2. However, reminiscent of EWI-2wint inhibition of HCV entry, EWI-F/CD9P-1 was found to negatively affect *P. yoelii* infection (Charrin et al. 2009b). Indeed, silencing of EWI-F/CD9P-1 in Hepa1-6 or HepG2/CD81 cells results in enhanced infection, and conversely over-expression of CD9P-1 in Hepa1-6 cells inhibited infection. Importantly, a EWI-F/CD9P-1 mutant that no longer interacts with CD81 has no effect on infection, demonstrating that EWI-F/CD9P-1 exerts its inhibitory action by interacting with CD81 (Charrin et al. 2009b). It is unlikely that EWI-F/CD9P-1 competes with a sporozoite receptor for association with CD81, because silencing of CD9P-1 enhances infection despite the fact that EWI-F/CD9P-1 is expressed at much lower levels than CD81 on hepatocytic cells, and because over-expression only partially blocks infection, even in cells expressing high levels of EWI-F/CD9P-1 after transfection. Furthermore, interaction with EWI-F/CD9P-1 involves the fourth transmembrane domain of CD81, which lies opposite to the critical A-B junction in the extracellular region (Seigneuret 2006; Yalaoui et al. 2008b). Based on these data, it is therefore more likely that EWI-F/CD9P-1 forms a complex with both CD81 and another still unidentified partner, the function of which would be altered in the presence of EWI-F/CD9P-1. A chimeric EWI-F/CD9P-1 where the cytoplasmic region has been substituted by that of HLA retains its ability to associate with CD81 and inhibit infection, ruling out a role for the C-terminus of EWI-F/CD9P-1 and potential involvement of ERM proteins in inhibition of infection (Charrin et al. 2009b).

14.8 Conclusion

Infections by HCV and *Plasmodium* sporozoites share some common features. Both pathogens selectively infect hepatocytes for replication. They both use liver HSPGs as docking sites in the liver, and both critically depend on the tetraspanin CD81 to enter hepatocytes. However, these two pathogens enter cells through very different mechanisms: HCV entry is mediated by clathrin-dependent endocytosis, whereas *Plasmodium* sporozoites invade host cells actively using their own actomyosin-based motor machinery. In addition, although both HCV and *Plasmodium* rely on CD81 to enter cells, they diverge in the way they use the tetraspanin. First, there is evidence for direct interaction of CD81 with HCV particle but not with *Plasmodium*. Second, although in both cases CD81LEL is critical, the LEL molecular determinants are different for HCV and *Plasmodium*. Third, although both infections can be inhibited by cholesterol depletion, localization of CD81 in tetraspanin-enriched microdomains is favorable to *Plasmodium* but not to HCV.

Fourth, interaction with its molecular partners affects CD81 differently, EWI-2wint and EWI-F/CD9P-1 selectively blocking HCV and *Plasmodium* sporozoites, respectively.

Additional entry factors have been identified that are critical for HCV entry, including SR-BI, Claudin and Occludin. Although Claudin is not necessary for *P. yoeli* infection (Yalaoui et al. 2008a) and the role of Occludin still needs to be characterized, SR-BI has been shown to be involved during sporozoite invasion of host cells. However, whereas SR-BI directly interacts with HCV, SR-BI function in *Plasmodium* infection seems to be mostly indirect, through the regulation of CD81 expression and organization at the cell surface. Altogether, these data show that although CD81 plays a pivotal role during liver cell infection by HCV and *Plasmodium* sporozoites, the molecular mechanisms involved may differ.

Interaction between HCV and *Plasmodium* has not been studied so far, but clearly deserves further attention, not only on a molecular level to understand the mechanisms involving CD81 in the context of infection, but also more generally from an epidemiological point of view, as a large part of the human population is exposed to both HCV and *Plasmodium* parasites.

References

Agnello V, Abel G, Elfahal M, Knight GB, Zhang Q-X (1999) Hepatitis C virus and other flaviviridae viruses enter cells via low density lipoprotein receptor. Proc Natl Acad Sci USA 96: 12766–12771

Akazawa D, Date T, Morikawa K, Murayama A, Miyamoto M, Kaga M, Barth H, Baumert TF, Dubuisson J, Wakita T (2007) CD81 expression is important for the permissiveness of Huh7 cell clones for heterogeneous hepatitis C virus infection. J Virol 81:5036–5045

Albecka A, Belouzard S, de Beeck AO, Descamps V, Goueslain L, Bertrand-Michel J, Terce F, Duverlie G, Rouille Y, Dubuisson J (2012) Role of low-density lipoprotein receptor in the hepatitis C virus life cycle. Hepatology 55:998–1007

Allander T, Forns X, Emerson SU, Purcell RH, Bukh J (2000) Hepatitis C virus envelope protein E2 binds to CD81 of tamarins. Virology 277:358–367

Aly AS, Vaughan AM, Kappe SH (2009) Malaria parasite development in the mosquito and infection of the mammalian host. Annu Rev Microbiol 63:195–221

Amino R, Thiberge S, Martin B, Celli S, Shorte S, Frischknecht F, Menard R (2006) Quantitative imaging of Plasmodium transmission from mosquito to mammal. Nat Med 12:220–224

Andre P, Komurian-Pradel F, Deforges S, Perret M, Berland JL, Sodoyer M, Pol S, Brechot C, Paranhos-Baccala G, Lotteau V (2002) Characterization of low- and very-low-density hepatitis C virus RNA-containing particles. J Virol 76:6919–6928

Andre P, Perlemuter G, Budkowska A, Brechot C, Lotteau V (2005) Hepatitis C virus particles and lipoprotein metabolism. Semin Liver Dis 25:93–104

Andreo U, Maillard P, Kalinina O, Walic M, Meurs E, Martinot M, Marcellin P, Budkowska A (2007) Lipoprotein lipase mediates hepatitis C virus (HCV) cell entry and inhibits HCV infection. Cell Microbiol 9:2445–2456

Bankwitz D, Steinmann E, Bitzegeio J, Ciesek S, Friesland M, Herrmann E, Zeisel MB, Baumert TF, Keck ZY, Foung SK, Pecheur EI, Pietschmann T (2010) Hepatitis C virus hypervariable region 1 modulates receptor interactions, conceals the CD81 binding site, and protects conserved neutralizing epitopes. J Virol 84:5751–5763

Bano N, Romano JD, Jayabalasingham B, Coppens I (2007) Cellular interactions of Plasmodium liver stage with its host mammalian cell. Int J Parasitol 37:1329–1341
Barth H, Schafer C, Adah MI, Zhang F, Linhardt RJ, Toyoda H, Kinoshita-Toyoda A, Toida T, Van Kuppevelt TH, Depla E, Von Weizsacker F, Blum HE, Baumert TF (2003) Cellular binding of hepatitis C virus envelope glycoprotein E2 requires cell surface heparan sulfate. J Biol Chem 278:41003–41012
Barth H, Ulsenheimer A, Pape GR, Diepolder HM, Hoffmann M, Neumann-Haefelin C, Thimme R, Henneke P, Klein R, Paranhos-Baccala G, Depla E, Liang TJ, Blum HE, Baumert TF (2005) Uptake and presentation of hepatitis C virus-like particles by human dendritic cells. Blood 105:3605–3614
Barth H, Liang TJ, Baumert TF (2006) Hepatitis C virus entry: molecular biology and clinical implications. Hepatology 44:527–535
Bartosch B, Cosset FL (2006) Cell entry of hepatitis C virus. Virology 348:1–12
Bartosch B, Bukh J, Meunier JC, Granier C, Engle RE, Blackwelder WC, Emerson SU, Cosset FL, Purcell RH (2003a) In vitro assay for neutralizing antibody to hepatitis C virus: evidence for broadly conserved neutralization epitopes. Proc Natl Acad Sci USA 100:14199–14204
Bartosch B, Dubuisson J, Cosset F-L (2003b) Highly infectious hepatitis C pseudo-viruses containing functional E1E2 envelope protein complexes. J Exp Med 197:633–642
Bartosch B, Vitelli A, Granier C, Goujon C, Dubuisson J, Pascale S, Scarselli E, Cortese R, Nicosia A, Cosset FL (2003c) Cell entry of hepatitis C virus requires a set of co-receptors that include the CD81 tetraspanin and the SR-B1 scavenger receptor. J Biol Chem 278:41624–41630
Bartosch B, Verney G, Dreux M, Donot P, Morice Y, Penin F, Pawlotsky JM, Lavillette D, Cosset FL (2005) An interplay between hypervariable region 1 of the hepatitis C virus E2 glycoprotein, the scavenger receptor BI, and high-density lipoprotein promotes both enhancement of infection and protection against neutralizing antibodies. J Virol 79:8217–8229
Basu A, Kanda T, Beyene A, Saito K, Meyer K, Ray R (2007) Sulfated homologues of heparin inhibit hepatitis C virus entry into mammalian cells. J Virol 81:3933–3941
Baum J, Gilberger TW, Frischknecht F, Meissner M (2008) Host-cell invasion by malaria parasites: insights from Plasmodium and Toxoplasma. Trends Parasitol 24:557–563
Benedicto I, Molina-Jimenez F, Bartosch B, Cosset FL, Lavillette D, Prieto J, Moreno-Otero R, Valenzuela-Fernandez A, Aldabe R, Lopez-Cabrera M, Majano PL (2009) The tight junction-associated protein occludin is required for a postbinding step in hepatitis C virus entry and infection. J Virol 83:8012–8020
Benga WJ, Krieger SE, Dimitrova M, Zeisel MB, Parnot M, Lupberger J, Hildt E, Luo G, McLauchlan J, Baumert TF, Schuster C (2010) Apolipoprotein E interacts with hepatitis C virus nonstructural protein 5A and determines assembly of infectious particles. Hepatology 51:43–53
Berditchevski F, Odintsova E, Sawada S, Gilbert E (2002) Expression of the palmitoylation-deficient CD151 weakens the association of alpha 3 beta 1 integrin with the tetraspanin-enriched microdomains and affects integrin-dependent signaling. J Biol Chem 277:36991–37000
Bertaux C, Dragic T (2006) Different domains of CD81 mediate distinct stages of hepatitis C virus pseudoparticle entry. J Virol 80:4940–4948
Blanchard E, Belouzard S, Goueslain L, Wakita T, Dubuisson J, Wychowski C, Rouille Y (2006) Hepatitis C virus entry depends on clathrin-mediated endocytosis. J Virol 80:6964–6972
Bollinger CR, Teichgraber V, Gulbins E (2005) Ceramide-enriched membrane domains. Biochim Biophys Acta 1746:284–294
Bonnafous P, Perrault M, Le Bihan O, Bartosch B, Lavillette D, Penin F, Lambert O, Pecheur EI (2010) Characterization of hepatitis C viral pseudoparticles by cryo-electron microscopy using functionalized magnetic nanobeads. J Gen Virol 91(Pt8):1919–1930
Boo I, Tewierik K, Douam F, Lavillette D, Poumbourios P, Drummer HE (2012) Distinct roles in folding, CD81 receptor binding and viral entry for conserved histidine residues of hepatitis C virus glycoprotein E1 and E2. Biochem J 443:85–94
Boucheix C, Rubinstein E (2001) Tetraspanins. Cell Mol Life Sci 58:1189–1205

Brazzoli M, Helenius A, Foung SK, Houghton M, Abrignani S, Merola M (2005) Folding and dimerization of hepatitis C virus E1 and E2 glycoproteins in stably transfected CHO cells. Virology 332:438–453

Brazzoli M, Bianchi A, Filippini S, Weiner A, Zhu Q, Pizza M, Crotta S (2008) CD81 is a central regulator of cellular events required for hepatitis C virus infection of human hepatocytes. J Virol 82:8316–8329

Buck M (2008) Direct infection and replication of naturally occurring hepatitis C virus genotypes 1, 2, 3 and 4 in normal human hepatocyte cultures. PLoS One 3:e2660

Cai Z, Cai L, Jiang J, Chang KS, van der Westhuyzen DR, Luo G (2007) Human serum amyloid A protein inhibits hepatitis C virus entry into cells. J Virol 81:6128–6133

Callens N, Ciczora Y, Bartosch B, Vu-Dac N, Cosset FL, Pawlotsky JM, Penin F, Dubuisson J (2005) Basic residues in hypervariable region 1 of hepatitis C virus envelope glycoprotein e2 contribute to virus entry. J Virol 79:15331–15341

Calvo-Calle JM, Moreno A, Eling WM, Nardin EH (1994) In vitro development of infectious liver stages of P. yoelii and P. berghei malaria in human cell lines. Exp Parasitol 79:362–373

Carloni G, Iacovacci S, Sargiacomo M, Ravagnan G, Ponzetto A, Peschle C, Battaglia M (1993) Susceptibility of human liver cell cultures to hepatitis C virus infection. Arch Virol Suppl 8:31–39

Catanese MT, Graziani R, von Hahn T, Moreau M, Huby T, Paonessa G, Santini C, Luzzago A, Rice CM, Cortese R, Vitelli A, Nicosia A (2007) High-avidity monoclonal antibodies against the human scavenger class B type I receptor efficiently block hepatitis C virus infection in the presence of high-density lipoprotein. J Virol 81:8063–8071

Catanese MT, Ansuini H, Graziani R, Huby T, Moreau M, Ball JK, Paonessa G, Rice CM, Cortese R, Vitelli A, Nicosia A (2010) Role of scavenger receptor class B type I in hepatitis C virus entry: kinetics and molecular determinants. J Virol 84:34–43

Chang KS, Jiang J, Cai Z, Luo G (2007) Human apolipoprotein e is required for infectivity and production of hepatitis C virus in cell culture. J Virol 81:13783–13793

Charrin S, Le Naour F, Oualid M, Billard M, Faure G, Hanash SM, Boucheix C, Rubinstein E (2001) The major CD9 and CD81 molecular partner. Identification and characterization of the complexes. J Biol Chem 276:14329–14337

Charrin S, Le Naour F, Labas V, Billard M, Le Caer JP, Emile JF, Petit MA, Boucheix C, Rubinstein E (2003a) EWI-2 is a new component of the tetraspanin web in hepatocytes and lymphoid cells. Biochem J 373:409–421

Charrin S, Manie S, Thiele C, Billard M, Gerlier D, Boucheix C, Rubinstein E (2003b) A physical and functional link between cholesterol and tetraspanins. Eur J Immunol 33:2479–2489

Charrin S, le Naour F, Silvie O, Milhiet PE, Boucheix C, Rubinstein E (2009a) Lateral organization of membrane proteins: tetraspanins spin their web. Biochem J 420:133–154

Charrin S, Yalaoui S, Bartosch B, Cocquerel L, Franetich JF, Boucheix C, Mazier D, Rubinstein E, Silvie O (2009b) The Ig domain protein CD9P-1 down-regulates CD81 ability to support Plasmodium yoelii infection. J Biol Chem 284:31572–31578

Ciccaglione AR, Costantino A, Marcantonio C, Equestre M, Geraci A, Rapicetta M (2001) Mutagenesis of hepatitis C virus E1 protein affects its membrane-permeabilizing activity. J Gen Virol 82:2243–2250

Ciczora Y, Callens N, Montpellier C, Bartosch B, Cosset FL, Op de Beeck A, Dubuisson J (2005) Contribution of the charged residues of hepatitis C virus glycoprotein E2 transmembrane domain to the functions of the E1E2 heterodimer. J Gen Virol 86:2793–2798

Clark KL, Zeng Z, Langford AL, Bowen SM, Todd SC (2001) PGRL is a major CD81-associated protein on lymphocytes and distinguishes a new family of cell surface proteins. J Immunol 167:5115–5121

Cocquerel L, Meunier JC, Pillez A, Wychowski C, Dubuisson J (1998) A retention signal necessary and sufficient for endoplasmic reticulum localization maps to the transmembrane domain of hepatitis C virus glycoprotein E2. J Virol 72:2183–2191

Cocquerel L, Duvet S, Meunier JC, Pillez A, Cacan R, Wychowski C, Dubuisson J (1999) The transmembrane domain of hepatitis C virus glycoprotein E1 is a signal for static retention in the endoplasmic reticulum. J Virol 73:2641–2649

Cocquerel L, Wychowski C, Minner F, Penin F, Dubuisson J (2000) Charged residues in the transmembrane domains of hepatitis C virus glycoproteins play a major role in the processing, subcellular localization, and assembly of these envelope proteins. J Virol 74:3623–3633

Cocquerel L, Meunier JC, Op de Beeck A, Bonte D, Wychowski C, Dubuisson J (2001) Coexpression of hepatitis C virus envelope proteins E1 and E2 in cis improves the stability of membrane insertion of E2. J Gen Virol 82:1629–1635

Cocquerel L, Op De Beeck A, Lambot M, Roussel J, Delgrange D, Pillez A, Wychowski C, Penin F, Dubuisson J (2002) Topological changes in the transmembrane domains of hepatitis C virus envelope glycoproteins. EMBO J 21:2893–2902

Cocquerel L, Kuo CC, Dubuisson J, Levy S (2003a) CD81-dependent binding of hepatitis C virus E1E2 heterodimers. J Virol 77:10677–10683

Cocquerel L, Quinn ER, Flint M, Hadlock KG, Foung SK, Levy S (2003b) Recognition of native hepatitis C virus E1E2 heterodimers by a human monoclonal antibody. J Virol 77:1604–1609

Cocquerel L, Voisset C, Dubuisson J (2006) Hepatitis C virus entry: potential receptors and their biological functions. J Gen Virol 87:1075–1084

Coller KE, Berger KL, Heaton NS, Cooper JD, Yoon R, Randall G (2009) RNA interference and single particle tracking analysis of hepatitis C virus endocytosis. PLoS Pathog 5:e1000702

Connelly MA, Williams DL (2004) Scavenger receptor BI: a scavenger receptor with a mission to transport high density lipoprotein lipids. Curr Opin Lipidol 15:287–295

Coppi A, Pinzon-Ortiz C, Hutter C, Sinnis P (2005) The Plasmodium circumsporozoite protein is proteolytically processed during cell invasion. J Exp Med 201:27–33

Coppi A, Tewari R, Bishop JR, Bennett BL, Lawrence R, Esko JD, Billker O, Sinnis P (2007) Heparan sulfate proteoglycans provide a signal to Plasmodium sporozoites to stop migrating and productively invade host cells. Cell Host Microbe 2:316–327

Cormier EG, Tsamis F, Kajumo F, Durso RJ, Gardner JP, Dragic T (2004) CD81 is an entry coreceptor for hepatitis C virus. Proc Natl Acad Sci USA 101:7270–7274

Coyne CB, Bergelson JM (2006) Virus-induced Abl and Fyn kinase signals permit coxsackie virus entry through epithelial tight junctions. Cell 124:119–131

Crotta S, Stilla A, Wack A, D'Andrea A, Nuti S, D'Oro U, Mosca M, Filliponi F, Brunetto RM, Bonino F, Abrignani S, Valiante NM (2002) Inhibition of natural killer cells through engagement of CD81 by the major hepatitis C virus envelope protein. J Exp Med 195:35–41

Crotta S, Ronconi V, Ulivieri C, Baldari CT, Valiante NM, Abrignani S, Wack A (2006) Cytoskeleton rearrangement induced by tetraspanin engagement modulates the activation of T and NK cells. Eur J Immunol 36:919–929

Dhillon S, Witteveldt J, Gatherer D, Owsianka AM, Zeisel MB, Zahid MN, Rychlowska M, Foung SK, Baumert TF, Angus AG, Patel AH (2010) Mutations within a conserved region of the hepatitis C virus E2 glycoprotein that influence virus-receptor interactions and sensitivity to neutralizing antibodies. J Virol 84:5494–5507

Dorner M, Horwitz JA, Robbins JB, Barry WT, Feng Q, Mu K, Jones CT, Schoggins JW, Catanese MT, Burton DR, Law M, Rice CM, Ploss A (2011) A genetically humanized mouse model for hepatitis C virus infection. Nature 474:208–211

Dreux M, Boson B, Ricard-Blum S, Molle J, Lavillette D, Bartosch B, Pecheur EI, Cosset FL (2007) The exchangeable apolipoprotein ApoC-I promotes membrane fusion of hepatitis C virus. J Biol Chem 282:32357–32369

Dreux M, Dao Thi VL, Fresquet J, Guerin M, Julia Z, Verney G, Durantel D, Zoulim F, Lavillette D, Cosset FL, Bartosch B (2009) Receptor complementation and mutagenesis reveal SR-BI as an essential HCV entry factor and functionally imply its intra- and extra-cellular domains. PLoS Pathog 5:e1000310

Drummer HE, Wilson KA, Poumbourios P (2002) Identification of the hepatitis C virus E2 glycoprotein binding site on the large extracellular loop of CD81. J Virol 76:11143–11147

Drummer HE, Maerz A, Poumbourios P (2003) Cell surface expression of functional hepatitis C virus E1 and E2 glycoproteins. FEBS Lett 546:385–390

Drummer HE, Wilson KA, Poumbourios P (2005) Determinants of CD81 dimerization and interaction with hepatitis C virus glycoprotein E2. Biochem Biophys Res Commun 328:251–257

Drummer HE, Boo I, Maerz AL, Poumbourios P (2006) A conserved Gly436-Trp-Leu-Ala-Gly-Leu-Phe-Tyr motif in hepatitis C virus glycoprotein E2 is a determinant of CD81 binding and viral entry. J Virol 80:7844–7853

Drummer HE, Boo I, Poumbourios P (2007) Mutagenesis of a conserved fusion peptide-like motif and membrane-proximal heptad-repeat region of hepatitis C virus glycoprotein E1. J Gen Virol 88:1144–1148

Dubuisson J, Rice CM (1996) Hepatitis C virus glycoprotein folding: disulfide bond formation and association with calnexin. J Virol 70:778–786

Duvet S, Cocquerel L, Pillez A, Cacan R, Verbert A, Moradpour D, Wychowski C, Dubuisson J (1998) Hepatitis C virus glycoprotein complex localization in the endoplasmic reticulum involves a determinant for retention and not retrieval. J Biol Chem 273:32088–32095

Evans MJ, von Hahn T, Tscherne DM, Syder AJ, Panis M, Wolk B, Hatziioannou T, McKeating JA, Bieniasz PD, Rice CM (2007) Claudin-1 is a hepatitis C virus co-receptor required for a late step in entry. Nature 446:801–805

Falkowska E, Kajumo F, Garcia E, Reinus J, Dragic T (2007) Hepatitis C virus envelope glycoprotein E2 glycans modulate entry, CD81 binding, and neutralization. J Virol 81:8072–8079

Flint M, McKeating JA (1999) The C-terminal region of the hepatitis C virus E1 glycoprotein confers localization within the endoplasmic reticulum. J Gen Virol 80:1943–1947

Flint M, Maidens C, Loomis-Price LD, Shotton C, Dubuisson J, Monk P, Higginbottom A, Levy S, McKeating JA (1999a) Characterization of hepatitis C virus E2 glycoprotein interaction with a putative cellular receptor, CD81. J Virol 73:6235–6244

Flint M, Thomas JM, Maidens CM, Shotton C, Levy S, Barclay WS, McKeating JA (1999b) Functional analysis of cell surface-expressed hepatitis C virus E2 glycoprotein. J Virol 73:6782–6790

Flint M, von Hahn T, Zhang J, Farquhar M, Jones CT, Balfe P, Rice CM, McKeating JA (2006) Diverse CD81 proteins support hepatitis C virus infection. J Virol 80:11331–11342

Fofana I, Krieger SE, Grunert F, Glauben S, Xiao F, Fafi-Kremer S, Soulier E, Royer C, Thumann C, Mee CJ, McKeating JA, Dragic T, Pessaux P, Stoll-Keller F, Schuster C, Thompson J, Baumert TF (2010) Monoclonal anti-claudin 1 antibodies prevent hepatitis C virus infection of primary human hepatocytes. Gastroenterology 139(3):953–964, 964 e951-954

Forns X, Allander T, Rohwer-Nutter P, Bukh J (2000a) Characterization of modified hepatitis C virus E2 proteins expressed on the cell surface. Virology 274:75–85

Forns X, Thimme R, Govindarajan S, Emerson SU, Purcell RH, Chisari FV, Bukh J (2000b) Hepatitis C virus lacking the hypervariable region 1 of the second envelope protein is infectious and causes acute resolving or persistent infection in chimpanzees. Proc Natl Acad Sci USA 97:13318–13323

Fournier C, Sureau C, Coste J, Ducos J, Pageaux G, Larrey D, Domergue J, Maurel P (1998) In vitro infection of adult normal human hepatocytes in primary culture by hepatitis C virus. J Gen Virol 79:2367–2374

Frevert U, Sinnis P, Cerami C, Shreffler W, Takacs B, Nussenzweig V (1993) Malaria circumsporozoite protein binds to heparan sulfate proteoglycans associated with the surface membrane of hepatocytes. J Exp Med 177:1287–1298

Frevert U, Sinnis P, Esko JD, Nussenzweig V (1996) Cell surface glycosaminoglycans are not obligatory for Plasmodium berghei sporozoite invasion in vitro. Mol Biochem Parasitol 76:257–266

Frevert U, Engelmann S, Zougbede S, Stange J, Ng B, Matuschewski K, Liebes L, Yee H (2005) Intravital observation of Plasmodium berghei sporozoite infection of the liver. PLoS Biol 3:e192

Gantt S, Persson C, Rose K, Birkett AJ, Abagyan R, Nussenzweig V (2000) Antibodies against thrombospondin-related anonymous protein do not inhibit Plasmodium sporozoite infectivity in vivo. Infect Immun 68:3667–3673

Garry RF, Dash S (2003) Proteomics computational analyses suggest that hepatitis C virus E1 and pestivirus E2 envelope glycoproteins are truncated class II fusion proteins. Virology 307:255–265

Gastaminza P, Cheng G, Wieland S, Zhong J, Liao W, Chisari FV (2008) Cellular determinants of hepatitis C virus assembly, maturation, degradation, and secretion. J Virol 82:2120–2129

Germi R, Crance JM, Garin D, Guimet J, Lortat-Jacob H, Ruigrok RW, Zarski JP, Drouet E (2002) Cellular glycosaminoglycans and low density lipoprotein receptor are involved in hepatitis C virus adsorption. J Med Virol 68:206–215

Goffard A, Callens N, Bartosch B, Wychowski C, Cosset FL, Montpellier C, Dubuisson J (2005) Role of N-linked glycans in the functions of hepatitis C virus envelope glycoproteins. J Virol 79:8400–8409

Gottwein JM, Scheel TK, Jensen TB, Lademann JB, Prentoe JC, Knudsen ML, Hoegh AM, Bukh J (2009) Development and characterization of hepatitis C virus genotype 1–7 cell culture systems: role of CD81 and scavenger receptor class B type I and effect of antiviral drugs. Hepatology 49:364–377

Grove J, Huby T, Stamataki Z, Vanwolleghem T, Meuleman P, Farquhar M, Schwarz A, Moreau M, Owen JS, Leroux-Roels G, Balfe P, McKeating JA (2007) Scavenger receptor BI and BII expression levels modulate hepatitis C virus infectivity. J Virol 81:3162–3169

Haid S, Pietschmann T, Pecheur EI (2009) Low pH-dependent hepatitis C virus membrane fusion depends on E2 integrity, target lipid composition, and density of virus particles. J Biol Chem 284:17657–17667

Harris HJ, Farquhar MJ, Mee CJ, Davis C, Reynolds GM, Jennings A, Hu K, Yuan F, Deng H, Hubscher SG, Han JH, Balfe P, McKeating JA (2008) CD81 and claudin 1 coreceptor association: role in hepatitis C virus entry. J Virol 82:5007–5020

Harris HJ, Davis C, Mullins JG, Hu K, Goodall M, Farquhar MJ, Mee CJ, McCaffrey K, Young S, Drummer H, Balfe P, McKeating JA (2010) Claudin association with CD81 defines hepatitis C virus entry. J Biol Chem 285:21092–21102

Hartsock A, Nelson WJ (2008) Adherens and tight junctions: structure, function and connections to the actin cytoskeleton. Biochim Biophys Acta 1778:660–669

Hauri AM, Armstrong GL, Hutin YJ (2004) The global burden of disease attributable to contaminated injections given in health care settings. Int J STD AIDS 15:7–16

Heiskala M, Peterson PA, Yang Y (2001) The roles of claudin superfamily proteins in paracellular transport. Traffic 2:93–98

Helle F, Goffard A, Morel V, Duverlie G, McKeating J, Keck ZY, Foung S, Penin F, Dubuisson J, Voisset C (2007) The neutralizing activity of anti-hepatitis C virus antibodies is modulated by specific glycans on the E2 envelope protein. J Virol 81:8101–8111

Helle F, Vieyres G, Elkrief L, Popescu CI, Wychowski C, Descamps V, Castelain S, Roingeard P, Duverlie G, Dubuisson J (2010) Role of N-linked glycans in the functions of hepatitis C virus envelope proteins incorporated into infectious virions. J Virol 84:11905–11915

Higginbottom A, Quinn ER, Kuo CC, Flint M, Wilson LH, Bianchi E, Nicosia A, Monk PN, McKeating JA, Levy S (2000) Identification of amino acid residues in CD81 critical for interaction with hepatitis C virus envelope glycoprotein E2. J Virol 74:3642–3649

Hollingdale MR, Leland P, Schwartz AL (1983) In vitro cultivation of the exoerythrocytic stage of Plasmodium berghei in a hepatoma cell line. Am J Trop Med Hyg 32:682–684

Hollingdale MR, Nardin EH, Tharavanij S, Schwartz AL, Nussenzweig RS (1984) Inhibition of entry of Plasmodium falciparum and P. vivax sporozoites into cultured cells; an in vitro assay of protective antibodies. J Immunol 132:909–913

Hsu M, Zhang J, Flint M, Logvinoff C, Cheng-Mayer C, Rice CM, McKeating JA (2003) Hepatitis C virus glycoproteins mediate pH-dependent cell entry of pseudotyped retroviral particles. Proc Natl Acad Sci USA 100:7271–7276

Huang H, Sun F, Owen DM, Li W, Chen Y, Gale M Jr, Ye J (2007) Hepatitis C virus production by human hepatocytes dependent on assembly and secretion of very low-density lipoproteins. Proc Natl Acad Sci USA 104:5848–5853

Iacovacci S, Sargiacomo M, Parolini I, Ponzetto A, Peschle C, Carloni G (1993) Replication and multiplication of hepatitis C virus genome in human foetal liver cells. Res Virol 144: 275–279

Imai T, Yoshie O (1993) C33 antigen and M38 antigen recognized by monoclonal antibodies inhibitory to syncytium formation by human T cell leukemia virus type 1 are both members of the transmembrane 4 superfamily and associate with each other and with CD4 or CD8 in T cells. J Immunol 151:6470–6481

Ishino T, Chinzei Y, Yuda M (2005) Two proteins with 6-cys motifs are required for malarial parasites to commit to infection of the hepatocyte. Mol Microbiol 58:1264–1275

Janse CJ, Ramesar J, Waters AP (2006) High-efficiency transfection and drug selection of genetically transformed blood stages of the rodent malaria parasite Plasmodium berghei. Nat Protoc 1:346–356

Jiang J, Luo G (2009) Apolipoprotein E but not B is required for the formation of infectious hepatitis C virus particles. J Virol 83:12680–12691

Jones CT, Catanese MT, Law LM, Khetani SR, Syder AJ, Ploss A, Oh TS, Schoggins JW, MacDonald MR, Bhatia SN, Rice CM (2010) Real-time imaging of hepatitis C virus infection using a fluorescent cell-based reporter system. Nat Biotechnol 28:167–171

Kapadia SB, Barth H, Baumert T, McKeating JA, Chisari FV (2007) Initiation of hepatitis C virus infection is dependent on cholesterol and cooperativity between CD81 and scavenger receptor B Type I. J Virol 81:374–383

Kappe S, Bruderer T, Gantt S, Fujioka H, Nussenzweig V, Menard R (1999) Conservation of a gliding motility and cell invasion machinery in Apicomplexan parasites. J Cell Biol 147:937–944

Keck ZY, Saha A, Xia J, Wang Y, Lau P, Krey T, Rey FA, Foung SK (2011) Mapping a region of hepatitis C virus E2 that is responsible for escape from neutralizing antibodies and a core CD81-binding region that does not tolerate neutralization escape mutations. J Virol 85: 10451–10463

Kitadokoro K, Bordo D, Galli G, Petracca R, Falugi F, Abrignani S, Grandi G, Bolognesi M (2001) CD81 extracellular domain 3D structure: insight into the tetraspanin superfamily structural motifs. EMBO J 20:12–18

Koutsoudakis G, Kaul A, Steinmann E, Kallis S, Lohmann V, Pietschmann T, Bartenschlager R (2006) Characterization of the early steps of hepatitis C virus infection by using luciferase reporter viruses. J Virol 80:5308–5320

Koutsoudakis G, Herrmann E, Kallis S, Bartenschlager R, Pietschmann T (2007) The level of CD81 cell surface expression is a key determinant for productive entry of hepatitis C virus into host cells. J Virol 81:588–598

Krey T, d'Alayer J, Kikuti CM, Saulnier A, Damier-Piolle L, Petitpas I, Johansson DX, Tawar RG, Baron B, Robert B, England P, Persson MA, Martin A, Rey FA (2010) The disulfide bonds in glycoprotein E2 of hepatitis C virus reveal the tertiary organization of the molecule. PLoS Pathog 6:e1000762

Krieger SE, Zeisel MB, Davis C, Thumann C, Harris HJ, Schnober EK, Mee C, Soulier E, Royer C, Lambotin M, Grunert F, Dao Thi VL, Dreux M, Cosset FL, McKeating JA, Schuster C, Baumert TF (2010) Inhibition of hepatitis C virus infection by anti-claudin-1 antibodies is mediated by neutralization of E2-CD81-claudin-1 associations. Hepatology 51:1144–1157

Labaied M, Harupa A, Dumpit RF, Coppens I, Mikolajczak SA, Kappe SH (2007) Plasmodium yoelii sporozoites with simultaneous deletion of P52 and P36 are completely attenuated and confer sterile immunity against infection. Infect Immun 75:3758–3768

Labonte P, Begley S, Guevin C, Asselin MC, Nassoury N, Mayer G, Prat A, Seidah NG (2009) PCSK9 impedes hepatitis C virus infection in vitro and modulates liver CD81 expression. Hepatology 50:17–24

Lavie M, Voisset C, Vu-Dac N, Zurawski V, Duverlie G, Wychowski C, Dubuisson J (2006) Serum amyloid A has antiviral activity against hepatitis C virus by inhibiting virus entry in a cell culture system. Hepatology 44:1626–1634

Lavie M, Goffard A, Dubuisson J (2007) Assembly of a functional HCV glycoprotein heterodimer. Curr Issues Mol Biol 9:71–86

Lavillette D, Tarr AW, Voisset C, Donot P, Bartosch B, Bain C, Patel AH, Dubuisson J, Ball JK, Cosset FL (2005) Characterization of host-range and cell entry properties of the major genotypes and subtypes of hepatitis C virus. Hepatology 41:265–274

Lavillette D, Bartosch B, Nourrisson D, Verney G, Cosset FL, Penin F, Pecheur EI (2006) Hepatitis C virus glycoproteins mediate low pH-dependent membrane fusion with liposomes. J Biol Chem 281:3909–3917

Lavillette D, Pecheur EI, Donot P, Fresquet J, Molle J, Corbau R, Dreux M, Penin F, Cosset FL (2007) Characterization of fusion determinants points to the involvement of three discrete regions of both E1 and E2 glycoproteins in the membrane fusion process of hepatitis C virus. J Virol 81:8752–8765
Lazaro CA, Chang M, Tang W, Campbell J, Sullivan DG, Gretch DR, Corey L, Coombs RW, Fausto N (2007) Hepatitis C virus replication in transfected and serum-infected cultured human fetal hepatocytes. Am J Pathol 170:478–489
Lemon SM, Walker C, Alter MJ, Yi M (2007) Hepatitis C virus. In: Knipe DM, Howley PM (eds) Fields virology, 5th edn. Lippincott Williams & Wilkins, Philadelphia, pp 1253–1304
Levy S, Shoham T (2005a) Protein-protein interactions in the tetraspanin web. Physiology (Bethesda) 20:218–224
Levy S, Shoham T (2005b) The tetraspanin web modulates immune-signalling complexes. Nat Rev Immunol 5:136–148
Li HF, Huang CH, Ai LS, Chuang CK, Chen SS (2009) Mutagenesis of the fusion peptide-like domain of hepatitis C virus E1 glycoprotein: involvement in cell fusion and virus entry. J Biomed Sci 16:89
Lindenbach BD, Evans MJ, Syder AJ, Wolk B, Tellinghuisen TL, Liu CC, Maruyama T, Hynes RO, Burton DR, McKeating JA, Rice CM (2005) Complete replication of hepatitis C virus in cell culture. Science 309:623–626
Lindenbach BD, Meuleman P, Ploss A, Vanwolleghem T, Syder AJ, McKeating JA, Lanford RE, Feinstone SM, Major ME, Leroux-Roels G, Rice CM (2006) Cell culture-grown hepatitis C virus is infectious in vivo and can be recultured in vitro. Proc Natl Acad Sci USA 103: 3805–3809
Lindenbach BD, Thiel HJ, Rice CM (2007) Flaviviridae: the viruses and their replication. Lippincott Williams & Wilkins, Philadelphia
Liu S, Yang W, Shen L, Turner JR, Coyne CB, Wang T (2009) Tight junction proteins claudin-1 and occludin control hepatitis C virus entry and are downregulated during infection to prevent superinfection. J Virol 83:2011–2014
Lopez D (2008) PCSK9: an enigmatic protease. Biochim Biophys Acta 1781:184–191
Maecker HT, Levy S (1997) Normal lymphocyte development but delayed humoral immune response in CD81-null mice. J Exp Med 185:1505–1510
Maillard P, Huby T, Andreo U, Moreau M, Chapman J, Budkowska A (2006) The interaction of natural hepatitis C virus with human scavenger receptor SR-BI/Cla1 is mediated by ApoB-containing lipoproteins. FASEB J 20:735–737
Manns MP, Wedemeyer H, Cornberg M (2006) Treating viral hepatitis C: efficacy, side effects, and complications. Gut 55:1350–1359
Masciopinto F, Freer G, Burgio VL, Levy S, Galli-Stampino L, Bendinelli M, Houghton M, Abrignani S, Uematsu Y (2002) Expression of human CD81 in transgenic mice does not confer susceptibility to hepatitis C virus infection. Virology 304:187–196
Mazier D, Landau I, Druilhe P, Miltgen F, Guguen-Guillouzo C, Baccam D, Baxter J, Chigot JP, Gentilini M (1984) Cultivation of the liver forms of Plasmodium vivax in human hepatocytes. Nature 307:367–369
Mazier D, Beaudoin RL, Mellouk S, Druilhe P, Texier B, Trosper J, Miltgen F, Landau I, Paul C, Brandicourt O et al (1985) Complete development of hepatic stages of Plasmodium falciparum in vitro. Science 227:440–442
McKeating JA, Zhang LQ, Logvinoff C, Flint M, Zhang J, Yu J, Butera D, Ho DD, Dustin LB, Rice CM, Balfe P (2004) Diverse hepatitis C virus glycoproteins mediate viral infection in a CD81-dependent manner. J Virol 78:8496–8505
Mee CJ, Grove J, Harris HJ, Hu K, Balfe P, McKeating JA (2008) Effect of cell polarization on hepatitis C virus entry. J Virol 82:461–470
Mee CJ, Harris HJ, Farquhar MJ, Wilson G, Reynolds G, Davis C, van ISC, Balfe P, McKeating JA (2009) Polarization restricts hepatitis C virus entry into HepG2 hepatoma cells. J Virol 83:6211–6221

Meertens L, Bertaux C, Dragic T (2006) Hepatitis C virus entry requires a critical post internalization step and delivery to early endosomes via clathrin-coated vesicles. J Virol 80:11571–11578

Meertens L, Bertaux C, Cukierman L, Cormier E, Lavillette D, Cosset FL, Dragic T (2008) The tight junction proteins claudin-1, -6, and -9 are entry cofactors for hepatitis C virus. J Virol 82:3555–3560

Menard R, Janse C (1997) Gene targeting in malaria parasites. Methods (San Diego, Calif) 13:148–157

Meola A, Sbardellati A, Bruni Ercole B, Cerretani M, Pezzanera M, Ceccacci A, Vitelli A, Levy S, Nicosia A, Traboni C, McKeating J, Scarselli E (2000) Binding of hepatitis C virus E2 glycoprotein to CD81 does not correlate with species permissiveness to infection. J Virol 74:5933–5938

Meuleman P, Hesselgesser J, Paulson M, Vanwolleghem T, Desombere I, Reiser H, Leroux-Roels G (2008) Anti-CD81 antibodies can prevent a hepatitis C virus infection in vivo. Hepatology 48:1761–1768

Meunier JC, Fournillier A, Choukhi A, Cahour A, Cocquerel L, Dubuisson J, Wychowski C (1999) Analysis of the glycosylation sites of hepatitis C virus (HCV) glycoprotein E1 and the influence of E1 glycans on the formation of the HCV glycoprotein complex. J Gen Virol 80: 887–896

Meunier JC, Engle RE, Faulk K, Zhao M, Bartosch B, Alter H, Emerson SU, Cosset FL, Purcell RH, Bukh J (2005) Evidence for cross-genotype neutralization of hepatitis C virus pseudo-particles and enhancement of infectivity by apolipoprotein C1. Proc Natl Acad Sci USA 102: 4560–4565

Meunier JC, Russell RS, Engle RE, Faulk KN, Purcell RH, Emerson SU (2008) Apolipoprotein c1 association with hepatitis C virus. J Virol 82:9647–9656

Michalak JP, Wychowski C, Choukhi A, Meunier JC, Ung S, Rice CM, Dubuisson J (1997) Characterization of truncated forms of hepatitis C virus glycoproteins. J Gen Virol 78:2299–2306

Molina S, Castet V, Fournier-Wirth C, Pichard-Garcia L, Avner R, Harats D, Roitelman J, Barbaras R, Graber P, Ghersa P, Smolarsky M, Funaro A, Malavasi F, Larrey D, Coste J, Fabre JM, Sa-Cunha A, Maurel P (2007) The low-density lipoprotein receptor plays a role in the infection of primary human hepatocytes by hepatitis C virus. J Hepatol 46:411–419

Molina S, Castet V, Pichard-Garcia L, Wychowski C, Meurs E, Pascussi JM, Sureau C, Fabre JM, Sacunha A, Larrey D, Dubuisson J, Coste J, McKeating J, Maurel P, Fournier-Wirth C (2008) Serum-derived hepatitis C virus infection of primary human hepatocytes is tetraspanin CD81 dependent. J Virol 82:569–574

Monazahian M, Bohme I, Bonk S, Koch A, Scholz C, Grether S, Thomssen R (1999) Low density lipoprotein receptor as a candidate receptor for hepatitis C virus. J Med Virol 57:223–229

Montpellier C, Tews BA, Poitrimole J, Rocha-Perugini V, D'Arienzo V, Potel J, Zhang XA, Rubinstein E, Dubuisson J, Cocquerel L (2011) Interacting regions of CD81 and two of its partners, EWI-2 and EWI-2wint, and their effect on hepatitis C virus infection. J Biol Chem 286:13954–13965

Morikawa K, Zhao Z, Date T, Miyamoto M, Murayama A, Akazawa D, Tanabe J, Sone S, Wakita T (2007) The roles of CD81 and glycosaminoglycans in the adsorption and uptake of infectious HCV particles. J Med Virol 79:714–723

Mota MM, Rodriguez A (2000) Plasmodium yoelii: efficient in vitro invasion and complete development of sporozoites in mouse hepatic cell lines. Exp Parasitol 96:257–259

Mota MM, Pradel G, Vanderberg JP, Hafalla JC, Frevert U, Nussenzweig RS, Nussenzweig V, Rodriguez A (2001) Migration of Plasmodium sporozoites through cells before infection. Science 291:141–144

Mueller AK, Camargo N, Kaiser K, Andorfer C, Frevert U, Matuschewski K, Kappe SH (2005) Plasmodium liver stage developmental arrest by depletion of a protein at the parasite-host interface. Proc Natl Acad Sci USA 102:3022–3027

Nakajima H, Cocquerel L, Kiyokawa N, Fujimoto J, Levy S (2005) Kinetics of HCV envelope proteins' interaction with CD81 large extracellular loop. Biochem Biophys Res Commun 328:1091–1100

Nakano I, Fukuda Y, Katano Y, Hayakawa T (1999) Conformational epitopes detected by cross-reactive antibodies to envelope 2 glycoprotein of the hepatitis C virus. J Infect Dis 180: 1328–1333

Nielsen SU, Bassendine MF, Burt AD, Martin C, Pumeechockchai W, Toms GL (2006) Association between hepatitis C virus and very-low-density lipoprotein (VLDL)/LDL analyzed in iodixanol density gradients. J Virol 80:2418–2428

Op De Beeck A, Montserret R, Duvet S, Cocquerel L, Cacan R, Barberot B, Le Maire M, Penin F, Dubuisson J (2000) The transmembrane domains of hepatitis C virus envelope glycoproteins E1 and E2 play a major role in heterodimerization. J Biol Chem 275:31428–31437

Op De Beeck A, Voisset C, Bartosch B, Ciczora Y, Cocquerel L, Keck Z, Foung S, Cosset FL, Dubuisson J (2004) Characterization of functional hepatitis C virus envelope glycoproteins. J Virol 78:2994–3002

Oren R, Takahashi S, Doss C, Levy R, Levy S (1990) TAPA-1, the target of an antiproliferative antibody, defines a new family of transmembrane proteins. Mol Cell Biol 10:4007–4015

Owen DM, Huang H, Ye J, Gale M Jr (2009) Apolipoprotein E on hepatitis C virion facilitates infection through interaction with low-density lipoprotein receptor. Virology 394:99–108

Owsianka A, Clayton RF, Loomis-Price LD, McKeating JA, Patel AH (2001) Functional analysis of hepatitis C virus E2 glycoproteins and virus-like particles reveals structural dissimilarities between different forms of E2. J Gen Virol 82:1877–1883

Owsianka AM, Timms JM, Tarr AW, Brown RJ, Hickling TP, Szwejk A, Bienkowska-Szewczyk K, Thomson BJ, Patel AH, Ball JK (2006) Identification of conserved residues in the E2 envelope glycoprotein of the hepatitis C virus that are critical for CD81 binding. J Virol 80: 8695–8704

Owsianka AM, Tarr AW, Keck ZY, Li TK, Witteveldt J, Adair R, Foung SK, Ball JK, Patel AH (2008) Broadly neutralizing human monoclonal antibodies to the hepatitis C virus E2 glycoprotein. J Gen Virol 89:653–659

Pacheco B, Gomez-Gutierrez J, Yelamos B, Delgado C, Roncal F, Albar JP, Peterson D, Gavilanes F (2006) Membrane-perturbing properties of three peptides corresponding to the ectodomain of hepatitis C virus E2 envelope protein. Biochim Biophys Acta 1758:755–763

Patel AH, Wood J, Penin F, Dubuisson J, McKeating JA (2000) Construction and characterization of chimeric hepatitis C virus E2 glycoproteins: analysis of regions critical for glycoprotein aggregation and CD81 binding. J Gen Virol 81:2873–2883

Patel J, Patel AH, McLauchlan J (2001) The transmembrane domain of the hepatitis C virus E2 glycoprotein is required for correct folding of the E1 glycoprotein and native complex formation. Virology 279:58–68

Perez-Berna AJ, Moreno MR, Guillen J, Bernabeu A, Villalain J (2006) The membrane-active regions of the hepatitis C virus E1 and E2 envelope glycoproteins. Biochemistry 45: 3755–3768

Perez-Berna AJ, Bernabeu A, Moreno MR, Guillen J, Villalain J (2008a) The pre-transmembrane region of the HCV E1 envelope glycoprotein: interaction with model membranes. Biochim Biophys Acta 1778:2069–2080

Perez-Berna AJ, Guillen J, Moreno MR, Gomez-Sanchez AI, Pabst G, Laggner P, Villalain J (2008b) Interaction of the most membranotropic region of the HCV E2 envelope glycoprotein with membranes. Biophysical characterization. Biophys J 94:4737–4750

Perez-Berna AJ, Pabst G, Laggner P, Villalain J (2009) Biophysical characterization of the fusogenic region of HCV envelope glycoprotein E1. Biochim Biophys Acta 1788:2183–2193

Petracca R, Falugi F, Galli G, Norais N, Rosa D, Campagnoli S, Burgio V, Di Stasio E, Giardina B, Houghton M, Abrignani S, Grandi G (2000) Structure-function analysis of hepatitis C virus envelope-CD81 binding. J Virol 74:4824–4830

Pileri P, Uematsu Y, Campagnoli S, Galli G, Falugi F, Petracca R, Weiner AJ, Houghton M, Rosa D, Grandi G, Abrignani S (1998) Binding of hepatitis C virus to CD81. Science 282:938–941

Pinzon-Ortiz C, Friedman J, Esko J, Sinnis P (2001) The binding of the circumsporozoite protein to cell surface heparan sulfate proteoglycans is required for plasmodium sporozoite attachment to target cells. J Biol Chem 276:26784–26791

Ploss A, Evans MJ, Gaysinskaya VA, Panis M, You H, de Jong YP, Rice CM (2009) Human occludin is a hepatitis C virus entry factor required for infection of mouse cells. Nature 457:882–886

Ploss A, Khetani SR, Jones CT, Syder AJ, Trehan K, Gaysinskaya VA, Mu K, Ritola K, Rice CM, Bhatia SN (2010) Persistent hepatitis C virus infection in microscale primary human hepatocyte cultures. Proc Natl Acad Sci USA 107:3141–3145

Podevin P, Carpentier A, Pene V, Aoudjehane L, Carriere M, Zaidi S, Hernandez C, Calle V, Meritet JF, Scatton O, Dreux M, Cosset FL, Wakita T, Bartenschlager R, Demignot S, Conti F, Rosenberg AR, Calmus Y (2010) Production of infectious hepatitis C virus in primary cultures of human adult hepatocytes. Gastroenterology 139:1355–1364

Ponnudurai T, Meuwissen JH, Leeuwenberg AD, Verhave JP, Lensen AH (1982) The production of mature gametocytes of Plasmodium falciparum in continuous cultures of different isolates infective to mosquitoes. Trans R Soc Trop Med Hyg 76:242–250

Popescu CI, Dubuisson J (2009) Role of lipid metabolism in hepatitis C virus assembly and entry. Biol Cell 102:63–74

Potel J, Rassam P, Montpellier C, Kaestner L, Werkmeister E, Tews BA, Couturier C, Popescu CI, Baumert T, Rubinstein E, Dubuisson J, Milhiet P-E, Cocquerel L (2013) EWI-2wint promotes CD81 clustering that abrogates Hepatitis C Virus entry. Cell Microbiol (in press)

Roccasecca R, Ansuini H, Vitelli A, Meola A, Scarselli E, Acali S, Pezzanera M, Ercole BB, McKeating J, Yagnik A, Lahm A, Tramontano A, Cortese R, Nicosia A (2003) Binding of the hepatitis C virus E2 glycoprotein to CD81 is strain specific and is modulated by a complex interplay between hypervariable regions 1 and 2. J Virol 77:1856–1867

Rocha-Perugini V, Montpellier C, Delgrange D, Wychowski C, Helle F, Pillez A, Drobecq H, Le Naour F, Charrin S, Levy S, Rubinstein E, Dubuisson J, Cocquerel L (2008) The CD81 partner EWI-2wint inhibits hepatitis C virus entry. PLoS One 3:e1866

Rocha-Perugini V, Lavie M, Delgrange D, Canton J, Pillez A, Potel J, Lecoeur C, Rubinstein E, Dubuisson J, Wychowski C, Cocquerel L (2009) The association of CD81 with tetraspanin-enriched microdomains is not essential for Hepatitis C virus entry. BMC Microbiol 9:111

Rodrigues CD, Hannus M, Prudencio M, Martin C, Goncalves LA, Portugal S, Epiphanio S, Akinc A, Hadwiger P, Jahn-Hofmann K, Rohl I, van Gemert GJ, Franetich JF, Luty AJ, Sauerwein R, Mazier D, Koteliansky V, Vornlocher HP, Echeverri CJ, Mota MM (2008) Host scavenger receptor SR-BI plays a dual role in the establishment of malaria parasite liver infection. Cell Host Microbe 4:271–282

Rothwangl KB, Manicassamy B, Uprichard SL, Rong L (2008) Dissecting the role of putative CD81 binding regions of E2 in mediating HCV entry: putative CD81 binding region 1 is not involved in CD81 binding. Virol J 5:46

Rubinstein E, Le Naour F, Lagaudriere-Gesbert C, Billard M, Conjeaud H, Boucheix C (1996) CD9, CD63, CD81, and CD82 are components of a surface tetraspan network connected to HLA-DR and VLA integrins. Eur J Immunol 26:2657–2665

Rumin S, Berthillon P, Tanaka E, Kiyosawa K, Trabaud MA, Bizollon T, Gouillat C, Gripon P, Guguen-Guillouzo C, Inchauspe G, Trepo C (1999) Dynamic analysis of hepatitis C virus replication and quasispecies selection in long-term cultures of adult human hepatocytes infected in vitro. J Gen Virol 80(Pt 11):3007–3018

Russell RS, Kawaguchi K, Meunier JC, Takikawa S, Faulk K, Bukh J, Purcell RH, Emerson SU (2009) Mutational analysis of the hepatitis C virus E1 glycoprotein in retroviral pseudoparticles and cell-culture-derived H77/JFH1 chimeric infectious virus particles. J Viral Hepat 16:621–632

Sala-Valdes M, Ursa A, Charrin S, Rubinstein E, Hemler ME, Sanchez-Madrid F, Yanez-Mo M (2006) EWI-2 and EWI-F link the tetraspanin web to the actin cytoskeleton through their direct association with ezrin-radixin-moesin proteins. J Biol Chem 281:19665–19675

Santos JM, Lebrun M, Daher W, Soldati D, Dubremetz JF (2009) Apicomplexan cytoskeleton and motors: key regulators in morphogenesis, cell division, transport and motility. Int J Parasitol 39:153–162

Scarselli E, Ansuini H, Cerino R, Roccasecca RM, Acali S, Filocamo G, Traboni C, Nicosia A, Cortese R, Vitelli A (2002) The human scavenger receptor class B type I is a novel candidate receptor for the hepatitis C virus. EMBO J 21:5017–5025

Schwarz AK, Grove J, Hu K, Mee CJ, Balfe P, McKeating JA (2009) Hepatoma cell density promotes claudin-1 and scavenger receptor BI expression and hepatitis C virus internalization. J Virol 83:12407–12414

Seigneuret M (2006) Complete predicted three-dimensional structure of the facilitator transmembrane protein and hepatitis C virus receptor CD81: conserved and variable structural domains in the tetraspanin superfamily. Biophys J 90:212–227

Selby MJ, Glazer E, Masiarz F, Houghton M (1994) Complex processing and protein: protein interactions in the E2:NS2 region of HCV. Virology 204:114–122

Sharma NR, Mateu G, Dreux M, Grakoui A, Cosset FL, Melikyan GB (2011) Hepatitis C virus is primed by CD81 protein for low pH-dependent fusion. J Biol Chem 286:30361–30376

Shaw ML, McLauchlan J, Mills PR, Patel AH, McCruden EA (2003) Characterisation of the differences between hepatitis C virus genotype 3 and 1 glycoproteins. J Med Virol 70:361–372

Silver DL, Wang N, Xiao X, Tall AR (2001) High density lipoprotein (HDL) particle uptake mediated by scavenger receptor class B type 1 results in selective sorting of HDL cholesterol from protein and polarized cholesterol secretion. J Biol Chem 276:25287–25293

Silvie O, Rubinstein E, Franetich JF, Prenant M, Belnoue E, Renia L, Hannoun L, Eling W, Levy S, Boucheix C, Mazier D (2003) Hepatocyte CD81 is required for Plasmodium falciparum and Plasmodium yoelii sporozoite infectivity. Nat Med 9:93–96

Silvie O, Franetich JF, Charrin S, Mueller MS, Siau A, Bodescot M, Rubinstein E, Hannoun L, Charoenvit Y, Kocken CH, Thomas AW, Van Gemert GJ, Sauerwein RW, Blackman MJ, Anders RF, Pluschke G, Mazier D (2004) A role for apical membrane antigen 1 during invasion of hepatocytes by Plasmodium falciparum sporozoites. J Biol Chem 279:9490–9496

Silvie O, Charrin S, Billard M, Franetich JF, Clark KL, van Gemert GJ, Sauerwein RW, Dautry F, Boucheix C, Mazier D, Rubinstein E (2006a) Cholesterol contributes to the organization of tetraspanin-enriched microdomains and to CD81-dependent infection by malaria sporozoites. J Cell Sci 119:1992–2002

Silvie O, Greco C, Franetich JF, Dubart-Kupperschmitt A, Hannoun L, van Gemert GJ, Sauerwein RW, Levy S, Boucheix C, Rubinstein E, Mazier D (2006b) Expression of human CD81 differently affects host cell susceptibility to malaria sporozoites depending on the Plasmodium species. Cell Microbiol 8:1134–1146

Silvie O, Franetich JF, Boucheix C, Rubinstein E, Mazier D (2007) Alternative invasion pathways for Plasmodium berghei sporozoites. Int J Parasitol 37:173–182

Slater-Handshy T, Droll DA, Fan X, Di Bisceglie AM, Chambers TJ (2004) HCV E2 glycoprotein: mutagenesis of N-linked glycosylation sites and its effects on E2 expression and processing. Virology 319:36–48

Snounou G, Gruner AC, Muller-Graf CD, Mazier D, Renia L (2005) The Plasmodium sporozoite survives RTS, S vaccination. Trends Parasitol 21:456–461

Stipp CS, Kolesnikova TV, Hemler ME (2001a) EWI-2 is a major CD9 and CD81 partner and member of a novel Ig protein subfamily. J Biol Chem 276:40545–40554

Stipp CS, Orlicky D, Hemler ME (2001b) FPRP, a major, highly stoichiometric, highly specific CD81- and CD9- associated protein. J Biol Chem 276:4853–4862

Strebel K, Luban J, Jeang KT (2009) Human cellular restriction factors that target HIV-1 replication. BMC Med 7:48

Tham TN, Gouin E, Rubinstein E, Boucheix C, Cossart P, Pizarro-Cerda J (2009) Tetraspanin CD81 is required for Listeria monocytogenes invasion. Infect Immun 78:204–209

Timpe JM, Stamataki Z, Jennings A, Hu K, Farquhar MJ, Harris HJ, Schwarz A, Desombere I, Roels GL, Balfe P, McKeating JA (2008) Hepatitis C virus cell-cell transmission in hepatoma cells in the presence of neutralizing antibodies. Hepatology 47:17–24

Trager W, Jensen JB (1976) Human malaria parasites in continuous culture. Science 193:673–675

Troesch M, Meunier I, Lapierre P, Lapointe N, Alvarez F, Boucher M, Soudeyns H (2006) Study of a novel hypervariable region in hepatitis C virus (HCV) E2 envelope glycoprotein. Virology 352:357–367

Tscherne DM, Jones CT, Evans MJ, Lindenbach BD, McKeating JA, Rice CM (2006) Time- and temperature-dependent activation of hepatitis C virus for low-pH-triggered entry. J Virol 80:1734–1741

van Dijk MR, Douradinha B, Franke-Fayard B, Heussler V, van Dooren MW, van Schaijk B, van Gemert GJ, Sauerwein RW, Mota MM, Waters AP, Janse CJ (2005) Genetically attenuated, P36p-deficient malarial sporozoites induce protective immunity and apoptosis of infected liver cells. Proc Natl Acad Sci USA 102:12194–12199

van Schaijk BC, Janse CJ, van Gemert GJ, van Dijk MR, Gego A, Franetich JF, van de Vegte-Bolmer M, Yalaoui S, Silvie O, Hoffman SL, Waters AP, Mazier D, Sauerwein RW, Khan SM (2008) Gene disruption of Plasmodium falciparum p52 results in attenuation of malaria liver stage development in cultured primary human hepatocytes. PLoS One 3:e3549

Vieyres G, Thomas X, Descamps V, Duverlie G, Patel AH, Dubuisson J (2010) Characterization of the envelope glycoproteins associated with infectious hepatitis C virus. J Virol 84: 10159–10168

Voisset C, Callens N, Blanchard E, Op De Beeck A, Dubuisson J, Vu-Dac N (2005) High density lipoproteins facilitate hepatitis C virus entry through the scavenger receptor class B type I. J Biol Chem 280:7793–7799

Voisset C, Lavie M, Helle F, Op De Beeck A, Bilheu A, Bertrand-Michel J, Terce F, Cocquerel L, Wychowski C, Vu-Dac N, Dubuisson J (2008) Ceramide enrichment of the plasma membrane induces CD81 internalization and inhibits hepatitis C virus entry. Cell Microbiol 10:606–617

von Hahn T, Lindenbach BD, Boullier A, Quehenberger O, Paulson M, Rice CM, McKeating JA (2006) Oxidized low-density lipoprotein inhibits hepatitis C virus cell entry in human hepatoma cells. Hepatology 43:932–942

Wack A, Soldaini E, Tseng C, Nuti S, Klimpel G, Abrignani S (2001) Binding of the hepatitis C virus envelope protein E2 to CD81 provides a co-stimulatory signal for human T cells. Eur J Immunol 31:166–175

Wakita T, Pietschmann T, Kato T, Date T, Miyamoto M, Zhao Z, Murthy K, Habermann A, Krausslich HG, Mizokami M, Bartenschlager R, Liang TJ (2005) Production of infectious hepatitis C virus in tissue culture from a cloned viral genome. Nat Med 11:791–796

Washburn ML, Bility MT, Zhang L, Kovalev GI, Buntzman A, Frelinger JA, Barry W, Ploss A, Rice CM, Su L (2011) A humanized mouse model to study hepatitis C virus infection, immune response, and liver disease. Gastroenterology 140:1334–1344

Weiner AJ, Brauer R, Rosenblatt J, Richman KH, Tung J, Crawford K, Bonino F, Saracco G, Choo Q-L, Houghton M, Han JH (1991) Variable and hypervariable domains are found in the regions of HCV corresponding to the flavivirus envelope and NS1 proteins and the pestivirus envelop glycoproteins. Virology 180:842–848

Witteveldt J, Evans MJ, Bitzegeio J, Koutsoudakis G, Owsianka AM, Angus AG, Keck ZY, Foung SK, Pietschmann T, Rice CM, Patel AH (2009) CD81 is dispensable for hepatitis C virus cell-to-cell transmission in hepatoma cells. J Gen Virol 90:48–58

World Health Organization, Geneva (2009) Malaria Report 2009

Xu Z, Choi J, Yen TS, Lu W, Strohecker A, Govindarajan S, Chien D, Selby MJ, Ou J (2001) Synthesis of a novel hepatitis C virus protein by ribosomal frameshift. EMBO J 20: 3840–3848

Yagnik AT, Lahm A, Meola A, Roccasecca RM, Ercole BB, Nicosia A, Tramontano A (2000) A model for the hepatitis C virus envelope glycoprotein E2. Proteins 40:355–366

Yalaoui S, Huby T, Franetich JF, Gego A, Rametti A, Moreau M, Collet X, Siau A, van Gemert GJ, Sauerwein RW, Luty AJ, Vaillant JC, Hannoun L, Chapman J, Mazier D, Froissard P (2008a) Scavenger receptor BI boosts hepatocyte permissiveness to Plasmodium infection. Cell Host Microbe 4:283–292

Yalaoui S, Zougbede S, Charrin S, Silvie O, Arduise C, Farhati K, Boucheix C, Mazier D, Rubinstein E, Froissard P (2008b) Hepatocyte permissiveness to Plasmodium infection is conveyed by a short and structurally conserved region of the CD81 large extracellular domain. PLoS Pathog 4:e1000010

Yamauchi LM, Coppi A, Snounou G, Sinnis P (2007) Plasmodium sporozoites trickle out of the injection site. Cell Microbiol 9:2093

Yang W, Qiu C, Biswas N, Jin J, Watkins SC, Montelaro RC, Coyne CB, Wang T (2008) Correlation of the tight junction-like distribution of claudin-1 to the cellular tropism of HCV. J Biol Chem 283(13):8643–8653

Zeisel MB, Koutsoudakis G, Schnober EK, Haberstroh A, Blum HE, Cosset FL, Wakita T, Jaeck D, Doffoel M, Royer C, Soulier E, Schvoerer E, Schuster C, Stoll-Keller F, Bartenschlager R, Pietschmann T, Barth H, Baumert TF (2007) Scavenger receptor class B type I is a key host factor for hepatitis C virus infection required for an entry step closely linked to CD81. Hepatology 46:1722–1731

Zhang J, Randall G, Higginbottom A, Monk P, Rice CM, McKeating JA (2004) CD81 is required for hepatitis C virus glycoprotein-mediated viral infection. J Virol 78:1448–1455

Zhang YY, Zhang BH, Ishii K, Liang TJ (2010) Novel function of CD81 in controlling hepatitis C virus replication. J Virol 84:3396–3407

Zheng A, Yuan F, Li Y, Zhu F, Hou P, Li J, Song X, Ding M, Deng H (2007) Claudin-6 and claudin-9 function as additional coreceptors for hepatitis C virus. J Virol 81:12465–12471

Zhong J, Gastaminza P, Cheng G, Kapadia S, Kato T, Burton DR, Wieland SF, Uprichard SL, Wakita T, Chisari FV (2005) Robust hepatitis C virus infection in vitro. Proc Natl Acad Sci USA 102:9294–9299

Chapter 15
Tetraspanins as Facilitators of Viral and Cellular Information Transfer

Markus Thali

Abstract Cells that are infected by a virus can infect other cells by passing on the viral genome. Such transfer of viral genetic information can occur by fusion of infected and uninfected cells, or through extracellular particles that are shed from infected cells and which carry the viral genome together with viral structural proteins and/or enzymes to uninfected target cells. The transfer of viral information, either by cell-cell fusion or via viral particles, requires the coordination of numerous membrane-based functions, several of which are regulated by tetraspanins.

In many ways, viral particles resemble cell-derived vesicles such as, for example, exosomes, which are known to transport (structural, biochemical, and genetic) information through extracellular space. With this brief essay, in addition to reviewing what we currently know about various tetraspanin functions during the replication of enveloped viruses, I will also discuss some of the similarities between viral and cellular information transfer processes. Emphasis will be placed on how, in either case, tetraspanins can facilitate short- and long-range transmission as well as transfer via cell-cell fusions.

As is true for any biological function, and as so elegantly expressed almost four decades ago by the geneticist Theodosius Dobzhansky, who, referring more generally to living things, wrote that "nothing in biology makes sense except in the light of evolution" (Dobzhansky 1964), a true understanding of how viruses (here discussed initially as genetic entities, not pathogens) are propagated, and how cells communicate with each other, requires that we put things into evolutionary perspective. An introductory, more general discussion of the relationship between viruses and cells seems necessary within the context of this review article for these two

---

M. Thali (✉)
Department of Microbiology and Molecular Genetics, College of Medicine and CALS, University of Vermont, Burlington, VT 05405-0084, USA
e-mail: markus.thali@uvm.edu

F. Berditchevski and E. Rubinstein (eds.), *Tetraspanins*, Proteins and Cell Regulation 9, DOI 10.1007/978-94-007-6070-7_15,

reasons: (a) enveloped viruses obviously interact with many different cellular proteins, and an understanding of how and why viruses use tetraspanins for their successful propagation will be facilitated if we compare viral and cellular functions of these proteins; (b) the Human Genome Project and other recent sequencing efforts revealed that viruses and virus-like genetic entities have an enormous presence, and the latter are tightly intertwined with the non-viral genome, likely reflecting (as briefly discussed below) that they predate cellular genomes and indeed provided the basis for non-viral, i.e. cellular genomes (Koonin et al. 2006; Holmes 2011). Nevertheless, many biomedical researchers and most textbooks unfortunately still adhere to the idea that viruses are cellular genes that "escaped" from cells, or even that they are the remnants of formerly independent cellular microbes and thus that they constitute, in the immunological sense, "non-self" genetic entities. Acknowledging that present day viruses and cells have the same root(s), and that viruses and virus-like genetic elements, overall, more appropriately should be seen as genetic symbionts, does affect how we interpret their interactions with cells. In turn, such interpretations are likely to lead to the development of more sophisticated treatments of diseases caused by pathogenic viruses.

What, then, do we know about the origin of viruses, about their evolution? And what is known about the interaction of cells in early life?

15.1 Origin of Viruses

While there is still quite some debate about the early history of the main groups of viruses, particularly about when and how DNA viruses entered the picture (e.g. Forterre 2005), it is now generally accepted among evolutionary biologists that viruses, (again, in this introduction discussed as genetic entities, and also including so-called endogenous viruses, i.e. viruses that never leave cells), are as old as cells, or that they even predate cells (for a very recent review, e.g. Holmes 2011). This notion may seem at odds with the textbook definition of viruses, which describes them as "obligate intracellular parasites". However, the textbook definition typically is used in a context where present day viruses are described as (pathogenic) invaders of cells, and thus clearly not with an eye on their origin and thus the true nature of these genetic entities. Most biologists who study early life however, would probably agree with the notion that early viruses, or what can be considered being the ancestors of present day viruses, were RNA-based, self-replicating genetic entities. Part of these self-replicating elements, which together formed the primordial gene pool, evolved and "stabilized" to become cellular genomes (still RNA-based first, and eventually transformed into DNA-based genomes), while others evolved to become present day viruses (described e.g. in Koonin et al. 2006; Villarreal and Witzany 2010). Because the latter, i.e. the viruses, continue to use cellular resources for their replication, one can describe them as "parasitic" entities, although given their very significant presence in e.g. the human genome (see below), and given their enormous contribution to the evolution of present day species, and, perhaps, even an involvement in ontogeny (e.g. Muotri et al. 2010; Studer 2010), the term still seems too narrow and ultimately inappropriate, as it introduces a biased view of things which is detrimental to a real understanding of the virosphere and to some extent also of pathogenic viruses.

15.2 Cell-Cell Communication

Another important aspect of our current understanding of early life is the interdependence of cells, i.e. of what are generally considered to be the basic units of life. Early cells, and also what were likely their predecessors, assemblages of primitive genetic elements, relatively freely shared genes (RNA- and later DNA-based), proteins and metabolites (e.g. see Woese 2002). While it is known for quite some time that present day bacteria still rather freely exchange genetic information, and that horizontal gene transfer (HGT) is ongoing and prevalent to the point that it is difficult to define bacterial species, only since the advent of the Human Genome Project, and the sequencing of additional, non-human genomes, are we beginning to understand that eukaryotic genomes are also far from being stable entities (Lander 2011). We now know that about half of the human genome consists of virus-like (often repetitive) sequences, and while some of these may be the result of infections by exogenous viruses that have crossed the species barrier at some earlier point in time, the majority of them most likely are remnants and expansions of the early genetic make-up of cells, as discussed above. Even more importantly, though we are only at the very beginning of understanding their complexity, it is already clear that these and other sequences, including non-coding genetic elements, are responsible for the evolution of novel classes of animals, or of specific species, including humans (Lander 2011). Viviparous mammals, for example, are thought to have evolved as a consequence of the infection of germ line cells of pre-mammals by retroviruses, whose envelope glycoproteins (which can trigger the fusion of plasma membranes of adjacent cells) were subsequently adapted by cells for the formation of syncytiotrophoblast, and thus placentas (e.g. Mi et al. 2000).

Finally, and directly relevant for the discussion of specific tetraspanin functions in this chapter, it is becoming clear that the flux of genetic information is not restricted to the evolutionary timescale, or to HGT between bacteria, but that transfer of genetic (and of course also biochemical) information also takes place within multicellular organisms, and between the myriad symbionts that cohabit multicellular organisms (for a further discussion of this, e.g. see (Goldenfeld and Woese 2007), also see below).

15.3 Tetraspanins and Other Scaffold Proteins as Mediators of Information Transfer

Cellular membranes are central to the above-described exchange of information between cells, whether viral or cellular in nature. Also, in order to process, i.e. transmit, biochemical and genetic information properly, membrane-based regulatory

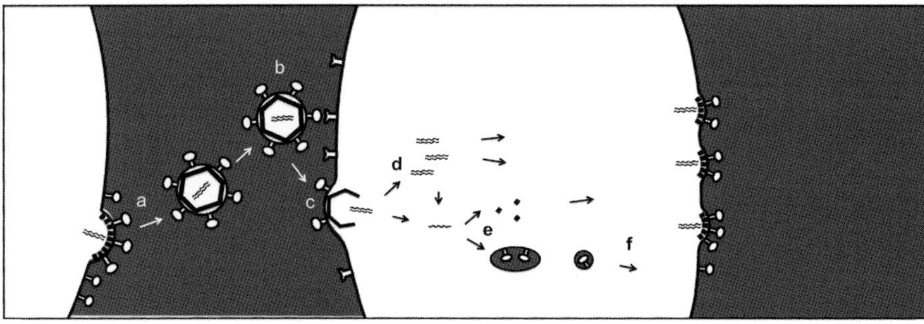

Fig. 15.1 Transmission and multiplication of an enveloped virus. Tetraspanins have been implicated in the regulation of viral assembly and release (**a**), particle attachment (**b**) and entry (**c**), genome replication (**d**) and expression (**e**), and transport (**f**) of viral components

units had to evolve, and it appears reasonable to assume that families of scaffold proteins, such as tetraspanins, did indeed evolve for that reason. While several chapters in this volume review their function, i.e. describe mechanism by which distinct members of the tetraspanin family of proteins control specific processes, in the remainder of this chapter I will discuss how some of these cellular regulatory activities are also used to control the transmission of viral genetic information. However, our understanding of how tetraspanins function during the replication of various viruses, with the exception of HCV (see the chapter by Cocquerel and Silvie), is still very much in its infancy, and we are thus only starting to comprehend similarities between tetraspanin activities that regulate cellular functions and the role of these proteins in life cycles of various viruses (Martin et al. 2005; van Spriel and Figdor 2010; Thali 2009, 2011). In Fig. 15.1, I provide a scheme for the transmission event and replication cycle of a generic enveloped virus. The figure illustrates that tetraspanins have functions at different steps of viral life cycles. Indeed, this multifunctionality complicates the analysis of their roles to some extent, as e.g. ablation of a specific tetraspanin can at the same time lead to repression of one step but enhancement of another step in the viral replication, resulting in unclear phenotypes.

15.4 Long-Range Information Transfer Via Viral Particles, Exosomes and Other Cellular Microvesicles

Decades ago, and thus long before results of the Human Genome Project told us that viruses and virus-like genetic entities make up a very significant fraction of the genome of cells, Baltimore, Campbell, Darnell, and Luria in the last paragraph of the textbook General Virology (Luria et al. 1978), declared: “A virus is essentially part of a cell”. Despite this (in retrospect) visionary statement, many scientists, including many virologists, think of viruses primarily as entities that exist outside of cells, i.e. of particles. Therefore, I will first discuss what we know about the involvement of tetraspanins in the transport of viral information (genes, proteins) via extracellular particles.

15.4.1 Tetraspanins and Release of Viral Particles

The first virus that was shown to be associated with a tetraspanin was HIV-1. Almost two decades ago, two groups reported that HIV-1 particles are enriched in CD63 (Meerloo et al. 1992, 1993; Orentas and Hildreth 1993), and numerous additional reports have since confirmed that finding (e.g. see (Chertova et al. 2006), reviewed in (Thali 2009)). Though it was demonstrated that the incorporation was specific, only recently several groups tested if this tetraspanin, and some other members of the family (CD9, CD81 and CD82) which are also acquired by virions, play a functional role in the formation of HIV-1 particles. Conflicting results have been reported about CD81, with one group showing a positive role for CD81 in the release of virus from Molt T cells, but two other groups not finding evidence for HIV-1 particle release enhancement by this tetraspanin in 293T cells or in Jurkat T cells (Sato et al. 2008; Krementsov et al. 2009; Grigorov et al. 2009). Further, none of the studies that measured if altering of CD63 levels affected virus particle release found a positive effect of that tetraspanin (Sato et al. 2008; Krementsov et al. 2009; Ruiz-Mateos et al. 2008). Similarly, despite the presence of CD9 at exit sites of another retrovirus, feline immunodeficiency virus (FIV) (de Parseval et al. 1997, and references therein) and at the release site of canine distemper virus (CDV) (Singethan et al. 2008), (Loffler et al. 1997) this tetraspanin also does not appear to influence release of viral particles (Schneider-Schaulies and Thali, unpublished observations). This latest observation regarding FIV was particularly surprising, as the paper by Elder and colleagues (de Parseval et al. 1997) very clearly documented that the treatment of FIV-producing cells with an anti-CD9 mAb resulted in reduced amounts of released virus, something that was subsequently confirmed by us in a study in which we also showed inhibition of HIV-1 release by another anti-CD9 antibody (Khurana et al. 2007). More recent work however revealed that the antibodies used in those two studies (Vpg15 and K41, respectively) were unique, as they both cluster CD9 and other associated tetraspanins at the plasma membrane of closely aligned cells. This may have lead to an unspecific interference with the viral budding process and/or a trapping of viral particles in a dense network of newly formed microvilli (which we called "microvilli zippers") at the cell-cell junction (Singethan et al. 2008). (Indeed, further studies of how binding of K41 to CD9 leads to the formation of microvilli may help revealing the mechanisms of microvilli formation per se, a process in which CD9 is known to play a key role (Runge et al. 2007)).

In contrast to the lack of support by tetraspanins of retroviral release (with the exception of the reported CD81-induced enhancement of HIV-1 shedding from certain T cells, see (Grigorov et al. 2009)), and as revealed by more recent investigations, members of the tetraspanin family may function as promoters of particle release for two other viruses: Tetraspanin 7, through an interaction with the viral capsid protein HP26, was recently shown to enhance the release of herpes simplex virus type 1 (HSV-1) (Wang et al. 2010), and CD9 and CD81 have been shown to be specifically incorporated into influenza virus particles (Shaw et al. 2008), suggesting the possibility that these tetraspanins promote particle shedding. In both cases, the (potential) mechanisms by which these tetraspanins (might) enhance particle release have yet to be elucidated. Further, a very recent report (Verweij et al. 2011) suggests that the tetraspanin CD63 can regulate the release of a viral component: while not addressing yet molecular mechanisms either, it documents that the viral oncogene latent membrane protein 1 (LMP1) of Epstein Barr virus associates with CD63, thus allowing LMP1 to be secreted via exosomes. This downregulation of LMP1 (due to its CD63-mediated secretion) leads to diminished NF-κB-induced cell activation, which is thought to favor virus persistence.

15.4.2 Tetraspanins in Viral Particles Can Modify Their Infectivity

Given that some tetraspanins are clearly enriched in HIV-1 particles, yet do not appear to play a role in release of that virus, several groups have asked if their presence makes the virions more infectious, e.g. by increasing their fusogenicity. So far results do not support this idea: indeed, the opposite is true, as overexpression of tetraspanins in producer cells and thus their enhanced incorporation into HIV-1 particles, while not affecting particle binding to target cells (Sato et al. 2008), renders them less infectious. Conversely, ablation of tetraspanins and thus reduced amounts of particle-associated tetraspanins correlates with increased infectivity (Sato et al. 2008; Krementsov et al. 2009). The situation might be different though for influenza virus: as discussed above, this virus specifically incorporates CD9 and CD81, and given that these tetraspanins promote oocyte-spermatozoa fusion (see the chapter by Boucheix), they might also play positive roles in the entry process of influenza virus.

15.4.3 Tetraspanins in Exosomes

Interestingly, shortly after we learned about the presence of tetraspanins in virions, it was also reported that exosomes are enriched in these scaffold proteins (Escola et al. 1998). Exosomes are extracellular vesicles of 30–100 nm that are secreted by most of the cells. Biogenesis of exosomes involves membrane budding into early/late endosomes, thus forming multivesicular bodies or multivesicular endosomes (MVBs/MVEs). Upon fusion of MVBs/MVEs with the plasma membrane, exosomes are shed into extracellular space (for an early review of these vesicles, particularly of their composition, see (Thery et al. 2002), see also the chapter by Ashman and Zoller for more information on exosome functions). At first sight, viral particles and exosomes, besides their acquisition of tetraspanins, do not appear to have much in common, indeed even their sizes do not really match (Pelchen-Matthews et al. 2004). Nevertheless, probably also because both use the same machinery for their formation (e.g. see Morita and Sundquist 2004), it was hypothesized that, at least functionally, exosomes and HIV-1 could be related (Nguyen et al. 2003; Gould et al. 2003), and more recent reports showed that, like viruses, exosomes can transport genetic information (in the form of mRNAs and microRNAs) (e.g. see Valadi et al. 2007). Further, and even more impressively, the glycome of HIV-1 released from T cells is identical to that of exosomes released from the same cells (Valadi et al. 2007; Krishnamoorthy et al. 2009), lending very strong support to the idea that viruses and exosomes can use the same cellular pathways for their generation. Exosomes are not the only microvesicles that can be produced by cells. Indeed we learn more and more about transfer of material or signals from cell-to-cell via small vesicles (e.g. Mack et al. 2000; Gillette et al. 2009), and it is already well established that some of these vesicles play important roles in various physiological processes (e.g. Cocucci et al. 2009). Importantly, such vesicles, whether they carry mRNA and/or microRNAs or merely proteins, can clearly modify the behavior of cells that are distant from the cell that released the vesicles, and in that regard they are biological entities that behave like viral particles.

What roles do tetraspanins play in information transfer via exosomes? To the best of our knowledge, no mechanistic studies have addressed that question yet. However, given that some tetraspanins can promote the release of certain viral particles, as discussed above, it would seem conceivable that they also can enhance exosome formation. Indeed, CD9 and CD82 expression was recently shown to augment the release of β-catenin-containing exosomes (Chairoungdua et al. 2010). Interestingly, like the above-mentioned CD63-induced release of EBV's LMP1, such enhancement of β-catenin-containing exosome release by CD9 and CD82 leads to reduced cell signaling. Do tetraspanins also play a role once they have been incorporated into exosomes? Based on the fact that tetraspanin incorporation can render viral particles less fusogenic and thus can inhibit their entry into target cells, it would appear likely that they can also regulate exosome-plasma membrane fusion processes. Indeed in one case, they are known to do that: for exosomes that are released from eggs before fertilization, a member of the tetraspanin family (CD9) was shown to act as a positive fusion regulator, as will be briefly discussed below (Sect. 15.6) because of similarities to virus-induced cell-cell fusion.

15.4.4 Tetraspanins and Virus Entry into Cells

So far I have discussed roles that tetraspanins play (or not) in the release of viral particles and cellular membrane vesicles, and their role in mediating the release of the viral genome into target cells. How about tetraspanins that are present at the surface of the target cells, do they have functions there as well, similar to e.g. how one of them (CD81) acts as co-receptor for HCV (see the chapter by Cocquerel and

Silvie)? While no information is available yet regarding attachment and entry of cellular vesicles, such as exosomes, several studies have shown that tetraspanins on target cells regulate virus attachment and entry. Two independent studies have suggested that CD63 and other tetraspanins can enhance HIV-1 uptake by macrophages (von Lindern et al. 2003; Ho et al. 2006), whereas one other study showed that their presence at the surface of T lymphocytes prevents fusion of viral and (T) cellular membranes (Gordon-Alonso et al. 2006). While nothing is known yet about how the presence of tetraspanins enhances virus uptake into macrophages, data presented in a recent study indirectly suggest that they could do so e.g. by negatively regulating virus-T cell fusions by re-organizing the receptors for HIV-1, i.e. CD4 and chemokine receptors (Barrero-Villar et al. 2009). Also, CD63 is engaged in the trafficking of CXCR4, one of the co-receptors for HIV-1, to the cell surface, and may thus co-regulate its surface levels and thus the permissiveness for virus entry (Yoshida et al. 2008). Perhaps comparably, CD63 and CD151 have been implicated in organizing the entry site for human papillomavirus (HPV) (Spoden et al. 2008), and a recent siRNA screen revealed that the presence of CD81 at the surface of target cells is critical for influenza virus entry steps (Konig et al. 2010). It should be pointed out though that in neither of these two cases (HPV and influenza virus) do tetraspanins act as co-receptors for the respective viruses.

15.5 Short-Range Information Transfer at Immunological and Virological Synapses

Cellular microvesicles, including exosomes, by using the vasculature as gateway, can travel long distances within an organism. This means that, in principal, just like viruses, certain microvesicles can even be transferred from one organism to another. However, as far as we know, many of them do not travel far at all, again resembling viral particles (as will be discussed below). Rather, they deliver their message (in the form of proteins or nucleic acids) to cells in the near vicinity of cells releasing them.

Interestingly, and as first documented in a seminal study almost three decades ago (Rodriguez-Boulan et al. 1983), it appears as if viruses fall into either one of two categories: some viruses release their newly formed particles at surface areas where the producer cell contacts a potential target cell, while others seem to be released almost exclusively at free surfaces. Delivery to target cells that are aligned with the producer cell appears to be a major route for the transmission of the human retroviruses HTLV-1 and HIV-1 and indeed for many other viruses that cause systemic infections of the host, such as HSV, measles virus etc. Research over the past two decades has established that these viruses are released from infected (and thus virus producing) cells at specialized cell-cell surface areas. Because these sites are used for the transmission of information, and also because of certain structural similarities with different types of synapses, relatively recently they have been dubbed "virological synapse" (VS) (e.g. (Jolly et al. 2004)). At the VS, the newly released viral particles need to travel only a few nanometers, if that, before reaching the surface of the uninfected (target) cell. While this transfer mode is very efficient, it is not without risk for both the producer and the target cell: the presence of the viral envelope glycoprotein on the pre-synaptic plasma membrane together with the presence of the viral receptors on the post-synaptic plasma membrane, in principal would allow the two cells to fuse with each other. Such fusion of producer and target cells, except for under certain circumstances (see below), would probably be detrimental to virus spread because it leads to the formation of so-called syncytia. Syncytia, however, while being able to produce progeny virus, are relatively short-lived entities and can thus be viewed as dead-ends for virus dissemination. Various tetraspanins have been shown to inhibit fusion (for a recent review, see (Fanaei et al. 2011), also, see the chapter by Hemler) and it thus appears likely that viruses such as HTLV-1 and HIV-1 evolved to recruit them to the viral pre-and post synapse for exactly that reason, i.e. because they inhibit cell-cell fusion (Weng et al. 2009). I write "evolved to recruit" because recent data by our group (Krementsov et al. 2011) and also by Ono and colleagues (Hogue et al. 2011), demonstrate the non-randomness of this process: HIV-1 Gag multimerization, which precedes viral budding, leads to a trapping of various tetraspanins at the future viral exit site, thus, tetraspanin accumulation at the pre-synapse is orchestrated by viral components. Such clustering of tetraspanins at the budding site comes at a cost, however. As mentioned above, tetraspanin incorporation into newly formed viral particles also renders them less fusogenic. However, given that those newly formed particles, if released at the VS, are placed right next to the target cell, even a reduced fusogenicity is likely enough to secure infection of the target cells and thus continued spread of the virus. In that regard it is interesting to see that the aforementioned transfer of certain cellular vesicles (Gillette et al. 2009) takes place also at sites that are enriched in CD63: perhaps the presence of this tetraspanin also allows the cells to separate (without fusion) upon vesicle transfer.

Tetraspanins are recruited not only to the viral budding site, i.e. to the viral pre-synapse, but, as mentioned above, they are also present at the target cell surface, where they may associate with viral receptors (e.g. CD4, a receptor for HIV-1, Imai et al. 1995). Similar to their anti-fusogenic function at the pre-synapse, receptor-associated tetraspanins may inhibit membrane fusions from taking place (likely by different mechanisms). While such fusion prevention precludes the entry of some particles, as mentioned before, given the intimate physical association with the producer cell, this will probably not hinder the virus from infecting the cell. Indeed, just as what was hypothesized regarding tetraspanins at the pre-synapse, we reason that their presence at the post-synapse also ultimately supports virus transmission from cell-to-cell and thus virus spread. It seems even plausible that the distribution of tetraspanins at the surface of both producer and the target cell, i.e. at the pre- and the post-synapse, is fine-tuned such that they can function primarily as inhibitors of cell-cell fusion, while not interfering too much with the fusion of viral and cellular membranes. Such a sophisticated arrangement of synapse molecules is not without precedent: neural synapses clearly are spatially organized (e.g. Gerrow and El-Husseini 2006), and research over the past decade has shown an equally sophisticated organization of yet another synapse, the so-called immune synapse (IS). In the following I will briefly discuss the relationship between the IS and the VS, also because of the similarities between the IS and the VS formed between uninfected cells and cell infected by primate retroviruses such as HIV-1 and HTLV-1 (for a recent review, see also e.g. Sattentau 2008).

ISs are formed between antigen-presenting cells (APCs) and T lymphocytes as well as between killer T cells and their target cells (which in principal can include also other T cells). Likely because APC-T cell synapses are better characterized, VSs are typically compared with these synapses rather than killer-target cell synapses, even when virus transmission between infected and uninfected T cells are described. What does the (T cell-T cell) VS have in common with the ISs formed between APC and T cells? Evidently the composition of the target cell surface is at least partially the same: the post-synaptic plasma membrane contains T cell-specific membrane proteins including CD4 and certain chemokine receptors as well as adhesion molecules such as the lymphocyte function-associated antigen 1 (LFA1). In addition to these factors, which interact with partner molecules in trans, i.e. receptors/ligands situated at the pre-synapse, the synapses include tetraspanins (e.g. CD81) which were reported to act as co-organizers. For example, studies have shown that CD81, which can act as a co-stimulator in T cells (see Levy and Shoham 2005 for a review), was shown to be redistributed during antigen presentation (e.g. (Mittelbrunn et al. 2002)). That co-stimulatory potential of CD81 may also bear fruit at the VS: it has been demonstrated that treatment of HIV-1-infected T cells with an anti-CD81 antibody leads to enhanced expression of the integrated viral genome (Tardif and Tremblay 2005). In addition, and as mentioned above, it appears likely that this tetraspanin, and possibly other members of the family, are involved in the reorganization of CD4, thus co-regulating viral entry at the post-synapse (Gordon-Alonso et al. 2006).—At first sight, the pre-synapses of the APC-T cell IS and the VS have less in common than the post-synapses. At the viral pre-synapse there are no peptide-loaded MHC complexes that would trigger an activation of the post-synaptic T cell (via cognate T cell receptors). Rather, the newly produced envelope glycoproteins at the pre-synapse engage, in the case of HIV-1, CD4 and a member of the chemokine family, e.g. CXCR4 or CCR5, and in some cases also $\alpha 4\beta 7$ integrin (Arthos et al. 2008), leading to the above mentioned activation events in the post-synaptic T cell. Two other cellular elements in the pre-synaptic cell play active roles in the formation of the VS: the kinase ZAP-70 and the microtubule-organizing center (MTOC). In the APC-T cell IS, ZAP-70-induced signaling in the post-synaptic cell promotes cytoskeleton and MTOC reorganization, and T cell activation, while ZAP-70-induced cytoskeleton reorganization and MTOC polarization (in the virus-producing cell) towards the pre-synapse of the VS has been reported to be necessary for efficient cell-to-cell transmission of HIV-1 (Sol-Foulon et al. 2007). It is possible, though remains to be investigated, that tetraspanins act as co-regulators of ZAP-70 initiated events at the pre-synapse. If so, this will require that tetraspanins be properly distributed at the pre-synapse. Arguably the most important result of numerous studies of the IS over the past decade is that this synapse has a sophisticated architecture, that it is subdivided into discrete zones where certain activities are initiated (or terminated). Further, and also of direct relevance to the VS, the synapse undergoes changes over time (Fooksman et al. 2010), and future studies of the VS, and of the role that tetraspanins play in its formation, thus should take such dynamics into account.

15.6 Information Transfer Via Fusion of (Virus-Infected and/or Uninfected) Cells

I should start this final section by pointing out that the term "uninfected cell" in this subtitle is an obvious misnomer, given that, as described above, all cells carry virus-like genetic elements that are remnants of pre-cellular life forms, and/or because at some point in time they have been infected by exogenous viruses, which, even if they are no longer active, still contribute to the overall genetic inventory of the cells. Perhaps a more appropriate term perhaps would thus be "not newly infected cell"—however, for obvious practical reasons I will continue to use the term "uninfected cells".

Uninfected cells can receive viral genetic information either when a viral particle binds to it, and releases the viral genome into the cell, or, and this arguably is the most straightforward way, when they fuse with infected cells. Fusion of virus-infected and uninfected cells leads to the formation of so-called syncytia. As mentioned above however, such syncytia formation, for many viruses, is not desirable, as syncytia are relatively short-lived entities. Nevertheless, under certain circumstances they may help spreading the virus, as they are motile and may exert force that allows them invade spaces that are not accessible to single infected cells (e.g. see Sylwester et al. 1998). And since, upon infection, i.e. here now upon cell-cell fusion, they acquire the properties of the "infecting cell", i.e. the fusion partner, as part of the new entity, the syncytia, can produce and shed progeny virus, which, again, makes them potentially useful agents of virus dissemination within their limited lifetime.

Because syncytia, under most circumstances, are detrimental to efficient spread of viruses such as HIV-1, this virus has evolved several mechanisms that tightly control the fusogenicity of its envelope glycoprotein (see Thali 2011, for a detailed discussion), and as discussed above, the recruitment of tetraspanins to the virus release site is one of them. As also discussed above, this negative regulation of membrane fusion processes is an important function of these proteins. Tetraspanins, however, can also do the opposite, i.e. they can promote membrane fusion. The most prominent case is that of CD9, whose expression in oocytes is absolutely required for fusion with spermatozoa to take place (see the chapter by Boucheix). However, it is important to point out that CD9 (and probably all the other members of the tetraspanin family) is not a fusogen itself, i.e. it does not trigger the fusion of membranes. Rather, this tetraspanin is thought to organize other membrane proteins, including cellular fusogens that mediate the fusion of the oocyte and the spermatozoa (Ziyyat et al. 2006). Because of the limited availability of human research material necessary for investigations of this particular fusion process, our knowledge of the molecular details is not yet very advanced. We know much more about how CD81, a closely related tetraspanin, which, through its interactions with specific other membrane proteins, promotes the fusion of the hepatitis C viral (HCV) membrane with the target cell membrane (see chapter by Cocquerel and Silvie). What we do know though is potentially interesting (Miyado et al. 2008), though remains unconfirmed so far (Gupta et al. 2009): to promote sperm-egg fusion, CD9 reportedly does not need to be present at the surface of oocytes, but can be released on exosomes (Miyado et al. 2008), the cellular membrane vesicles discussed above. How CD9 on exosomes could ultimately promote sperm-egg fusion remains to be seen. Such exosomes could form bridges between sperm and egg. Because of the more curved surfaces of exosomes (relative to the surfaces of sperm and egg), the energy barrier that needs to be overcome by a fusogen to trigger mixing of two opposing lipid bilayers is lower. Such a situation is reminiscent of how viral particles, when incubated with uninfected cells, can form bridges and induce the fusion of two uninfected cells (a phenomenon called "fusion from without" (Bratt and Gallaher 1969)). It has also been suggested, however, that the exosomes may deliver CD9 to the sperm, where they could reorganize the surface of those cells such that the fusogen there may be activated (Barraud-Lange et al. 2007). Such delivery to other cells of molecules involved in the fusion process would be reminiscent how exosomes have been shown to transfer CCR5, one of the co-receptors for HIV-1 to cells, thus rendering them infectable by HIV-1 (Mack et al. 2000).

Finally, it should be pointed out that information transfer by cell-cell fusion, whether virus-induced or virus-independent, is probably an underappreciated phenomenon, and thus that tetraspanins are likely to play important regulatory roles that have not yet been detected. Besides sperm-egg fusion, and e.g. the fusion of myoblasts, there are probably many more cell-cell fusions taking place, during ontogeny and also later on (for example, see Ying et al. 2002). However, as fused cells can divide again, in many instances we may not realize that they previously received genetic information from another cell (their fusion partner).

15.7 Perspectives

Research on tetraspanin functions in virus replication, with a few exceptions, is still very much in its infancy. Nevertheless, as outlined in this chapter, these membrane proteins are clearly involved in numerous steps of virus spread. Exogenous viruses, in contrast to endogenous viruses, can function as semi-independent entities, and because of this, many fundamental principals of cellular and molecular biology were revealed through the study of these genetic entities. Given that the transmission of enveloped viruses includes, and is controlled by, numerous membrane-based processes, and because these processes, such as fusion of lipid bilayers, signaling, etc., are of obvious importance also for cellular functions, it can be expected that further analyses of tetraspanin functions in virus replication will continue to shed light also on important cellular mechanisms.

Acknowledgements Work in my laboratory is supported by the NIH (RO1 AI080302-02). I thank current members of my group for critical reading of the manuscript, and Jane McKeating and Peter Monk, with whom I organized the 2011 FASEB meeting "membrane organization by molecular scaffolds", for discussions.

References

Arthos J, Cicala C, Martinelli E, Macleod K, Van Ryk D et al (2008) HIV-1 envelope protein binds to and signals through integrin alpha4beta7, the gut mucosal homing receptor for peripheral T cells. Nat Immunol 9:301–309

Barraud-Lange V, Naud-Barriant N, Bomsel M, Wolf JP, Ziyyat A (2007) Transfer of oocyte membrane fragments to fertilizing spermatozoa. FASEB J 21:3446–3449

Barrero-Villar M, Cabrero JR, Gordon-Alonso M, Barroso-Gonzalez J, Alvarez-Losada S et al (2009) Moesin is required for HIV-1-induced CD4-CXCR4 interaction, F-actin redistribution, membrane fusion and viral infection in lymphocytes. J Cell Sci 122:103–113

Bratt MA, Gallaher WR (1969) Preliminary analysis of the requirements for fusion from within and fusion from without by newcastle disease virus. Proc Natl Acad Sci USA 64:536–543

Chairoungdua A, Smith DL, Pochard P, Hull M, Caplan MJ (2010) Exosome release of beta-catenin: a novel mechanism that antagonizes Wnt signaling. J Cell Biol 190:1079–1091

Chertova E, Chertov O, Coren LV, Roser JD, Trubey CM et al (2006) Proteomic and biochemical analysis of purified human immunodeficiency virus type 1 produced from infected monocyte-derived macrophages. J Virol 80:9039–9052

Cocucci E, Racchetti G, Meldolesi J (2009) Shedding microvesicles: artefacts no more. Trends Cell Biol 19:43–51

de Parseval A, Lerner DL, Borrow P, Willett BJ, Elder JH (1997) Blocking of feline immunodeficiency virus infection by a monoclonal antibody to CD9 is via inhibition of virus release rather than interference with receptor binding. J Virol 71:5742–5749

Dobzhansky T (1964) Biology, molecular and organismic. Am Zool 4:443–452

Escola JM, Kleijmeer MJ, Stoorvogel W, Griffith JM, Yoshie O et al (1998) Selective enrichment of tetraspan proteins on the internal vesicles of multivesicular endosomes and on exosomes secreted by human B- lymphocytes. J Biol Chem 273:20121–20127

Fanaei M, Monk PN, Partridge LJ (2011) The role of tetraspanins in fusion. Biochem Soc Trans 39:524–528

Fooksman DR, Vardhana S, Vasiliver-Shamis G, Liese J, Blair DA et al (2010) Functional anatomy of T cell activation and synapse formation. Annu Rev Immunol 28:79–105

Forterre P (2005) The two ages of the RNA world, and the transition to the DNA world: a story of viruses and cells. Biochimie 87:793–803

Gerrow K, El-Husseini A (2006) Cell adhesion molecules at the synapse. Front Biosci 11:2400–2419

Gillette JM, Larochelle A, Dunbar CE, Lippincott-Schwartz J (2009) Intercellular transfer to signalling endosomes regulates an ex vivo bone marrow niche. Nat Cell Biol 11:303–311

Goldenfeld N, Woese C (2007) Biology's next revolution. Nature 445:369

Gordon-Alonso M, Yanez-Mo M, Barreiro O, Alvarez S, Munoz-Fernandez MA et al (2006) Tetraspanins CD9 and CD81 modulate HIV-1-induced membrane fusion. J Immunol 177:5129–5137

Gould SJ, Booth AM, Hildreth JE (2003) The Trojan exosome hypothesis. Proc Natl Acad Sci USA 100:10592–10597

Grigorov B, Attuil-Audenis V, Perugi F, Nedelec M, Watson S et al (2009) A role for CD81 on the late steps of HIV-1 replication in a chronically infected T cell line. Retrovirology 6:28

Gupta S, Primakoff P, Myles DG (2009) Can the presence of wild-type oocytes during insemination rescue the fusion defect of CD9 null oocytes? Mol Reprod Dev 76:602

Ho SH, Martin F, Higginbottom A, Partridge LJ, Parthasarathy V et al (2006) Recombinant extracellular domains of tetraspanin proteins are potent inhibitors of the infection of macrophages by human immunodeficiency virus type 1. J Virol 80:6487–6496

Hogue IB, Grover JR, Soheilian F, Nagashima K, Ono A (2011) Gag induces the coalescence of clustered lipid rafts and tetraspanin-enriched microdomains at HIV-1 assembly sites on the plasma membrane. J Virol 85:9749–9766

Holmes EC (2011) What does virus evolution tell us about virus origins? J Virol 85:5247–5251

Imai T, Kakizaki M, Nishimura M, Yoshie O (1995) Molecular analyses of the association of CD4 with two members of the transmembrane 4 superfamily, CD81 and CD82. J Immunol 155:1229–1239

Jolly C, Kashefi K, Hollinshead M, Sattentau QJ (2004) HIV-1 cell to cell transfer across an Env-induced, actin-dependent synapse. J Exp Med 199:283–293

Khurana S, Krementsov DN, de Parseval A, Elder JH, Foti M et al (2007) Human immunodeficiency virus type 1 and influenza virus exit via different membrane microdomains. J Virol 81:12630–12640

Konig R, Stertz S, Zhou Y, Inoue A, Hoffmann HH et al (2010) Human host factors required for influenza virus replication. Nature 463:813–817

Koonin EV, Senkevich TG, Dolja VV (2006) The ancient virus world and evolution of cells. Biol Direct 1:29

Krementsov DN, Weng J, Lambele M, Roy NH, Thali M (2009) Tetraspanins regulate cell-to-cell transmission of HIV-1. Retrovirology 6:64

Krementsov DN, Rassam P, Margeat E, Roy NH, Schneider-Schaulies J, Milhiet PE, Thali M (2011) HIV-1 assembly differentially alters dynamics and partitioning of tetraspanins and raft components. Traffic 11:1401–1414

Krishnamoorthy L, Bess JW Jr, Preston AB, Nagashima K, Mahal LK (2009) HIV-1 and microvesicles from T cells share a common glycome, arguing for a common origin. Nat Chem Biol 5:244–250

Lander ES (2011) Initial impact of the sequencing of the human genome. Nature 470:187–197

Levy S, Shoham T (2005) The tetraspanin web modulates immune-signalling complexes. Nat Rev Immunol 5:136–148

Loffler S, Lottspeich F, Lanza F, Azorsa DO, ter Meulen V et al (1997) CD9, a tetraspan transmembrane protein, renders cells susceptible to canine distemper virus. J Virol 71:42–49

Luria SE, Darnel JE, Baltimore D, Campbell A (1978) General virology, 3rd edn. Wiley, New York

Mack M, Kleinschmidt A, Bruhl H, Klier C, Nelson PJ et al (2000) Transfer of the chemokine receptor CCR5 between cells by membrane-derived microparticles: a mechanism for cellular human immunodeficiency virus 1 infection. Nat Med 6:769–775

Martin F, Roth DM, Jans DA, Pouton CW, Partridge LJ et al (2005) Tetraspanins in viral infections: a fundamental role in viral biology? J Virol 79:10839–10851

Meerloo T, Parmentier HK, Osterhaus AD, Goudsmit J, Schuurman HJ (1992) Modulation of cell surface molecules during HIV-1 infection of H9 cells. An immunoelectron microscopic study. AIDS 6:1105–1116

Meerloo T, Sheikh MA, Bloem AC, de Ronde A, Schutten M et al (1993) Host cell membrane proteins on human immunodeficiency virus type 1 after in vitro infection of H9 cells and blood mononuclear cells. An immuno-electron microscopic study. J Gen Virol 74(Pt 1): 129–135

Mi S, Lee X, Li X, Veldman GM, Finnerty H et al (2000) Syncytin is a captive retroviral envelope protein involved in human placental morphogenesis. Nature 403:785–789

Mittelbrunn M, Yanez-Mo M, Sancho D, Ursa A, Sanchez-Madrid F (2002) Cutting edge: dynamic redistribution of tetraspanin CD81 at the central zone of the immune synapse in both T lymphocytes and APC. J Immunol 169:6691–6695

Miyado K, Yoshida K, Yamagata K, Sakakibara K, Okabe M et al (2008) The fusing ability of sperm is bestowed by CD9-containing vesicles released from eggs in mice. Proc Natl Acad Sci USA 105:12921–12926

Morita E, Sundquist WI (2004) Retrovirus budding. Annu Rev Cell Dev Biol 20:395–425

Muotri AR, Marchetto MC, Coufal NG, Oefner R, Yeo G et al (2010) L1 retrotransposition in neurons is modulated by MeCP2. Nature 468:443–446

Nguyen DG, Booth A, Gould SJ, Hildreth JE (2003) Evidence that HIV budding in primary macrophages occurs through the exosome release pathway. J Biol Chem 278:52347–52354

Orentas RJ, Hildreth JE (1993) Association of host cell surface adhesion receptors and other membrane proteins with HIV and SIV. AIDS Res Hum Retroviruses 9:1157–1165

Pelchen-Matthews A, Raposo G, Marsh M (2004) Endosomes, exosomes and trojan viruses. Trends Microbiol 12:310–316

Rodriguez-Boulan E, Paskiet KT, Sabatini DD (1983) Assembly of enveloped viruses in madin-darby canine kidney cells: polarized budding from single attached cells and from clusters of cells in suspension. J Cell Biol 96:866–874

Ruiz-Mateos E, Pelchen-Matthews A, Deneka M, Marsh M (2008) CD63 is not required for production of infectious human immunodeficiency virus type 1 in human macrophages. J Virol 82:4751–4761

Runge KE, Evans JE, He ZY, Gupta S, McDonald KL et al (2007) Oocyte CD9 is enriched on the microvillar membrane and required for normal microvillar shape and distribution. Dev Biol 304:317–325

Sato K, Aoki J, Misawa N, Daikoku E, Sano K et al (2008) Modulation of human immunodeficiency virus type 1 infectivity through incorporation of tetraspanin proteins. J Virol 82:1021–1033

Sattentau Q (2008) Avoiding the void: cell-to-cell spread of human viruses. Nat Rev Microbiol 6:815–826

Shaw ML, Stone KL, Colangelo CM, Gulcicek EE, Palese P (2008) Cellular proteins in influenza virus particles. PloS Pathog 4(6):e1000085

Singethan K, Muller N, Schubert S, Luttge D, Krementsov DN et al (2008) CD9 clustering and formation of microvilli zippers between contacting cells regulates virus-induced cell fusion. Traffic 9:924–935

Sol-Foulon N, Sourisseau M, Porrot F, Thoulouze MI, Trouillet C et al (2007) ZAP-70 kinase regulates HIV cell-to-cell spread and virological synapse formation. EMBO J 26:516–526

Spoden G, Freitag K, Husmann M, Boller K, Sapp M et al (2008) Clathrin- and caveolin-independent entry of human papillomavirus type 16–involvement of tetraspanin-enriched microdomains (TEMs). PLoS One 3:e3313

Studer L (2010) Neuroscience: excessive mobility interrupted. Nature 468:383–384

Sylwester A, Daniels K, Soll DR (1998) The invasive and destructive behavior of HIV-induced T cell syncytia on collagen and endothelium. J Leukoc Biol 63:233–244

Tardif MR, Tremblay MJ (2005) Tetraspanin CD81 provides a costimulatory signal resulting in increased human immunodeficiency virus type 1 gene expression in primary CD4+ T lymphocytes through NF-kappaB, NFAT, and AP-1 transduction pathways. J Virol 79:4316–4328

Thali M (2009) The roles of tetraspanins in HIV-1 replication. Curr Top Microbiol Immunol 339:85–102

Thali M (2011) Tetraspanin functions during HIV-1 and influenza virus replication. Biochem Soc Trans 39:529–531

Thery C, Zitvogel L, Amigorena S (2002) Exosomes: composition, biogenesis and function. Nat Rev Immunol 2:569–579

Valadi H, Ekstrom K, Bossios A, Sjostrand M, Lee JJ et al (2007) Exosome-mediated transfer of mRNAs and microRNAs is a novel mechanism of genetic exchange between cells. Nat Cell Biol 9:654–659

van Spriel AB, Figdor CG (2010) The role of tetraspanins in the pathogenesis of infectious diseases. Microb Infect 12:106–112

Verweij FJ, van Eijndhoven MA, Hopmans ES, Vendrig T, Wurdinger T et al (2011) LMP1 association with CD63 in endosomes and secretion via exosomes limits constitutive NF-kappaB activation. EMBO J 30:2115–2129

Villarreal LP, Witzany G (2010) Viruses are essential agents within the roots and stem of the tree of life. J Theor Biol 262:698–710 von Lindern JJ, Rojo D, Grovit-Ferbas K, Yeramian C, Deng C et al (2003) Potential role for CD63 in CCR5-mediated human immunodeficiency virus type 1 infection of macrophages. J Virol 77:3624–3633

Wang L, Liu L, Che Y, Jiang L, Dong C et al (2010) Egress of HSV-1 capsid requires the interaction of VP26 and a cellular tetraspanin membrane protein. J Virol 7:156

Weng J, Krementsov DN, Khurana S, Roy NH, Thali M (2009) Formation of syncytia is repressed by tetraspanins in human immunodeficiency virus type 1-producing cells. J Virol 83: 7467–7474

Woese CR (2002) On the evolution of cells. Proc Natl Acad Sci USA 99:8742–8747

Ying QL, Nichols J, Evans EP, Smith AG (2002) Changing potency by spontaneous fusion. Nature 416:545–548

Yoshida T, Kawano Y, Sato K, Ando Y, Aoki J et al (2008) A CD63 mutant inhibits T-cell tropic human immunodeficiency virus type 1 entry by disrupting CXCR4 trafficking to the plasma membrane. Traffic 9:540–558

Ziyyat A, Rubinstein E, Monier-Gavelle F, Barraud V, Kulski O et al (2006) CD9 controls the formation of clusters that contain tetraspanins and the integrin alpha 6 beta 1, which are involved in human and mouse gamete fusion. J Cell Sci 119:416–424

Index

F. Berditchevski and E. Rubinstein (eds.), *Tetraspanins*, Proteins and Cell Regulation 9,
DOI 10.1007/978-94-007-6070-7,

N

O

Printed by Printforce, the Netherlands